STATE OF NEW HAMPSHIRE. MERRIMACK COUNTY.

# HEARING

IN THE MATTER OF

# CONCORD RAILROAD CORPORATION

VS.

# GEORGE CLOUGH AND TRUSTEES,

BEFORE

HON. E. L. CUSHING, HON. H. A. BELLOWS.
HON. WILLIAM HAILE, *Referees.*

FOR PLAINTIFFS:

JOHN H. GEORGE, C. W. STANLEY.

FOR DEFENDANTS:

MASON W. TAPPAN, H. P. ROLFE, J. Y. MUGRIDGE.

Reported by Jay Read Pember, Law Stenographer.

THE PEOPLE STEAM PRESS, CONCORD.
1869.

# PREFACE.

In October, 1865, as will hereafter appear in the following pages, on account of the several times repeated assertion of Mr. Joseph A. Gilmore, the then superintendent of the Concord Railroad, that "the conductors were stealing fifty thousand dollars a year," the directors put a detective force upon the cars, and continued them for the space of about six weeks; and caused some seventeen hundred dollars in money, more or less, to pass into the hands of the conductors for tickets and fares paid for in the cars.

The result of the labors of that detective force was made known to John H. George, counsel for, and then clerk of the road, and to no one else in the government of the road. His deductions, in the shape of a report, were given to the directors,—the report itself being concealed from them. Upon this representation, made by Mr. George, the directors voted to discharge all the conductors on the Concord road, and on the different roads connected with it, as all were represented as having embezzled the funds of the road, but no action was commenced against any one of the conductors, though all were immediately discharged. This happened about the middle of January, 1866.

February 10, 1866, Mr. Gilmore and Mr. George found on the person and in the possession of one James Whitcher, of Hooksett, N. H., several hundred tickets, which he said were given him by the conductors of the Concord Railroad, in exchange for brooms and pails, and that the largest share of them were given him by George Clough. Thereupon suits were at once commenced against all the conductors who had any property, including Robert N. Corning, who had not then been in the employ of the railroad for six years; and by a vote of the directors, at a meeting subsequently held, the action of the clerk, Mr. George, and of the superintendent, Mr. Gilmore, in this matter, was approved.

Mr. Everett, the conductor on the Manchester and North Weare road, was soon put back and restored to his place, which he has occupied since. James M. Jones has been continued in the employ of the road as conductor till the present time, although the original suit stands at the present time against him on the docket. Mr. Kendrick, with whom George Clough, in this trial, has been accused of dealing in bogus tickets, has been voted, by the present board of directors, as guiltless of any fraud or irregularity while conductor on the road. The suit against Henry Eaton was discontinued by the railroad upon their own motion, with no previous knowledge of the defendant.

It is now three years and a half since these suits were commenced, and no single effort has been made to bring Mr. Corning's case to trial; although he has been dead over three years, no effort has been made to summon in the executor of Mr. Corning's estate.

The basis of these suits was the testimony of said James Whitcher and the detective, James G. Carney, of Lowell, and his corps of assistants, and yet it will be seen, by an examination of the testimony, that said Whitcher, although living within nine miles of the place of hearing, was not brought in as a witness at all, and the tickets which were found in his possession were proved not to have been furnished him by Mr. Clough; while the testimony of Mr. Carney and his assistants did not show a single fare taken in the cars by Mr. Clough that was not returned on his way-bills and paid over to the general ticket agent.

The entire evidence in this case has been taken down in short-hand—questions and answers—and has been reported at an expense of eleven hundred dollars, and the printing has involved an expense of thirteen hundred dollars more. All this expense has been incurred by Mr. Clough, in addition to the expense occasioned him by this trial, in order that any one and every one who has either an interest in the road or feels any interest either in the defendant or the trial, may have a full and fair opportunity of judging for themselves of the true state of this controversy. The public ear was burdened, for three years previous to the trial, with all sorts of false and malicious stories, and now all the facts as they are can be seen and read of all men.

On the 28th of May following the commencement of this suit, Mr. George, then clerk of the road, went before the stockholders, at their annual meeting, and gave a detailed *ex parte* statement of his agency in the discoveries made by himself and Mr. Carney. The stockholders gave him and his board of directors an unconditional discharge, with the exception of Mr. Stickney.

Soon after this, as will be seen, Mr. Carney was employed in his capacity as detective by the United States Treasury Department, but was soon detected in taking a thousand dollar bribe from one John Leighton, whose name is familiar to many, and he received a discharge from the service of the United States, which, if not honorable, was unconditional.

To the public generally, and more especially to those connected with railroads, a careful perusal of this trial is very respectfully commended. The judgment of the referees has been given. On that no comments are made—none are necessary. These pages contain every word of the evidence on the trial. As will be seen, it covers the entire history of Mr. Clough's life during the time he was a conductor on the Concord Railroad; and the details of every one of his business transactions, from the time when, a mere boy, he started out from home with all his worldly effects tied up in a pocket handkerchief, down to the time of the commencement of this suit, have been gone into, and are embodied in the evidence. The way in which he has acquired his property is here stated; a comparison of the amounts returned by him from day to day, while he was conductor, with those of other conductors who were, from time to time, on the same trains, is given; and the testimony of the detectives who followed him like hounds on the track, day after day for six weeks, unknown to him, trying every method in their power to "spot" him, is set forth in their own language.

Mr. Clough feels that he has been unjustly and wrongfully pursued; and, conscious of his own innocence, he has felt compelled, by way of vindication, to place before the public all the facts, as they were elicited by the trial. To the candid and impartial judgment of that public he submits his cause.

Sept. 1, 1869.

# HEARING.

[First Day, July 21, 1868.]

Concord, N. H., July 21, 1868.

At a hearing in the matter of the Concord Railroad Corporation *vs.* George Clough and Trustees, before Hon. E. L. Cushing, Hon. H. A. Bellows, and Hon. William Haile, Referees, there appeared in behalf of the Plaintiff Corporation, Mr. John H. George, of Concord, and Mr. C. W. Stanley, of Manchester; and in behalf of the Defendant, Mr. Mason W. Tappan, of Bradford, Mr. H. P. Rolfe, of Concord, and Mr. J. Y. Mugridge, of Concord.

Before submitting the opening statement in behalf of the plaintiffs, Mr. George, counsel for the corporation, read the writ and the returns thereon, and the specifications, which are as follows:

## WRIT.

[L. S.] THE STATE OF NEW HAMPSHIRE.

Merrimack, ss. *To the Sheriff of any County in this State, or his Deputy.*

We command you to attach the goods or estate of George Clough, of Concord, in said County of Merrimack, Esquire, to the value of one hundred thousand dollars, and summon him (if to be found in your precinct,) to appear before the Supreme Judicial Court, to be holden at Concord, in said County of Merrimack, on the first Tuesday of April next, to answer to the Concord Railroad Corporation, a corporation established by law, and having its place of business at Concord aforesaid,

In a plea of the case, for that the said defendant, at said Concord, on the day of the date of this writ, being indebted to the plaintiffs in the sum of one hundred thousand dollars for goods, wares and merchandise bargained and sold by the plaintiffs to the said defendant; and in the further sum of the same amount, for goods, wares and merchandise sold and delivered by the plaintiffs to the said defendant; and in a further sum of the same amount for work done and materials for the same provided by the plaintiffs for the said defendant, and at the said defendant's request; and in a further sum of the same amount for money by the plaintiffs lent to the said defendant at the said defendant's request; and in a further sum of the same amount for money by the plaintiffs paid for the use of said defendant, at the said defendant's request; and in a further sum of the same amount of money had and received by the said defendant to the plaintiffs; and in a further sum of the same amount for interest, for the plaintiffs' forbearance, at the said defendant's request, of moneys due and owing from the said defendant to the said plaintiffs; and in a further sum of the same amount for money found to be due from the said defendant to the said plaintiffs, on account stated between them; in consideration thereof promised to pay the said several sums to the said plaintiffs on demand. Yet no part of said sums has ever been paid. To the damage of the plaintiffs, as they say, the sum of one hundred thousand dollars.

We also command you to attach the money, goods, chattels, rights and credits of the said Clough in the hands and possession of Henry P. Rolfe, of said Concord, Esquire, and also in the hands and possession of the Union Bank, a corporation existing by law in said State, and having its place of business in said Concord,

To the value of one hundred thousand dollars; and summon said trustees, if to be found in your precinct, to appear at said court and show cause, if any they have, why execution should not issue against them for the damages which may be received by said plaintiffs against said Clough.

And make return of this writ, with your doings therein.

Witness, IRA PERLEY, Esquire,
The tenth day of February, Anno Domini, 1866.

J. D. Sleeper, Clerk.

[Internal Revenue Stamp.]

Merrimack, ss., February 10, 1866.

I attached all the lands and tenements in the City of Concord, in said County, in which the within named defendant has any right, title, interest or estate; and on the same day I left at the dwelling house of Charles F. Stuart, the city clerk of said city, a true and attested copy of this writ, and of this, my return, endorsed thereon, at 10 of the clock, in the afternoon of said day. J. L. Pickering, Deputy Sheriff.

Fees: Service, .23; travel, .10; paid clerk, .20; copy, 2.00; attachment, 1.00; travel to clerk's office, .10; revenue stamp, .15—$3.78.

Merrimack, ss., February 10, 1866.

I attached, as the property of the within named George Clough, four hundred shares in the capital stock of the Concord Railroad Corporation, by giving in hand to John H. George, clerk of said corporation, an attested copy of this writ, with an attested copy of this, my return, thereon, at nine o'clock in the afternoon of said day.

J. L. Pickering, Deputy Sheriff.

Merrimack, ss, February 15, 1866.

I then attached the money, goods, chattels, rights and credits of the within defendant, in the hands and possession of the within named Union Bank and Henry P. Rolfe, trustees, and summoned said trustees to appear as within commanded, by giving in hand to A. C. Pierce, the cashier of the Union Bank, an attested copy of the within writ; and on the 19th day of February, by leaving at the usual place of abode of Henry P. Rolfe an attested copy of this writ. I have also made service on the within named Clough, by leaving at his usual place of abode an attested copy of the within writ.

Jonathan L. Pickering, Deputy Sheriff.

Fees: Three services, .69; travel, .30; three copies, etc., 3.00—$3.99.

Merrimack ss., February 19, 1866.

I attached all the land and tenements in the town of Dunbarton, in said county, in which the within named defendant has any right, title, interest or estate; and on the same day I left at the dwelling house of

Gilbert B. French a true and attested copy of this writ, and of this, my return, endorsed thereon, at ten of the clock in the forenoon of said day. J. L. PICKERING, Deputy Sheriff.

FEES: Service, .23; travel, .80; paid clerk, .20; copy, 2.00; attachment, 1.00; travel to clerk's office, 1.20; revenue stamp, .15—$5.58.

## SPECIFICATIONS.

MERRIMACK SS. SUPREME JUDICIAL COURT. April Term, 1866.

CONCORD RAILROAD CORPORATION *v.* GEORGE CLOUGH.

In the above action the plaintiff seeks to recover of the defendant the sum of one hundred thousand dollars ($100,000), being for moneys received by the defendant while acting as passenger conductor upon the plaintiff's railroad for the fares of passengers over the plaintiff's railroad and connecting railroads, not accounted for or paid over to the plaintiff.

Also for the value of passenger tickets, the property of the plaintiff, over the plaintiff's and connecting railroads, disposed of by the defendant and converted to his own use while acting as such conductor.

Also for the fares of passengers passed free by the defendant while acting as such conductor, over the roads aforesaid, without right and for the defendant's benefit.

Also for moneys of the plaintiff received by the defendant while in pursuance of his duties as conductor as aforesaid, and not accounted for or paid over to the plaintiff.

---

## TESTIMONY OF ALONZO H. WESTON.

Mr. Alonzo H. Weston, of Manchester, was called in behalf of the plaintiff, and was duly sworn and testified as follows:

Q. (*By Mr. George.*) Mr. Weston, will you be kind enough to state what is your business and where you live?

A. I live at Manchester.

Q. How long have you lived at Manchester and been in business there?

*Mr. Haile.* He has not stated what his business was.

Q. What did you say your business was? I don't know as you stated that.

A. Clothing business.

Q. How long have you been in that business?

A. About three years.

Q. Will you state whether you purchased any tickets of James Whitcher? If so, when, and to what extent?

[Here followed a long discussion between counsel as to the admissibility of evidence that Weston purchased tickets of Jim Whitcher, unless it was shown that in some way the tickets came from Clough. The referees ruled that the evidence might go in, but that unless it was shown by plaintiff's counsel that Clough furnished the tickets to Whitcher the evidence would be incompetent.]

Q. You stated that you were in the clothing business at Manchester, and had been for the last three years. Whom did you succeed in business ?

A. Jacob Morse.

Q. Will you state whether or not, if you ever purchased tickets? If so, of whom? If you ever purchased tickets of James Whitcher, or not? Give the times, dates, and amounts, so far as you are able, and the kinds of tickets.

*Mr. Tappan.* We shall be very glad to have all these tickets put in.

*Mr. George.* I cannot tell all these. These tickets came from Whitcher himself; those are entirely a different lot of tickets.

Q. You might state how many you purchased of Whitcher. State when you purchased them, as far as you recollect, and all the circumstances.

A. I could not say when the first was.

Q. Within how long? Was it after you went into business or before?

A. It was after I went into business; yes, sir. It was after Aug. 1st.

Q. After August 1st, 1865?

A. Yes, sir.

Q. Well, now state how much dealing you had with Mr. Whitcher; and what portion was paid in tickets, and what kind of tickets they were.

A. I could not tell anywhere near what portion in tickets, because I credited everything as cash, because I considered them as cash.

Q. Make the statement to the referees as near as you can. How much was your business with Mr. Whitcher?

A. It was sixty-seven dollars and fifty cents.

Q. Did you have any other trades besides what your books show?

A. No, sir: not that I know of.

Q. Now, what portion of that was paid in tickets, according to your best judgment and recollection?

A. Well, I couldn't tell. Sometimes he would come and pay me—— I see $7 twice, $10.50, $14, and $29, and some proportion of these were tickets. That is all I know.

Q. State as near as you can.

A. I have forgot all about it; I supposed the thing was dropped.

Q. State as near as you can. Your attention was called to it at the time of Whitcher's arrest?

A. It was.

Q. Did you make a statement at that time?

A. Yes, sir; I think I did; I don't recollect about it.

Q. Mr. Weston, did you make a statement at the time of Mr. Whitcher's arrest?

A. I did.

Q. Was that statement written down and read to you by me, with regard to the amount of business that you had done with Mr. Whitcher and the proportion paid for in tickets?

A. I made a statement. and it was written down; quite a long one, I think. I don't seem to recollect whether it was addressed to me or not; I know I made a statement.

Q. Did you make a statement in regard to the proportion of tickets?

A. I presume I did.
Q. Don't you know you did?
A. I think I did.
Q. Was that written down and read to you?
A. Not that I remember of.
Q. Was the statement that you made at that time true?
A. Yes sir; so far as I could remember, sir.
Q. Do you mean to say there was a long statement written?
A. I suppose——there was a man writing there all the time, I know.
Q. Who was it?
A. I don't know.
Q. Now, will you state, if you please, just as well as you can, the proportion of $67.50 that was paid you in tickets, that you received from Mr. Whitcher, according to your best recollection?
A. I couldn't say for any certainty.
Q. State as well as you can.
A. It might have been a quarter; and it might have been a half. That is as near as I can say.
Q. On what roads were these tickets?
A. Manchester to Boston.
Q. What did you do with these tickets?
A. On my way to Boston I rode on them.
Q. Did you sell any from your store?
A. I did, once or twice.
Q. When you sold them, for how much did you sell them?
A. A dollar and seventy-five cents to Boston.
Q. What was the regular fare from Manchester to Boston?
A. About two dollars, I think, at that time.
Q. How many tickets had you on hand at the time Mr. Whitcher was arrested?
A. I think it was either two or three. They were coupon tickets, I think. That is, they were—if I recollect right,—they were tickets from Boston to Concord or Hooksett. I guess one was Hooksett or Suncook. I know they were small—not worth much, you know. I think there were two or three.
Q. From Boston?
A. No, sir.
Q. You said from Boston.
A. I don't think there was any from Boston. I had rode on them up as far as Manchester, and they had been punched.
Q. You had rode on them?
A. Yes, sir.
Q. Were any tickets punched tickets—partially so; that is, punched a portion of the way?
A. Not that I——I don't seem to remember that they were.
Q. Don't remember how that fact is?
A. No, I do not.
Q. How was it with regard to your predecessor, if you know?
A. I don't know anything about it. I had heard——[Objected to.]
Q. Did you ever see any tickets in their hands? [Objected to.]
*The Chairman.* It seems to me that is hardly admissible. [Excepted to by plaintiff's counsel.]
Q. (*By Mr. George.*) How happened you to be trading in tickets with Mr. Whitcher?

A. I knew that Mr. White here, my neighbor, had tickets now and then; and I think I got, perhaps, one of him; and I think I must have asked him where he got them. That is the only way how I happened to——

Q. Did you go to Mr. Whitcher, or Mr. Whitcher come to you?

A. I could n't say.

Q. I want to ask you, sir, if Mr. Morse had any Concord Railroad tickets—had any tickets on hand, at the time you bought him out, to your knowledge; and if you, yourself, knew that he had them of Mr. Whitcher? [Objected to.]

Q. State what you know about Mr. Morse's having railroad tickets, and where he obtained them?

A. I think Mr. Morse had some tickets to Boston. That was before I bought him out. And that is all I know about it. I might, possibly, have had one of them. I could not tell certainly.

Q. From whom did he have them?

A. I don't know, sir.

Q. Did you see Mr. Morse sell tickets, or know of his selling tickets from Manchester to Boston? If so, at what rate?

A. I do not know.

CROSS EXAMINATION.

Q. (*By Mr. Mugridge.*) Just a single question, Mr. Weston. I understood you to say that at the time you were called upon, at Manchester, you had two or three tickets. What were these tickets? Where did they go?

A, They went from Boston. My impression is that one went from Boston to Suncook, or Hooksett, and the other one from Boston to Concord. I know I rode on them as far as Manchester—as far as I wanted to.

Q. Were these coupon tickets?

A. Well, I think they were card tickets like that [referring to tickets upon table]. Each run through; and of course as I rode to Manchester they would be punched.

Q. Were they coupon tickets?

A. I could not tell.

Q. Had the Boston and Lowell or Nashua and Lowell part been taken off from the tickets?

A. I could not say.

Q. You don't know how that was?

A. No.

Q. And you think you had some two or three of them at the time?

A. Yes, sir.

Q. (*By Mr. George.*) And these, you have stated, you had of Whitcher?

A. Yes, sir.

Q. (*By Mr. Mugridge.*) Over which road did these tickets go? Over the Manchester and Lawrence, or the Nashua?

A. I had some on both.

## TESTIMONY OF WILLIAM WHITE.

Q. (*By Mr. George.*) Mr. White, you are a trader at Manchester?

A. I have been, sir.

Q. For how long?

A. Twenty years.

Q. And you were in trade there at the time Mr. Whitcher was arrested, in February, 1866.

A. Yes, sir.

Q. Will you state what your trade was?

A. Dry goods.

Q. Have you bought tickets of Mr. James Whitcher, over the Concord Railroad? Had you prior to that time?

A. I have had dealings with Mr. Whitcher, and I have had tickets of him occasionally.

Q. To what extent?

A. It would be impossible for me to tell.

Q. Tell as nearly as you can.

A. I don't know that I have any means whereby I could tell. I could tell the amount of business I have had in a year or two, or four or five years. It was some four hundred dollars.

Q. Did you give any credit?

A. I had orders on his grocery store for merchandise, to offset his account as cash, the whole thing. Therefore I couldn't state.

Q. Give as near as you are able.

A. The fact is, the whole thing would have to be guess work.

Q. Give the whole amount.

A. From a third to one half.

Q. What was the whole amount?

A. Four hundred and thirty dollars; and I commenced September, 1864.

Q. (*By Judge Bellows.*) From what time to what time?

A. September, 1864, to January, 1866.

Q. (*By Mr. George.*) Did you have some tickets on hand at the time? Are these some of the tickets that you had? [Showing some tickets to witness.]

A. I could not state. I don't recollect as you took any tickets, sir.

Q. Don't you recollect of any?

A. You might have done so; I don't recollect.

Q. Are these similar?

A. I should say, of course, some of them resemble these.

Q. You cannot tell whether you handed me those two tickets?

A. I could not say.

Q. Did you ever deal with Whitcher before 1864 in tickets?

A. I should think not; but I could not be certain about it.

Q. What did you allow Whitcher?

A. I think twenty-five cents less than the regular fare. I should say, as a general thing, that that was the case.

Q. Did you sell them from the store?

A. I may have disposed of some of them.

Q. What did you sell them for?

A. At the same price.

Q. Did you sell some to Mr. Cutter, or his clerk? Did you sell some to Mr. John D. Patterson?

A. I have no recollection of selling any to Mr. Patterson. I don't think I ever did.

Q. Won't you allow me to refresh your recollection if I can? Have you any recollection of any conversation with me at the time of Mr. Whitcher's arrest, and saying that you had bought the tickets and paid for them, and that if I took them you wanted your pay?

A. It seems to me that conversation must have been with somebody else; I don't recollect it, upon my word.

Q. Won't you examine these tickets?

A. I might have said that; but it has gone from my mind now. I forgot it; I have had so much else which has shoved it out, I suppose.

Q. I will ask you the same question I asked Mr. Weston: whether your neighbors were dealing in these same tickets with Mr. Whitcher?

A. Well, there was a time when the package ticket was sold by Mr. Merrill and the other broker there. It was common talk about there.

Q. I am asking now about these tickets purchased of Mr. Whitcher. I want to know if you have any personal knowledge in regard to your neighbors dealing in this same kind of ticket with Mr. Whitcher. In the first place, if you know anything about their having dealings?

A, I don't know about other parties having tickets.

Q. How is it? You have rode in the cars, I suppose, with Mr. Clough?

A. I have.

Q. Ever passed you free? [Objected to.]

[SECOND DAY, July 22, 1868.]

*The Chairman.* We rule it to be inadmissible, unless the plaintiff further shows that, as the result of this transaction, money, or something which as between the parties ought to be considered money, came to the defendant's hands.

*Mr. George.* Now, then, supposing I show that he passed people free?

*The Chairman.* We rule that that is not admissible.

Q. (*By Mr. George.*) Did you know Mr. Starkey, who was a brakeman upon Mr. Clough's train.

A. I did.

Q. Will you state whether a dress was bought for his wife at your store? [Objected to.]

Mr. George then submitted, in writing, what it was proposed by plaintiff's counsel to show, viz:—

We propose to show that Mrs. Starkey, wife of a brakeman upon Mr. Clough's train, purchased a dress of the witness: and that Starkey, in consideration of such purchase, agreed to pass him over the road; and that Mr. Clough, as the conductor of the train, did so pass him at Starkey's request.

*The Chairman.* I believe we all think that this testimony now stands on the same ground, so far as this is concerned, as what we have already ruled upon. We do not conclusively rule it out, but we rule it out for the present, until some evidence is offered to show that Mr. Clough derived some equivalent of money for it. We consider this to stand on the same ground as the last ruling. As a matter of discretion, we do not think this evidence is now admissible. We think there should be some further evidence tending to show that Mr. Clough derived a benefit from the transaction.

CROSS EXAMINATION.

Q. (*By Mr. Mugridge.*) I understand, Mr. White, that you commenced your transactions in the ticket line with Whitcher, some time in 1864, or about that time—was it?

A. Well, to the best of my recollection; I am not positive on that point.

Q. You know Starkey?

A. Yes, sir.

Q. Let me ask you whether he was upon the road in the capacity of a brakeman, at that time?

A. Well, I really couldn't fix it in my mind whether he was, or not.

Q. Don't you recollect how that was?

A. It seems to me that I didn't know Starkey as early as that, as connected with the road; I am not sure on that point.

Q. You say you received tickets of Whitcher from time to time—did you receive tickets from other persons between the times you have mentioned, in '64 and '66.

A. I did.

Q. Did you get them of Starkey?

A. I had tickets of Starkey, but it wasn't in 1864. It was along just before this.

Q. I say from '64 to '66?

A. Those I had of him was along sometime previous to this affair.

Q. It was between '64 and '66?

A. Yes, sir.

Q. Did you have them of other people?

A. I did.

Q. Did you send to Boston and get them of parties there?

A. I have bought tickets in Boston for parties coming from Montreal, and sold to parties coming up that would make me some discount.

Q. Which you got North?

A. I did not send there. I have had parties come to me; for instance, coming from Ogdensburgh or Potsdam.

Q. Have you not procured a good many tickets in that way?

A. I have—a great many. At that time the fare from Ogdensburgh to Boston was nine dollars, and the fare from there to Manchester was nine dollars. Therefore people would buy their tickets to Boston, because it was the same price as it was to Manchester, and sell them for whatever they could get.

Q. Did you not procure a good many tickets in that way, from persons who purchased their tickets up North to go to Boston?

A. As I said before, I have, a good many.

Q. And, as you said, you have had them from parties going North?

A. Not as many.

Q. You have had some that way?

A. Yes, sir.

Q. You say you had tickets from other parties. Did you ever have a ticket, in your life, from George Clough?

A. I never did, to my recollection.

Q. Now, sir, you have spoken of tickets that you had from time to time of Mr. Whitcher; will you be kind enough to state what kind of tickets those were, and from what stations to what stations, did those tickets that you had of Whitcher go, as a general thing?

A. Well, a portion of the tickets were package tickets.

Q. What proportion of them, do you think, were package tickets, and between what stations upon the road did these package tickets run?

A. I should not want to undertake to say what proportion; because I could not tell.

Q. Give us your best recollection upon that subject.

A. It would only have to be from recollection.

Q. Let us have that.
A. Thirty-three per cent., or a third of them.
Q. Then you think that a third of the tickets that you had from Whitcher were package tickets? Now, sir, between what stations and on what lines of road were those? On what line of road were these tickets to go—speaking now of package tickets?
A. Understand that one-third is more or less.
Q. Yes, sir; I understand that is your best recollection.
A. These were on the road from Manchester to Boston, by the way of Lowell.
Q. Now we will take the portion remaining. You may explain what package tickets are.
A. Well, they are issued by the company in packages, at a discount; I suppose a discount from the regular——
Q. Tariff?
A. Yes, sir; I take it so.
Q. After taking out one-third that you received from Whitcher, what kind of tickets were those that were left?
A. Well, a portion over the other road.
Q. What road?
A. The Lawrence road.
Q. Between what stations?
A. Manchester and Boston.
Q. Manchester and Boston—Manchester, Lawrence and Boston, on the Maine road?
A. Yes, sir.
Q. Now, will you be kind enough to state what number of the tickets that you had of Whitcher went that way, over that road to Boston?
A. My recollection could only be this, that it was less than the other.
Q. Will you be kind enough to indicate your best recollection on that subject?
A. It was less, as I said.
Q. Then you mean those going to Boston that way, were less than those going to Boston by Lowell?
A. Yes, sir.
Q. How much less? Have you an idea?
A. Not so many by considerable.
Q. What were the balance that you had of Whitcher, I mean. After taking out the package tickets, and the tickets to Boston by way of Lawrence, where were the balance of tickets that you had of Whitcher, and what kind were they?
Q. (*By the Chairman.*) Do you mean that the tickets by way of Lawrence were package tickets?
A. Miscellaneous—varying.
Q. (*By Mr. Mugridge.*) Between what stations?
A. Well, they were generally, I think, from Manchester to Boston. There were a few from Concord to Boston.
Q. But over the line of the Lowell road, generally?
A. I should think, as a general thing—to a large amount—as near as my memory would serve me.
Q. State whether you got any tickets from Whitcher, other than those you have indicated in kind, at any time?
A. I could not say that I recollect positively.

Q. Aside from those tickets that you have spoken of, the package tickets; tickets from Manchester to Boston, by way of Lawrence; tickets from Manchester to Boston, by way of Lowell; and some few, as you say, from Concord to Boston, by way of Lowell, did you get other tickets from Whitcher?

A. Going anywhere—any direction?

Q. Yes, sir.

A. I think I had one or two from Manchester to Portsmouth, or down to the Junction, I can't tell which—not to any amount.

Q. Was there any one time more than another when you received more tickets from Mr. Whitcher? That is, did you receive more tickets, at or about some particular time, than you did at any other time during your transactions with Whitcher? If so, when? If at any time between 1864 and 1866 you were receiving more tickets than at any other time, won't you be kind enough to indicate that time?

A. Well, my mind is——I could not state whether there was any particular time.

Q. Could you refresh your recollection by your book?

A. I might have got more out of him at one time than another.

Q. Was there any particular time, in reference to this arrest of Whitcher, that you received more tickets than at any other time?

A. Well, it may be. It is an impression in my mind. He would send orders, and come in and get orders, and left some tickets. I can't tell you.

Q. In how large numbers were you accustomed to receive tickets from Whitcher? How many did you ever receive at any one time from him?

A. I couldn't tell you.

Q. Give us your best recollection.

A. Sometimes a single trip, and sometimes three or four.

Q. Was there ever a time when you received more than three or four?

A. I couldn't say, so that it would be of any use. I might say fifteen dollars at a time.

Q. You say you had tickets of Starkey. How many did you ever have of Starkey at any one time?

A. Well, about all. The larger part that I had was just before this Whitcher affair came up.

Q. How many did you receive from him at that time?

A. I really couldn't tell.

Q. Give us some kind of an impression.

A. My impression is nothing unless it comes pretty near the mark, you know.

Q. Give us the best impression you can of the number of tickets that you had from Starkey at any one time.

A. I don't like to go on guess work and impression very well.

Q. Did you receive ten from him at any one time?

A. Might be ten of them.

Q. Did you receive from him, at some one time, more than twenty-five tickets?

A. I should say not.

Q. Will you say that you didn't receive from him twenty tickets?

A. I shouldn't want to say it; but I should think not.

Q. At the time he let you have these ten tickets, what kind of tickets were they, and where did they go?

A. Manchester to Boston.

Q. Over what road?

A. By the way of Lowell.

Q. State whether all the tickets you had of Starkey were tickets to Boston over the Lowell road.

A. Those that I had of Starkey were.

Q. Those that Starkey furnished you were not over the Lawrence road, but all over the Lowell road?

A. I think so.

RE-DIRECT EXAMINATION.

Q. (*By Mr. George.*) You have been speaking about Starkey furnishing you tickets. About how many times did he furnish you tickets?

A. Not many times; two or three times. I was passed on the cars. [Objected to.]

*The Chairman.* I do not understand that there is any objection yet made to proving that Starkey let him have tickets coming from Clough. The referees do not see any connection between that and the fact that Mr. Starkey may have passed him over the road free. They do not see how that is one of the collateral circumstances.

Q Did Mr. Starkey's wife buy a dress at your store? [Objected to and ruled in.]

(*The Chairman.*) The counsel brought out, on cross examination, the fact that this witness had tickets from Starkey. I believe the referees are of opinion that the plaintiff is entitled now to inquire as to the circumstances connected with the giving of these tickets; for instance, to show how he paid for them, and all those circumstances. We don't think, if I understand the opinion of the referees, that it opens the case to show that Starkey got him passed over the road free sometimes; but that he proposes to show that Starkey was paid for them.

Q. Will you then answer the question, whether Mrs. Starkey purchased a dress?

A. I not only say that she purchased a dress, but that she had been in the habit of trading with us, and paying for her goods, up to a certain point.

Q. How was it after that point?

A. There was an order came in, brought by Hill's express, for some goods; I think a dress pattern.

Q. (*By Mr. Haile.*) Did you state what time this was?

A. No; I could not state.

Q. (*By Mr. George.*) About what time was it, say? How long before the arrest of Whitcher?

A. Oh, this must have been some little time before that.

Q. Some time before the arrest of Whitcher?

A. Yes, sir; of this dress we are now talking of.

Q. Was it while Mr. Clough was running as conductor?

A. I presume it was.

Q. And at that time was Mr. Starkey running as Mr. Clough's brakeman?

A. I think so.

Q. Well, sir, how was that dress pattern paid for, if at all?

A. Well, I was going on to state.

Q. I only want to know if it was paid for by tickets; and, if so, the circumstances under which it was paid?

A. I could not tell you, sir. This order that I have spoken of came in enclosing a sample that his wife had previously taken, stating that he would pay for it in a few days. *I want it understood that the dress was not sold expecting railroad tickets for it; but it was sold expecting the cash in a few days; nothing said about any other kind of pay for it but money.*

Q. Well, go on and state how that was paid for.

A. I will state how he procured it in the first place. I talked with Mr. Hill to know if he trusted Mr. Starkey. I wasn't very anxious to sell the fellow, anyway.

Q. Go on, and state how that dress was paid for.

A. I couldn't tell whether he did pay for that dress or not; but either that, or else he got another dress afterwards that he didn't pay for.

Q. I want to know if he paid for it in tickets?

A. Whether that is the last charge or not—for the last one all I ever got I got in tickets.

Q. Now, won't you tell how you got them in tickets and why you got them? Tell us all the circumstances.

A The simple fact is, that Starkey handed them to me. That is all there was about it.

Q. How did Mr Starkey get the tickets?

A. That is all I know; I don't know where he got them.

Q. Hadn't these tickets been used before? That is, didn't they have indications of use upon them, either stamp or punch?

A. I cannot tell, sir.

Q. What is your recollection?

A. It isn't clear enough to tell you. They were tickets that were good from Manchester to Boston.

Q. I want to know if they were tickets that had indications of being once used?

A. My memory isn't good enough to tell about that.

Q. Where did he hand you these tickets?

A. I suppose his idea was——

Q. What point? What place?

A. Oh, I should think at the depot in Manchester once, and in the cars.

Q. In the cars as you were going down?

A. Coming up, probably.

Q. Of whom did he obtain these tickets that he handed you in the cars, and who was the conductor on board the train?

A. I tell you, sir, the last tickets that I had was when he was boss and all hands himself.

Q. My question was of whom he obtained these tickets.

A. I couldn't tell you, to save my soul.

Q. Who was the conductor on the train at the time he gave you these tickets?

A. I don't know what conductor ran on the train.

Q. Did ever Mr. Starkey go to Mr. Clough and get a ticket to pass you over the road?

A. Not to my knowledge.

Q. Did you ever state that he did?

A. Well, my statements you have got. I can only say that from this point, to-day, I have no recollection of saying that Mr. Starkey ever got a ticket from Mr. Clough.

Q. You cannot tell whose train Mr. Starkey ran on as brakeman?

A. I don't know; I don't know as I ever knew.

Q. What did you pay Mr. Starkey for those tickets that you bought? How did you pay him, and at what rate?

A. I suppose Mr. Starkey intended that there was a balance due, and called it an offset, to settle his account. There was no talk about it. I supposed it was so.

Q. You paid him no money for the ticket? Did you credit him for the value of the tickets on his account?

A. Well, sir, you was asking how that first dress was paid for——

Q. I want to know whether you credited the tickets on Mr. Starkey's account?

A. I think very likely I did.

Q. (*By the Chairman.*) Is your book here in which that account is charged?

A. There are more things than that dress. But still I could not tell you the items there without another book.

Q. (*By Mr George.*) The book shows, prior to October 10, 1865?

A. It shows in 1865, July 25th, merchandise, $10
August 26th, 9
October 10th, 13

Cash is credited, July 25th, $5.25
27th, 1.75
August 1st, there is no amount carried out. Page 173.
August 21st, 3.00
September 20th, 1.75
September 25th, 3.50
October 4th, 10.00
October 8.00
February 27th, 1866, 12.13
February 7th, 1866, merchandise, 4.75
" 7th, 5.25
" 8th,

Q. This shows all your dealings with Starkey?

A. No, not all our dealings; I had some cash.

Q. All your ticket dealings?

A. That is all the ticket dealing. That shows that the tickets are in there somewhere. Whatever he has had that he did not pay cash for is there.

Q. And you never bought tickets of him and paid cash for them?

A. No, sir.

Q. Mr. White, have you any recollection of Mr. Clough being on the train when Starkey was brakeman?

A. I have no special recollection about it.

Q. Don't you know the fact?

A. Did Mr. Starkey run on a particular train? If he did, I don't know.

Q. Have you any recollection of Mr. Clough's being on any particular train when Starkey went on that particular train to Boston?

A. I have no special recollection of it. He might have been on, and might not. If you ask me what I had for dinner last Thanksgiving, I could probably tell you something about it.

Q. (*By Mr. Mugridge.*) Do you know whether Starkey used to go over the Manchester road to Boston rather than over the Nashua and Lowell?

A. All I know about it is that he used to run down on the Nashua and Lowell road.

Q. Did you ever notice Starkey, when he was brakeman on the road, commencing at the end of the train to collect tickets himself; that is, to aid the conductor, while Mr. Clough, who was conductor, was going through the train from the front?

A. It seems to me as though I had an impression of seeing him, once or twice, with a lantern, taking them up.

Q. He would commence at the rear of the train and go towards the front?

A. I never thought of it before, but it does seem as though I had seen him in the evening coming up with a lantern.

Q. How frequently have you seen him doing this?

A. I should think the cases were rare, and exceptions to the rule; but it seems to me that I have seen him.

Q. Do you recollect how far he would go, how many cars on a train, or how many tickets he would take up in this way? Have you such a recollection?

A. Nothing more than that of seeing him pass along.

Q. Will you state whether this Starkey acted as conductor for a while on the road after Mr. Clough left it?

A. I have an impression that he did for a few days.

Q. You don't know when he was appointed, and how long he ran the train on the road as conductor?

A. No, I do not; my recollection does not show me.

Q. Here seems to be, on your books, transactions with Starkey upon the debit side, amounting to some eighty or eighty-one dollars. Now the amount is not on the bills. You do not mean to give the referees to understand that these payments, either of them, were made by tickets?

A. Not much.

Q. Are you able to state what proportion of the credits which appear upon your books were made from the price of tickets?

A. Not more than I have stated of the price of tickets.

Q. You don't mean that these tickets here resulted from the transactions between you and him?

A. No, sir.

Q. Would you be able to come to a conclusion that would approximate the truth as to how much of these debit charges were paid by these tickets?

A. I should have to take my memory along back, and do the best I could. Still, I don't like to do the thing, because I may hit very far from it. These things are matters that have dropped out of my mind.

Q. Yes, but we will take it as your best recollection. Ten dollars?

A. More than ten dollars.

Q. Twenty?

A. Well, I should set it along in the neighborhood of twenty-five dollars. You don't take that only you know for what it is worth.

Q. (*By Mr. George.*) You were asked if you had tickets of Starkey and others, Mr. White. Wasn't part of these tickets that were bought of individuals? Whether you had any from officers of the road? [Objected to.]

*The Chairman.* The referees think that this matter may be followed up just as far as it was opened by counsel for the defence. We understand that the counsel for the defence put a general question.

Q. (*By Mr George.*) Will you tell me what others? I don't mean the Ogdensburg tickets. Have you had tickets given you?

A. Well, that is pretty much a matter of my——that is, I would say individuals that I got them of.

Q. I mean, have you had any tickets from conductors upon the road directly? And if so, of whom?

A. None unless I made——I don't know as I have, unless I was passed.

Q. It is pretty essential that it should be stated correctly.

A. I may possibly have had a ticket or two of Noyes. You see I do not want to do any injustice in that direction.

Q. Was Mr. Noyes conductor upon that road?

A. Yes, sir.

Q. Did Mr. Noyes's wife buy a dress and give you tickets in payment for it, or part payment?

A. His wife has traded at the store more or less. I can't say that he ever paid for dresses.

Q. Has his wife bought a dress and had it charged, and given you tickets in part payment? [Objected to, and withdrawn.]

Q. Mr. White, you have spoken of purchasing tickets thus far of Mr. Whitcher going to Boston, principally, as you recollect, of people from Canada, Ogdensburg and the West, and of Mr. Noyes; I will now ask you of what other persons and to what amount you have purchased these tickets over the Concord Railroad?

A. Well, I can tell you as far as I can recollect the persons; but as to the amount, that will be impossible.

Q. Give the amounts as near as you can.

A. I have had some tickets from Charles Morse (I think his name was Charles).

Q. Now I want you to state about how many?

A. It was so long ago it would be impossible to tell exactly. Perhaps if I set it at ten dollars I should not be within bounds, you know.

Q. When was it, as near as you are able to state?

A. Well, it was in war time, That is about as nigh as I can give it.

Q. Now you say about ten dollars. What did you pay him for them?

A. Well, the amount I would not state certain.

Q. What per centage of their par value? How much did you pay for the tickets?

A. I should say about the same discount as those I stated here.

Q. That is twenty-five cents on a ticket from Manchester to Boston?

A. Yes, sir; that's my impression.

Q. Did you buy any tickets—if so, at what time—of Jacob Morse?

A. Yes, sir; I had a few of Jacob Morse.

Q. Who was Jacob Morse?

A. He was a brother of Charles Morse, a trader at Manchester.

Q. And at what time did you buy them, and to what extent?

A. It was only during this year or two; not a large amount. I have no means of stating it.

Q. As near as you can?

A. I want to tell you as near as I can. I cannot say. Not a large amount.

Q. I want you to tell the referees as near as you can.

A. Well, sir, I might shoot very wide of the mark, therefore let it be so understood. I should say twelve or fifteen dollars.

Q. Did you buy any tickets of Herman Hershfield?

A. Yes, sir.

Q. When?

A. I could not give the time.

Q. He was trading there?

A. Yes, sir.

Q. To what extent did you buy?

A. It was a small matter.

Q. About how much?

A. I was trying to get hold of something or other to go from. It was a small amount.

Q. Who was Mr. Hershfield, and what was his business, and where did he live?

A. I think at that time he was in the store with Mr. Morse.

Q. He was a partner?

A. I think he was not a partner; he was in his employ.

Q. Was the brother, Jacob Morse, employed in his clothing store?

A. Yes sir; at any rate there is this much about it—he went to Lowell after that; I know after he went to Lowell he asked me for a ticket. I do not know what was said. I think I refused him.

Q. These tickets you used for yourself, and sold them?

A. Used them for myself and family

Q. What did you do with the balance of them?

A. Sold a few.

Q. To whom did you sell them?

A. Well, let me see; there was, I recollect—I disposed of a few to E. K. Chandler.

Q. Well?

A. D, F. Straw.

Q. Well, any to William B. Johnson?

A. Johnson had some tickets of me. I have sometimes bought a package of tickets myself in Boston. Yes, I guess Johnson did have a few; I think he did.

Q. I mean—not talking package tickets. How was it about John S. Folsom?

A. I think Folsom had some to Boston once or twice.

Q. S. L. Hastings?

A. I recollect he had a ticket down to Boston once.

Q. Do you mean to say that that is all he had? Did you give your deposition? Will you not look at your deposition and see what you then sold?

A. E. L. Hastings had some; I don't know how many. [Objected to.]

Q. How was it about Mr. A. H. Webb?

A. He must have had some.

Q. Mr. William H. Elliott?
A. He had some.
Q. Mr. Stanton, insurance agent?
A. Yes, sir.
Q. How many did you sell to Mr. Stanton?
A. I cannot tell.
Q. How many do you think?
A. I haven't any idea; he had more than any other one.
Q. Mr. Thomas P. Pierce—sell him any?
A. Well, I don't know what I told you at that time. My mind is not clear on that point to-day.

[Counsel for the plaintiff proposes to show a deposition to the witness which is objected to.]

*The Chairman.* We think that is a kind of paper that may be submitted to the witness, if he wants to see it to refresh his memory.

Q. (*By Mr. George.*) I will now ask you if you sold these tickets to Thomas P. Pierce, or such tickets?

A. It says so there, I think. My recollection is not clear on the Pierce matter; my recollection was better, then.

*The Chairman.* I suppose the true way is to ask him to refresh his memory.

Q. (*By Mr. George.*) Looking at that paper, is your memory refreshed?

A. I don't seem to remember; I should not want to say; not at this time.

Q. Now I want to ask you whether all these tickets that were purchased by you—Concord Railroad tickets in packages, were purchased at twenty-five cents discount—whether that is the price?

A. I take it for granted that that is the fact of the case.
Q. In what did you pay?
A. Well, as I before stated, tickets were paid for in trade.
Q. In goods, you mean?
A. Yes, sir.

Q. When you went to Boston in the cars, how often did you go to Boston from Manchester, say for six or eight years, or for five or six years previous to this controversy?

A. Well, sometimes I went down once a week, or twice a week, and then again not for two weeks.

Q. That is, you mean it to be understood that you went all the way from once in two weeks to once or twice a week?

A. Seldom twice a week—not as often; may possibly have been times that we ran oftener than that.

Q. And you paid money in the cars to the conductors in all these cases?

A. Well, I should say that I very rarely paid money in the cars. It was provided for otherwise; but still there is no doubt, but that occasionally ——

Q. Did you pay money to Mr. Clough?

A. I have no distinct recollection now of paying to Mr. Clough; it would be very strange, the number of times I rode on the cars, if I did not pay something to Mr. Clough.

Q. Is that the only answer you can give?
A. Well, I guess that perhaps covers the ground.

Q. Does it refresh your recollection? Did you pay money in the cars? And was it to Mr. Clough?

A. I have rarely paid money.

Q. Did you at any time pay money to Mr. Clough in the cars, without regard to time?

A. Well, to-day, this hour, and this minute, I have no recollection of taking out my purse to pay any money to Mr. Clough. There may be such an occasion, and I have no doubt that there is.

Q. Have you any general recollection of that fact?

A. I have nothing more than—I should be quite likely sometimes to be not provided.

Q. What do you mean by that?

A. Having no ticket about me.

Q. That you would not have a ticket with you?

A. Not having a ticket about me, and not having been to the office.

Q. That is the only way? Do you mean to swear that is the only way you had to get to Boston?

A. Only these, or else Starkey passed me.

Q. That is only three ways.

A. I have stated that Mr. Starkey had passed me.

Q. Have you had any conversation with the counsel for the conductors since you have been at Concord?

A. No, sir.

Q. With any other persons since you gave this deposition?

A. Not that I recollect of.

CROSS EXAMINATION *resumed.*

Q. (*By Mr. Mugridge.*) You have not had any conversation with me, have you?

A. No, sir.

Q. You have not had any with Mr. Tappan?

A. No, sir. I do not recollect that I did.

Q. Have you had any with Mr. Rolfe?

A. I have not.

Q. Any with Mr. Clough?

A. I never spoke with him on the subject in the world.

Q. You have been inquired of as to the different persons to whom you sold these tickets. I suppose you sold them all save what you needed for yourself and family?

A. Yes, sir.

Q. That is, you used what you had occasion to use for your own purposes, and you got them for twenty-five cents less?

A. Just that and nothing else.

Q. You have been inquired of as to Mr. Charles F. Morse; was he a man who lived in Concord—a clothing dealer?

A. Yes, sir.

Q. Was he a German?

A. Yes, sir.

Q. And Hershfield was a German?

A. Yes, sir.

Q. And you got all these tickets that you had during the war?

A. Yes, sir.

Q. They were Germans, and you had them of these men during the war?

A. That is my impression that I had them.

Q. At the time the substitute business was carried on in Manchester?

A. Of course that was carried on during the war. I wouldn't say that I did not have any from them before the war.

Q. The bulk of them were during the war?

A. Yes, sir.

Q. Did you know that Charles F. Morse and Mr. Hershfield—or that Mr. Charles Morse dealt in substitutes?

A. I didn't know it particularly.

Q. You were not intimately acquainted with them?

A. No, sir.

Q. Did you know whether his brother dealt in substitutes during the war?

A. Not to my knowledge.

Q. Now I want to know, first, as to the tickets that you had of these persons; to what stations and upon what lines of road were they good? Let me inquire particularly of those you had of Charles F. Morse.

A. As far as my recollection is of the matter, I should think mainly what I had, went to Boston by way of Lowell I should say.

Q. And those that you had of——

A. Possibly there might be some the other way.

Q. Were those you had of his brother and Hershfield by the way of Lowell?

A. I should say they were both ways, by Lawrence and by Nashua. That would be my opinion.

Q. Did they run from Manchester to Boston and from Boston to Manchester, both ways?

A. Such is my impression. If I had known this thing was coming up, I would have had these things down.

Q. How many tickets did you have in your possession at the time of this controversy. Do you remember?

A. No, sir.

Q. Have you any sort of an idea as to how many?

A. I could not have many.

Q. Half-a-dozen?

A. What, do you mean half-a-dozen trips?

Q. Half-a-dozen trips?

A. I should not think I had half-a-dozen down and back.

Q. Half-a-dozen trains, do you think?

A. I should think I had.

## TESTIMONY OF DR. JUSTUS BLAISDELL.

Q. (*By Mr. George.*) Please state where you live?

A. I reside at Concord.

Q. Will you state whether you formerly kept your office in the Masonic building, owned by Mr. Clough and Mr. Corning?

A. Yes, sir.

Q. At what time, Doctor?

A. I hired of them; it was October, I think, 1859.

Q. And how long did you continue to occupy?

A. I was there seven years.

Q. While you were in the occupancy of that office, will you state whether you were passed over the Concord road by Mr. Clough? and if so, how many times, and under what circumstances?

A. Well, I went over the road a few times. Clough passed me, or gave me a pass.

Q. What kind of a pass, if you please, Doctor?

A. It was a pass with his name signed to it, "Please pass Dr. Blaisdell from Concord to Boston."

Q. From Concord to Boston, or to Boston and back?

A. I think from Concord to Boston.

Q. Will you state now under what circumstances?

A. Well, it was a favor. I had asked Clough to pass me, and he done it as a favor.

Q. Dr. Blaisdell, you gave your deposition in this suit on the 27th February, 1866, did you not?

A. Yes, sir.

Q. Will you please look at your deposition? [Objected to.]

*Mr. Tappan.* Wouldn't it be competent to inquire and see what he recollects now?

*Mr. Mugridge.* He is your own witness.

*Mr. George.* I understand that—certainly. I believe that I am in the regular order.

*Mr. Mugridge.* It don't appear that his memory is at fault. If the witness desires his recollection to be refreshed; if he exhibits any intimation that he cannot testify, and desires to look at a proper memorandum for that purpose, it is regular, and of course, proper, that he should; but he has answered every question that has been put, and has not expressed any desire to refresh his recollection by a deposition which they took some time in the past.

*The Chairman.* We think that the deposition should not be shown to the witness, or any other papers, for the purpose of refreshing his memory until it becomes apparent that his memory is at fault, and that he desires to refresh it.

*Mr. George.* I move for leave to cross-examine this witness on the ground that he has stated entirely different from what he stated before, and in order to lay a foundation. I simply desire to pass the deposition to the court, for the purpose of cross-examining the witness.

*Mr. Tappan.* I should like to inquire whether it is usual and just and fair to such a gentleman as Dr. Blaisdell—or anybody, after parties have gone on such an excursion as they have, and taken depositions in the way these have been taken, and when the witnesses are produced on the stand, and do not come exactly up to the scratch, that they are to be put in the position of culprits

*The Chairman.* I believe the referees do not think at this time there is a case made for putting leading questions. That is, they do not think that the witness has at present shown any reluctance to testify, such as could justify the counsel in putting leading questions. But I do understand that the counsel may refresh his memory in any matter. If, upon inquiring about anything, the witness don't remember, his memory might be refreshed as to what he did testify.

*Mr. George.* Suppose the witness testified that he applied to Mr. Clough for tickets, and supposing that Mr. Clough offered them to him voluntarily as a consideration for repairs.

*Judge Bellows.* We think you might show him the deposition and ask him to say, under the circumstances.

The deposition was shown to the witness.

Q. (*By Mr. George.*) Well, Mr. Blaisdell, I have shown you your deposition?

A. Yes, sir.

Q. I want to know whether it refreshes your recollection as to the instances in which you received passes?

A. It does.

Q. Now I want you to state how it was about you receiving passes from Mr. Clough to Boston and back; and how many times you received such passes; and under what circumstances, and all the circumstances of receiving such passes?

A. Well, I asked Mr. Clough if he would pass me over the road; and I supposed that he had a right to do so. He said he would, and did so. Well, that answers that question, I believe?

Q. How many times did Mr. Clough give you a written pass to Boston and back, if at all?

A. I could not say as to the number of times; but few times for myself.

Q. Did these written passes extend from Concord to Boston and back?

A. Well, I couldn't say positive. I think a proportion or some of them were to Boston and return, or something like that; not on all of them. I think they were only one way, some of them.

Q. Now I will ask you the same question as to your deposition, when I read it to you, "How happened Mr. Clough to furnish you with such passes?"

A. By my request—I requested him to do it, I suppose. I asked it as a favor. I supposed he had a right to do so, or else I should not have asked him.

Q. Will you look at your answer to that very question, and see whether it refreshes your recollection? That is the third.

A. Yes, sir; I understand. I recollect of answering as they state then.

Q. Is that answer true?

A. Yes, sir. To explain that, I would say I went on to suggest——

Q. I will put the question to you, and I want you to answer it to the referees. I will put this question. How happened Mr. Clough to furnish you with such passes? State fully.

A. That may be the question that you proposed; I do not recollect.

Q. I propose it now.

A. I do not remember that question in that form.

Q. Will you read that question?

A. The question that I understood was, if there had been anything said about repairs — that these passes were to go for repairs. My recollection of what I said was that Mr. Clough went into the office; I asked him to go in and asked him about repairs. He asked me what the rent was, and I told him. Then he said that — well, I said to him that — or he said to me that I might see Mr. Corning, and talk with him. He thought they could not afford it, but whatever Mr. Corning said, he would agree to.

Q. (*By Mr. Mugridge.*) That was about the repairs?

A. Yes, sir. That is all I recollect of saying. I might have said what it states here, but I have no recollection of it, or Mr. Clough's saying so to me.

Q. Did Mr. Clough offer you free passes?
A. No, sir; he never did.
Q. Did Mr. Clough say to you that he could not afford to make any repairs, and if you would fix it up, he would make it all right?
A. I have no recollection; he might have said something of that kind, but I have no recollection of his saying anything of that kind.
Q. (*By Mr. George.*) Dr. Blaisdell, I understood you that you occupied a room in Clough's and Corning's building?
A. Yes, sir.
Q. Did you apply to Mr. Clough to make repairs on that room?
A. I did.
Q. What did he say about it?
A. I told him what needed to be done; that it needed whitening and painting. Well, he asks me what rent I paid. I told him, and he said that he could not afford to make any repairs.
Q. Did he say anything about your going on and making repairs, and he would make it all right with you?
A. No, sir.
Q. Did you go on and make the repairs?
A. I painted it myself. Painting myself cost me about two dollars. If I had hired it, it would have cost about six.
Q. Did you at different times make repairs on that room that would amount to about fifteen dollars? [Objected to.]
Q. State whether you made repairs on that room at different times amounting to about fifteen dollars? [Objected to and ruled out.]
Q. To what extent did you make repairs on that room? To what amount in money?
A. I would say that when I moved in there, I said to Mr. Corning that I wanted repairs made, and to have another room, and suggested that I had been paying out considerable money and that he might put it on to the rent. That was when I first moved in there, before the building was completed, and his workmen were there, and they put up a partition. The rent was to be seventy-five dollars, and he put it up to one hundred dollars the first year. That was in another room. I moved into the front room. That was in the back room, or on Pleasant street. Well, sir, I moved the partition in there. Then I spoke of these other repairs. That is all that I recollect of laying out on the room. And when I moved out — it was after Mr. Corning died — I spoke to Mr. Taylor about moving the partition; he objected, and said he had no right to remove it, and I let it remain there. It cost about eight dollars to remove the partition. It cost me twenty-five.
Q. Now, Dr. Blaisdell, I wish to ask you the direct question, how much did you expend in repairs; in the repairs that you had a conversation with Mr. Clough about, about which you have testified?
A. Well, I have stated all the repairs and the amount, as near as I could judge of what it would cost.
Q. I ask you directly, sir, how much did you expend for the repairs, about which you spoke to Mr. Clough, when he said he could not afford to do it?
A. Well, I could not give the exact amount. It was at different times, as I said. I had it colored and whitened overhead, and the casing painted. It might have been ten or fifteen dollars. I could not tell the whole amount, what it would be.

Q. Did you ever have any compensation — I am speaking of the repairs that you spoke to Mr. Clough about — if so, what was it?

A. No, sir.

Q. How long after these repairs were made before you received a pass to Boston?

A. Well, I could not say. I think they were before; it might have been that he passed me once after the repairs.

Q. Did Mr. Clough offer you tickets, or did you ask him for the tickets?

A. I asked him for the tickets.

Q. He did not offer them to you?

A. No, sir; not without my asking him.

Q. Were you ever in the cars when Mr. Clough came to you and told you he would get you along, and bought you tickets and gave them to you from Concord to Boston and back — before you got to Nashua — tickets or passes?

*Mr. Mugridge.* That is a pretty leading question, I think.

A. I have no recollection.

Q. Did ever I or any one else speak to you on the subject of these suits prior to your giving your deposition?

A. No, sir.

CROSS EXAMINATION.

Q. (*By Mr. Rolfe.*) Has ever Mr. Clough spoken to you about this matter, either before or since you gave your deposition, in his life?

A. No, sir.

Q. Have you ever spoken to him?

A. No, sir.

Q. Has any one in Mr. Clough's interest or employment ever spoken to you at all about it?

A. No, sir.

Q. At the time you asked Mr. Clough to do certain repairs, and he stated he could not afford to, and referred you to Mr. Corning, was anything said about passing you over the railroad; or giving you any favors in that direction?

A. I have no recollection of it — there might have been; I would not say; I do not know but he said to me, when I asked him, that he would make it right, straight, or something of that kind; that he could not afford to make the repairs at that rent. He rented it low to me, lower than I could get rent anywhere else.

Q. Did he say anything at that time by which you could infer, or by which you did infer that he intended to grant you any favors by railroad, any free passes; if so, what did he say?

A. He never gave me any intimation, that I can recollect, that he would pass me over the road, only as I had asked him.

Q. And you say that these times you had been passed free was previous to your asking him to make these repairs?

A. Yes, sir; I think they were.

Q. Now, Dr. Blaisdell, had you, previous to your being passed over the railroad at all, stated to Mr. Clough your pecuniary condition?

A. Yes, sir; I had.

Q. Did you tell him in words that you were poor and had a hard time to get a living, and that you would like to be favored in that direction?

A. Yes, sir; I think I did.

Q. You stated to him what was then true, did you not?

A. Yes, sir.

Q. For what purpose did you want to go to Boston?

A. Well, go down there to see my friends; I had a brother and sister.

Q. Do you recollect, when you first went to Boston, whether you asked Mr. Clough to pass you before you got into the cars or after?

A. Before, I think; I think it was just before the train started; standing by the cars.

Q. And he told you he would pass you, and got you something to go to Boston, did he not?

A. Yes, sir.

Q. The next time that you passed down in the cars, did you speak to Mr. Clough before you got into the cars?

A. Yes, sir; I think I did.

Q. Did Mr. Clough finally have to refuse, or did he refuse to pass you?

A. He did.

Q. You came back without going to your journey's end, did you?

A. Yes, sir.

Q. What time in the year was this?

A. I could not tell.

Q. With reference to the time when this excitement was produced in relation to the conductors; how many months before that?

A. I do not know when the excitement was; I did not hear much about the excitement; I could not tell whether it was before or after.

Q. The time previous to the last time he passed you, did he suggest to you that he did not know as he ought to pass you, and did he finally consent to pass you?

A. No, sir; I have no recollection of his intimating anything of that kind. I always supposed that he had a right to. I always supposed that a conductor had a certain amount of passes given him, that he could give to friends. I never should have asked him for a pass if I had not supposed that he had that right.

Q. What person did you do that business with? To whom did you pay your rents while you were in the office?

A. Mr. Corning.

Q. And after Mr. Corning died, who to?

A. Mr. Taylor.

Q. You never hired the office of Mr. Clough?

A. No, sir.

Q. *(By Mr. George.)* He was owner, was he not?

A. His name was signed. I supposed he was.

Q. Mr. Taylor continued to do Mr. Corning's business after he died?

A. Yes, sir.

Q. *(By Mr. Tappan.)* Mr. Corning was postmaster, was he?

A. Yes, sir.

Q. Mr. Taylor was his clerk?

A. Yes, sir.

[Adjourned.]

[THIRD DAY, July 23, 1868.]

## TESTIMONY OF WILLIAM ROBY.

Q. (*By Mr. George.*) Mr. Roby, please state where you live and what your age is.

A. I live in this city, on Washington street.

Q. And what is your age, sir.

A. My age is seventy-two, nearly.

Q. Mr. Roby, will you state whether you and your wife had occasion to go down to Boston in the winter of 1865, and whether you did go?

A. I was down there, and she was with me, some time in 1865, if my memory serves me.

Q. Which way did you return when you came home?

A. By Nashua and Lowell.

Q. You left Boston and came to Concord by way of Lowell.

A. Yes, sir.

Q. Who was the conductor on the Concord railroad when you got to Nashua?

From Nashua up here?

Q. From Nashua to Concord; yes, sir.

A. Mr. Clough, I think.

Q. Now, if you noticed anything with regard to tickets you may state what tickets you bought and what you noticed with regard to tickets? [Objected to, and objection withdrawn.]

Q. Well, Mr. Roby, will you be kind enough to just state, when you got into the cars in Boston, what tickets you bought, and what transpired?

A. I did not buy any myself; my son stepped into a place where he bought the tickets, and brought me four tickets; two to go part way up. The first conductor took two and Mr. Clough took the other two.

Q. Now, with regard to punching the tickets; state what was done by the conductors?

A. Well, the first conductor punched his ticket, and the second time he called on me he punched them again and took them.

Q. That was below Nashua?

A. Yes, sir.

Q. State what happened when you got this side of Nashua?

A. Mr. Clough took the two tickets, and made, either in this corner or that,—he did not happen to hit them.

Q. Mr. Roby, now you may show, if you please, exactly the motion which Mr. Clough made.

A. The punch went by the end, either by that corner, or that, and did not strike the tickets, and he handed them back to me.

Q. Now go on, sir, and state what further happened with regard to these tickets.

A. When we got up this side, he took the tickets and carried them off.

Q. Did he punch them?

A. No, sir.

Q. When you got up this side of where, did he take the tickets and carry them off?

A. Between here and Manchester, of course.

Q. Before you got to Concord?

A. Before we got to Concord. It might have been on the edge of Concord. [No cross examination.]

## TESTIMONY OF ISAAC W. FARMER.

Mr. Isaac W. Farmer, of Manchester, N. H., was called by the plaintiff and duly sworn, and testified as follows:

Q. (*By Mr. George.*) Mr. Farmer, are you deputy sheriff?
A. I am, sir.
Q. Do you live at Manchester?
A. I do, sir.
Q. Will you state whether you went to make an examination at Mr. Whitcher's house, and if so, just state the circumstances, and what you found?
A. I went to Mr. Whitcher's house ——

*Mr. Mugridge.* I suppose the object of this testimony is to show the tickets found on Mr. Whitcher?

*Mr. George.* Yes, sir.

*Mr. Mugridge.* We would like to inquire whether the referees will constantly receive this testimony with regard to tickets found in the possession of Whitcher, until it is shown that a portion of them are found to be coming from Mr. Clough. There has been a large amount of evidence here, upon that point, upon the assurance that there would be evidence adduced hereafter, connecting these tickets with Mr. Clough. The counsel on the other side is now introducing cumulative testimony on that point, and the question is how far the court will receive evidence of this kind until it is shown to be connected in some way with Mr. Clough. We object to admission of the testimony.

*The Chairman.* We do not see any occasion to change our principle, or the course that we have adopted before, if the counsel states that he has evidence by which he expects to connect with Mr. Clough.

*Mr. George.* I shall prove it from Mr. Clough's own deposition.

Q. You say, Mr. Farmer, that you are deputy sheriff?
A. I am.
Q. Were you called upon to go and arrest Mr. Whitcher? Can you give the day?
A. February the tenth.
Q. Will you state the circumstances of his arrest?
A. Mr. Whitcher was arrested in Manchester, and I was sent up to his house on purpose to find the tickets at his house.
Q. What time of day was this, sir?
A. This was in the afternoon, I think.
Q. How did you go?
A. Upon an engine, or a special train. I think it was upon the engine.
Q. When you got there what did you find?
A. I found, in one room up stairs, in a small bureau, one hundred and seventeen railroad tickets.
Q. What was done with those tickets?
A. Those tickets I took and brought down to Hooksett, and gave them to Mr. George.
Q. What did Mr. George do?
A. He took them and marked them with the letter G on the face of them; marked them there in the depot.
Q. In the depot at Hooksett, on the desk?

A. Yes, sir.

Q. Won't you see if these are the tickets?

A. I should presume they were; they look like it.

Q. Were these tickets the one hundred and seventeen tickets that you found at Mr. Whitcher's?

A. Found at his house.

Q. Were they all together?

A. All together, in one place, in a small drawer in a bureau.

*Mr. George.* I want, in this connection, to call attention to the stamps on the backs.

*Mr. Mugridge.* That applies only to a portion of the tickets, with regard to being stamped.

CROSS EXAMINATION.

Q. (*By Mr. Mugridge.*) Substantially, is your testimony that you found one hundred and seventeen tickets?

A. Yes, sir.

Q. Do you know that these are the same tickets?

A. That is my impression, so far as they are marked in that way.

Q. I suppose you cannot say but what somebody else might have put on that G?

A. Yes, sir; if he counterfeited his hand.

Q. Are you able to say that a single one of these tickets was among the tickets that you found in Mr. Whicher's house?

A. Oh, I could not identify positively.

Q. Are you able to say that a single one of the tickets presented to you is a ticket you took from Whitcher?

A. I have no particular mark on them.

Q. (*By Mr. George.*) You say the general character of them is the same?

A. Yes, sir.

Q. Is there anything about the general character of them that would enable you to identify them?

A. No, sir.

Q. Or is there anything about the letter G that would enable you to identify them?

A. No, sir.

Q. (*By the Chairman.*) Mr. Farmer, I want to ask you one question, and that is whether or not you saw these tickets marked, or whether they were marked in your presence?

A. I did; yes, sir.

Q. Have you any knowledge of Mr. George's hand-writing?

A. Well, I've seen his hand-writing, but am not very familiar with it.

Q. These tickets that you have examined this morning, you can state what you think about them.

A. They have every appearance of being the same tickets.

Q. I mean the marks that you saw on them?

A. Yes, sir.

Q. (*By Mr. Mugridge.*) Do you mean that every ticket was marked at the time?

A. Yes, sir; I think they were.

Q. Now, let me ask if all the tickets contained in these packages are marked;

A. I think not.

Q. Your impression is that all the tickets that you saw taken from Mr. Whitcher were marked G by Mr. George?
A. That is my understanding. I think they were.
Q. There were one hundred and seventeen of them?
A. There were one hundred and seventeen tickets taken from the house.
Q. And they were tickets over definite roads? Tickets over the Manchester and Lawrence and Boston and Lowell?
A. Yes, sir; and the adjoining roads.
Q. Are you particular that all the tickets were marked?
A. I was not particular. I saw him there marking them.
Q. Was it made a point that they should all be marked?
A. Yes, sir; he asked me to see them marked.
Q. All the tickets?
A. That is my recollection.
Q. (*By Mr George.*) Do you recollect that you handed the tickets, one by one, to me while I marked?
A. I do not remember about that
Q. Who took the tickets from you? Who went with you?
A. I have the impression that there was somebody else there.
Q. Who carried them to the depot?
A. I took them to the depot. Mr. Whitcher went and showed me where they were.
Q. (*By Mr. Mugridge.*) You went up to Whitcher's on an engine that ran up from Manchester?
A. He was arrested in Manchester.
Q. To make search for the tickets?
A. I went with Mr. Prescott. My impression is that there was somebody else.
Q. Who was present at the house at the time Mr. Whitcher produced the tickets?
A. Mr. Whitcher himself.
Q. Colonel George?
A. No, sir.
Q. And you took them from Whitcher's house to the depot?
A. Yes, sir.
Q. Have you seen them until to-day?
A. Not until I came here to this court.
Q. And you are not able to say positively that they are the very same tickets that you took from Whitcher's house at that time?
A. I have no particular mark.
Q. Would you be able to say, with any degree of certainty, that either one of these was a ticket that you took from Whitcher's bureau?
A. I would not swear to it.
Q. Is there the slightest mark as being the tickets taken from Whitcher, and they have not been in your possession since?
A. They never have.
Q. Is there anything about that ticket by which you can identify it?
A. There was a mark that was put on. It has the general appearance of being the same ticket.
Q. Will you be kind enough to take one of those tickets that has got a "G" on it. There is a ticket marked G too. Are you able to swear that that G is, or not, Colonel George's?
A. No, I could not.

Q. Now there are two Gs; is there any resemblance?

A. There is not.

Q. Are you able to say that either of them is in Colonel George's hand-writing?

A. No, sir.

Q. Now take either of these tickets in that package, and are you able to say that Colonel George made a G on them?

A. No, I am not.

Q. Now, I ask you if there is anything connected with them which may enable you to identify them as taken from the drawer of Whitcher?

A. Not positively.

Q. (*By Mr. George.*) Any doubt about it?

A. I have not any doubt about it, but still I could not be positive.

Q. (*By Mr. Mugridge.*) Let me ask you if there is anything connected with the tickets that enables you to swear with any degree of certainty?

A. No, sir.

Q. If an equal number of tickets were laid before you, marked in the same way, would not you be just as positive?

A. If they were similar tickets to those, I would be.

Q. And you would have no more doubt about those than you would about these?

A. I might have.

Q. Supposing they were brought in an envelope like these and laid on the table before you?

A. If I had not known where they were, and where they were kept, or had any knowledge of it, I might be.

Q. (*By the Chairman.*) Mr. Farmer, will you be kind enough to look at that and tell us whether you have any knowledge of the writing?

A. I have not much knowledge of Mr. George's hand-writing, though I saw him mark the tickets at that time. I have seen his writing, but I could not identify it.

Q. What is your best impression about that hand-writing, whether it is or is not the hand-writing of Mr. George?

A. My impression is that it is, but I am not able to swear positively about it?

Q. Now, will you not tell me about that; whether it is or is not the hand-writing of Mr. George?

A. That is the ticket that was found there.

Q. That is not the question. The question is whether you think that is Mr. George's hand-writing; or what is your impression?

A. I do not know, I am sure.

Q. (*By Mr. Mugridge.*) There are two tickets with two Gs on them; do those tickets look alike?

A. Not much.

Q. Which of those Gs looks most like Colonel George's hand-writing?

A. That is the kind he generally makes, I guess; that is the kind that most of them are marked with.

Q. Is there anything about that that is like Colonel George's hand-writing in any single particular?

A. I do not know as there is.

Q. Do those two Gs look alike?

A. No.

Q. Now, assuming that one of those Gs was made by Col. George and one by Col. Tappan, could you say which of the two was made by Col. George?

A. No, I could not tell.

[The deposition of William B. Hardy was put in evidence, and read by Mr. Stanley. Mr. Mugridge objected to that part of the deposition which related to free passes. The chairman suggested that as there seemed to be a difference of opinion, it would be better to hear the deposition, and see if there was any evidence that Mr. Clough received any benefit from them.]

## TESTIMONY OF DANIEL R. PRESCOTT.

Mr. Daniel R. Prescott, of Manchester, N. H., was called in behalf of the plaintiff, and duly sworn, and testified as follows:

Q. (*By Mr. George.*) Mr. Prescott, you are a police officer? and you were in 1866?

A. I was deputy marshal of the City of Manchester,

Q. Would you state how it was on the tenth of February, 1866, about your arresting James Whitcher, of Hooksett, and state where you arrested him?

A. A warrant was put in my hands by Mr. Stanley. I went to find Mr. Whitcher, and found him very near the depot.

Q. That was in Manchester?

A. Yes, sir.

Q. About what time in the day?

A. I am not able to state that.

Q. Do you know whether it was in the forenoon or afternoon?

A. I think it was in the forenoon; I am sure it was.

Q. Well, sir; after the warrant was placed in your hands, where did you go?

A. I went to several places where I thought he would be, and at last I went to the depot, thinking he might be going on the train. I met him on the street, near the depot, very near the platform of the depot.

Q. Well, sir; you arrested him; where did you take him?

A. To the office of Messrs. Morrison & Stanley.

Q. Well, sir; when you got to the office of Messrs. Morrison & Stanley, you may state what he produced or took from his pockets?

A. He produced different tickets; tickets on various railroads, to the amount of two hundred and twenty-six.

Q. Have you a memorandum?

A. I have.

Q. What did you give the number at Hooksett?

A. One hundred and seventeen.

Q. You took these from his vest pockets?

A. Yes, sir.

Q. Will you state now what was done with the tickets as he produced them?

A. As he produced them from his pockets they were passed to Col. George, and he put the letter G on them in my presence.

Q. (*By Judge Bellows.*) You are speaking now of those you found at Manchester?

A. Yes, sir.

Q. (*By Mr. George.*) Whether was it in the front or back office?

A. It was in the back room.

Q. As the tickets were produced, will you state how it was that I put my name on them?

A. Well, sir; you put them on as I passed them to you.

Q. On the face or back?

A. On the back, I think.

Q. Won't you examine these tickets, sir?

A. I don't recognize the figure 2 on these. I don't know about that. The other ones—I should say these were the tickets.

Q. Any doubt about it?

A. No doubt about that package.

Q. Won't you look at them and see.

A. Yes, sir; I have no doubt; these with the extra figure I think I cannot tell. There is the figure 2 on some of them. There is another package, but that I have no doubt they are the ones.

[Seven packages identified and marked with the initial P.]

Q. After Mr. Whitcher had produced these, and you had passed them to me, and they had been marked, where then did you go?

A. In the afternoon I came with Mr. Farmer to Hooksett, to Mr. Whitcher's house.

Q. And what did you find there?

A. We there found one hundred and seventeen tickets.

Q. In what place?

A. In his chamber, in a bureau..

Q. What was done with those tickets?

A. Those tickets were delivered to Mr. Farmer.

Q. Were you present when they were marked?

A. I was not.

Q. Did you see the tickets yourself?

The one hundred and seventeen?

Yes, sir.

A. Yes; I had these in my hand.

Q. What was the age of these tickets?

A. They were various ages. Some very old, and some of them not so old. I note the tickets on the different roads.

Q. But you simply made a memorandum of the number of tickets?

A. That is all.

Q. Won't you read your memorandum, if you will?

A. Saturday, February the 10th, 1866. Arrested, for being concerned in the railroad robbery, A. H. Weston, William White and James Whitcher. Found upon Whitcher two hundred and twenty-six tickets, on different roads; also, one hundred and seventeen tickets at his house in Hooksett. That is all.

#### CROSS EXAMINATION.

Q. (*By Mr. Mugridge.*) Can you state how many of these packages there are that you do not identify? You have given seven that you do; now how many are there that you do not identify?

A. Ten.

Q. Now, out of seventeen packages you can identify the contents of seven. I understand you to say that you arrested Whitcher at the depot and took him to Morrison's office immediately?

A. Yes, sir.

Q. You were there present in Mr. Morrison's office after you had got Mr. Whitcher there?

A. In a private room.

Q. Who was there?

A. Colonel George and Mr. Stanley and, I think, Squire Morrison.

Q. Mr. Gilmore?

A. I am not able to state.

Q. I want you to tax your recollection and see if Mr. Gilmore was there.

A. It is my impression that Mr. Gilmore was not there.

Q. Are you certain?

A. I would not say.

Q. Mr. Farmer there?

A. No, sir.

Q. What was done in the back office after they got Whitcher in there? Or, in the first place, who was with Whitcher in the back office?

A. Colonel George, Mr. Stanley and myself.

Q. Mr. Morrison in there?

A. I think not.

Q. What was done by these gentlemen, with reference to Whitcher, after they got him in the back office?

A. Nothing; only he was asked to give up the tickets.

Q. And did he give them up?

A. He did, after awhile.

Q. Now, without going into circumstances, was there a large amount of talk between Mr. Whitcher, Colonel George and Mr. Stanley, in reference to this matter?

A. There was a long consultation about giving up the tickets?

Q. How long were they in the office there together?

A. I should think thirty minutes or more before he gave up all the tickets.

Q. Was there evidence written down there in the office by anybody?

A. Well, sir, I did not see any, except putting the initial on these tickets.

Q. Did you leave before Colonel George and Whitcher and Morrison left?

A. I think not.

Q. Were there any statements of Whitcher's taken down there?

A. I think not.

Q. None at all?

A. Not to my knowledge.

Q. Now, I understand you to say that you saw Colonel George mark some tickets. Let me ask you if there is anything in the seven packages that enables you positively to identify them?

A. They correspond to these tickets.

Q. Is there anything else?

A. The hand-writing corresponds, as I saw him put them on.

Q. Do you know Col. George's hand-writing?

A. I know that he wrote on them.

Q. That is not the question. Did you ever see his writing before?

A I never did.

Q. Were you then, or are you now, familiar with his hand?

A. His general hand-writing I am not.

Q. Now, I want to ask you if you can swear that that is Colonel George's hand-writing?

A. I have no doubt.

Q. Will you look at these Gs and see if they are in Col. George's hand-writing, any of them?

A. I should think they were not like these on the back of the tickets.

Q. See if you can find a G there that resembles at all any G that he put on the back of any of these tickets?

A. I suppose there might be quite a difference in a man's hand-writing. As he put them on there, he put them on very fast.

Q. Can you find a G that he made?

A. Yes, sir; there are several that are similar to these on the tickets.

Q. Will you pick out a G that you are able to say he wrote?

A. I am not able to say that he wrote any of them. There is one there similar to the ones that he made on the tickets.

Q. Now find any one that you say he made.

A. I don't say that he made any.

Q. Find a single one that you think he made.

A. He may have made the whole of them.

Q. Find one that you think he made.

A. There is another that looks similar to those. I suppose that there are very many different letters on the tickets in the same hand.

Q. Now, are you able, from the examination of the hand-writing on these tickets, and with your knowledge of the hand-writing of Colonel George, to point to a single one on that paper that was made by him?

A. Let me point out to you a difference; I do not pretend to state that Colonel George made these; I state that there is a resemblance; but if you look at that you will find that the point is the other way.

Q. Well, without going into the reasons, point out any of those that are his?

A. I do not say that any are his.

Q. Are you able to swear, from having seen him mark on the ticket, and from your own knowledge of his hand-writing, that either of those Gs were made by Colonel George?

A. I would not swear that there was one of them made by Colonel George.

Q. Are you able to tell, from your knowledge of his hand-writing, and from the tickets, that a single one of the tickets were marked by Colonel George?

A. I do swear that I saw him mark two hundred and twenty-six with the letter G, and these resemble them.

Q. You say you saw him mark two hundred and twenty-six tickets. Will you be kind enough to count those tickets and see whether there are two hundred and twenty-six tickets in them that are marked that are in the seven packages?

*Mr. George.* Excuse me; there is another package that you have not seen. [Question withdrawn.]

Q. I want you to point out the differences that there are between the letters G on the tickets and the letters G on the lower line [handing papers to witness], and explain the differences that you notice between the letters on the ticket and the letters on that lower line?

A. In finishing the point at the lower end of the letter G in this line, it is brought this way, and in finishing these it is brought to the right. On the ticket the point is carried to the left, and on the paper to the right.

Q. Now then, sir, from your knowledge of Colonel George's hand, and your opinion of the tickets, did he make any of the Gs there on that lower line?

A. Well, in deciding, I should decide on the turning of that point altogether.

Q. Now did he, in your opinion, make either of those Gs on the lower line?

A. I should hardly want to say that this was both the same hand-writing.

Q. Did he, in your opinion, make either of those Gs on that line [calling attention to another line on the same sheet]?

A. I should hardly think that it was the same hand-writing.

Q. From your knowledge of his hand-writing, independently of the fact that you saw him mark any tickets at any time—supposing these tickets were shown to you—and independently of the fact that you saw him mark on any tickets at Manchester at any time, let me ask you if you are able to say that either of these Gs were made by Colonel George?

A. These are exactly similar to those that I saw him put on the tickets. I should say that they were made by him.

Q. Aside from the fact that you saw him put the letter on the tickets?

A. Aside from that fact I am not able to state. If I may be allowed, there is a difference in these three; these are straight, and the others turn to the left.

[Mr. Tappan stated that at some period of the case he should call for a description of each one of these tickets.]

Q. You say that you arrested two others besides Mr. Whitcher. Who were they?

A Mr. White and Mr. Weston.

Q Are they the gentlemen who were examined here yesterday?

A. I understood that they came here; I did not see them here.

Q. (*By Mr. George.*) Have you identified the others?

A. Yes, sir.

Q. (*By Mr. Mugridge.*) Now let me ask you if you had under arrest White and Weston?

A. I had a warrant for each of them, at Stanley's office.

Q. Did you have them there at the same time, or before?

A. Before.

Q. Did you have them there at the same time, or did you arrest them and bring them in separately?

A. I think they were both in there at the same time.

Q. What did you arrest them for? What offence? Have you got the warrant?

A. No, sir.

Q. What offence did you understand, or did you give them to understand that you were arresting them for?

A. I supposed the warrant sufficient authority, and I took them at Mr. Stanley's request.

Q. Did you understand the nature of the offence charged against them in the warrant?

A. I am not able to say now whether I did or not.

Q. Did you understand it?
A. I did afterwards.
Q. What was it?
A. In regard to railroad matters.
Q. Did you understand what they were arrested for?
A. Mr. Stanley gave me warrants.
Q. Were you present when they were made?
A. I was not.
Q. How long did you have them in custody?
A. Not a great while.
Q. Did you have them in custody elsewhere than in Stanley's office?
A. No, sir: I did not.
Q. Were you present all the time they were in Stanley's office?
A. I was not.
Q. You don't know that during the time that they were under arrest they made any statements in writing?
A. No, sir; I did not.
Q. How long a time did you have them under arrest?
A. Long enough to get from the stores to Stanley's office.
Q. You arrested them and took them to Mr. Stanley's office at the request of Mr. Stanley?
A. Yes, sir.
Q. At Stanley's request?
A. Yes, sir.
Q. Have you had them under arrest since then?
A. No, sir.
Q. Who directed you to bring them to Stanley's office and leave them?
A. He did.
Q. Did you go and arrest them and bring them to Stanley's office and leave them there? What did you leave them for?
A. Because I done all that I had to do.
Q. Is that the way you do ordinarily?
A. Not always.
Q. Why did you abandon them?
A. Because I supposed it had been fixed according to Mr. Stanley's satisfaction.
Q. Were you told anything to that effect?
A. I may have been told that it was all right.
Q. Can you account for it in any other way?
A. I cannot account for it in any other way than I was satisfied that it was all right.
Q. (*By the Chairman.*) Mr. Prescott, I understood you to say that you took the warrants, and went and invited these men to go up to Mr. Stanley's office. Did you inform them that you had warrants?
A. No, sir.
Q. Did you show the warrants?
A. No, sir.
Q. In point of fact, you did not give them any notice that they were under arrest?
A. No, sir; I merely asked them to step up to Mr. Stanley's.
Q. (*By Mr. Mugridge.*) Didn't you say that you wished them to go up to Mr. Stanley's office, and wasn't that according to your usual custom?

A. That's what is called arresting them, if we have a warrant.
Q. It is the same way that you arrest half that you do arrest?
A. Yes, sir.
Q. And were they as completely in your custody as nine-tenths of the parties that you arrest?
A. If they had refused to go, I should have shown them the warrant.
Q. Did you not regard them as much under arrest as nine-tenths of the parties you do arrest?
A. Yes, sir.
Q. (*By Mr. George.*) You went for the purpose, if they had not come, of arresting them?
A. Yes, sir.
Q. So far as you know, did either White or Weston know that you had warrants against them?
Q. (*By Mr. Mugridge.*) Didn't they know very well that you were the assistant marshal of the City of Manchester? By the way, what day of the week was it that you arrested Whitcher?
A. Saturday.
Q. Let me ask you if you had him in your keeping all the time?
A. I put him in charge of a keeper and another officer.
Q. Who was the other officer?
A. His name was Griffin.
Q. What was done with him?
A. He was taken to Concord.
Q. On Sunday?
A. Yes, sir.
Q. What time in the day was he taken to Concord?
A. I think it was in the afternoon.
Q. How was he taken?
A. Mr. Gilmore sent a special train.
Q. Did you come up with him on Sunday?
A. Yes, sir; Mr. Farmer, also.
Q. You came up with him on Sunday?
A. I came up with Mr. Farmer. Farmer was sheriff of the county; I was not.
Q. Where was he taken to after he got here?
A. He was taken into the second story of the depot down here.
Q. Who was present in the second story of the depot there, on Sunday, after you got him up here?
A. I think that Col. George and Gov. Gilmore took him into another room from where we were. I am very sure.
Q. He was taken on a special train from Manchester, on Sunday. What time did you arrive in the city here?
A. I am not able to state.
Q. About what time?
A. My impression is that it was about two o'clock.
Q. About two o'clock in the afternoon? Now who came up in the train with him, besides you and Mr. Farmer? Anybody?
A. I think not.
Q. He was taken by you into the second story of the depot, and there you found Gilmore and Col. George?
A. Yes, sir.
Q. How long was he there before he was taken away by Col. George and Mr. Gilmore?

A. I can't remember; it was some time ago.
Q. They took him in another room?
A. They took him in another room.
Q. How long did they keep Whitcher in that private room?
A. I should think it was two hours or more.
Q. Well, sir; when he came from the room——he was under arrest at the time he came here?
A. Yes, sir.
Q. Did you take him into custody after he came from Col. George and Gov. Gilmore?
A. Yes, sir.
Q. What did you do with him then? Did you let him get off at Hooksett?
A. Yes, sir.
Q. Did the train run back that day?
A. Yes, sir.
Q. Did you take him into custody after he came from Col. George and Gov. Gilmore, and take him on board the special train running back that day, and let him get off the train at Hooksett?
A. Yes, sir.
Q. Has he been in your custody since then?
A. No, sir.
Q. Do you know whether he locked the door?
A. I do not.
Q. Did they come out from the time they went in until they finally came out?
A. I am not able to state.
Q. Did anybody go in during that time?
A. I am not able to state.
Q. So far as you remember, were they in consultation with Whitcher two hours or more?
A. It is my impression that they were.
Q. What time did you start with Whitcher to go back to Hooksett? Do you remember that?
A. I do not remember. We got home, I think, about dark—a little dusky.
Q. (*By the Chairman.*) Mr. Prescott, what day of the month was this—this Sunday?
A. The eleventh day of February.
Q. The day after his arrest?
A. Yes, sir.
Q. Where was Whitcher during that time, between the time of his arrest and the time you brought him here?
A. He was in charge of an officer; I think they stayed at the City Hotel.
Q. (*By Mr. George.*) At Manchester?
A. Yes, sir.
Q. Mr. Prescott, I would ask you whether Col. Cheney was there?
A. Yes, sir.
Q. (*By Mr. Mugridge.*) Do you know whether they came up at the request of Gov. Gilmore?
A. Yes, sir. I think it was on the train. I had forgotten that.
Q. (*By the Chairman.*) Was he before any police court, or magistrate, before that? Was he before any tribunal?

A. Yes, sir.

Q. Who directed you, if anybody, to release the custody that you had over Mr. Whitcher?

A. I am not able to state whether it was Col. George or Mr. Gilmore.

Q. Was it either of them?

A. I think it was.

Q. Was the warrant put into your hands the same time as the warrant for White and Weston?

A. I think not the same.

Q. Did they give you orders to release them, after the interview with Col. George and Gov. Gilmore?

A. I think they gave us these instructions on the platform in the depot.

Q. After the interview? After you got ready to start?

A. Yes, sir.

Q. Were you directed by anybody not to take Whitcher before any tribunal—not to return the warrant.

A. No, sir; I was not.

## TESTIMONY OF F. M. CRANE.

Mr. F. M. Crane, of Concord, N. H., was called in behalf of the plaintiff, and duly sworn, and testified as follows:

Q. (*By Mr. George.*) Please state whether you were a clerk to Mr. Gilmore at the time of the arrest of Mr. Whitcher, and of the finding of the tickets upon him?

A. I was.

Q. You were confidential clerk?

A. I was.

Q. Will you state whether these tickets came into your possession, and you sorted them?

A. I did.

Q. Whether these were the tickets, and you sorted them? These are the tickets given you by Mr. Gilmore? You have examined them?

A. Yes, sir.

Q. Whether they were put together in one package or separate when they were given you?

A. They were all together.

Q. You assorted them?

A. Yes, sir.

Q. Are these packages in your hand-writing?

A. Yes, sir.

Q. Are these the same tickets? Can you identify them?

A. I can only identify them by the G that is on them.

Q. Well, you know my hand-writing?

A. Yes, sir; they were all marked by you at the time.

Q. Have you any doubt that that is my hand-writing G on the back?

A. No, sir.

Q. Now, you must state when Mr Gilmore gave you those tickets, and where you assorted them, and what you did?

A. I do not recollect how long after; it was certainly a fortnight.

Q. Certainly a fortnight after the arrest of Whitcher?

A. Yes, sir.

Q. What directions did he give you? and where were you when he gave them to you? and what did you do in pursuance of those directions?

A. He gave them to me in my office, and told me to assort them—on the different roads—and mark on them where they were from, and who they were taken from, and so forth.

Q. Well, sir; you did it, did you?

A. I did it.

Q. And you put them in the same envelopes they are in now?

A. These are the envelopes. After I assorted them he requested me to turn them over to the cashier.

Q. The cashier? Mr. Webster?

A. Yes, sir.

Q. And have you seen them from that time till now?

A. Yes, sir.

Q. Where were they kept?

A. In the safe in the cashier's office.

Q. How long did they remain there? How long did you know of them remaining there until the governor's death? Or did you leave before that?

A. I left about the fifth of September.

Q. Were they there when you left?

A. I think they were.

Q. Are you familiar with the tickets of the Concord Railroad?

A. Yes, sir.

Q. Won't you look at these, and see if they are not genuine tickets of the Concord Railroad—tickets in general use on the Concord Railroad?

A. They are; but some of them are dated back, and were out of use at that time.

Q. That is, they were not the kind of tickets that were used at that time?

A. I recollect at the time they were some two or three years old.

Q. Mr. Crane, did you make an abstract of the various tickets, and the amounts, and so forth?

A. Yes, sir.

Q. Do you recollect what was done with that?

A. It was handed over to the cashier with those tickets, I think. I made several copies of it.

Q. For whom did you make the several copies?

A. I believe for Mr. Gilmore.

Q. Do you know whether any abstracts were made and given to the conductors?

A. No, sir.

Q. Do you know how many abstracts were made?

A. I think I made three.

Q. Did you make any abstracts of those found at one place, and those found at the other places: that is, those found on Mr. Whitcher, and those found at his house?

A. Yes, sir; I made two different abstracts. If I am not mistaken they are here.

*Mr. Mugridge.* That is of no consequence, as a mere memorandum made at the suggestion of Mr. Gilmore.

*The Chairman.* The suggestion is correct, perhaps, that he made it at the suggestion of Mr. Gilmore. but he will testify as to the different marks.

Q. (*By Mr. George.*) You made a separation, just as you were instructed.

A. Yes, sir.

*Mr. Mugridge.* No man has sworn that they were marked G 2.

*Mr. George.* That is where the back and face were just alike.

Q. (*By Mr. George.*) How did you mark those found on his person, as distinguished from those found at his house?

A. They were given to me in different packages; those on his person in one package, and those at his house in another.

Q. (*By the Chairman.*) What is your knowledge?

A. Simply from what Mr. Gilmore told me.

CROSS EXAMINATION.

Q. (*By Mr. Mugridge.*) Mr. Crane, are you able to testify, positively, that these are the tickets?

A. Only by the G on the back.

Q. Supposing other tickets were marked on the back?

A. I could not testify.

Q. Was there any distinguishing mark pointed out to you to designate those found at the house, from those found on his person?

A. No, sir.

Q. There was no mark on the back of those found on his person that was not found on those found at the house?

A. No, sir; neither of them were marked; that is, designating where they were found. Mr. Gilmore gave me both packages.

Q. But was there any figure 2 on them?

A. I did not notice.

Q. Was figure 2 pointed out to you as being a distinguishing mark upon those tickets?

A. No, sir; it was not.

Q. Now, then, are you perfectly confident that the tickets handed to you at that time had any figure 2 upon them?

A. I can say that there was no figure 2 on them.

Q. Are you pretty confident that the figure 2 did not appear on every one of the tickets that were handed to you at that time?

A. I should have noticed it if there had been.

Q. Now then, sir, if the figure 2 appears upon any of these tickets here (there is one with the figure 2 on it), did you see any such tickets handed to you by Mr. Gilmore?

A. No, sir; there was nothing said about it.

Q. Did you see any?

A. I did not; simply because I was not requested.

Q. Wouldn't you have noticed the 2 as well as the G?

A. I probably should

Q. Do you remember how many tickets you marked at that time?

A. I do not.

Q. Have you any idea with regard to the number?

A. I don't think I have.

Q. Could not you tell. I mean marked on the envelopes? Have you any idea of the number of tickets passed to you to sort, at that time?

A. I should say perhaps there was in the vicinity of two hundred.

Q. And I understand you that these were not handed to you to be sorted until some two weeks after Mr. Whitcher was arrested?

A. I think it was two weeks.

Q. Then he brought them along to you in these envelopes?

A. Yes, sir.

Q. That is, he brought to you a lot of tickets that he wanted you to sort and mark in the way you did, and that these were sorted in the envelopes?

A. Yes, sir.

Q. Let me ask you if it was not four weeks?

A. I could not say; I know it was some time.

Q. From your present recollection can you say that it was not two months?

A. No, sir; I cannot.

Q. Was it more than three months? Would you say that it was not more than three months?

A. No, sir. That is after Mr. Whitcher's arrest.

Q. You would not say that it was not?

A. No, sir; I should not.

Q. Let me ask you whether you were down at the depot?

A. I was.

Q. Was there a meeting of the officers of the Concord Railroad, with regard to railroad matters? and if so, who was present at that conference? And how long a time was spent?

A. I think they arrived between half past two and three.

Q. In the afternoon?

A. Yes, sir.

Q. Who was present?

A. Mr. Gilmore and Col. George and Mr. Pecker.

Q. Mr. Pecker, a correspondent of the Boston Journal?

A. Yes, sir.

Q. Was Mr. Pecker, the correspondent of the Boston Journal there, with Col. George and Mr. Gilmore?

A. No, sir; he was not.

Q. Where was he?

A. He was in the depot Pecker attempted to go into Gilmore's office, and Gilmore told him he would kick him down stairs if he did not get out of the way.

Q. Let me ask you if this Pecker is a special correspondent, living here in Concord?

A. He was at the time.

Q. He is a man who telegraphs his effusions from here over the signature of "F. F."?

A. Yes, sir.

Q. Whether Pecker had been down there before on that Sunday?

A. I couldn't say.

Q. Let me ask you whether a communication had been handed to Pecker and he had telegraphed it to the Boston Journal?

A. Yes, sir.

Q. There had been?

A. Yes, sir.

Q. Who prepared the article to be telegraphed by Pecker to the Boston Journal next morning, with regard to the railroad robbery?

A. I could not say.

Q. Was it prepared by Colonel George and Gilmore and Pecker, altogether?

A. I could not say by whom it was prepared.

Q. Were they all present at the time it was prepared?

A. I could not say whether Pecker was there or not.

Q. Did Col. George and Mr. Gilmore prepare that article themselves, in the absence of Mr. Pecker?

A. That I could not say.

Q. Did you see Mr. Gilmore and Col. George at work upon an article that was to be telegraphed to the Boston Journal?

A. I did.

Q. What day of the week was this communication prepared by Col. George and Gov. Gilmore?

A. That I could not say.

Q. Was it Sunday? Was it the day they came up from Manchester?

A. Yes, it was, sir; I think it was Sunday evening.

Q. Did you see the communication? and who did the writing and who did the dictating in the preparation of the article?

A. I thought at the time it was Col. George.

Q. He did the writing and the dictating. Should you recognize the article?

A. I copied the article myself.

Q. Should you recognize the article if you saw it again.

A. I think I should.

Q. At whose suggestion was the article prepared?

A. It was handed me by Col. George.

Q. So far as you know, did anybody but Col. George have anything to do with getting it up?

A. I could not say.

Q. So far as you know, could anybody else have?

A. Nobody except Mr. Gilmore.

Q. Did Mr. Gilmore make any suggestion as to what should be incorporated in it?

A. I do not know.

Q. So far as you know, did not Col. George himself write that article? Do you know that there was a single sentiment that came from Mr. Gilmore?

A. I could not say that there was. I recollect that Colonel George came from Mr. Gilmore's room and handed it to me, and asked me to copy it. [Intermission.]

Q. Will you be kind enough to look at that article in the Boston Journal and see if that is the article prepared by Col. George?

A. The latter part was something that was added after I copied it.

Q. Let me ask you if you have had any conversation with any one in regard to this since the adjournment?

A. I have; with Col. George.

Q. What did Col. George say to you?

A. He simply asked if I understood why this was sent. I told him I did; and he asked me what I understood from him, and I simply told him it was sent that the public might ascertain from this an idea of things, so that there would not be an exaggeration, or anything of that kind.

Q. Did Col. George say that there was?

A. He asked me if I knew that there was.

Q. Did he call your attention to the last part as added after it was copied by you by somebody?

A. No, sir; he didn't state that, but he asked me if I was aware of it.

Q. Will you be kind enough to say where that article ends, as copied by you?

A. I think it was—— The commencement of that that was added was that about the city clerk's office.

The counsel for the defendant proceeded to read the article in question, beginning with the headings of the article; whereupon the counsel for the plaintiff objected to the reading of the headings, as not forming a part of the article as originally sent.

*The Chairman.* We think it proper for the plaintiff's counsel to inquire about that matter, in order to determine what was actually stated.

Q. (*By Mr. George.*) Was there any heading to the article given to Mr. Pecker?

A. In speaking of this article I had no reference to the heading.

Q. (*By Mr. Mugridge.*) After you copied that article, what did you do with it?

A. I handed it to Col. George.

Q. For what purpose did you understand from Col. George that the copy you made was made?

A I understood it was to be given to Mr. Pecker.

Q. For publication?

A. Yes, sir.

Q. Now, after having copied it, you handed it to Col. George?

A. I did.

Now, we shall argue from that that it was prepared in that office?

*Mr. George.* I object that the testimony was not relevant.

*Mr. Tappan.* Now then, it appears that a portion of this article was prepared by Col. George, and that a portion of it was copied, and that it went into the hands of Col. George. And one part of our defence is this, may it please your Honors, that this whole affair against the conductors is a conspiracy to cover up the fraud and infamy of the railroad managers themselves, to make a scape-goat of these conductors. And it does not make any difference whose shoes are trodden upon in this investigation. This whole thing was concocted to raise a great smoke, under the cover of which they were to go out harmless. Suspicion had already attached to the managers of the Concord Railroad. There was a great deal said in regard to that management, and a great deal of talk going on everywhere. The annual meeting was soon to come off, and it was necessary for Mr. Gilmore, and perhaps others, to get up something that would draw public attention from themselves and what they had done, and would palm it off and shoulder it upon others. And we shall show in this investigation that not only this gentleman, but others, were innocent of the frauds which were the frauds of the principal, and not of the mere servants and agents of those who went over this road

*Mr. George.* The remark of my brother to Gov. Gilmore himself was not exactly like what he says here. "Sir, we do not accuse you of anything — any of these things; if you had had the management of these things everything would have been all right." I notice that his remarks of Gov. Gilmore now are very different from what they were.

The question here is whether Mr. George Clough has been guilty of such conduct that under this writ and action of assumpsit we are entitled to recover of him what he has taken from the Concord Railroad. If he has not taken anything from the Concord Railroad, why then, of course this suit cannot be maintained. If Mr. Gilmore has not stolen anything, and if Mr. Clough has not stolen anything, then this writ cannot be maintained. And Mr. Gilmore's conviction cannot increase nor lessen Mr. Clough's. We are here to try this case on the law and evidence applicable to this case. I have not objected in this matter, because when you talk about sensation, I should like to have an opportunity of showing the number of lawyers retained in this case, and the number of sensation people. I do not think it would be difficult to see the source from which the sensation came. If we cannot maintain our case, then it is our misfortune. Let me say this: if the present directors of the corporation do not unanimously insist that this suit shall be prosecuted to the end, then I am misinformed. And if the gentleman will satisfy me of the fact that they are not, I will go out of this case immediately. When the gentleman undertakes to say that these suits which were ordered to be commenced when the old board was in office, are not desired by the present board, I think he will find himself mistaken.

*Mr. Tappan.* I do not think it is necessary to prolong this discussion. I have only stated strongly what I thought necessary to state; and we think that all the circumstances surrounding this matter should go in. We say that the whole thing, to repeat, was gotten up by Mr. Gilmore—I won't name the others, they will appear in the course of the evidence — for the purpose, not so much of ever getting anything from this conductor, or that conductor, as for the purpose of getting up a smoke under which the guilty parties could go clear, and securing a re-election, which was not secured. Now gentlemen, I do not think it comes with a very good grace to talk about multiplicity of counsel. You see before you all the counsel in this case. I suppose Mr. Clough has a right to employ all the counsel he chooses. Mr. Noyes and Mr. Jones have a right to employ whom they choose. And I think considerable capital has been attempted to be made from the fact that Gen. Butler has been engaged to prosecute these men. And Gen. Butler has been conferred with, and he has been consulted. And therefore I do not think the array of counsel is greater on this side than on the other.—But this does not amount to anything here. We ask to put in this article as part of the *res gestæ* in order that the commission may understand where this originated, and the means and appliances used to carry it on.

*The Chairman.* There is no question made but that certain portions of it should be read.

*Mr. George.* I am willing that everything should be read, except the headings.

*Mr. Mugridge.* We regard the headings as a material part. I submit that the article, as he understood, was written by Col. George for publication, and it was presented to him to copy, and was handed back by him, and the next we see it is this article in the Boston Journal. And the presumption is that as it went so it was forwarded; that it was sent as it appeared in the Journal.

*The Chairman.* We think that, as far as the heading of this article is concerned, it would be better to pursue the inquiry a little further and ascertain, as well as can be done, what the condition of the article was, and how much of that article was handed to the correspondent. I understand the counsel do not object to that article being read.

Q. (*By Mr. Mugridge.*) Wont you be kind enough to look at that article, and state what part of it was made in the copy you made?

A. I copied to where it is turned——

Q. Where did you commence?

A. I commenced here [pointing to the beginning].

Q. You commenced, then, with the first line of the article and copied down to the words "city clerk's"? Now, having copied it in that way, you handed it back to Col. George, to be published? Didn't he say, at the time, that it was for publication?

A. No, sir. I think he was aware that I knew the fact that it was to be published.

Q. How did you learn the fact?

A. On account of Mr. Pecker being there to receive it.

Q. From what did you learn it?

A. I inferred it.

Q. Was Pecker present at the time you handed this article to Col. George?

A. Yes, sir; I think he was.

Q. Did you see Pecker when he left the room, or did you know whether this article had been handed to Pecker before he left the room?

A. He left the room with it.

Q. With the article?

A. Yes, sir.

Q. Do you know whether any alterations had been made in the article by Col. George, after it was handed to him?

A. I know that they added something, but whether it was Colonel George, or Mr. Gilmore, or Judge Upham, I don't know. After that was added to it, permission was given to him to have it published.

Q. Did you know that something was added to the article after it was copied by yourself, and returned to Col. George, before Pecker took it?

A. No, sir; I think not. All the addition that was made was made by Mr. Pecker himself.

Q. In the presence of these gentlemen?

A. No, sir.

Q. In the presence of Mr. Gilmore?

A. No, sir; I think not. I think he went into the telegraph office and brought it back.

Q. After he had been to the telegraph office and altered it, he then returned?

A. Yes, sir.

Q. And then permission was given him to publish it?

A. Yes, sir; from some authority. I don't know who.

Q. Now who was in Gilmore's office when Pecker returned with the article from the telegraph office?

A. I could not say.

Q. Mr. Gilmore was there, was he?

A. I couldn't say that.

Q. Upham?

A. Well, I couldn't say who was there.

Q. Who gave him permission to publish it?

A. It was one of the three, or from some one connected with the affair that had authority to do so.

Q. (*By the Chairman.*) Did you see that article after Pecker went in the hall?

A. Yes, sir.

Q. Do you know whether there was any heading to it?

A. No, sir; there was not. I simply read it over with the addition that he made. I think there was no heading on it.

Q. Then the article, as it was published, hadn't headings on it?

A. No, sir.

Mr. Mugridge offered in evidence the entire article, including the headings, as evidence of an alleged conspiracy against the conductors.

*The Chairman.* I think the opinion of the referees is, that they would not rule in any more than what is authorized to be published, as shown by the witness.

Mr. Mugridge then read the body of the article, which is as follows:

"In October last, on account of information officially communicated, the Board of Di- " rectors of the Concord Railroad directed Colonel John H. George, clerk and counsel for " the corporation, to make a thorough investigation as to the manner in which the passen- " ger conductors discharged their duties. Colonel George employed Major George J. Car- " ney, of Lowell, to act as a detective agent. Major Carney, as an excuse for residence " at Concord, and for constant riding in the cars, established a large claim agency here, " and for the following months spent most of his time passing over the various roads oper- " ated by the Concord Corporation, without exciting suspicion. It is understood that the " result of Major Carney's investigations show not only disregard of the general rules of " the road, but a failure to account to the corporation for money paid for tickets in the cars. " Upon the submission of Major Carney's report, all the conductors were immediately dis- " charged and new appointments made. Since that time a very searching investigation " has been going on in furtherance of the general object. Yesterday, Governor Gilmore " and Colonel George went to Manchester, having learned that tickets were being sold upon " the streets for far less than their value, and caused a careful examination, and also several " arrests to be made. One of the persons arrested had in his possession several hundred " tickets over the Concord and connecting roads, which he alleges he purchased of the " conductors. A further investigation in the same direction is still in progress. Last evening " writs were made against the conductors for embezzlement, and all their property attach- " ed; and there can be no doubt, from the character of those having charge of the investi- " gation, that this matter will be probed to the bottom. An examination of the City Clerk's " office shows that attachments have been made in the sum of three hundred thousand " dollars, from which it may be inferred that the defalcations have been of a large and " of an alarming character.

"Two of the parties arrested for alleged participation in the sale of illicit tickets are " merchants of high standing, whose names are not yet given to the public. The individ- " ual upon whom was found the large number of tickets, is James Whitcher of Hooksett, " a manufacturer of wooden ware, and formerly a Canterbury Shaker.

"There seems to be no doubt but that the alleged system of plunder had gained a mag- " nitude before detection, that will completely astonish the public.

"The discharge of the conductors, several weeks ago, led people to think that some- " thing was wrong, but they were not prepared for such startling disclosures as the above. " and their announcement in our city to-day is attended with great surprise, and no little " astonishment. It is now reported that further arrests have been made to-day and that " more disclosures are in progress. The result of further investigation will be looked for " with intense interest."

Q. (*By Mr. Mugridge.*) Now, Mr. Crane, do you know anything of the preparation of the article that appeared in the Boston Journal next day? See if you know anything about that article.

A. No, sir; I do not.

Q. You know whether that was prepared at Gilmore's office, by anybody in his office?

A. I don't know that it was.

Q. Do you recollect of another article being prepared the next day?

A. No, sir.

Q. Do you recollect of Pecker, the correspondent, being there with reference to obtaining information?

A. I saw him there.

Q. Let me ask you whether he made any inquiries and made minutes of the information he received?

A. I could not say.

Q. Do you know what his object in being there was?

A. I suppose it was to get information.

Q. Let me ask you whether there was any talk made in Gilmore's office in reference to the commencement of suits against the conductors?

A. I don't know that there was. I was generally present when he was.

Q. How long were they in consultation—Col. George and Mr. Gilmore—what part of the day—about this business ?

A. I should think in all from an hour and a half to two hours.

Q. (*By Mr. George.*) Mr. Crane, you remember, after you copied what has been read—do you remember, when I handed it to Mr. Pecker, the remark that Mr. Pecker made.

A. I heard no conversation.

[Plaintiff's counsel then offered in evidence the deposition of Martin Starkey, and a portion of the same was admitted and read.

Mr. Tappan stated that the defendant's counsel did not object to the testimony of Mr. Starkey in regard to free passes, but did not withdraw their general objection.

Judge Bellows stated that it had been distinctly understood that, although some evidence had gone in upon this point, the same objection applied generally.

Mr. Rolfe objected to testimony as to conversation between Mr. Whitcher and Mr. Starkey, in which Mr. Starkey attempted to state what Whitcher told him, and the same was ruled out.

Objection was made to the seventh interrogatory, as to Mr. Whitcher's having given some tickets away.

It was stated that in the view of the referees what was done was evidence, and what was said was not evidence.]

## TESTIMONY OF BARUCH BIDDLE.

Q. (*By Mr. George.*) What is your occupation ?

A. I have charge of the depot.

Q. You are what is called depot-master ?

A. I am not now.

Q. Have you had any such charge ?

A. I have, until within a year.

Q. During the war, Mr. Biddle, did you keep an account of the number of cars attached to each train ? If so, by whose direction ?

A. By Mr. Gilmore's.

Q. Did you keep such an account ?

A. I did, sir, at times ; I should say about six months ; short of a year.

Q. During the war ?

A. It was during the latter part of the war.

Q. Where is that record, if you know ?

A. I do not, sir. Mr. Gilmore took it from me, and I never saw it afterwards.

Q. You may state, from your recollection, as near as you can, how many cars a train—take the 10.15 and the 3.30 trains from Concord—how many cars a train ran out of Concord during this period that you kept this account ?

[Defendant's counsel objected to showing the amount of travel by the number of cars as bearing upon the question of stealing by the conductors. A long discussion ensued. Mr. Biddle was withdrawn, and was not afterwards put upon the stand.]

## TESTIMONY OF JAMES R. KENDRICK.

Q. (*By Mr. George.*) Mr. Kendrick, I want you to explain to the referees the manner of doing the passenger business on the Concord Railroad, so far as the sale of tickets is concerned, and the return of tickets made, and the number of passengers, and from what it is determined.

*The Chairman.* What you want to know is, what it was at that time.

Q. What is there — how did you determine the number of passengers from the Northern road?

A. They made a return monthly of the sales.

Q. And what account did you have?

A. We have a coupon from that road for each passenger that passes over it.

Q. The Concord Railroad takes up that coupon ticket?

A. Yes, sir.

Q. And these two should agree, if correct?

A. Yes, sir.

Q. They make a return to you and you take up the coupon ticket, and if they agreed, it was correct?

A. Yes, sir; that is the principle that we go upon.

Q. Now then, how did you make up the number of passengers on the Concord road? I mean the local business of the Concord Railroad.

A. They are made up by returns from the different ticket-sellers monthly.

Q. And the general ticket agent?

A. Makes up an account. They are rendered to me. The ticket sellers make a return to me of the sales monthly.

Q. What does the conductor do with the returns?

A. They are returned to the office.

Q. To whom are they returned?

A. To the general ticket agents, every trip.

Q. Suppose a man does not buy his ticket at the ticket office, but of the conductor?

A. The returns are made by the conductor to the general ticket office every trip.

Q. Just tell me how the number of passengers over the Concord Railroad was made out?

A. From the returns of the ticket sellers and the returns of the conductors. Those two are the only two that we have; the returns of the ticket sellers—the report of the tickets that they sold—and the returns of the conductors of those without tickets. I suppose, put both these together, they would make up the amount of travel upon the road.

Q. What do you call a trip on the road by a passenger conductor?

A. A trip down and back—from Concord to Nashua and back, or to different places.

Q. You would call from here to Nashua a train, and from here to Nashua and back a trip?

A. That is the usual way.

Q. Suppose a man runs to Nashua; at the end of the trip, where does the conductor take the tickets?

A. The tickets are carried into the general ticket office, and the money is paid to the cashier.

Q. Supposing, Mr. Kendrick, that ten men should get into the cars at Concord to go to Nashua, and they should pay ten fares to the conductor in the cars, and the conductor should not return these ten fares, would the road have any evidence whatever of the passage of these ten men?

A. If the conductor did not return the amount on his way-bill, I don't know as we could.

Q. These ten men would not be reckoned in the number of passengers?

A. Not if they were not returned.

Q. Supposing ten persons should buy their tickets at the ticket offices, and the conductor should take up the tickets going down, and then re-sell them, so they would be used six times, and at the end of the week returned, would there be any record or evidence of more than a single passage?

A. Of these ten tickets?

Q. Of these ten persons?

A. No, sir; these ten tickets were represented as the first ten fares that were taken, or rather the sale of the first ten tickets.

Q. And the others would not find any representation on the records of the road?

A. No, sir.

Q. (*By the Chairman.*) I was going to ask Mr. Kendrick this question: In the first place, under what circumstances, by the course of business, does the conductor have tickets to sell?

A. He has tickets to sell only to stations below our road. He has tickets to sell to Lowell, Boston, Lawrence, and those points that he doesn't run to himself. with which we connect ourselves. He has conductors' checks for that purpose himself. They are furnished by the general ticket agent and charged to the conductor, and he is accountable, the same as the ticket seller.

Q. These would be used in cases where he took fare on the train?

A. Yes, sir.

Q. What has been the usage in regard to date?—putting the date upon tickets sold? How is it, for instance—how has been the usage as to tickets sold by the Northern or the Vermont Central road? Are these tickets dated?

A. Some of these are dated, I think, and some are not. I am not certain in regard to them all. Mr. Sanborn could tell.

Q. How was it with regard to ours?

A. At our own office we furnished our own tickets—they are not dated.

Q. Suppose it should happen that a ticket is sold at your office in Concord—we will say the 22d day of July, and is dated on the 22d day of July—and is returned by the conductor on the 10th of August. What is the course of business in regard to that ticket?

A. We make no difference with the tickets—we take the tickets without regard to dates, so far as our road is concerned, at that time.

Q. Then, if I understand you, this date of the ticket does not furnish any check?

A. So far as the date is concerned it is no material check. We have now the consecutive checks, and in that way we keep track of the checks.

Q. Does that enable you to show whether a check has been used more than once?

A. I suppose it does not, unless it is punched by the conductor. It enables us to see that our tickets are collected. We can keep better track of our tickets in that way, than if the tickets were not numbered.

Q. If I understand you, there is no check. This dating of the ticket, as it is practiced. does not furnish any check to prevent the ticket being used over and over again a great many times, before it gets back to the office?

A. Well, no sir—that is—the ticket is presented to the conductor of an old date, and he very often questions the passenger in regard to it, and in some cases, I believe, where they are very old, they have been refused.

Q. The question is, when the conductor brings it back to the ticket office, whether it furnishes any check? Is there any check by which you can prevent, or by which you can know, whether one of your tickets is used more than once, supposing your conductor is violating his duty, is there any way by which you can ascertain, supposing the ticket is purchased one day and returned a week afterwards?

A. I should not be able to tell, as I know of, by the date.

Q. (*By Mr. George.*) Is there any sort of a check?

A. No, sir; I don't know as there is.

Q. (*By Mr. Cushing.*) Do you know anything about the practice of other roads?

A. I am acquainted with those that we connect with. The practice is pretty much the same, I expect. As we sell a ticket we advertise that the ticket is good only for that day, but it has been ruled by law that we cannot oblige a passenger to use a ticket on the day of sale, and as a practice we have done away with it.

Q. (*By Judge Bellows.*) Supposing, taking all passengers on one train, according to the course of your business, what ought the return of the conductor to show—that is, how far ought it to show the exact number of passengers who traveled—taking those who have passes and season tickets, or mileage tickets, or anything else?

A. The return of the conductor would show the travel with the exception of the tickets going on to the roads below. Those, I think, we did not take up—a coupon of our sales. I mean by that, a man buys a ticket from Concord to some place below our road—we did not take up the coupon; we punched the ticket, and it goes on to the other road. Tickets on the roads above us we took a coupon. Every passenger going over our road to Boston buys a ticket from here to Boston, and we have no coupon; that is returned to us at the close of the month, with their collection of tickets. But a passenger going from here to Boston, we should have no coupon returned to our office. Our local tickets are taken up. The ticket from here to Manchester would go on to the lower road, and be taken up there.

Q. How is it in regard to those who have passes?

A. From connecting roads, if they are going over our road, the conductor takes a record of them usually. That is his direction usually. Of the employees of our own road they are not taken; that is, our working hands—those doing our own work. Where a pass terminates on our road it is taken up on our road.

Q. In case a pass is given by other roads?

A. The conductor takes the name and the pass is returned.

Q. If it is a season pass?

A. There is a record taken of the person holding it. The conductor furnishes nothing more than this record.

Q. That is the same as a season ticket?

A. No, sir; we keep no record of the season ticket passengers.

Q. And the mileage tickets?

A. The mileage tickets—there isn't any record taken of them. The ticket shows itself when it is taken up.

Q. (*By Mr. George.*) The mileage tickets have come into use since you were on the road?

A. Yes, sir.

Q. The season tickets were used before?

A. The season tickets were before.

Q. You said the tickets were returned?

A. The tickets themselves are returned from the other roads at the end of the month.

Q. That is, these local tickets. They returned the tickets themselves, from which you ascertained the number of passengers?

A. Yes, sir. That is, we verify the amount, by these collections, at the close of the month.

Q. Is there any check whatever, or has there ever been—I mean a check, or any system by which, if a conductor takes fifty dollars on a day and returns only twenty-five, you would know it from the records which you have?

A. No, sir; I don't know that there is.

Q. Is their any way, by your statistical information, that you can show that fact at all?

A. No, sir.

CROSS EXAMINATION.

Q. (*By Mr. Mugridge.*) With regard to the manner of your business between the roads above you and yourself—to illustrate, between Plymouth and Boston—suppose a passenger buys a ticket at Plymouth to go to Boston, is there a record made at the Plymouth office of that fact?

A. I suppose there is.

Q. Now when that passenger gets into Concord, he gets in, having in his hand what is called a coupon ticket, from Plymouth to Boston, does he not?

A. Yes, sir.

Q. Now, then, is a coupon taken off on the Boston, Concord and Montreal Railroad?

A. I think that is their system.

Q. The next one is taken off on the Concord road by the conductor on the Concord Railroad train, is it not?

A. Yes, sir.

Q. And the others are taken off below Concord?

A. Yes, sir.

Q. Now, Mr. Kendrick, it is the duty of your conductor to return that coupon ticket, taken on the Concord road, to the general agent, is it not?

A. Yes, sir.

Q. Because by that return of the coupon your account for the travel is settled with the Boston, Concord and Montreal Railroad?

A. That we didn't settle by account.

Q. It represents your interest?

A. Yes, sir. But then, if we should lose the coupon we should get our pay.

Q. Because they have got a record of it at Plymouth—that is your evidence?

A. Yes, sir, that is our evidence; that is what we check our record by.

Q. Now, Mr. Kendrick, at the end of the month, your tickets being all returned, you make an account of all the tickets sold for the Boston, Concord and Montreal road?

A. Yes, sir.

Q. Now the coupons at your office should agree with the number of tickets charged at the Plymouth office?

A. The number that they reported to us? Yes, sir; if they were all taken up.

Q. At the end of every month you should have in your office a coupon for every passenger to whom they sold a ticket?

A. We should, if these passengers all went through. A passenger might start the last day of the month and come to Concord, and the next day we were to take it up, our record would not show it in the month in which it was sold. But take six months together it would show, or substantially show.

Q. Now we will suppose, Mr. Kendrick, that fifty tickets are sold at the Plymouth office to go to Boston on the first day of the month; if the plan is properly carried out there should be fifty coupons in your office at the end of that month?

A. Yes, sir.

Q. Suppose a conductor should take fifty coupons and turn over twenty-five of them, wouldn't that defalcation be manifest at the end of the month, when you came to make up your record?

A. It would, of course.

Q. Well, suppose a thousand tickets were sold north of Concord, upon the Boston, Concord and Montreal, Northern, and Concord and Claremont roads, all going to stations below Concord, and that the conductor takes up one hundred of those tickets on the Concord road—coupons, I mean—one hundred from the Boston, Concord and Montreal, and one hundred from the Concord and Claremont—making three hundred—and should return only these three hundred tickets to the general office, when you come to make your monthly statement it would be seen from the return that the conductor had abstracted the difference?

A. Without there might be an error in the account.

Q. It would be palpable, if there were any taken away, that the conductor had abstracted them?

A. It would look as if he had forgotten to return them, or had mislaid them some way.

Q. Now, if you will allow me, Mr. Kendrick—suppose that your plan worked perfectly, just as it is intended, isn't there a perfect check upon the conductor in regard to his stealing tickets, in the operation of that plan. That is, can the conductors improperly withhold any amount of these tickets without its being surely discovered at the time of the monthly settlement?

A. We have half a dozen conductors—we don't keep each conductor's returns in that shape. If there was one conductor, and the returns were kept in that way, it would show in that way, or else a good many passengers didn't travel that way.

Q. Now is there any possible way for a conductor to withhold, or keep back, coupon tickets sold to passengers over the upper roads going over your road to any amount, without that discrepancy or that lack being made manifest at the time of your settlement?

A. The amount of it is, I don't suppose we agree each month, because passengers go from one month to another.

[Adjourned.]

[FOURTH DAY. Friday, July 24th, 1868.]

## TESTIMONY OF G. G. SANBORN.

The hearing was resumed, and Mr. G. G. Sanborn, the general ticket agent of the Concord Railroad, was called by the plaintiff and sworn, and testified as follows:

Q. (*By Mr. George.*) You may state your residence.
A. I live in Concord.
Q. How old are you?
A. Forty years old.
Q. What is your business on the Concord Railroad?
A. General ticket agent.
Q. How long have you been general ticket agent?
A. About eleven years.
Q. How long have you been connected with the road in any capacity?
A. About twenty years.
Q. Before you were general ticket agent, what were you?
A. I sold tickets at the station.
Q. How long did you sell tickets at the station?
A. About six years.
Q. Now, Mr. Sanborn, I want you to explain, just as clearly as you can, and just as minutely, the whole ticket system of the Concord Railroad. I will just indicate generally what subjects I want you to explain, and then you can go on in your own way. As to how tickets are given out to the ticket sellers; and what rates, if any, you take; how much you charge; what tickets you give to the conductors; explain how the conductors make their returns, and to whom they are made; and how the conductors make their returns for sales of tickets in the cars; and then I want you to explain how the accounts are kept, as to the number of passengers carried on the trains, and as to the amounts received for passengers. Now, I will begin at the beginning of the whole thing. Who gets the tickets printed in the first instance?
A. I will state, so that you can see how it is done. The system is now different from the system that existed several years ago, and as I found it when I came on the road. The tickets were printed here by Morrill & Silsby, until within two years; all the tickets were printed alike, and could be used over and over again, without regard to their age, so long as they were legible; and after that the thin card was printed, which was used only once. Well, sir, I should think about four or five years ago the old card system was abandoned.
Q. (*By Mr. Mugridge.*) For the thin ticket that you speak of?
A. For the thin ticket.
Q. Now, then, that was about fifteen years ago. Who procured the printing of these card tickets? I mean, what officer of the road?
A. The general ticket agent.
Q. Did he keep a record of them?
A. Yes.
Q. Well, then, about fifteen years ago this system was abandoned; what was the next system that was adopted?
A. The next was the thin card with coupon card.
Q. The general agent procured the printing of these tickets?
A. Yes, sir.
Q. And what should have been done with the ticket when it was once used?

A. It was entered on the book and destroyed.

Q. Now, I am speaking about the conductor.

A. It was his duty to return it.

Q. And what with regard to mutilation?

A. To punch them.

Q. You procured the printing of these tickets at Morrill & Silsby's?

A. Yes, sir.

Q. What did you do to authenticate them? What mark did you put on to authenticate them?

A. I put a stamp on them.

Q. A raised character?

A. Yes, sir. You will see it better on the back.

Q. Now you procured these tickets to be printed; what check did you have on the printer, apart from this die? Any?

A. No, sir.

Q. This die was simply to prevent the use of any other tickets?

A. That was the object of it.

Q. That is the way you got the possession of the tickets in your office?

A. Yes, sir.

Q. Now, then, just go on and state how you supplied the ticket officers.

A. The ticket master at this time was supplied with a given number of tickets to each station, and their number was made good each month.

Q. You may, if you please, Mr. Sanborn, if you can make it any clearer, go on and state further the practice in this respect.

A. For instance, from Concord to Manchester, he was supplied with one thousand tickets, and when he made his report at the end of the month, that number that was taken out—the number indicating his sale—was returned to him.

Q. If he sold nine hundred tickets you returned nine hundred tickets to him, and he kept his thousand good every month?

A. Yes, sir; that was the case at every station. That was the system.

Q. Go on and explain, if you please.

A. If the ticket seller at Concord sold out five hundred, or at the first of the month he is charged with five hundred, and during the month sells five hundred and makes such a return to me at the end of the month, then I take that report and return him the five hundred that he has sold, and that makes the number good.

Q. Where is he charged with the number that he has?

A. At my office.

Q. He is, then, charged with the number of tickets that he has sold?

A. Yes, sir.

Q. Now, will you go on and state as to this ticket office system. If you supplied any tickets to the conductors, won't you state what kind of tickets they were? I mean under this system that was adopted four or five years ago.

A. We furnished the tickets to Lowell and Boston and Lawrence; and these tickets were charged to them.

Q. You say, to Boston, Lowell and Lawrence. From what points to Boston and Lowell?

A. From the Concord road.

Q. Without regard to stations?

A. Without regard to stations. On the back, however, of each check, was printed the various stations on the road, and the latter part —I think for the last five years—I cannot be certain about that date—my impression is that for the last five years they have printed the stations.

Q. Mr. Sanborn, tell the way in which you delivered these checks to the conductors; how many you delivered; what receipts you took, and what accounts you kept.

A. These checks are charged, perhaps from five to twenty-five, to the different conductors. Some of them used a great deal more than others. Those that had occasion to use more, we charged more to them. Whenever they returned any on their way-bills, that check was returned to them; at least a new check was furnished to them to make the number good. The checks were charged to them the same as they were to the station agents; precisely the same principle.

Q. Now, what return did the conductors make to you as general ticket agent? You were the person, you said, whom the conductors made their returns to?

A. Yes, sir; they returned to me a way-bill. Originally they returned to me the money; but for the last few years, I think, the money has not been returned to me. For two years the money was returned to me. Since then—I think for eight or nine years—the money was returned to the cashier, and the way-bills and tickets were returned to me.

Q. (*By the Chairman.*) What do you say the way-bill shows?

A. It shows the number which they sold in the cars.

Q. (*By Mr. Haile.*) The amount of tickets sold?

A. Yes; that was carried out and the amount put down on the way-bill.

Q. (*By the Chairman.*) All the tickets were returned to you?

A. The collections in the cars.

Q. (*By Judge Bellows.*) That is, they were accounted for?

A. They were returned to me.

Q. (*By Mr. George.*) Were any tickets furnished to the conductors between stations on the Concord road—I mean between local stations—to be sold by them in the cars?

A. No, sir; not until recently. I think not until within two years.

Q. Since the new conductors were on?

A. Yes, sir. We had checks which were issued, which might be used for that purpose; but I think they were never adopted.

Q. (*By Mr. Bellows.*) You say not until recently. How recently?

A. Not until the spring of 1866.

Q. (*By Mr George.*) Then prior to 1866 the money paid in the cars to the conductors, for passengers between local stations, simply appeared on the way-bill, and not by a ticket in addition?

A. No, sir.

Q. Is there anything further in regard to the ticket system?

A. I think not.

Q. (*By the Chairman.*) I should like to know at this time how minutely the way-bills were filled up?

*Mr. George.* I am coming to the way-bills, which will show for themselves.

Q. (*By Mr George.*) Have you the way-bills here?

A. Yes, sir.

Q. Won't you show them? [Way-bills shown and explained to the referees.]

Q. The prices are all printed in, with ten cents additional for the amounts taken in the cars?

A. Yes, sir.

Q. Mr. Clough used to go from Concord to Nashua, and return one of these way-bills to you?

A. Yes, sir.

Q. (*By Mr. Stanley*) If he went two trips on the same day there would be two way-bills?

A. Yes, sir.

Q. (*By Mr. George.*) Well, Mr. Sanborn, I will ask you, in order that it may appear what the rule was in regard to the terms in the cars; that is, how much was required to be paid in the cars more than for tickets sold in the office; and how much at various times?

A. The first advance from the regular fare must have been as much as seventeen years ago; and the next was about 1857, the spring of 1857, perhaps in January, 1857, an addition of five cents more, which made ten cents

Q. Then, originally, five cents more were added, and after 1857 ten cents more, and that was the rule of the road?

A. Yes, sir.

Q. And the way-bills were printed with this addition?

A. Yes, sir.

Q You may state, if you know, the object of this additional charge?

A. The object was to induce passengers to buy their tickets at the office. That is, so far as I know as to the object.

Q Now, you spoke of a further change in the matter of the tickets; won't you just tell what that further change was, and when it went into operation?

A. The change was principally in the numbering of the tickets consecutively, so that we might know the number.

Q. When was that done?

A. October, 1866.

Q. Now, won't you explain. Where were these tickets printed?

A. They were printed in New York.

Q. Supposing you sent on for a hundred thousand tickets,—we will say between the various stations on the Concord road—won't you explain how these tickets are printed?

A. They commenced with a cypher, and so on, one and two.

Q. There would be no two tickets of the same number?

A. No. Those tickets are kept in the office.

Q. The only difference, then, as I understand you——you may state if there is any particular difference now, except the mere numbering of the tickets.

A. There is no difference, only in the effect of the system. The effect of it is that we are better able to understand about the number of tickets that are out. For instance, if the number 99 is out, we know that that particular ticket is gone, and whenever that ticket comes in we know who brings it in; we know on which train it is missing, and who brings it in.

Q. How do you know on which train?

A. Every conductor's checks, as they are brought in, are examined, and we know if there is a missing one.

Q. (*By the Chairman.*) There is one thing I do not understand; and that is how you know, if you do know, what tickets are sold for each train?

A. Tickets are sold for the numbers as they come along. That is the duty of the ticket seller, to sell them in order.

Q. And he keeps the sale on record?

A. When the conductors bring them, we know the highest and lowest number; and if the middle one is gone, the inference is that it is gone on his train.

Q. You say that applies to all the stations on the road?

A. That applies to all except "Robinson's Ferry;" and they don't sell enough there to make it an object.

Q. I want to know how the two thousand tickets are numbered?

A. They are numbered just alike; the difference is the station where they are.

Q. Then the number of the tickets commences new from each particular station?

A. Yes, sir.

Q. Supposing it coming out to one month, then do you begin again?

A. If we come up to 999 in June, we commence in July with 1000.

Q. Have you any stop until you get to the end?

A. We have not.

Q. (*By Mr. Tappan.*) That is to say, you get a certain number of tickets printed, and you go on until these are all used up?

A. Yes, sir.

Q. (*By Mr. George.*) Supposing, Mr. Sanborn, that I should buy a ticket at the Concord ticket office to day, and should not use it for a week; won't you explain how any numbering or dating, or anything else, would show on what train that was?

A. There would be nothing to show any further than this: if the number was within the first and last number which that particular conductor had, it would be charged to him. If it was previous or after—for instance, if the first ticket was sold, the next date that the conductor brought, it would not be charged, but when it was brought in it would be known.

Q. Supposing I buy a ticket to-day, and on going to the train find a friend, and don't go; how does any number show when that was used?

A. It does not tell us where the ticket is; but the ticket itself, if it is stamped as it should be when it comes in, of course we know the train it comes on. As to the stations, the Boston tickets we do not collect.

Q. All the tickets are stamped on the day they are sold?

A. Yes, sir.

Q. Then the stamp shows the date of their sale?

A. Yes, sir.

Q, Is there any way—if there is you may state it—by which you can tell by the stamping of the ticket, or in any other way, when it is used? I simply want to ask this question: I will suppose that a man buys tickets from Concord to Nashua on the first day of the month; I will suppose that the conductor on the road takes that ticket, and does not return it; and he sells that ticket over and over again; and that it is used every day of the month, and he returns it at the end of the month; I want to know if there is any way of telling that that ticket has not been used over and over again?

A. No, sir.

Q. (*By Judge Bellows.*) That is an answer applicable to both your systems?

A. Yes, sir.

Q. Now, I want you to describe, if you will, Mr. Sanborn, the ticket system—if it is applicable to the joint ticket. I want you to explain, clearly as you can, to the referees, the ticket system as applicable to joint tickets from the upper and from the lower roads. We will first take the Northern road. Will that illustrate the whole system?

A. I should think so.

Q. Take that for example, and explain the general ticket system of passengers coming from the other roads.

A. Every ticket that is sold by the Northern road to come on to any part of the Concord road has a coupon which is left on our road; that is taken up and carried into the general ticket office.

Q. (*By Mr. George.*) Have you any specimens of full coupon tickets?

A. I have not; but I can get them.

Q. Now, won't you go on, exactly and minutely, and explain the ticket system in reference to joint tickets—what checks there are, to start with?

A. The tickets brought into our office by the conductors are credited to the Northern road.

Q. You mean by the Northern road?

A. Only so far as they are to be used as a check.

Q. That is, you keep an account of them?

A. We keep an account of them; and when they report the numbers sold we put these opposite and compare them.

Q. (*By Mr. Tappan.*) The tickets returned by the conductors are put opposite the tickets sold? When the Northern road reports at the end of the month—when they report the number of tickets—then what do you put opposite?

A. The collection.

Q. Have you a tabular sheet that will show the return for one month?

A. I have nothing except the whole book.

Q. Now if you will explain from your book how this is done: first, in regard to the upper roads, and then the others and all of them.

A. For instance, on the Northern road, in black is what is given out; in red is what we collect of them. To Manchester 56½—255, 257.—Those are the number of the tickets.

Q. How do you account for that little discrepancy?

A. Oh, they stop over. Generally — take it for six months — they will very generally agree. But the collections never exceed the sales. I never knew them to in the whole six months.

Q. (*By the Chairman.*) You don't absolutely expect that the check will come back on the same day that it is sold? What is there to prevent its being used several times?

A. There is nothing.

Q (*By Mr. George.*) We will say the tickets are sold on the 30th day of June, and they may not be used until the first of July; and, therefore, of course the returns for June would not show it.

A. No, sir; but perhaps at the end of six months or a year. I never have known but one instance where the collections have exceeded the sales.

Q. I want to ask you a question that is perhaps involved in the other, but I want it perfectly clear. I understand that the collections of coupon tickets from the Northern road, or any other road, as returned by the conductors, are entered in the book, and then at the end of the month the Northern Railroad makes a record of the sales. Now supposing on the first day of the month a ticket is sold from Fisherville to Boston; what is there, if anything, to prevent the conductor, if he chooses, from retaining that coupon until the end of the month, and using it every day, over and over again, and selling it?

A. There is nothing in our power.

Q. Suppose the conductor should, in the case I have cited, take up a coupon ticket on the first day of the month and use it every day during the month; and that the last day of the month he returned it, why then your account would agree with the account shown by the Northern Railroad exactly, would it not?

A. Yes, sir.

Q. Now, will you be kind enough to take, for instance, the Boston and Lowell Railroad. How is it?

A. The system is the same with that, precisely, only that they make their report earlier in the month than the Northern Railroad. The collections are the same, and the system is the same, except that it is earlier in the month.

Q. You mean by earlier in the month that they report earlier than the Northern Railroad?

A. Yes, sir.

Q. Supposing the ticket was sold over the Boston & Lowell road from Boston to Concord, or from Boston to any other station, and the conductor should take up that ticket, and not return it until the end of the month, what is there to show that use?

A. There is nothing to do it.

Q. It rests entirely upon the honesty of the conductor?

A. Yes, sir.

Q. Supposing the conductor should take a local ticket—take any of those tickets that are furnished to him to sell in the cars, for example, or any other tickets that are sold at the ticket office; and supposing it is sold on the first of the month, and he continued to use it over and over again, what is there to show that he used them more than once?

A. The tickets that are furnished him he could not use so.

Q. You may give all the reasons, if you please, in reference to that and the others?

A. The reason why he could not use these is because when they are once used they go on to another road. He don't use these except to go on another road; and without collusion with passengers he could not; and even if that was so, he could not, because they go on to the road.

Q. Supposing a ticket was sold in the cars by the conductor to go from Manchester to Lowell; and supposing the conductor on the Concord road should neglect to punch it, and the conductor on the Nashua and Lowell road should neglect to punch it, what is there to prevent that ticket being used over and over again?

A. Nothing to prevent it.

Q. Supposing a man gets in the cars at Concord and pays in the cars to Nashua, and the conductor fails to return upon his way-bill, or in any other way, the amount paid, what is there in the rules and regulations of the road, or what is there in any way to make that fact apparent?

A. There is nothing to prevent it.

Q. You may state the entire evidence that the road has, or has had, of the amount received by conductors in the cars; upon what the entire evidence rests. State the whole evidence.

A. There is none at all, only their report, their way-bill.

Q. You have spoken of the return of tickets to you, and the return of money paid in the cars. What further report, if any, are the conductors required to make to you, and have they been always?

A. For the last seven or eight years—seven years, I should think—they have been required to report the number of free passes. I think that is all in addition.

Q. And how are these reports made?

A. These are on a sheet by itself.

Q. Will you state how that report was made? I mean the contents of the report?

A. Just the man's name, and where from and to, and by whose authority he was passed. I believe that is the order in which the report is made. Those lists we do not use. It has been several years that way.

Q. (*By Mr. Haile.*) No account made of annual passes?

A. There has been at times, and at times not; there is no account made to me. There was a record made by the superintendent at his own office.

Q. But does the conductor report that a gentleman having an annual pass—that he passes?

A. They have at some times, and some times not.

CROSS EXAMINATION.

Q. (*By Mr. Mugridge.*) You spoke of furnishing a certain number of tickets to the ticket seller, and then at the end of the month taking his report of the number of tickets sold, and then supplying enough to make his number. At that time do you require him to return the tickets that he has left to return?

A. No, sir.

Q. What check have you upon the ticket-seller? How do you know that he may not sell the tickets that he has and not return?

A. We have a list which shows that the collection is in excess; that is, that he had a number to go around, and counting up their tickets will show if they hold out.

Q. Now, Mr. Sanborn, having explained the manner in which business is done, I want you to give before the referees the receipts from passengers on the Concord Railroad, as far back, say, as 1857. Have you condensed a table?

A. No, sir; we have not as far back as that. I think the reports we have here, if that is what you refer to—the reports from 1860—we have each year, from 1860 to 1868 inclusive.

Q. Now you may give the receipts from passengers. That is what we want.

A. For the year ending 1860, commencing

| | | | |
|---|---|---|---|
| April, 1859, | $11,738 06 | October, | 14,564 74 |
| May, | 11,693 67 | November, | 12,246 41 |
| June, | 11,888 70 | December, | 9,442 76 |
| July, | 15,235 21 | January, 1860, | 9,935 64 |
| August, | 19,343 00 | February, | 8,371 27 |
| September, | 23,113 45 | March, | 12,361 78 |
| For the year, | | | $160,573 42 |
| April, 1860, | $12,691 00 | October, | 20,285 80 |
| May, | 12,011 30 | November, | 13,976 39 |
| June, | 14,367 50 | December, | 11,932 83 |
| July, | 16,032 18 | January, 1861, | 10,360 49 |
| August, | 20,782 91 | February, | 9,633 92 |
| September, | 20,830 99 | March, | 12,468 20 |
| For the year, | | | $175,144 42 |
| April, 1861, | $13,588 62 | October, | 14,560 91 |
| May, | 12,814 42 | November. | 11,924 39 |
| June, | 12,340 32 | December, | 10,434 97 |
| July, | 13,820 33 | January, 1862, | 8,717 91 |
| August, | 23,090 79 | February, | 6,240 25 |
| September, | 18,458 96 | March, | 11,411 53 |
| For the year, | | | $155,065 43 |
| April, 1862, | $9,113 50 | October, | 17,574 83 |
| May, | 9,817 03 | November, | 16,345 37 |
| June, | 11,376 26 | December, | 11,736 40 |
| July, | 16,193 56 | January, 1863, | 10,758 16 |
| August, | 20,461 50 | February, | 9,474 84 |
| September, | 23,390 70 | March, | 12,549 97 |
| For the year, | | | $167,794 32 |
| April, 1863, | $12,881 44 | October, | 24,701 05 |
| May, | 12,711 04 | November, | 20,498 63 |
| June, | 16,224 01 | December, | 15,823 76 |
| July, | 20,005 60 | January, 1864, | 13,342 33 |
| August, | 23,664 56 | February, | 10,512 43 |
| September, | 24,875 34 | March, | 10.660 04 |
| For the year, | | | $223,900 24 |
| April, 1864, | 66,442 01 | October, | 33,813 36 |
| May, | 20,940 76 | November, | 29,857 71 |
| June, | 25,498 15 | December, | 25,337 95 |
| July, | 34,197 33 | January, 1865, | 22,797 09 |
| August, | 51,770 10 | February, | 23,513 79 |
| September, | 44,943 20 | March, | 25,128 97 |
| For the year, | | | $361,442 94 |

| | | | |
|---|---|---|---|
| April, 1865, | $21,746 75 | October, | 28,763 07 |
| May, | 23,539 19 | November, | 23,652 84 |
| June | 33,443 49 | December, | 29,596 96 |
| July, | 37,077 26 | January, 1866, | 21,704 08 |
| August, | 46,306 62 | February, | 23,091 30 |
| September, | 45,512 50 | March, | 30,056 21 |
| For the year, | | | $362,490 27 |

| | | | |
|---|---|---|---|
| April, 1866, | $29,924 47 | October, | 37,159 22 |
| May, | 28,890 03 | November, | 31,008 24 |
| June, | 33,476 69 | December, | 25,648 83 |
| July, | 36,748 41 | January, 1867, | 23,224 67 |
| August, | 46,206 58 | February, | 23,229 60 |
| September, | 40,632 64 | March, | 26,525 04 |
| For the year, | | | $381,726 82 |

| | | | |
|---|---|---|---|
| April, 1867, | $30,573 99 | October, | 27,270 57 |
| May, | 27,159 65 | November, | 23,252 45 |
| June, | 31,160 17 | December, | 18,447 20 |
| July, | 39,764 53 | January, 1868, | 16,210 45 |
| August, | 37,726 27 | February, | 14,834 80 |
| September, | 33,819,05 | March, | 19,019 56 |
| For the year, | | | $319,238 99 |
| From Manchester to Lawrence to be added to last amount, | | | 47,653 79 |

This takes the Concord Railroad alone. It is divided.

That goes with the last year, 1868.

Q. (*By Mr. George.*) Now, I want you to give the referees the rates of fare?

*Mr. Mugridge.* And now I want you to produce, at the same time, the receipts from all the ticket offices on your line of road, from 1861 to 1868, as distinguished from the amount received for tickets.

Q. (*By Mr. George.*) I want to ask you the rates of fare on the Concord Railroad, say from 1848 to 1868. Is this abstract the same that you have drawn?

A. This is the abstract that I made of the rates of fare.

Q. Now, I want you to read that, and I want the referees to take down the prominent parts of it. And, Mr. Sanborn, while they are reading, I wish you to make out the per centage and decrease of fares between the highest fares and the present fares.

*Mr. Mugridge.* Now, when that is offered, we have something to say; at least until some further statistics are put in. This runs from 1848 to 1865; and I want the per centage made between these two times. We want the receipts of the road put in from 1848 to 1868.

*Mr. Stanley.* You shall have them.

Q. (*By Mr. Mugridge.*) Now then, won't you make the average between 1861 to 1868?

| | Manchester. | Nashua. | Lowell. | Boston. | Lawrence. |
|---|---|---|---|---|---|
| From Concord, 1848 to 1850, - - | 35 | 70 | 1 00 | 1 50 | 1 00 |
| From Manchester, - - - - - - | | 35 | 65 | 1 15 | 65 |
| From Concord, 1850 to 1851, - - - | 40 | 80 | 1 10 | 1 50 | 1 00 |
| From Manchester, - - - - - - | | 40 | 75 | 1 15 | 65 |
| From Concord, 1851 to 1854, - - - | 45 | 90 | 1 25 | 1 75 | 1 25 |
| From Manchester, - - - - - - | | 45 | 80 | 1 30 | 80 |
| From Concord, 1854 to 1857, - - - | 50 | 1 10 | 1 50 | 2 00 | 1 50 |
| From Manchester, - - - - - - | | 45 | 80 | 1 30 | 80 |
| From Concord, 1857 to 1862, - - - | 60 | 1 20 | 1 50 | 2 25 | 1 50 |
| From Manchester, - - - - - - | | 60 | 1 00 | 1 75 | 1 00 |
| From Concord, 1862 to 1864, - - - | 65 | 1 25 | 1 55 | 2 35 | 1 55 |
| From Manchester, - - - - - - | | 65 | 1 05 | 1 85 | 1 05 |
| From Concord, 1864 to 1865, - - - | 80 | 1 60 | 2 10 | 2 90 | 2 10 |
| From Manchester, - - - - - - | | 90 | 1 40 | 2 25 | 1 45 |
| From Concord, 1865 to 1866, - - - | 75 | 1 50 | 2 00 | 2 75 | 2 00 |
| From Manchester, - - - - - - | | 75 | 1 25 | 2 00 | 1 25 |
| From Concord, 1866 to 1867, - - - | 70 | 1 40 | 1 85 | 2 60 | 1 85 |
| From Manchester, - - - - - - | | 70 | 1 15 | 1 90 | 1 15 |
| From Concord, 1867 to 1868, - - - | 65 | 1 30 | 1 75 | 2 50 | 1 65 |
| From Manchester, - - - - - - | | 65 | 1 10 | 1 85 | 1 00 |
| From Concord, 1868 to ——, - - - | 55 | 1 10 | 1 45 | 2 20 | 1 45 |
| From Manchester, - - - - - - | | 55 | 90 | 1 65 | 90 |

[Adjourned.]

[FIFTH DAY. Tuesday, July 28th, 1868.]

MR. SANBORN'S TESTIMONY CONTINUED.

Q. (*By Mr. George.*) Does this contain all the returns made by Mr. Clough?

A. Yes, sir; that is the number of passengers carried.

Q. You made it in the manner that I requested? You made up all the tickets?

A. Yes, sir.

*Mr. George.* Now my point is this; that we propose to show that Mr. Clough carried on the trains run by him more than the returns on all the trains.

*The Chairman.* If I recollect right it was thought when Mr. Biddle was testifying, that we would suspend this a while to show the conduct of the road. I understand that it is not laid aside yet.

*Mr. Mugridge.* I wish to inquire of Mr. Sanborn if there is any way of getting at the number of passengers each day?

A. No, sir.

Q. (*By Mr. George.*) Mr. Sanborn, you have examined this statement, haven't you?

A. Yes, sir, I suppose that is right.

Q. You examined that at the time?

A. Yes, sir.

Q. You may look at them again, if you please. I want to know whether these are genuine tickets or otherwise?

A. Yes, sir, I should say these were genuine.

Q. These are genuine tickets of the road, are they?

A. Yes, sir.

Q. I want to know if these tickets show any indications, and if so what, that they have been through your office, and also other ticket offices of the road?

A. They have a stamp on them which was put on them in my office. That is so far as we had gone. There are some of them old tickets before we had begun to use these. And these are dated, showing that they have been through the ticket offices. Sometimes in the rush they would get out without being stamped; but they have the stamp of the general ticket office.

Q. You spoke of having examined these before. When were they examined by you before?

A. I think it was in February, 1866.

Q. At the time of their being found on Whitcher?

A. Yes, sir, I think sometime between the 15th and 23d. I have somewhere a list.

Q. Of the tickets found on Whitcher?

A. Yes, sir.

Q. Who gave you this?

A. Mr. Crane brought them into my office.

Q. Now, Mr. Sanborn, have you made an abstract of the amounts returned by Mr. Clough from the way-bills, returned as taken in the cars by him, from the way-bills?

A. Yes, sir.

Q. And from what time to what time?

A. From 1857, I think; December 1st, 1857.

*Mr. George.* I propose to put in the amounts returned by Mr. Clough from December 1st, 1857, until the time when he left the road.

*Mr. Mugridge.* Why can't you go behind that?

*Mr. George.* The books are all burned up; burned in 1859.

*Mr. Mugridge.* The records that were burned were back of 1857.

*Mr. George.* Yes, sir.

Q. (*By Mr. George.*) You said you had been connected with the road how long?

A. Twenty years—nearly twenty years.

Q. Mr. Clough went on the road in 1842?

A. That is about the time.

Q. What train did Mr. Clough run? and how many conductors were then upon the road?

A. Mr. Clough generally ran the train that left Concord Monday at half past ten.

Q. (*By the Chairman.*) What period does this cover?

A. Since I have been on the road?

Q. How long is that?

A. About twenty years—nineteen years. I think the rule has generally been that he has run one train on Monday and two Tuesday.

Q. Going down one day and back the next?

A. That has been the rule; there might have been exceptions for a month or two perhaps.

Q. What time did the three trains leave Concord, and what time did they return?

A. The first train left at half past five (the same as they do now), quarter past ten, and half past three; and reached here at half past ten, three and eight.

Q. Now, for how long a period of time were these the only three trains; that is, these three trains down and three back? You may state the date.

A. From November, 1859, there was an express train put on and run for months.

Q. A fourth train you mean?

A. A fourth train run by Mr. Jones.

Q. Then up to 1859, how many trains had there been?

A. Three trains.

Q. And how many conductors?

A. Two.

Q. Now you may go on and state how the trains have run from that time to this.

A. There has been part of the time, for six months, but three trains, and then again they commenced with four, and then they had three trains. Up to January 31, 1860, for four months they ran four trains, and from November 1 to January 31, 1959 and '60.

Q. What time of day did that go? That went down in the evening; that is, it was an extra train down and not an extra train up?

A. Yes, sir. Then for September and October of the same year, 1860, four trains were run.

Q. Well, then go on to the next.

A. Then next April, 1861, they ran six months, four trains. Then they commenced with January, 1862, and ran a fourth train regularly, and it has run ever since.

Q. (*By Judge Bellows.*) At what time does that run?

A. That left at half past seven in the morning. The "express" runs separate now. That is still another train.

Q. When was the train put on that goes from here at eight o'clock in the evening, in connection with the Montreal express?

A. That was put on in January, 1865, according to my list. That is what is now called the "express."

Q. Mr. Sanborn, which is the heaviest train on the road?

A. Half past three in the afternoon, down. That is the heaviest train now.

Q. Has it been for twenty years?

A. The half past ten, down, in the morning, and the half past three, up, in the afternoon.

Q. And the third?

A. The evening train, up. The half past five train in the morning is always the lightest train.

Mr. Clough's way-bills were then read, showing the amounts returned from December, 1857, to December, 1865, inclusive, as follows: (December, 1857, the amount returned was $69.55.)

| | 1858. | 1859. | 1860. | 1861. | 1862. | 1863. | 1864. | 1865. |
|---|---|---|---|---|---|---|---|---|
| January, ............... | 74 00 | 115 85 | 108 40 | 240 20 | 237 25 | 174 25 | 255 70 | 337 95 |
| February, ............... | 110 60 | 122 35 | 81 55 | 212 97 | 207 30 | 159 20 | 303 65 | 254 00 |
| March, ................... | 139 20 | 120 25 | 117 70 | 228 15 | 211 75 | 209 65 | 359 40 | 254 10 |
| April, ................... | 98 35 | 120 35 | 113 25 | 209 50 | 200 85 | 223 05 | 395 20 | 277 35 |
| May, ..................... | 73 80 | 144 80 | 108 70 | 202 85 | 168 90 | 218 95 | 393 15 | 225 90 |
| June, .................... | 85 50 | 118 30 | 109 80 | 174 95 | 250 60 | 204 95 | 367 80 | 253 60 |
| July, .................... | 138 90 | 138 25 | 158 30 | 164 75 | 214 95 | 193 60 | 276 75 | 252 30 |
| August, .................. | 111 60 | 158 10 | 180 55 | 347 35 | 212 60 | 181 20 | 334 30 | 277 90 |
| September, ............... | 86 95 | 231 55 | 302 60 | 193 25 | 242 30 | 246 55 | 355 60 | 267 10 |
| October, ................. | 138 10 | 181 40 | 344 40 | 351 40 | 278 40 | 322 20 | 251 95 | 274 45 |
| November, ................ | 134 75 | 133 95 | 223 15 | 252 20 | 160 90 | 247 30 | 244 60 | 414 25 |
| December, ................ | 108 85 | 96 00 | 210 55 | 245 25 | 201 40 | 369 95 | 344 60 | 422 25 |
| | 1300 60 | 1681 15 | 2058 95 | 2822 82 | 2593 20 | 2750 85 | 3881 70 | 3511 15 |

Q. (*By Mr. George.*) What time did Mr. Clough leave the road?
A. Left in January, 1866, I think.
Q. Do you remember what day of January?
A. I think about the middle of the month.
Q. (*By Mr. Rolfe.*) The fourteenth?
A. I could tell if I were at the office.
Q. (*By Mr. George.*) Have you a computation how much this average is per train for Mr. Clough?
A. Yes, sir.
Q. Won't you be ready with that in the morning, and exactly the day Mr. Clough left, and the amount which he returned for the fourth day of January? [Adjourned.]

[SIXTH DAY. Wednesday, July 29, 1868.]

The hearing was continued at at nine o'clock, A. M., and the examination of Mr. Sanborn resumed, after putting in the following documentary evidence:

CONCORD, MANCHESTER & LAWRENCE, CONCORD & PORTSMOUTH AND MANCHESTER & NORTH WEARE RAILROADS.

## INSTRUCTIONS TO CONDUCTORS.

On and after the first day of June, 1864, NO PERSON WHATEVER, except as hereinafter provided, will be allowed to pass over this road, without a ticket or pass signed by the president, superintendent or assistant superintendent of the Concord, Manchester & Lawrence Railroad, in their own hand-writing.

1—Pass presidents, managing agents, superintendents and passenger conductors of railroads named in article 4.

2d—Pass the following persons connected with the Concord, Manchester & Lawrence Railroad:—Directors, clerk, treasurer, master of transportation, road master, ticket master, overseer of repairs in iron shop, overseer of repairs in wood shop, wood agent, regular freight conductors, depot master at Concord, cashier, superintendent's clerk, ticket seller at Concord, freight agent at Concord, station agent at Manchester, station agent at Nashua, regular mail agents in their car with their mails, also regular express men in charge of their express matter, who have a business contract with the Concord, Manchester & Lawrence Railroad.

3d—If necessary for persons employed by the company to pass over the Concord, Manchester & Lawrence road, *strictly on business of the corporation*, the following persons will furnish written passes to the men employed in their respective departments, viz:—Master of transportation, general ticket agent, cashier, superintendent's clerk, overseer of repairs in iron shop, overseer of repairs in wood shop, road master, wood agent, and station agents at Manchester and Nashua, good for one trip only over the C., M. & L. road, and must be taken up every trip by the conductor and returned to the ticket master.

4th—Conductors will receive written passes signed by the president, superintendent, agent, or managing agent, in their own hand-writing, of the following roads:—Boston & Maine, Eastern, Boston & Lowell and Nashua & Lowell, Worcester & Nashua, Northern, Passumpsic, Boston, Concord & Montreal, Providence & Worcester, Cheshire, Vermont Central, Ogdensburgh, Concord & Claremont, Contoocook River, Concord & Portsmouth, and Norwich & Worcester Railroads. Also, agent Norwich & Worcester Steamboat Company. Such passes will be limited to one trip each way over the road.

5th—The conductors will mark all free passes given by the Concord, Manchester & Lawrence road, or by a connecting road, over the Concord, Manchester & Lawrence road, by punching a hole through them when presented, each way over the road, that they may know they have been used; and after they have been so marked, the conductor will allow no person to ride on them.

6th—In cases of manifest poverty or inability to pay, conductors can exercise their discretion in allowing a free pass; and whenever such a free pass is granted, the name of the person passed, and the attending circumstances, will be noted by the conductor, which, with the names of all persons having passes from connecting roads, will be reported every trip to the ticket master.

7th.—The ticket master will give free passes in cases of evident inability to pay; but in all such cases the pass will be given for one trip only, and the name of the person passed, and the word "charity" must be written plain across the pass, and it will not extend over connecting roads.

8th.—All free passes over the Concord, Manchester & Lawrence road will be taken up by the conductor and returned to the ticket master every trip, excepting those given by and extending to and from connecting roads, which will be marked and reported as provided by articles 5th and 6th.

9th.—In all cases, conductors will demand and collect of the passengers ten cents more for tickets sold or fares taken in the cars than when purchased at the ticket office. All money so taken must be entered *at the time* on a way-bill, and a return made thereof, with all tickets collected, *each trip*, to the general ticket master.

10th.—All outstanding passes, dated prior to January 1, 1864, will be collected by the conductors, and returned to the superintendent, except the regular season tickets paid for by the holders.

11th—All tickets over the Concord, Manchester & Lawrence Railroad shall be dated on the day of sale; said tickets shall only entitle the holder thereof to a passage on the day of their date, provided that joint tickets shall be good for such further time as may be necessary to enable the holders, by the regular trains, of the road to reach the stations to which such tickets are sold.

12th—Conductors, when on duty, will wear the badge prescribed by the company, and must be at their head station at least fifteen minutes prior to the departure of the train, attending to the business of the train and accommodation of the passengers, and remain after the arrival of the train a suitable time for this purpose.

13th—The conductors are required to punch all coupons and tickets in the up and down trains between Concord and Manchester, and Manchester and Nashua. Also, between Manchester and Derry, and Derry and Lawrence. Between Manchester and Candia, Epping and New Market, and New Market and Portsmouth, tickets, coupons and passes required to be punched *are to be punched when they are presented by the holders, in the holders' presence,* so as to be seen by them.

14th—Tickets or coupons of other roads shall not be sold, used or retained by the conductors, and they shall report the names of any persons using them or offering them, when they believe or the date indicates that they were not received in the regular course of business. No tickets or checks shall be sold, used or retained by the conductors, except those received by them from the general ticket agent at Concord, and charged to them in account.

By order of the Directors of the Concord Railroad,
JOHN H. GEORGE, Clerk.

N. G. UPHAM, Pres't.
J. A. GILMORE, Sup't.

MR. SANBORN'S TESTIMONY CONTINUED.

Q. (*By Mr. George.*) I asked you last night to state the largest amounts returned by Mr. Clough at any one train, with the date. Have you the list?

A. Yes, sir; that is the list.

Q. Here is the date and amounts returned by George Clough exceeding fifteen dollars a trip? A trip is from here to Nashua and back?

A. Yes, sir.

Mr. Tappan inquired the purpose of the proposed statement.

*Mr. George.* We propose to show—and that we have already—what he received on the trains during these years. The amounts shown here are on individual trains. All that he returned, except in these individual instances, simply shows that he did not return the amounts received in the cars.

*Mr. Tappan.* Those commenced when?

*Mr. George.* In 1862. We have no list farther back than that.

The following statement was then read;

| | | | | | |
|---|---|---|---|---|---|
| 1862. | June 19, | $83 25 | 1864. | December 31, | $23 40 |
| | October 25, | 65 00 | 1865. | June 3, | 16 90 |
| | December 27, | 20 95 | | June 12, | 15 60 |
| 1863. | October 16, | 15 10 | | June 31, | 19 65 |
| | October 19, | 18 05 | | April 29, | 17 00 |
| | December 19, | 17 65 | | June 30, | 29 40 |
| | December 22, | 17 00 | | July 31, | 24 80 |
| | December 29, | 15 90 | | August 14, | 17 35 |
| 1864. | March 29, | 16 05 | | September 4, | 15 40 |
| | April 9, | 15 05 | | October 25, | 17 40 |
| | April 19, | 20 95 | | October 26, | 17 50 |
| | May 30, | 22 80 | | November 9, | 19 35 |
| | May 31, | 16 85 | | November 14, | 28 25 |
| | June 1, | 16 45 | | November 18, | 20 25 |
| | July 30, | 36 05 | | November 21, | 15 70 |
| | August 31, | 18 90 | | November 22, | 15 95 |
| | November 30, | 16 75 | | November 23, | 30 40 |
| | December 29, | 21 10 | | November 24, | 15 90 |

| | | | | | |
|---|---|---|---|---|---|
| 1865. | November 25, | $25 10 | 1865. | December 23, | 16 55 |
| | December 2, | 16 45 | | December 27, | 16 20 |
| | December 6, | 17 65 | | December 28, | 15 85 |
| | December 11, | 21 25 | 1866. | January 4, | 15 40 |
| | December 18, | 18 65 | | January 6, | 18 65 |
| | December 22, | 19 10 | | January 13, | 16 10 |

Q. (*By Mr. George.*) Mr. Sanborn, will you now be kind enough to show how much Mr. Clough's returns show on the average?

Average amounts returned by George Clough each trip and each train, from 1858 to January 15, 1866:

| YEAR. | TRIP. | TRAIN. |
|---|---|---|
| 1858 | 2 77 | 1 38½ |
| 1859 | 3 58 | 1 79 |
| 1860 | 4 39 | 2 19 |
| 1861 | 6 00 | 3 00 |
| 1862 | 5 53 | 2 76 |
| 1863 | 5 85 | 2 92 |
| 1864 | 8 27 | 4 13 |
| 1865 | 7 47 | 3 73 |
| *1866 | 8 78 | 4 39 |

*January, including half month by Mr. Le Bosquet.

Average for the whole trip $5 46; for the train, $2 73.

Q. (*By the Chairman.*) Have you any average made at any time when the fares were the same?

*Mr. George.* I propose to have that prepared.

From 1857 to 1862 the prices were all the same.

FARES FROM CONCORD.

| | Manchester. | Nashua. | Lowell. | Boston. |
|---|---|---|---|---|
| 1857–62 | 60 | 1 20 | 1 50 | 2 25 |
| 1862–64 | 65 | 1 25 | 1 55 | 2 35 |
| 1864–65 | 80 | 1 60 | 2 10 | 2 90 |
| 1865–66 | 75 | 1 50 | 2 00 | 2 75 |
| 1866–67 | 70 | 1 40 | 1 85 | 2 60 |

Q. (*By Mr. George.*) The rules and regulations require a return of the free passes to you. Were these returned to you by Mr. Clough?

A. They were returned by him.

Q. And did Mr. Clough make his returns?

A. The free passes as they were collected were turned in to the office, and will show, at least for the last few years of his returns, where it was required that the report should be made of all the free passes, except annual passes, of which there was no return made; unless it might have been the last few months that he returned.

Q. In what did he make the return?

A. On a blank sheet for that purpose. Another exception too, was the passes given to workmen on the road, by the overseers of the different departments. These were not reported. I believe these they didn't require, and also the annual passes.

Q. They were not made on the same sheet of paper as the way-bill?

A. No, sir.

Q. What became of these returns?

A. Destroyed as waste paper; no account made of them after making the annual report. There had to be an annual report to the Legislature.

Q. How many free passes did Mr. Clough return?

A. I couldn't tell; no idea except general estimation of the returns; should judge all the way from one to half a dozen on a train.

Q. What do you mean? A train?

A. I mean one trip. There would be times, perhaps in June, during the Legislature, when there would be more passes returned.

Q. (*By the Chairman.*) Mr. Sanborn, did I understand you that the superintendent undertook to abrogate these rules?

A. I couldn't say what he said to the conductors; I know it was by his consent that some of these things were done.

Q. Of course I only ask from what you know, whether you gave you instructions?

A. I have asked him myself whether such were to be returned, and he said that it was all right, and need not be done. The matter of annual passes was spoken of, and I found that the conductors did not report them; and I was required to report any deviation. And he said that he had seen the conductors.

Q. That was at the time that the rule was to return all of them?

A. Yes, sir.

Q. (*By Mr. George.*) Those are the annual passes and the passes of the workmen that you refer to?

A. Yes, sir.

Q. And no others?

A. No others.

Q. I want you to tell the referees, on these long western tickets, what the proportion to the road was. For instance, take a ticket to Chicago, from Boston, west. How much did the Concord road receive from a passenger?

A. These tickets were reported to us from the Vermont Central. We got, I think, the lowest was twenty-eight cents, and from that up to the regular rates, fifty-two cents.

Q. How much was the Concord Railroad's proportion of the fare of passengers on these western through tickets?

A. All the way from twenty-eight cents up to the regular price, which was fifty-two cents at that time.

Q. The regular price on the Concord Railroad, from Concord to Nashua, was $1.20. Supposing a man purchased a ticket on the Northern Railroad for Nashua, how much did the Concord Railroad receive?

A. Fifty-two cents.

Q. Now, sir, supposing you bought a ticket from Concord to Boston, that would take you from Concord over the Nashua and Lowell and the Boston and Lowell Railroad—take it in 1857, for example—when the fare was $2.25, how much did the Concord road receive of that $2.25? That is, what did they pay out to the lower roads?

A. They paid out sixty-eight cents and received one dollar and fifty-seven cents.

Q. Which was thirty-seven cents more than their local fare. Now we take it in 1864. The fare to Boston was $2.90. How much did the Concord Railroad receive of that $2.90?

A. Two dollars and twelve cents; they then paid seventy-eight cents.

Q. And their local fare was at that time?

A. A dollar and sixty.

Q. Then they received fifty-two. With regard to 1864, on these western tickets, how much did you receive?

A. The average was about the same. There has been no change in these, because it has always been settled.

Q. Now, what was the average amount received by the Concord Railroad on these western tickets?

A. I don't know as I could say.

Q. State as near as you can.

A. Perhaps it might have been forty-five cents. There have been cases, as I said before, that we have got more when they were divided *pro rata*.

Q. Now, supposing a man substituted one of these western tickets for the regular ticket from Concord to Boston. Out of how much would the Concord Railroad be defrauded? Take it in 1864.

Q. (*By Mr. George.*) Mr. Sanborn, then you may, if you please, state what the difference would be to the Concord Railroad in the first place, and we will take it between Concord and Boston for the local ticket. Take it from 1864 to 1865 for example. The fare was to Boston $2.90, and the local ticket that you would receive would be how much?

A. Two dollars and twelve cents.

Q. On one of these western coupons, how much did they receive, calling it 52 at the ordinary price $1.66. How much would they lose at the highest price?

A. One dollar and sixty cents.

Q. How much would they lose at the lowest rate, 28 cents?

A. One dollar and eighty-four cents.

Q. How much would they lose on the average?

A. One dollar and sixty-seven cents.

Q. You say these settlements were made with the Vermont Central road; were they made from the sales, or from tickets taken up in the cars?

A. From the sales.

Q. During the fall, when a regiment of soldiers was sent on, how did it go, on the regular trains, or special?

A. The special trains, almost invariably; I do not recollect a single instance where they went on the regular trains; it might have been.

Q. Generally, how many did they carry?

A. Generally as many as three or four hundred; they had on special trains.

Q. How was an account kept of those?

A. There was an order given by the quartermaster for their transportation.

Q. And were tickets issued?

A. No tickets issued for it. That order was presented to the government for payment.

Q. How was it, were there any tickets taken with reference to the government transportation?

A. If a small squad of men, they were sent, and a general order used to come down to the ticket office for tickets.

Q. State what peculiarity they had?

A. These tickets were marked "military tickets," and were not good for a citizen to use.

Q. (*By the Chairman.*) A man could not use them without brass buttons on his clothes?

A. No, sir.

Q. (*By Mr. Tappan.*) Did I understand that a small squad went on a regular train?

A. Yes, sir; the small squads generally went on the regular trains.

Q. To whom were those tickets returned?

A. Returned the same as any regular tickets, by the conductors.

Q. And how were these vouched for?

A. By the ticket for transportation which they left at the office.

Q. Supposing soldiers came home on government transportation; if there were such cases, explain that. Explain the tickets, if any?

A. They usually had orders if they came home either in squads or single individuals. They had an order from the government from Washington to New York, and beyond there they exchanged their orders. We had a coupon of such tickets to take up on this road.—Such tickets were marked "military tickets."

Q. And those were returned to you?

A. Yes, sir.

Q. For how long a time were these sold through New York, and from what stations?

A. From Concord to Manchester only.

Q. For how long a period of time—when did you sell to New York again?

A. I don't know as I could state exactly. I should say 1859 or 1860, and then, perhaps for two years they were cut off, and would not sell. They have always sold here both ways. both by rail and by boat, but we have not sold on these all the time.

Q. You commenced in 1859, or 1860. and sold for two years?

A. I should think so.

Q. How long a time was that sale discontinued?

A. I should not think more than a year.

Q. Have you means of getting exactly the date?

A. I might probably.

Q. I wish you to state to the referees—you say you will get all the exact dates at the office?

A. I will do so if I can.

Q. How has it generally been after that?

A. After that we commenced again to New York by boat, but by rail I think it was not commenced until about a year ago? I think not until last fall.

Q Then from 1860, till 1861, there was no connection, and no sale of tickets over the Concord Railroad to New York by rail, until since Mr. Clough's discharge?

A. No, sir.

Q. I want you to state, from what other points on the road, the Worcester and Nashua roads, to and beyond there, was any sale of tickets, and to what extent it continued?

A. At the same time we commenced with New York, we commenced with Springfield and Albany.

Q. This last time?

A. No, the first time; they were discontinued the same time they were at New York.

Q. And have they ever been resumed?

A. No, sir.

Q. Then during the war you had no business connection with those?

A. No, sir, only military tickets; those were to anywhere.

Q. Have you now mentioned all the points between which there have been sales of tickets, between Worcester and Nashua and the points beyond?

A. Groton Junction, this side of Worcester, and from Groton Junction to Keene and Bellows' Falls. To Keene and Bellows' Falls we sell now. They all commenced at the same time in the first place, and after being discontinued they were renewed again.

Q. When were they renewed?

A. Perhaps a year and a half; it may be more.

Q. Were they discontinued at the time you discontinued the New York tickets?

A. Yes, sir.

Q. And those to Keene and Bellows' Falls were renewed at the same time?

A. Those to Keene and Bellows Falls were renewed the same time that those by boat were. Groton Junction has never been renewed.

Q. How was it with Worcester?

A. Worcester has always been so.

Q. That commenced in 1859 or 1860?

A. Whenever it did commence.

Q. And have you now stated all the points to which you take tickets from passengers on that line?

A. Yes, sir.

Q. How were the settlements made with these roads? That is, how was the settlement made with the Worcester and Nashua road?

A. We reported everything to the Worcester and Nashua roads, deducting our proportion, and paid the whole over to the Worcester and Nashua road.

Q. Was the Worcester and Nashua road the intermediate road that settled with the roads beyond?

A. Yes, sir.

Q. When you say that the account was with the Worcester and Nashua road, upon what basis was the account made up?

A. Made up on sales.

Q. Sales made by the Concord road, do you mean?

A. Yes, sir.

Q. How did you settle, supposing the tickets came the other way?

A. It was settled by their sales.

Q. Made mutual returns; you to them and they to you?

A. Yes, sir.

Q. Supposing the conductor on the Concord road sold ten tickets from Concord to New York, we will say; supposing he took that money and put it into his pocket, and did not make any returns whatever, and you settled by the amount of sales; won't you tell me ——

A. He had no such tickets to sell.

Q. Did he have any tickets to sell on other roads?

A. He did to Boston and Lowell and Lawrence.

Q. That is all? He didn't have any on the Worcester road?

A. No, sir.

Q. Did the conductor have tickets from Concord to Boston?

A. Yes, sir.

Q. Supposing the conductor sold ten tickets from Concord to Boston; and supposing, having sold ten tickets from Concord to Boston, and put the money in his pocket, and did not make any return to the road, of money; and the next day he took up ten tickets on the road, and did not punch them, and used them over again, and returned these, I want to know what means you had of knowing it?

A. He could not take up ten Boston tickets without collusion with the conductor below.

Q. I want you to explain this whole matter fully, just what could be found out, and what could not be found out. Mr. Sanborn, supposing the conductor took out five tickets a day, out of the amount collected; and supposing there were five hundred passengers, and he took out five tickets a train, and did not return them, and used those tickets over and over again, until he had got out five hundred tickets; is there any method of discovering it?

A. If he took that five hundred in one month, I think it would be noticeable, but if he was five months or a year, I do not think it would be noticed particularly. *Still, we should notice the difference between the sales and returns, and we should expect to look to the following months to see it come in. For instance, the months of July, August and September are large months, in which the sales usually exceed the collections; I mean on through tickets; and as the sales are greater than the number of tickets returned, there are many sales, reports of which do not come in until succeeding months.*

Q. Supposing a conductor should take two tickets a train, and he ran three trains a day (there would be six tickets a day), and he took them out every day for a month (there would be one hundred and odd tickets), and they were used over and over again, and they were returned next month; that is, he returned them gradually, as they were taken. Won't you tell me if there is any way of discovering it?

A. No way of discovering them by ordinary notice.

Q. Well, sir, after the conductor had taken these tickets—supposing the conductor chose to keep those as his stock in trade—is there anything to show that will prevent his using them over, at the rate of one hundred a day. Take the case of the tickets found upon Mr. Whitcher, and sold elsewhere, was there a discovery of that fact until the tickets were found?

A. No, sir; there was not.

Q. Mr. Sanborn, are you able to state what Mr. Clough's salary was?

A. I should not be able to state with any accuracy.

*Mr. Mugridge.* Mr. Clough commenced with five hundred dollars a year, and along at the last of his service he had seventy dollars a month.

Q. (*By Mr. George.*) Mr. Sanborn, did you ever make any per cent. of the sales returned of these western coupon tickets?

A. Yes, sir; when the tickets first began to sell, perhaps the first few years, at any rate as early as 1856 or 1857, the collections overrun the sales very heavily. It was in the fall of the year they first began to overrun so heavily.

Q. (*By the Chairman.*) That, you say, was when the system first began?

A. Yes, sir.

Q. (*By Mr. George.*) At the very beginning of it?

A. Yes, sir; for the first two or three years, until the matter was investigated.

Q. Where were these tickets sold?

A. At all points west, wherever they sold any tickets to Boston.

Q. Were they sold by agents?

A. By agents principally, I should judge.

Q. To what extent did the collections overrun the sales?

A. Being a long time ago I should not recollect, but I should judge from fifteen hundred to twenty-five hundred in one year.

Q. When was that leak stopped?

A. When the matter was investigated. That was, I think, in the summer of 1857. It was inquired into, and I know I made out the figures for the sales and collections for one year. I was requested to inquire into it, and I did so.

Q. What was the result as to the defrauding of the road? [Objected to.]

*The Chairman.* If the counsel had been permitted to put in, without any objection, the fact that from fifteen hundred to twenty-five hundred tickets had been returned more than were sold, it seems to me that it does not make much difference whether you get a man to state what the result is, or whether you let the witness make the calculation, or leave the referees to do it.

Q. (*By Mr. George.*) Now, Mr. Sanborn, was or was not the road defrauded? [Objected to and ruled out.]

A. I inquired in regard to every collection of tickets; comparing with them where the tickets were sold at a less price than what they should have been, many of them at various prices; and they didn't wish to make any report on account of the contract then existing. They didn't wish to make any report, and for that reason lumped the amount and divided them into tickets at the regular price.

Q. (*By the Chairman.*) Who were *they?*

A. The interest of the Vermont Central road.

Q. The effect of that was what, upon the number of tickets?

A. To reduce the number of sales, in some cases, one-half; or they went for twenty-eight cents when we should have received fifty-two.

Q. (*By Judge Bellows.*) You carried more passengers than——

A. Than we had made sales.

Q. (*By the Chairman.*) How long did that go on before you investigated it?

A. It went on a year before it was changed.

Q. (*By Mr. George.*) When did this state of things cease?

A. In the season of 1857.

Q. Then what was the condition of things?

A. It was required of them to report the rates received, and settle *pro rata.*

Q. (*By Mr. George.*) What do you mean by *pro rata?*

A. The distance; the whole distance is taken and divided into so many miles for each.

Q. (*By the Chairman.*) I should like to make one inquiry of Mr. Sanborn; and I limit my inquiry to his own knowledge, but I should like to inquire of Mr. Sanborn whether or not the Vermont Central had any authority from the Concord Railroad to sell tickets under the regular price, and then allow a *pro rata?*

A. So far as my knowledge went, they had no right to sell, though they reported to us at less rates than that. That was only so far as I was told by the superintendent; that is all.

*The Chairman.* I don't think that would be competent.

Q (*By Mr. George.*) The result was that they afterwards did so return?

A. They returned them *pro rata* after that.

Q. (*By the Chairman.*) Well, in that case, Mr. Sanborn, did they return, or allow any returns, that came within your knowledge, where the tickets were sold under price?

A. Yes, sir; that is common, almost invariable, in western tickets.

*Mr. Rolfe.* I desire to raise a little objection, although the referées asked the question. And I do it for the purpose of making a suggestion to the referees that at that time the object was to get the travel over this road; and the Concord Railroad was just as much interested in getting the travel; and that the upper roads did not violate any understanding; but the difficulty was in reporting wrong, so as to deceive the lower road.

*The Chairman.* It is difficult, so far as the difference between the collection and sales is concerned, to say that there was any fault about that.

*Mr. George.* The result of the matter was this: It was understood that the Vermont Central line might sell these tickets for a price as low as that, but not below a given sum. The Vermont Central road did sell in violation of that understanding; and then, instead of returning the actual number of tickets sold, they simply took the actual amount received and divided it by the full price.

Q. (*By Mr. George.*) And how has the western ticket business continued from that time on?

A. They have sold at reduced rates ever since. I mean by reduced rates, by less than local rates.

Q. What do you mean by local rates? Does the Concord road get over fifty-two cents on any western ticket?

A. Yes, sir, they have done so.

Q. Well, I want you to give the ——

A. I could not state instances, but I know I have figured it up, and it gives more *pro rata* to us, and I have, from curiosity, figured it up, and in some cases we have got as high as ninety cents, when our proposition is only seventy.

Q. How many instances of that kind are there?

A. I do not know; not many. It is not often that I figure them.

Q. (*By the Chairman.*) If I understand you, the ticket is sold for so much, and you get so much for the distance of your road?

A. Yes, sir.

*Mr. George.* Deducting rent and risk.

Q. (*By Mr. George.*) Do you know what the fare is from Boston to Montreal?

A. My idea is, ten dollars and a half.

Q. Did they sell at all sorts of prices?

A. They did in the summer season.

Q. Is that the regular fixed price, or do they sell just as they run?

A. That is the regular price.

Q. How much of that comes to the Concord road?

A. I could not tell.

Q. Will you ascertain?

A. I will ascertain.

Q. And also from Boston to Chicago, and how much the *pro rata* is?

A. I do not know as I could do that, because I did not have the distances. That is done by the Northern road, and I should not know what the car risk is.

*The Chairman.* What is the length of the Concord road?

*Mr George.* It is between thirty-four and thirty-five miles.

CROSS EXAMINATION.

Q. (*By Mr. Mugridge.*) I understood you to say that when regiments left Concord for the seat of war, they ordinarily went in trains, furnished by the road, by themselves, and the transportation was charged to the government?

A. Yes, sir.

Q. But that when they were sent off in squads or companies they went in the regular trains. Supposing, Mr. Sanborn, that a squad of soldiers numbering fifty, recruited here to be sent to the different regi. ments, was sent on some particular morning, were they not put into a separate car which was in connection with the regular train?

A. Yes, sir.

Q. Supposing that a squad of a hundred were recruited here, to make up regiments that had been cut up in the field, were they not put in a car ordinarily and run out with the regular train?

A. Yes, sir.

Q. Did you ever, during the war, furnish special trains, to take soldiers away from Concord, unless a regiment went at a time. And did you not invariably run them with the regular trains?

A. My impression is that we did once or twice, but I cannot now be definite. When a regiment, or perhaps two or three hundred, or perhaps three or four hundred, there has been a special train run.

Q. Did you, as a general rule, send out a special train, unless it was to take away a full regiment?

A. No, sir; not as a general rule.

Q. A full regiment of soldiers was one thousand men, was it not?

A. Yes, sir.

Q. How many cars did it take to send a full regiment from Concord?

A. About twenty or twenty-four.

Q. Now, then, every time you sent a full regiment, you sent them by a train of from twenty to twenty-four cars, which made up a full train by itself; did you not, sir?

A. Yes, sir.

Q. How many cars would it take to send out a squad of two hundred men?

A. About four cars.

Q. About fifty in a car?

A. That is about what they calculate.

Q. But where you had a squad, you annexed them to the regular train?

A. Yes, sir.

Q. How many times did that occur? We will take squads numbering all the way from twenty-five to two hundred men.

A. The number from twenty-five to fifty is common.

Q. But it was very seldom that you had one that would reach the amount of two hundred?

A. I do not recollect of any such case where there was as many as that.

Q. I would put in squads of from fifty to one hundred men?

A. That was quite common.

Q. Do you remember the fact that in the early part of the war the soldiers refused to pay their fare, and were left this side of Manchester?

Q. (*By Mr. George.*) Were you there?

A. No, sir.

Q. (*By Mr. Mugridge.*) Didn't it occur almost every week that squads of soldiers, numbering from twenty-five to a hundred, were sent off?

A. No, I should think not, sir.

Q. Wasn't it as often as every week?

A. It is my impression that it was not.

Q. Now, these soldiers had military tickets, had they not?

A. Yes, sir.

Q. I want you, sir, this afternoon, to take the two hundred or three hundred Whitcher tickets, and the schedule that you have prepared here, look them all over and be able to state to the referees, how many of these tickets could possibly have got into the hands of George Clough in the regular course of business.

A. It won't need any preparation, because I shall know every ticket. [Intermission.]

Q. (*By Mr. Mugridge.*) Mr. Sanborn, I was inquiring of you in regard to soldiers riding in the cars. Let me ask you whether they were in the habit of riding more or less when they were not sent off in squads. Was there a large number of soldiers going home?

A. Yes, sir.

Q. Now then, where regiments came here to Concord, and were discharged here, did not the soldiers go over the road a great deal?

A. Yes, sir.

Q. Were there not many who came here to vote?

A. Yes, sir.

Q. Were they not constantly going back and forward from their rendezvous?

A. Yes, sir.

Q. And there was continual travel back and forward during the time of the war, was there not?

A. Yes, sir.

Q. Now, while they were rendezvousing here preparatory to being sent off, were they not riding backwards and forwards constantly?

A. Yes, sir.

Q. From here to their home and backward again. And was there a day passed but what there was any number passing over the road in this way?

A. I could not state as to that, but I think there might possibly not have been a day, that is, I mean during the active part of the war.

Q. How many passengers does an ordinary passenger car contain?

A. They hold fifty I believe, but the cars they make now hold sixty; fifty I suppose would be the average.

Q. Supposing you take the coupon tickets returned by the conductor to the general ticket office; add to them the local tickets sold here at the stations; add to that amount the fares returned by conductors as paid in the cars; and divide that number by fifty, will it not show the number of cars that made up the trains here ordinarily? I mean the regular ordinary travel on the road?

A. I cannot say as to that; I never reckoned.

Q. Let us see now. You take the number of coupon tickets returned by the conductor, that would show the number of passengers that came from the Northern road, would it not?

A. Yes, sir.

Q. Now then, take the tickets sold here at this station?

A. The tickets that were sold at this station you could not reckon on that day from the returns of the conductor, because a ticket sold from here to Boston has a single coupon, and that is punched on this road, and goes through.

Q. There is a record kept here of the tickets sold at this station?

A. Monthly; yes, sir.

Q. Take the number of tickets returned for a month; then take the local tickets sold at this station for a month; and the fares returned by the conductors in the trains for a month; add them up; then calculate the number of passengers to a car; and can you, with those facts, get at the number of cars which made up a train that left the city of Concord each day?

A. You get the number it was necessary to carry.

Q. I understood you to say that the soldiers travel was represented by a peculiar kind of ticket, marked as a military ticket, which they had of you directly, and which they did not receive from the local ticket office?

A. Yes, sir.

Q. (*By the Chairman.*) Did you intend to have it understood that in all cases where the soldiers traveled singly, where they had a furlough, for instance; they had to use these military tickets?

A. They did in the latter part of the war; at first they were obliged to pay.

Q. (*By Mr. Mugridge.*) That was the rule?

A. Yes, sir.

Q. Now, I want you to prepare these statistics, during 1862 to 1865. Take the number of coupon tickets taken within the month; the number of tickets sold at the local stations here; the number of fares returned by conductors, as sold in the cars; and upon the basis of fifty persons to a car, calculate the number of cars that left Concord during the years 1861, '62, '63, '64 and '65.

A. [Schedule shown.]

Q. That is not what I want. I want the number of cars that made up the train here?

A. The number of cars that made up the trains I cannot give you. If I understand you, you want the number of cars that would have been necessary.

Q. Now, Mr. Sanborn, I desire you to take the tickets found at the house, and upon the person of Mr. Whitcher, and ascertain the number of tickets found at each place, that could, according to the regular course of business, by any possibility, come into the possession of George Clough.

A. Of those found at his house, eighteen.

Q. And how many of those found on his person?

A. Those found on his person, as returned to me, one hundred and fifty-one.

Q. They would come through his hands, in the ordinary course of business?

A. They might have.

Q. Now, I want you to give the list. What are they. First, those found at his house?

A. Concord to Manchester, 4; Manchester to Concord, 2; Manchester to Concord (again), 1; New Market Junction to Concord, 2; Manchester to Concord (again), 4; Manchester to Suncook, 1; Suncook to Concord, 1; Concord to Lawrence, 1.

Q. Wouldn't he have had to part with that?

A. Being in his hands——

Q. I mean that he could have retained?

A. Then that would have to be left according to that. That would be all on that then.

Q. Now give me those found on his person?

A Concord to Manchester, 7; Manchester to Concord, 9; Lawrence to Concord, 2; Manchester to Nashua, 14; the same package (one more), 1; Manchester to Hooksett, 7; Manchester to Suncook, 1; Nashua to Manchester, 10; Hooksett to Manchester, 6; Suncook to Manchester, 1; Goff's to Manchester, 1; Robinson's to Manchester, 1; Thornton's to Nashua, 3; Concord to Manchester, 3; Manchester to Nashua, 6; Concord to Nashua, 14; Manchester to Concord (again), 1; Nashua to Manchester, 19; Manchester to Concord (again), 1; Manchester to Concord (again), 8; Concord to Suncook, 2; Concord to Nashua, 3; Concord to Manchester (again), 4; Concord to Manchester (again), 8; Concord to Nashua (again), 4; Manchester to Nashua, 3; Concord to Hooksett, 1. Total, 143.

Q. Speaking of the age of these tickets—I want you to look them over and see how long some of those have been on the road, those that he might have retained. I want to show the age of the tickets; how long they had been out of use, and the number of those that are the most ancient.

A. There seems to be forty-eight of these.

Q. Now those have been in use how long a time?

A. There are some of them marked 1865; can't always tell. They were in use, perhaps, all the way from 1860.

Q. Now give us the most remote date of these forty-eight tickets; and then the most recent date.

Q. (*By Judge Bellows.*) Is it your statement that that style of tickets has been in use ten years?

A. Yes, sir; this style. I think '64 and '65 are the only two years; and there is one Feb. '66—that which is stamped.

Q. (*By Mr. Mugridge.*) When you spoke of 1864 and '65, you spoke of the time it was printed?

A. Yes, sir.

Q. Tell when it was sold.

A. Well, that tells on the back here.

Q. Well, give us the most remote date.

A. The dates are very indistinct.

*Judge Bellows.*. How long prior to the seizure of these tickets did Mr. Clough leave?

*Mr. George.* He left the road January 14th, I think, 1866; and these tickets were seized February 10th.

Q. (*By Mr. Mugridge.*) We want you to go over these tickets one by one, and find the date of these tickets,—of all the tickets that Mr. Clough could have used.

A. Yes, sir.

Q. Let me ask you whether you have ever known Joseph A. Gilmore, the superintendent of the Concord Railroad, to direct Mr. Clough to pay back money for fares that he had collected in the cars; and if so, in how many instances?

A. I haven't known any such cases.

Q. Understand the question. When Mr. Clough has come in and made returns of money for tickets sold in the cars, if you have known Gilmore to direct him to pay back the fares that he has collected; and in how many instances you have known that to occur?

A. I have known him to come in while he as well as others were there, and direct him to pay back money to a gentleman that he had with him; but I wouldn't know who the gentleman was.

Q. The gentleman that Gilmore had with him?

A. Yes, sir.

Q. That person would be a person that rode to Concord with Clough in his train, and Clough had collected his fare in the cars, did he not?

A. Yes, sir; I should judge so.

Q. And after Mr. Clough had taken his fare, Gilmore would come with that person and direct Mr. Clough to pay him back his fare, collected in the cars. Now I want to ask you, Mr. Sanborn, in how many instances you have known that to occur, in the case of George Clough alone?

A. Perhaps three or four times.

Q. Have you stated to Mr. Clough that you have known it to occur "times without number,"—using just these terms?

A. I don't think I have.

Q. Have you stated that you have known it to occur a great many times?

A. I shouldn't think that I ever made any such statement.

Q. Have you known that to occur a great many times?

A. I should think not a great many times, if I understand your question.

Q. Well, the question is simply: haven't you known Mr. Gilmore to come in with a person and direct Mr, Clough to pay him back his fare; haven't you known it to occur a great many times?

A. No, I have not.

Q. Who were these men?

A. I don't know any of their names.

Q. Do you know whether Mr. Clough hasn't frequently paid money to you, that Mr. Gilmore has ordered you to pay back to the man?

A. There may have been cases of that kind, when Mr. Clough had passed out, and he came in to see, and if Mr. Clough was there he got it, but if Mr. Clough wasn't there, he didn't.

Q. Have you done that?

A. I have done that; I should think I had with him.

Q. Under the order of Gilmore?

A. Yes, sir.

Q. I want to ask you, sir, when Mr. Gilmore was a candidate for the office of governor, how it was about persons being passed over the road by his direction, without anything whatever to show, and in large numbers?

A. I couldn't tell about that.

Q. Haven't any positive knowledge on that point?

A. No positive knowledge; because I never ran as conductor.

Q. Whether you ever knew anything about the issuing of tickets by Mr. Spalding, of Nashua?

A. You said "without anything to show."

Q. Let me ask you about his passing persons over the road, when he was candidate for senator, without having the regular tickets at the office? I mean tickets that would not be represented at all in the general office?

A. Yes, sir.

Q. Was that his custom?

A. There was considerable done of that kind, I should think.

Q. Did you ever see a ticket like that, Mr. Sanborn [showing ticket to witness]?

A. I should think I had.

Q. Let me ask you if the tickets of that kind had any representation in the accounts of the Concord Railroad?

A. Not on the accounts.

Q. Whether any tickets (I mean on account of any other person save Isaac Spalding) were in circulation, or in use on the road, save that?

A. Fred. Smyth had a lot of them for two or three years, while he was governor.

Q. How many did Fred. Smyth have?

A. I should think I sent fifty or sixty in one day, and took them up on special trains. That was to come up on special trains, election day, I think.

Q. (*By the Chairman.*) By running on a special train, you mean a train specially to bring them up?

A. Yes, sir.

Q. (*By Mr. Mugridge.*) Whether you sent to Mr. Smyth on other occasions?

A. I don't recollect whether I ever did, or not.

Q. Was he governor at that time?

A. It might have been when he was to be inaugurated. I presume that was the case.

Q. In the case of Mr. Gilmore, to what extent did he send tickets to voters to come home and vote for him?

A. If you mean for the purpose of election, I can't tell.

Q. How is it?

A. I know he ordered a good many tickets to send out; at least, got them of me at different times.

Q. While he was a candidate for office?

A. Yes, sir.

Q. Did they find any representation in your office?

A. They did not.

Q. How many tickets, while Mr. Gilmore was a candidate for any office, or while he was in office, did he procure from you, to send off for political purposes?

A. I could not tell.

Q. Won't you give your best estimate?

A. Oh, there might have been, perhaps, in one year, three or four hundred.

Q. I want to ask you if there were not more than twenty-five hundred tickets passed by you to Joseph A. Gilmore, to be used for political purposes, that never found any representation in your office?

A. I shouldn't want to make that statement without the figures.

Q. Have you any sort of statistics, any memoranda of any kind, to enable you to state any correct number of tickets sent out from that office, for political purposes, through Joseph A. Gilmore alone, that found no representation in your office?

A. No, sir.

Q. Let me ask you if you recollect the time when Gov. Harriman was nominated as a third candidate?

A. I remember the time.

Q. How many were there sent out from your office on that particular day?

A. I could not tell.

Q. Were there not three hundred and fifty?

A. I should not want to state; I have no recollection of any particular day. I know that he would come into the office and want so many tickets to such a place; and I was requested to keep no account of it.

Q. When Mr. Gilmore came to your office to get tickets for political purposes in this way, what direction did he give you in reference to those tickets?

A. Not to make any account of them.

Q. Did he direct you to make no account of them?

A. Yes, sir.

Q. And did you obey those directions?

A. I tried to.

Q. (*By Mr. Haile.*) Did he do this all the time?

A. He did at first; afterwards there were special tickets for political purposes.

Q. What year was this?

A. For that matter it has always been so, more or less.

Q. (*By Mr. Mugridge.*) Will you state, from the best information that you have got on that subject, that Mr. Gilmore hasn't taken from the general ticket office, to be used for political purposes (as you have said), two thousand tickets?

A. I shouldn't want to state.

Q. How many will you state as your best recollection?

A. It would be a pretty hard matter. I shouldn't want to state, because I should be afraid that it wouldn't be very near.

Q. While Mr. Gilmore was candidate for office—have you any sort of an idea that you can give that can be reliable on that subject?

A. I should think that one year there might have been from three to four hundred, and other years they were smaller. My impression is that 1864 was the time that I got the most. The other years it was a great deal smaller.

Q. Let me ask you if that wasn't one year when Harriman was run for governor—if there wasn't three hundred of these taken from your office at that time, or about that number?

A. No, sir; I couldn't say that I do recollect. Perhaps some facts connected with it might refresh my mind—the individual that got them, if he got them.

Q. Do you recollect the time when Mr. Gilmore and Mr. Harriman were run as candidates for governor?

A. I recollect that.

Q. Do you recollect, in connection with that time, of any large number of tickets being taken out of the office?

A. I don't recollect that there was two or three hundred taken out in that year, if I understand you, by Mr. Gilmore.

Q. You say that Mr. Gilmore used to come and get these tickets. Let me ask you if he used to send men for them?

A. It wasn't a common thing. If there was an arrangement made with parties to come for special tickets, they would come with written orders.

Q. From the republican headquarters?

A. Yes; and the democratic.

Q. Let me ask you if Joseph A. Gilmore while he was a candidate for governor, gave you instructions to answer orders of persons who came?

A. Yes, sir.

Q. And did you obey these orders?

A. I did.

Q. Did he give you instructions to keep these tickets secret?

A. The special tickets were represented to a certain extent; but the regular tickets were not represented at all.

Q. And did you keep these secret, under his instruction?

A. I think I did; I always endeavored to, at any rate.

Q. Did Jim Whitcher ever come there and get tickets?

A. Not for that purpose.

Q. For any purpose?

A. I don't think he ever did.

Q. Do you mean to say that he never came there to get tickets for any purpose?

A. He had regular season tickets.

Q. Did he ever come and get tickets for political purposes?

A. Yes, I think he did; it comes to my mind; I think he did.

Q. Don't it come to your recollection that he came two or three times?

A. No, sir; I don't recollect of his coming there but once; and that was when he said he came to see Mr. Gilmore, and he told me the story (for Mr. Gilmore had gone away,) that he had so many men that he had got at Hooksett, and that he was going to keep there until town-meeting day.

Q. I want to ask you how many men Whitcher had there at Hooksett, and how many he expected there for which he wanted tickets?

A. If I recollect right, he said he had three or four there, and he was going to have three or four more there; at any rate he was going to get ten or fifteen there.

Q. Did you keep them a secret?

A. Just the same as I did any of them. I have told you all I know about it.

Q. Let me ask you if you have ever known of his sending tickets to Whitcher?

A. He never told me what he wanted them for.

Q. Let me ask you if Natt. Head, who lives in Hooksett, used to come up to get tickets for political purposes?

A. He did; that is, he said they were for that.

Q. How many tickets did Natt Head get there for political purposes?

A. I can't tell how many he has had for political purposes. He has had them almost every year when Gilmore was candidate for governor; and I think one year when Natt. Head was candidate for some office.

Q. Did Natt. Head get them there in large numbers?

A. Oh, not many; perhaps twenty-five.

Q. Twenty-five at a time?

A. Yes, sir; sometimes half-a-dozen and sometimes twenty-five.

Q. How many times did he get them?

A. Perhaps three or four times during these years.

Q. Do you know Jesse Gault, of Hooksett, sir?

A. I think I do, sir.

Q. Was he a candidate for railroad commissioner at the same time Gilmore was candidate for governor?

A. I don't recollect about that; I didn't know him then.

Q. He lives in Hooksett?

A. I think so; I did not know him until recently.

Q. Did he ever come there to get tickets for political purposes?

A. I can't say that he did, for I didn't know him.

Q. Did ever Mr. Gilmore come in with him?

A. I say I couldn't tell, because I didn't know the man until after that.

Q. Do you know Joseph Goss, of Hooksett?

A. Yes, sir.

Q. Did he ever come there to get tickets?

A. I can't say that he did.

Q. Let me ask you if you will state that you have given them to Mr. Goss, of Hooksett?

A. I don't know.

Q. He was one of Jim Whitcher's neighbors?

A. Yes, sir; I believe so.

Q. Let me ask you if he did not come there and get tickets for political purposes?

A. I shouldn't want to take my oath to that.

Q. Would you take your oath that he didn't?

A. No, sir.

Q. Who is the station agent at Hooksett?

A. Marshall.

Q. Did he ever come there and get tickets for political purposes?

A. I don't think he ever did.

Q. Will you say that he did not?

A. No, sir.

Q. Do you know Thomas Wattles?

A. Yes, sir.

Q. Did Tom Wattles come there and get tickets?

A. Yes, sir; I should think he had; I shouldn't want to take my oath, but I think he did.

Q. Let me ask you how many tickets Wattles had?

A. I couldn't tell.

Q. How many times did he ever come?

A. I couldn't tell that.

Q. Mr. Jim Whitcher lives in the town of Hooksett?

A. I think he does.

Q. Let me ask you if particular men came from Pembroke, from the city of Manchester, and from other places, to get tickets, aside from those that Mr. Gilmore got, regular tickets? I am not speaking of the tickets dealt out by the republican state committee, but did particular men from Manchester, and Portsmouth, and Nashua, and Pembroke, from all the large towns, come there to get tickets to be used for political purposes?

A. I recollect of a man coming there from Manchester.

Q. Who was he?

A. His name was Harrington.

Q. How many did he have?

A. He had twenty-five, I think.

Q. (*By the Chairman.*) Were these special tickets?

A. No; those were not.

Q. (*By Mr. Mugridge.*) Did Jim Cheney ever come up there for any?

A. I think he was there at the same time with Harrington; and I think Harrington took the tickets.

Q. How many did Harrington take?

A. Twenty-five.

Q. How many times did Harrington ever come there for tickets?

A. I think that was the only time that he got them.

Q. Did ever you hear or see men ask Gilmore for any?

A. No, sir.

Q. Did Jim Cheney ever come there except then?

A. I don't think he did. I should be very much inclined to say that I don't think he did.

Q. Did men come there from Nashua and Portsmouth that you didn't know, with orders from Gilmore for tickets?

A. Come with him?

Q. Yes, come with him for tickets for political purposes?

A. That was'nt a very common thing, however, beyond what you have said.

Q. Did instances of that kind happen where men came there with Gilmore, and he would say, "Mr. Sanborn, let us have so many tickets"; and take them and carry them off?

A. There was a few such cases.

Q. While Gilmore was candidate for office—we will say four weeks before election, each year, for the four years—was there scarcely a day but what men came, either with Gilmore, or under orders from Gilmore, for tickets?

A. That would be a hard question to answer.

Q. What is your impression?

A. I should think there might be a good many days, perhaps.

Q. Were not the days when they didn't come exceptional days?

A. I should think that they were, rather.

Q. You think they were rather exceptional days?

A. Yes, sir.

Q. Now, when these men come in this way, were you not accustomed to deal out the tickets from the general stock of tickets in your possession, all the way from five to two hundred?

A. I should think two hundred was very high.

Q. Well, I will put it in that way.

A. I should like to alter that statement.

Q. Well, you may.

A. From one to a dozen.

Q. I ask you, sir, if, when these men came for tickets, you didn't deal them out in numbers ranging from five to two hundred?

A. I should say from one.

Q. From one?

A. I don't recollect any case of two hundred.

Q. Have you not known one hundred tickets to be sent out at a time?

A. You mean regular tickets, do you?

Q. Yes, sir.

A. I don't recollect any particular occasion. I recollect once of giving Mr. Gilmore fifty tickets at one time.

Q. Is your recollection so clear that you will state, positively, that you haven't dealt out a hundred?

A. No, sir.

Q. You won't state that?

A. No, sir.

Q. I suppose you don't know whether any of these tickets got into Jim Whitcher's hands, do you?

A. No, sir.

Q. I suppose you wouldn't be able to say whether any of these tickets were taken there by Mr. Whitcher?

A. Those tickets that I dealt out to Mr. Gilmore had no stamp of the ticket seller at all; they had no mark, except, may be, the date.

Q. I want to ask you if there has been a single year that Joseph A. Gilmore has been superintendent of the Concord Railroad, but what he has been in the habit of going to your office and getting tickets to send away, more or less each year?

A. I couldn't say as to the first part of the time.

Q. Well, what do you mean by the first part of the time?

A. The first two or three years that I was there; I could not say as to that because it was sometime ago.

Q. Now, has there been a year since that time that Mr. Gilmore has not been there to get tickets, either to give to persons or to send off?

A. There was a time when he gave them to individuals just as they were going down on the train. That was a common thing.

Q. It was a common thing?

A. Yes, sir.

Q. (*By the Chairman.*) And these tickets, you say, were not dated?

A. No, sir.

Q. Didn't have any seller's mark upon them?

A. No, sir.

Q. (*By Mr. Mugridge.*) Now I want to ask you, Mr. Sanborn, in how large numbers—(I will lay his political career entirely out of account)—Gilmore has been in the habit, during these other years, of taking tickets from you, at any one time, for the purpose of sending away?

A. I recollect, one time, of carrying twenty-five into his office, between Manchester and Nashua,—twenty-five regular tickets; or twenty-four, perhaps the number was.

Q. (*By Judge Bellows.*) These were not stamped with the mark of the seller?

A. No, sir.

Q. These were between Manchester and Nashua?

A. Yes, sir; one-half up and the other half down.

Q. (*By Mr. Mugridge.*) Were you directed by Mr. Gilmore to keep the fact that you gave them to him, a secret?

A. Not in so many words; I was directed to make no account of it.

Q. (*By the Chairman.*) Mr. Sanborn, did you, as general ticket agent, ever see any of these tickets again?

A. Yes, sir; tickets unstamped came in.

Q. (*By Mr. Mugridge.*) Now, you say you recollect giving him, at one time, twenty-four or twenty-five tickets? Did you, on other occasions, give him a less number?

A. It was a very common thing for him to come in and get tickets for persons going down.

Q. (*By Mr. Haile.*) I understand you to say that was while he was not in political office?

A. Yes, sir.

Q. Before or after?

A. Both.

Q. (*By Mr. Mugridge.*) Well now, sir, let me ask you if it was not a common thing for him, and if there was hardly a day passed over his head but what he came to your office and wanted one or more tickets to be sent away under the circumstances you have described?

A. Perhaps it would average two or three times a week, that he would take tickets for so and so to Boston.

Q. During all the time?

A. Well, it was a common thing; I could not designate.

Q. When the weeks occurred that these facts did not occur, were they not exceptional ones?

A. I think they were.

Q. Now I want to ask you sir, where these tickets went that he took?

A. I couldn't tell in all cases.

Q. Did they go in all directions?

A. Generally they would go to individuals that were there present.

Q. Did some go to Boston and Portsmouth, and some over the Boston and Lawrence road?

A. Yes, sir.

Q. To all stations—to Boston and this side of Boston?

A. Yes, sir.

Q. Now, Mr. Sanborn, did you ever see any of these tickets back again?

A. I think I did.

Q. What did you do with them?

A. Threw them out of the account.

Q. What did you do with them?

A. Turned them out the same as I did the other tickets.

Q. You took them and turned them out, and never made any return of them whatever?

A. Yes, sir.

Q. Did Mr. Gilmore ever ask you if you were destroying those tickets?

A. I don't know as he did. He kept repeating to me to do so; not always when he got the tickets, understand, but only occasionally.

Q. Did you ever hear Mr. Gilmore give Mr. Clough verbal orders to pass men over the road, without giving a written pass?

A. I should think I had, some of the conductors; I couldn't say that it was Mr. Clough.

Q. You couldn't say that it was not Mr. Clough?

A. No, sir.

Q. When that party to whom that order was given got into the cars, he didn't have anything to represent his passage? [Objected to unless the witness knew.]

*The Chairman.* I do not understand that this objection is well-taken, because he has described the course of business.

Q. (*By Mr. Mugridge.*) I want to ask you this: taking the course or usages of business as it was done there, with which you are familiar, I understand: supposing Mr. Gilmore met Mr. Clough on the platform, and says, "Here, pass this man to Manchester;" and he did; was his passage represented in any return that the conductor could make to the office?

A. Not necessarily. Perhaps I should say, though, when the rules were adopted for the conductor to report all free passes, it would come in.

Q. Now, let me ask you, how many instances, how frequently did you see Gilmore give instructions of that kind to Clough?

A. Not but a very few times.

Q. Do you know how many times?

A. No, sir, for the reason that I wasn't there.

Q. And you do not know how many instances of that kind occurred in Mr. Clough's case?

A. No, sir, I couldn't tell.

Q. Did you ever hear Gilmore tell Clough to disregard the instructions of the road with reference to making a return of free passes?

A. No, sir.

Q. Now, I want to inquire with regard to trip passes. Did Mr. Gilmore frequently issue trip passes?

A. Yes, sir.

Q. Now it was the duty of the conductor to return these to the general ticket office.

A. Yes, sir.

Q. Let me ask you, if these came into the office and you destroyed them?

A. They were all returned to me every month.

Q. Let me ask you if there were occasions when there was no record taken of annual passes?

A. There was one time that the conductors were required to return every free pass to the superintendent; they didn't come into my office at all.

Q. What year was that?

A. I could not tell.

Q. As near as you can?

A. I wouldn't undertake; I should think it was as far as 1863.

Q. Now, then, by the usage of the road, the conductors for a time in 1863 were required to return every free pass to the superintendent; and they did not come into the general ticket office at all?

A. Yes, sir.

Q. Now for how long a time did that practice or usage of the road prevail?

A. I think only two or three months.

Q. Was Gilmore a candidate for office at that time?

A. I don't recollect.

Q. Now you say that you think in 1863, if that was the right year, when that was the usage of the road, was there any possible way for the number of free passes that were issued by Gilmore to be detected by any general account kept at your ticket offices?

A. No, sir.

Q. Now when the trip passes were, under the usage of the road, to be sent to your office, let me ask you if Gilmore didn't come to your office and take these passes away before they were reckoned in the accounts of the road, and destroyed?

A. As I told you before, the passes were all returned to me at the end of the month.

Q. When they were returned to you at the end of the month, they ought to have been put into the account?

A. Reckoned in the list of free passes.

Q. Now, before they were entered on the list of free passes, hasn't he come to your office and taken them, and destroyed them?

A. Not frequently.

Q. Hasn't he ever done it?

A. I couldn't say that he had.

Q. Do you recollect on one occasion after he was governor, of his coming to your office and calling for the free passes?

A. I don't recollect now.

Q. Will you say that he didn't do it?

A. I wouldn't say that he didn't do it; I might have forgotten. It wouldn't be just according to my recollection. I should rather say that he didn't—until they had been counted up.

Q. Has he ever given you directions to modify them in any way?

A. I don't think he has.

Q. Are you able to state the number of free trips that were made upon passes used by Gilmore the last year that he was superintendent of the Concord Railroad?

A. No, sir; I have no means of knowing.

Q. Did you see an article, made up and published in the Concord Monitor?

A. I read it; but that was annual passes.

Q. I want to know, sir, if you know, as general ticket agent of the Concord Railroad, how many free trips were made over the Concord Railroad, upon passes issued by Gilmore the last year he was superintendent?

A. No, sir; I couldn't tell.

Q. Did you ever see a record of it?

A. Yes, sir; I think I have.

Q. What did that record show?

A. I could not tell.

Q. What statement did ever you see, I mean such annual passes as given by Mr. Gilmore for that year?

A. I could'nt tell; it was quite a number of thousands.

Q. Were there ten thousand?

A. I couldn't state.

Q. Was there five thousand?

A. I should judge there might have been three or four pages of names; I don't know how many names one page would come to; perhaps forty or fifty; perhaps more; that is the list that I saw.

Q. How large a book was it?

A. As large again as that [referring to book on table]. Mr. Nutter always kept that book. It was a book that I never had occasion to go to.

Q. Do you know Seth Eastman?
A. Yes, sir.
Q. Did Seth Eastman have free passes?
A. I couldn't swear to that.
Q. Do you know his son?
A. I couldn't swear.
Q. Do you know Samuel C. Eastman?
A. Yes, sir.
Q. Did he have them?
A. I couldn't say as to that.
Q. (*By the Chairman.*) What do you mean by annual passes?
A. Passes given for a year, or to end with the year.
Q. Mr. Sanborn, did you see the article published in the Daily Monitor?
A. I might.
Q. Did you, shortly before that, see a record made up at the Concord Railroad office?
A. No, sir.
Q. Did you know who did make it up?
A. No, sir.
Q. Did anybody come to your office for information?
A. I think so.
Q. You don't know who did it?
A. No, sir; I did not.
Q. You saw the article?
A. I think I saw the article.
Q. Do you know that the statistics that went into that article were prepared at the Concord Railroad office?
A. I could not say.
Q. Now, Mr. Sanborn, I want to turn your attention to another matter here, that I want to go over briefly.
Q. (*By Mr. Haile,*) Will you allow me to ask one question; you said that at one time Mr. Gilmore called upon you for tickets, and you distributed twenty-four or twenty-five between Concord and Manchester. I ask you if there was any unusual circumstance for that?
A. No, sir; only there was a person there which I understood from the conversation was to receive the tickets.
Q. (*By Mr. Mugridge.*) Who was that person?
A. It was a woman.
Q. Do you know Starkey's sister?
A. I don't think I do.
Q. Did you ever see the sister of Martin T. Starkey?
A. I have had such a person pointed out to me.
Q. I want to ask you if you ever saw tickets given to her.
A. That person wasn't the person pointed out to me.
Q. Who was the woman that was there?
A. I don't know now, sir.
Q. What kind of a looking woman?
A. She was a shortish, medium-sized, good looking lady.
Q. Did you ever see her in Gilmore's company?
A. No, sir.

Q. Do you know where she lived?
A. My impression was that she lived at Manchester.
Q. Did she live at Hooksett?
A. I don't know, only I know she used to go to Manchester.
Q. Did you ever see her with Jim Whitcher?
A. I don't know as I have.
Q. Won't you say?
A. No; I couldn't say that.
Q. Do you know of his giving tickets to Jim Whitcher besides this time?
A. Yes, sir.
Q. How many times has Gilmore given tickets to females in that office?
A. I couldn't state that.
Q. How many times has Gilmore given tickets to females?
A. I should think the males would predominate.
Q. How many times has he given tickets to females?
A. I cannot tell.
Q. Has he many and many a time?
A. Well, that is a little too strong.
Q. Well, I ask you the question. I want to ask you how many times you have known Gilmore to give tickets to females in that office?
A. I could not say.
Q. Will you say that he has done so fifty times?
A. I don't believe that he has.
Q. Twenty-five?
A. He may have done that twenty-five times.
Q. How large a number at a time?
A. Generally two, but oftener perhaps not but one.
Q. Did you ever know Jim Whitcher to come with a female to get tickets?
A. No, sir.
Q. Did you know of Jim Whitcher being in Gilmore's office with a female?
A. No, sir.
Q. Was Whitcher frequently at Gilmore's office?
A. I don't think he was.
Q. Was he ever?
A. I wouldn't say that he was.
Q. Has he ever been around your office?
A. No, sir.
Q. Was he around the office of the Concord Railroad? was he occasionally there, sir?
A. Occasionally; saw him in the depot.
Q. Did you ever ride on the cars between Concord and Hooksett, and not see him?
A. Oh yes, sir.
Q. Did ever you see Mr. Whitcher with any of these women to whom Gilmore has given tickets?
A. I don't know as I have.
Q. Can you say that you never have?
A. No, sir; that I don't know.
Q. Did you use to see females in there?
A. Once in a while.

Q. Pretty often?

A. Perhaps some might call it often.

Q. How were they dressed?

A. Generally pretty well.

Q. Young ladies, ordinarily?

A. All ages.

Q. Do you know Starkey's sister?

A. I couldn't tell her; I have had her pointed out to me.

Q. Did you ever see her in there with Mr. Gilmore?

A. I can't say that I ever did.

Q. Did Gilmore ever get you to issue passes, and sign his, Gilmore's, name to them?

A. No, sir.

Q. Did you ever sign Gilmore's name to a pass?

A. I might have done so, once or twice. He used to give passes, by his order, in my name.

Q. How many did you give?

A. I could not tell, sir; a great many.

Q. Has it been your practice, while Gilmore was there, to issue passes in that way?

A. While he was there; the last four years; certainly.

Q. And you did it under specific directions from Gilmore?

A. Yes, sir.

Q. Whether he gave you specific instructions to keep those passes secret?

A. No, sir; I don't know that he did.

Q. Did you keep any account?

A. In the report of free passes, I did.

Q. To what persons were they given?

A. To individuals put down here, and people who came up on the train.

Q. Who were some of them, sir?

A. There was Seth Eastman, Sam. Eastman and Edson Eastman.

Q. And their wives?

A. Yes, sir.

Q. Who else, Mr. Sanborn?

A. Benjamin Grover.

Q. All the members of the firm of Barron, Dodge & Co.?

A. Yes, sir, I think I have; not much to them.

Q. Colonel Grover's family?

A. I think I have to Colonel Grover and his wife.

Q. Give us another?

A. Mr. Hart.

Q. Did you ever see Mr. Gilmore give Christopher Hart's daughters passes?

A. No, I don't think I have.

Q. Tom, his son?

A. Tom? I guess they never gave to him.

Q. Who else?

A. I don't know, I'm sure; most everybody; I could not designate.

Q. Is there a bank president or bank director of any company in the County of Merrimack that didn't have a pass over the road?

A. I think the cashier of the Warner Bank did.

Q. Did George Jones have one?

A. I don't recollect as he did.
Q. Franklin Simonds?
A. Yes, sir.
Q. Dr. Ames, of Bradford?
A. I think not.
Q. Ruel Dirkey?
A. Yes, sir; I think occasionally; not very often for him.
Q. Take the State Capital Bank here for instance; Hall & Roberts; did you give to them?
A. Yes, sir.
Q. To Mr. Samuel Butterfield?
A. I don't recollect that he ever did.
Q. Abraham Bean?
A. Yes, sir.
Q. Enos Blake?
A. Yes, sir.
Q. Theodore French?
A. Yes, sir.
Q. His family?
A. I don't recollect of ever giving his family one.
Q. Now take the Union Bank; A. C. Pierce?
A. Yes, sir.
Q. John H. Pearson?
A. I don't recollect about him.
Q. Didn't he have annual passes?
A. I think he did.
Q. Woodbridge Odlin?
A. I think I have, once or twice.
Q. Isaac Elwell?
A. Yes, sir.
Q. His son and family?
A. Yes, sir.
Q. Jacob C. Dunklee?
A. I think I never gave him any.
Q. Benjamin Dunklee?
A. Yes, sir.
Q. John Kimball?
A. I don't recollect.
Q. Frank Dunklee and John Dunklee?
A. Yes, sir.
Q. S. Dumas and family?
A. Yes, sir; I should think I had.
Q. B. F. Martin and family, of Manchester?
A. Yes, sir, I presume so.
Q. Joe Goss?
A. Yes, sir.
Q. E. A. Straw?
A. Yes, sir.
Q. Asa Fuller, wife and daughter and two boys?
A. I could not tell about that.
Q. How was it about passing his entire family?
A. I couldn't say that I have ever given anything more than self and wife, perhaps.
Q. Don't you recollect of giving the boys a pass?

A. I don't recollect that I have.

Q. Colonel Norton, of the army?

A. Yes, sir.

Q. Lieutenant Graham?

A. I could not say.

Q. Colonel Graham?

A. I could not say as to him.

Q. Think the thing over by days, and say how many you gave in this way, during the last four years of Gilmore's stay on the road?

A. Perhaps it would average three a week; sometimes it was more than three in a day, and then some days there wouldn't be any.

Q. Now I want to ask you if Mr. Gilmore didn't ask you to keep that sly?

A. No, sir, I don't recollect as he ever did; for I always turned them in, the same as I would any one's else.

Q. Now, sir, there was a considerable rumpus, if you will allow that expression, raised about these free passes shortly before Gilmore left; and before George Clough left the road. Now let me ask you whether there were orders given by Gilmore to cut off these passes after that?

A. There was an order not to issue any of our own issue.

Q. (*By the Chairman.*) What do you mean by your own issue?

A. My own signature.

Q. (*By Judge Bellows.*) When did you say that was?

A. I could not say.

Q. (*By Mr. Mugridge.*) Was it a short time before George Clough left the road?

A. It might have been a year previous.

Q. You observed that order, did you?

A. Yes, sir.

Q. Do you know whether, after the annual passes expired January 1st, 1866, any annual passes were issued?

A. I had nothing to do with the annual passes.

Q. Now, I want to refer for a minute to another subject. Suppose a man buys a ticket at Franklin, on the line of the Northern road, to go to Boston. Is that ticket that he buys at Franklin charged to the Concord Railroad at the Franklin office?

A. There is an account of it made.

Q. Now, then, that passenger going from Franklin to Boston, should leave a coupon ticket upon the Concord road, which should be returned to your office?

A. Yes, sir.

Q. Now, is the settlement between the Northern Railroad and the Concord Railroad made by that charge and by that coupon?

A Yes, sir; by the charge. The coupon does not go into the account.

Q. The coupon should properly, if the business is regularly done on the Concord road, tally with the charge at Franklin?

A. Yes, sir.

Q. Now an account is taken of this charge at Franklin; and this is kept regularly for the month, is it, sir?

A. Yes, sir.

Q. And there should be an exact correspondence between the charges and the coupons?

A. Yes, sir.

Q. Now, sir, if any great number of coupons were abstracted by the conductor during that month, and not returned, would not that fact inevitably be noted at the time of the monthly return?

A. Yes, sir, the fact of that deficiency would.

Q. Now, supposing that some coupons were not returned that month. Supposing that the sales exceeded the collections, and it ran along into the next month. Supposing there was a great deficiency the second month, would not that be noticed?

A. Yes, sir.

Q. Now, is there a chance, Mr. Sanborn,—I do not speak now of individual instances—is there a chance for a conductor to appropriate any considerable number of tickets from the upper road without its being discovered, from the first or second or third month's settlement?

A. If he keept them out entirely, it would be discovered that there was a deficiency.

Q. If he kept them, it would be discovered and noted at the end of the month in the settlement.

A. It would be noted?

Q. Can the conductor keep back those tickets in that way, to any considerable amount, without the fact being detected—if there is any amount of them?

A. You mean a considerable number?

Q. I mean twenty, or thirty, or forty.

A. If it was that number we should inquire into it at once.

Q. Now then, sir, that is the way the business is done in connection with the upper roads. That I think isn't clear as the matter stands now, We will take a case of the sale of tickets at the Concord office, for instance. Is it possible for the conductor to abstract tickets sold at the Concord office, going to Boston?

A. Not by himself alone.

Q. I want to put this question to you: Supposing the ticket master should sell tickets here at the Concord office, to me, for instance, and I get on board the train to go to Boston. Is there any possible way that George Clough alone, without colluding with the conductor on the Boston & Lowell road, can abstract that ticket?

A. Not without an arrangement with you or the conductor.

Q. He has got to have an arrangement with me, and also with the conductor?

A. Either one or the other.

Q. Hasn't he got to collude with the conductor below here, to sell it, for instance?

A. I should think so; if I understand the whole question. The only point I see where he might get it is, if he had some other tickets that he might give you.

Q. And if I were to collude with him?

A. Yes, sir. He might give you a check.

Q. Now, then, in order to pass a western ticket, he has got to collude with the conductor below Nashua?

A. If the ticket is good, he has not.

Q. Supposing I go and buy a ticket, to-day, to Boston. Now that ticket has to go over the Boston & Lowell and Lowell & Nashua roads?

A. Yes, sir.

Q. Now how is it possible for Mr. Clough, unless he colludes with the conductors of the two roads, to get possession of the ticket?

A. I do not know as he could. I spoke of his keeping a ticket that he took honestly.

Q. Is there a possible chance? That ticket is finally taken up by the conductor on the Boston & Lowell road, is it not?

A. Yes, sir.

Q. Now then, he knows what kind of a ticket is sold over the Concord road?

A. Yes, sir.

Q. The conductor, also, on the Nashua & Lowell road understands what ticket I ought to have?

A. Yes, sir.

Q. Suppose I should present a ticket different from that sold by the Concord office, to the conductor over the Boston & Lowell road; wouldn't he notice it?

A. Yes, sir.

Q. If he was an honest man he wouldn't pass me?

A. No, sir.

Q. Now, in order to keep that ticket without discovery, has there not got to be collusion between him and the other conductors?

A. I say if he has one of these western tickets which were good from here to Boston.

Q. Was there ever a ticket issued by a western railroad which would pass a man from Concord to Boston?

A. Yes, sir; there has been such.

Q. How long ago?

A. It was several years ago. It was a mistake, however.

Q. Let me ask you if all these western ticket coupons are not taken off at each road?

A. Yes, sir.

Q. Now for a man to go out of the City of Concord with a western ticket, he has got to have a ticket made up of three coupons?

A. Two.

Q. Two coupons; one of which is taken on the Concord and Nashua, and one on the Boston and Lowell road—each to be taken up?

A. Yes, sir.

Q. Mr. Clough would take up the first coupon?

A. Yes, sir.

Q. Now, when he gets to the Boston and Lowell road, the other would be taken up?

A. Yes, sir.

Q. When a man buys a ticket here regularly from the City of Concord to Boston, he gets one ticket, does he not?

A. Yes, sir.

Q. And that ticket is known by Mr. Clough? It is known to the conductor on the Nashua and Lowell road, and to the conductor on the Boston and Lowell road?

A. Yes, sir.

Q. It is a kind of ticket which, whenever it is presented on either of the lower roads, is detected at once?

A. It might be.

Q. Now, when a man buys a western ticket which would take him, we will say to Boston, from any point west of here, that western ticket is a long one—ordinarily a long ticket?

A. It has only two coupons when it come here.

Q. And one takes the passenger from Concord to Nashua?
A. Yes, sir.
Q. And the other from Nashua to Boston?
A. Yes, sir.
Q. When he gets on the cars, it is the duty of Mr. Clough to take up his coupon?
A. Yes, sir.
Q. And that is an entirely different one from the one presented below Nashua?
A. So far as regards printing?
Q. So there is no sort of similarity between the coupons?
A. No further than the color of the ticket.
Q. Now, I want you to show how it is possible for a man to get into the cars here with a western coupon ticket, and go to Boston on it, without the ticket being taken up in just the way I have stated. Can it be done?
A. I should think not.
Q. (*By Mr. Cushing.*) Suppose that the conductor should have in his hands coupons of the western road, that he had obtained from somebody that had gone as far as Concord; and that their time has not expired; so that they are good for some number of days. Supposing a passenger gets in at Concord and buys a ticket there; and the conductor takes up the passenger's check which he gives him, and substitutes another for it. Is there any difficulty in that?
A. That is the point that I have been trying to make.
Q. (*By Mr. Mugridge.*) Now you say if Mr. Clough should take up that ticket and give him a coupon, that he might ride on that coupon ticket through to Boston?
A. Yes, sir; that was my idea.
Q. Now, Mr. Sanborn, I want to ask you whether these western tickets are sold in the coupon shape now?
You mean from the west to Boston?
Yes.
A. Yes, sir; we receive them constantly.
Q. Now let me ask you if the Boston tickets are all charged to George Clough?
A. What Boston tickets?
Q. Is it George Clough's duty to return any of the Boston tickets that are sold at this office?
A. No, sir; he don't return any of them; they are punched.
Q. Does the Lowell conductor return them?
A. He returns to the Boston & Lowell office.
Q. In the monthly settlement they are returned and taken into the account here at Concord?
A. Yes, sir.
Q. Now then, shouldn't the returns from that road agree with the sales here at the City of Concord?
A. They should.
Q. Now, supposing that any large number of tickets were abstracted in that way each month, by the conductor, wouldn't they inevitably be discovered?
A. Yes, I think they would. The discrepancy would be discovered.
Q. Supposing that twenty or thirty tickets a month were taken and abstracted in that way, or even fifteen?

A. It would be noticed.

Q. That is so in regard to all the stations this side of Boston?

A. That is the general rule.

Q. That is so in regard to stations this side of Nashua?

A. All the stations: that is, so far as returns are made.

Q. If, for instance, a man should buy a ticket here over the road to Nashua, there is an account made of it, not returned here but at Nashua?

A. No, sir.

Q. And that is returned to the general ticket office?

A. Yes, sir.

Q. Well, then, that applies to tickets sold here at the general office. Now we will take tickets put into the hands of the conductors to sell in the cars. Isn't every single ticket given to him to sell in the cars charged to him as so much money?

A. Yes, sir.

Q. And hasn't he got to account for it as so much money?

A. Yes, sir.

Q. Now, supposing there is put into George Clough's hands here, by you, ten tickets from Concord to Boston; they are charged to Mr. Clough as so much in tickets; and if he does not account to you for them, they are charged to him?

A. Yes, sir. The tickets are not from Concord to Boston; he can ticket from any station to Boston.

Q. Well, now is there any chance for a conductor to steal by these tickets, without cheating himself?

A. Yes, sir; there is a chance for him to steal with these tickets.

Q. How is it?

A. If a man gets in at Concord, and the fare is $2.20, while from Thornton's it is $1.30 or $1.20; and he reports that a ticket is sold from Thornton's to Boston instead of from Concord to Boston; in that case he makes a dollar, while he gets his check back.

Q. But he cannot sell the full ticket?

A. No, sir.

Q. Wouldn't there be any way to discover by the monthly returns?

A. No, sir: because if he put it on his way-bill as from Thornton's, it would be charged so on my account. That is, all I should know would be what he would return.

Q. But still, is there any great opportunity for a man to abstract money in that way?

A. I don't see how he could very much.

Q. Haven't you such checks against him as to prevent a very extravagant use of tickets in that way?

A. As regards checks, I don't think there are many, except that he has so many tickets, and can have these sold. Our present conductors —four—report, perhaps, sixty or seventy in a month, from all parts of the month. They used to report in the same proportion—perhaps twenty apiece.

Q. Is there a chance for a man to abstract much?

A. Not any further than that.

Q. Is there any chance for a man to do a large wholesale business in stealing, by the use of tickets over the lower roads, tickets sold from the local stations? Haven't you such checks and can't you make such checks as to be able to discover it?

A. I don't know as I could. Yes, I should think, taking your statement, that I might.

Q. I mean a hundred or a hundred and fifty thousand dollars ?

A. Yes, sir.

Q. Since 1842 ? Do you think a man could steal half a million of money by the use of these tickets without its being discovered ?

A. I should think he would have to be pretty shrewd.

Q. (*By the Chairman.*) In this matter of tickets which are sold at the office, which at some time or other should be returned to you, is it practicable for the ticket to be used several times over, and then returned within the month ?

A. Yes, sir.

Q. Suppose, from the ticket office there are sold a certain number of tickets in one day, and suppose that a proportion of them are retained. There may be a portion of them retained several times, and returned within the month ?

A. Under the old system.

Q. If that is done, what means have you of discovering ?

A. We have none.

Q. (*By Mr. Mugridge.*) Let me ask you if twenty tickets could be discovered ?

A. If they were all returned within the month they would not be discovered.

Q. Can they be abstracted and kept back entirely by the conductor ? If there was any considerable number of them, you would be sure to notice them ?

A. It would call our attention to them.

Q. (*By Mr. Bellows.*) These tickets that you deliver to the conductors to sell on the road, are they tickets from Concord to Nashua—Concord to Boston, rather ?

A. Over the Concord road to Boston.

Q. Not as from Concord ?

A. No, sir.

Q. (*By Mr. Haile.*) They are different from what you sell regularly ?

A. Yes, sir.

Q. Supposing an individual should get into the cars, and the conductor should not have any of these tickets, could he not take pay through to Boston, and take a check ?

A. These are checks.

Q. (*By Mr. Stanley*) The question is, if he didn't have any of these tickets ?

A. I presume that no conductor would refuse to take his pass on that account. But it is understood that sometimes he writes on a piece of paper, explaining to the conductor, who keeps it until he comes back and gives him the check.

Q. (*By Mr. Rolfe.*) Mr. Sanborn, is it the practice of the conductors, where they take fare on the cars, to give out a local ticket ?

A. There has been a time ; it has not all the time.

Q. Has it been so in six or eight years ?

A. Yes, sir ; in the commencement of 1866, I think, we had a system of checks for every station. Every single fare that was taken, the conductor was supposed to take one.

Q. What became of that system ?

A. That was as long as the conductors would do it.

Q. You sell a ticket, a local ticket from here to Nashua. Now, Judge Cushing asked you the question whether that might not be taken up and used over and over again?

A. Yes, sir.

Q. Now, what occasion has the conductor to take that, so long as he never gives any tickets to persons in the cars, when, if he used the tickets, he exposed himself to detection?

A. If I understand you—I spoke of using a ticket over—there is no necessity of that.

Q. (*By the Chairman.*) How is it about the tickets that these gentlemen from Manchester testify that they have sold?

A. They might be done in that way.

Q. (*By Mr. Mugridge.*) How was it with the old conductors taking fares from here to Nashua? Were they accustomed to give out a ticket for that fare?

A. There was times when they did.

Q. When was that?

A. I could not tell without looking; but I think the general practice was to let them get along without anything. There were no tickets except these checks; they could give them or not.

Q. Now, aside from this particular time, did they have anything furnished them that they could give?

A. No, sir.

Q. Could they give out anything as a representation of the fares?

A. No, sir.

Q. Now, can you see any occasion, suppose a man wished to steal, can you see any object that he would have in using a coupon ticket over and over again, when he made a sale, and thereby exposing himself to the risk of detection, when he might just as well take the fare of the passenger and put it into his pocket at once?

A. So far as that is concerned, there is no need of it.

Q. Supposing Mr. Clough wanted to steal a fare from Hooksett to Nashua, does this giving of a check furnish him any aid in doing it?

A. No, sir.

Q. Well, don't he take one chance of detection if he does this?

A. Yes, sir; I should think he would.

Q. Wouldn't it be much easier, if he wanted to steal, to take the fare itself at once?

A. Yes, sir; I should think it would.

[The further examination of Mr. Sanborn was deferred for the time being, at the request of Mr. Mugridge.]

## TESTIMONY OF HENRY P. LANE.

Q. (*By Mr. George.*) Mr. Lane, where do you live?

A. Worcester, Mass.

Q. Who is your father? Where did you formerly live?

A. Manchester.

Q. Son of whom?

A. Warren Lane.

Q. Were you engaged in the substitute business at Concord, and if so, at what time?

A. I was.

Q. About how long in all?

A. Six or nine months; seven months I think here.

Q. Will you state what your business was?

A. My business was to start from here and go to Worcester, and take the first train from Worcester in the morning, and fetch around substitutes, brokers, etc.

Q. What time would you arrive at Concord?

A. Half-past nine or ten—I don't know—half-past ten, I think.

Q. Ride by the morning train?

A. The morning train; the first.

Q. Will you state how many substitutes, brokers, etc., you brought on during the seven or nine months that you were engaged, nearly as you are able to state?

A. Perhaps from seven to nine thousand.

Q. That is the number in the aggregate?

A. I cannot state the number exactly; it was a number of thousand.

Q. Will you state now who paid the fares of these substitutes; how they were paid, and by whom they were paid, on the Concord Railroad?

A. Why, from Nashua to Concord we always paid the conductors inside fare, in the cars, invariably. We never bought a ticket, as I can recollect. And when I didn't have the direct charge, the brokers who had charge of them paid in the cars.

Q. With what conductor did you usually ride, and to what conductor did you pay?

A. Well, I rode with Mr. Clough. I always paid Mr. Clough when I rode with him, and the other ones. He was always on this early train. I paid him.

Q. How much have you ever paid? How many did you pay for? Give the best idea you can to the board of referees how many you paid for on the train; from what number to what number, and how much it would amount to.

A. I could not state exactly; about from five to thirty-five. Well, there might have been sometimes more or less.

Q. You mean the number of substitutes, or the amount of money?

A. The money I could not tell.

Q. How many times a week did you go from here to Worcester?

A. Every day, when I was able. Probably didn't miss but very few days in the seven months.

Q. You spoke of paying in the cars. From what point to what point did you pay in the cars?

A. From Nashua.

Q. Do you remember what the fare was from Concord to Nashua at that time; do you recollect?

A. I have forgot; I couldn't tell.

Q. Mr. Lane, how were the cars? How many cars ran on the train at that time from Nashua up, and how were they loaded?

A. Oh, there were extra long trains in those days

Q. (*By Mr. George.*) During the eight or nine months that you were engaged in the substitute business, won't you describe the trains on which you rode? You may describe the number of cars, and how many passengers there were. Give the best description you can during the period that you were running upon them every day.

A. I used to go from here to Worcester, to meet the boat train from New York; and from there to Nashua.

Q. I am talking about the cars of the Concord Railroad, from here to Nashua. How many cars ran on a train?

A. I couldn't tell, exactly; but I have seen, perhaps, eleven to fifteen cars.

Q. Describe the ordinary trains; about how many ran, and how full they were. Just give the best account you can.

A. They was always very full, what cars there were. They used to be very large trains; eleven to fifteen cars.

Q. How was it with regard to the seats of cars accommodating passengers? Give just as fully as you can how full the cars were.

A. They was always very full, so that we used to stand upon the platforms. There wasn't seats enough, very often, for the passengers.

Q. Were there any other substitute brokers, or runners, as they were called? You called them runners?

A. Yes, sir.

Q. Were there others besides yourself?

A. Yes, sir.

Q. About how many? How large a number?

A. Sometimes three to five used to run on the train which came from New York, clear through to Concord.

*Mr. George.* Mr. Tappan, do you make any question that Concord was the rendezvous, or head-quarters?

*Mr. Tappan.* No, sir.

Q. Assuming that Concord was the rendezvous of the central district of the state, will you state whether you have reason to know how it is, having seen money paid to Mr. Clough, and to what extent besides money paid yourself?

A. I could not tell exactly to what extent I have seen paying.

Q. Have you seen any paid, in the first place?

A. I have, sir.

Q. Now give the best idea that you can in reference to how much a train.

A. I could not state the amount exactly. I have seen from fifteen to twenty; fifteen, five, ten, all along to thirty-five.

Q. What do you mean, fares or dollars?

A. Fares.

Q. Have you seen him have money in his hand in the cars; and if so, in what form and to what extent?

A. I have seen him carrying money in his hands, the same as every conductor.

Q. To what extent?

A. I could not say to what amount; fives, ones, twos, threes, &c., &c., tens, &c.

Q. How large a bundle of money did you ever see him have in his hand?

A. I have seen him have a handful; what I call a handful.

Q. When you paid Mr. Clough, for instance, supposing you had twenty-five substitutes in the cars who paid in the cars on coming up, how was the practice?

A. I never paid for twenty-five myself. I would make an arrangement with the brokers for them.

Q. Did the conductor go and count them, or did you tell him?

A. He would go and count them himself; such men sat in these seats and all through the cars.

*The Chairman.* Do I understand that Mr. Lane is talking about any particular conductor.

Q. (*By Mr. George.*) Who were you speaking of?

A. I am speaking now of when that Mr. Clough was on. That is the only one that I suppose I am speaking about at all.

CROSS EXAMINATION.

Q. (*By Mr. Mugridge.*) Where do you live now, sir?
A. Worcester.
Q. What is your business, Mr. Lane?
A. The liquor business.
Q. Do you keep a liquor saloon?
A. I stop in a saloon.
Q. Whose saloon is it?
A. It is No. 17 Exchange street.
Q. Do you keep it yourself?
A. I stop there?
Q. Do you keep it?
A. I don't own it.
Q. What is your business there?
A. Liquor business.
Q. Do you sell liquor in the saloon yourself?
A. We sell there occasionally in the place.
Q. Have you an interest in the sale?
A I have an interest in the business there myself.
Q. Are you a partner?
A. No, sir.
Q. Now, what interest have you?
A. I have an interest because I am hired there.
Q. Do you work on hire as clerk there?
A. Yes, sir.
Q. What other business is done there besides the liquor business? Anything?
A. Well, a little.
Q. Any playing cards?
A. I couldn't say.
Q. Do you mean to say there are no cards played there?
A. There might occasionally be card playing.
Q. Any gambling done there?
A. If you call that gambling, there may be.
Q. Do you know whether there is any gambling done there?
A. Be I obliged to answer that question?
Q. No, sir; you can do just as you please. Let me ask you if the saloon where you stay is a gambling saloon?
A. It is not.
Q. Is there any gambling done there in the saloon?
A. Not that I know of.
Q. Do you mean to say that you do not know that there is any gambling done in that saloon at Worcester?
A. I don't gamble myself.
Q. The question I ask you is if there is any gambling done in the saloon at Worcester, where you are employed? Do you mean to say to the referees that you do not know whether there is any gambling done there, or not, sir?
A. Well, there might be some going on there.
Q. Well, is there, or is there not?

A. O, I have seen playing there.

Q. How many times have you seen any playing there for the last year?

A. I don't know as I am obliged to answer these questions.

*Mr. George.* You need not answer them unless you choose—if you decline.

Q. (*By Mr. Mugridge,*) How many times have you seen playing there for the last year?

A. I might have seen it twenty-five times, and I might have seen it twice.

Q. How long have you been connected with this place?

A. I have been connected with the saloon about ten weeks.

Q. Do you keep a house where Clark was murdered, in Worcester?

A. No, sir; not within two or three houses from there.

Q. Is your place similar to the one where Clark was?

A. I never saw his place.

Q. Were you never in Clark's place?

A. I never was.

Q. What kind of a place did he keep?

A. I understood he kept a room for playing—for gambling.

Q. And yours is in the same block?

A. I am on Exchange street, and he was on Main.

Q. You say you have been at this place ten weeks. What was your business before going there?

A. I have always been in the liquor business.

Q. Whereabouts was you before you were connected with this place?

A. I was in Manchester.

Q. With whom were you connected, and what was your principal business?

A. My business has always been the liquor business—agent for selling, and carrying on the liquor business.

Q. How old are you?

A. Thirty-eight years old.

Q. Will you state what business you have been employed in since you were eighteen years old, and where you were employed?

A. I have been in a number of businesses.

Q. Will you state what you did first after you were eighteen years old?

A. I worked in a machine shop three years.

Q. From what time to what time?

A. From the time I was eighteen until I was twenty-one?

Q. What was your business after you left the machine shop?

A. I tended bar.

Q. For whom?

A. For Henry Nichols?

Q. What was his business?

A. He kept a saloon and a stable.

Q. Any gambling done in that saloon?

A. No, sir.

Q. Where did you go from Nichols's?

A. I could not tell every place that I have tended bar in?

Q. You worked for Nichols how long?

A. A few months.

Q. How many months did you work for Nichols?

A. Probably seven or eight months.
Q. In whose employ did you go then?
A. I worked at gas work—making gas fixtures.
Q. How long did you work at gas work?
A. I couldn't say; three or four months, may be.
Q. What did you do then?
A. I went into the liquor business again.
Q. Into whose employ did you go then?
A. Well, I have worked in Boston.
Q. What place did you work at in Boston?
A. Harvard street.
Q. Whose place?
A. His name was Allen.
Q. What kind of a place did he keep?
A. A saloon.
Q. Any gambling done there?
A. Yes, sir.
Q. How long were you in Allen's employ?
A. Four or five months.
Q. Were you employed by anybody else in Boston?
A. No; I don't know that I was.
Q. Then you returned from Boston to Manchester?
A. Yes, sir.
Q. Into whose employ did you go then?
A. The Island Pond House.
Q. That is down to Massabesic?
A. Yes, sir.
Q. Any gambling done there?
A. Not that I know of.
Q. Do you mean to state that there was no gambling done at that place?
A. I mean to say that he didn't allow it.
Q. How many months were you there?
A. Three or four months.
Q. What did you do after you left the Island Pond House?
A. I went to——I have forgot where I did go to.
Q. Go to New York?
A. Yes, sir.
Q. Ever been in business in New York?
A. No, sir.
Q. How long have you stayed in New York at any one time?
A. From two to four weeks.
Q. Where have you put up?
A. Lovejoy's Hotel.
Q. What has been your business there?
A. Substitute business.
Q. Have you ever been there at any other time?
A. I have not, but once.
Q. When was that?
A. Three or four years ago.
Q. Where did you stay?
A. Brooklyn, N. Y.
Q. Who have you been employed for in Manchester, besides those you have stated?

A. I have worked for Collis.
Q. What does he do?
A. Keeps a liquor place.
Q. Any gambling there?
A. I didn't know of any.
Q. You don't mean to say that there wasn't any?
A. No, sir.
Q. How long were you there at a time?
A. A week at a time, and sometimes more.
Q. Who else have you worked for in Manchester?
A. I have worked for different people there.
Q. Give us the names of the different persons you have worked for. Is this Collis that you worked for "Hod Collis"?
A. That's the man.
Q. Who else did you work for beside Hod Collis?
A. For J. C. Ricker.
Q. What does he do?
A. Works in a machine shop.
Q. What other persons have you worked for?
A. I can't think of many that are there now.
Q. Mention another person that you have worked for there?
A. I have worked for Mr. Lang.
Q. How long?
A. Some little time.
Q. Tend bar?
A. Yes, sir; took charge of the house.
Q. Well, sir; have you been in the employ of any one man, since you were twenty-one years of age, more than three months at a time?
A. O, yes.
Q. Well, give us the name of any man in whose employ you have been for three months since you were twenty-one years of age?
A. O, I have been in the employ of men a good while.
Q. Give us a name?
A. Bancroft.
Q. What did he do for a living?
A. Kept an eating house.
Q. Ale room?
A. Yes, sir.
Q. Any gambling done at his place?
A. No, sir.
Q. None at all?
A. No, sir; he wouldn't allow it.
Q. How long were you in Bancroft's place?
A. Probably there most a year.
Q. Have you been in the employ of any one man more than a year at a time?
A. I guess not.
Q. Think.
A. I couldn't say that I have.
Q. More than eight months?
A. I can't say that I have.
Q. Have you lived in any other places than Manchester and Worcester since you were twenty-one years old?
A. Yes, sir.

Q. Where?
A. In Boston.
Q. How long have you lived in Boston at any one time?
A. O, a year, may be.
Q. What did you do?
A. Part of the time sick at the beach.
Q. What beach were you at?
A. Cohasset and Squantum.
Q. How long did you stay there?
A. Two months.
Q. What was your difficulty?
A. Rheumatism.
Q. What was your business in Boston?
A. I was in the liquor business; selling on commission, and such things.
Q. For whom were you selling on commission?
A. Edwards & Co.
Q. The persons were resident in Boston you were selling for?
A. Outside folks, in Manchester, etc.
Q. What was the smallest part of your business?
A. That was about the amount of business that I attended to.
Q. What did you mean to say, in reply to my question, that that was the biggest part of your business? If there was any other kind of business what was it?
A. There might have been a number of other.
Q. What do you mean by that?
A. That was my main business.
Q. If you had any collateral branches of business what were they?
A. Well, I can't tell.
Q. Haven't you any sort of an idea?
A. I can't tell, sir, anything particular.
Q. Did you play any there, sir?
A. Once in a while.
Q. How much of the time did you spend in gambling, sir, that you were in Boston?
A. Not a great deal.
Q. Will you state how much time, how many days, for the last seventeen years, have passed over your head without your being engaged in gambling?
A. I could not tell the number of days.
Q. Will you state how many days a week, upon an average, for the last sixteen years that have passed over your head, without your being engaged in gambling?
A. I cannot tell.
Q. Has there been a day a week?
A. Yes, sir.
Q. Has there been two days?
A. Yes, sir.
Q. Has there been three days?
A. Certainly.
Q. Do you mean to say that there have been three days a week that you have not been engaged in gambling, during the last sixteen years?
A. There has been six months that I have not seen a card. There has been times that I have been sick.

Q. When you were up and about your business has there been a day?
A. Yes, sir.
Q. How many?
A. I could not tell how many days; quite a number though.
Q. Isn't the main business upon which you rely for a living, that of gambling?
A. No, sir.
Q. Have you got any other honest occupation, or way of earning your living, and have you had for the last ten years? And if so, please name it?
A. Oh, yes; I have worked for a living.
Q. You have stated the way in which you have worked. Now if you have had any other business please name it.
A. No other business but the liquor business.
Q. Have you won large sums in gambling?
A. No large amounts.
Q. How large an amount at any one time?
A. I cannot recollect.
Q. Have you won a hundred dollars?
A. It seems to me I may.
Q. Have you won a thousand dollars?
A. I cannot say that I have or haven't won a thousand dollars.
Q. Do you mean to say that you cannot tell whether you have or not?
A. Betting on horse races, perhaps I have.
Q. Do you go to all the horse races that are got up?
A. Most always.
Q. Do you always bet?
A. No, sir.
Q. As a general rule, don't you always bet on horse races?
A. Well, it is according to how they are trotting.
Q. Do you go there for the purpose of betting?
A. Sometimes.
Q. Don't you attend the state fairs, etc., for that purpose?
A. I try to see them.
Q. For what purpose?
A. To see what is going on; for seeing the people.
Q. Do you go there to gamble?
A. I might do a little.
Q. Isn't your purpose in attending these places to gamble?
A. Sometimes I try to bet, and sometimes I do.
Q. Isn't that the main object?
A. No, sir.
Q. What else do you go for?
A. To see the cattle and horses.
Q. Now let me ask you if you haven't got your living, for the last seventeen years, as a gambler?
A. No, sir.
Q. How much money have you earned, put it all together, for the last seventeen years, by tending saloons in the manner you have described?
A. I could not say.
Q. Upon the average?
A. I could not say.

Q. Have you earned enough to buy your salt ?
A. I think likely I could have bought my salt.
Q. Have you earned enough in tending bars in this way to pay your board ?
A. Well, I can't tell exactly.
Q. For a hundredth part of the time ?
A. Perhaps so.
Q. Your expenses are considerable ?
A. Not very heavy.
Q Well, you live tolerably fast, generally ?
A. Well, I live as well as I can.
Q. Now let me ask you if the bulk of the money you get, you don't get from gaming in Massachusetts, New Hampshire and New York ?
A. I can't tell.
Q. You are not able to say how that is, upon your oath ?
A. I don't know, hardly.
Q. Upon your oath, are you able to say that the bulk of the money you have earned or got for the last seventeen years you have not got by gambling or gaming ?
A. I can't tell how much I have got.
Q. Isn't the bulk of it ? I mean the most of it ?
A. Well, probably it is ; the biggest part of it.
Q. Now, don't you regard yourself as a professional gambler ?
A. No, sir.
Q. Don't you make gambling your profession ? and are not your associates gamblers ? and have not they been for the last seventeen years ?
A. No, sir ; not exactly. There might have been men that gambled, and attended horse races, etc.
Q. Now, havn't your associates for the last seventeen years been mainly gamblers ? Haven't they been gamblers and their associates ?
A. Well, I wouldn't wonder if they did play once in a while.
Q. I want to ask you if some of the houses that you frequent in Manchester and Worcester and New York are not of the vilest and most infamous kind ?
A. I cannot say, sir.
Q. Won't you say, sir ?
A. I cannot say, sir.
Q. I put the question, if the houses you frequent in Manchester and Worcester and New York are not of the vilest and most infamous kind ?
A. I cannot say, sir ?
Q. Will you say that they are not ?
A. I cannot say that they are.
Q. Will you say that they are not ?
A. I say they are not.
Q. Will you name a respectable place in Boston that you have stopped at a night for the last seventeen years ?
A. Revere House.
Q. How many times ?
A. A number of times.
Q. Where do you ordinarily put up in Boston ?
A. One place and another. I have put up at the American House.
Q. Let me ask you if your associates have not been of the vilest and most infamous character ?
A. I don't know whether they are or not.

Q. Will you say that the men with whom you have associated are not of that character?

A. I will say that they are not.

Q. Haven't the men with whom you have associated intimately—your confidential friends—for the last seventeen years, all been gamblers and the associates of gamblers?

A. No, sir.

Q. Wasn't it a custom, when substitutes were passed through from New York, to ticket them to Concord, because the fare was half a dollar less?

A. It was sometimes, sir.

Q. Now, sir, couldn't you in these times save a half a dollar on the fare by buying a ticket right through to Concord?

A. I never did the thing; I couldn't tell.

Q. Didn't you know that the ticket was half a dollar less?

A. I never bought one through to Concord.

Q. So you didn't know?

A. No, sir.

Q. Didn't you ever ticket a man through from New York to Concord, as a substitute, in your life?

A. No, sir.

Q. Didn't know anything about that?

A. No, sir.

Q. Now, sir, when you met substitutes at Worcester that you designed to take to the Concord market, you did not close your transactions with the brokers, but you let it remain open until they were taken to Concord, and until it was seen whether they passed?

A. I used to make a bargain for these men at the depot, and take their names.

Q. You met the broker at Worcester?

A. Yes, sir.

Q. Now, you didn't take the man from the broker there. You made an arrangement there with the broker, and the conductor simply took fare through from Worcester?

A. Yes, sir.

Q. Now then, sir, you had no responsibility connected with the men until they arrived at Concord? Or, in other words, they remained in the hands of the broker until they were actually delivered here in Concord?

A. Sometimes.

Q. Wasn't that ordinarily the case?

A. Yes, sir, ordinarily.

Q. You took and delivered the men from the broker here at Concord?

A. Aboard the cars at Concord.

Q. And the broker had the charge and entire responsibility of the men until you did take them at Concord?

A. Yes, sir, generally.

Q. The broker paid the fare of the men, didn't he?

A. Yes, sir.

Q. Generally?

A. Yes, sir, if they had money enough.

Q. Well, ordinarily when the substitute broker came on with his men, you noted the men and then the broker came on with them, and paid the fare himself; and you came here to Concord and took the men here?

A. Yes, sir.

Q. You say the fare was paid by the broker, if he had money enough? If he didn't you would advance the money to him, and have the money allowed on your account for your men?

A. Yes, sir.

Q. Now, Mr. Lane, as a general thing, had you anything to do with the fare of the substitutes from Worcester to Concord?

A. Not always; no.

Q. I mean as a general thing, as a rule, the broker took charge of the substitute and paid his fare to Concord.

A. Yes, sir.

Q. Exactly; and you had nothing to do with it? Now, sir, so far as getting the men, the business that you did in the substitute line was done in the way that you have last described?

A. What? In regard to the ——

Q. The manner of taking them?

A. Well, I have taken a good many myself.

Q. But the principal part was with the brokers?

A. Yes, sir.

Q. But when through the brokers, you have done it in the way you have described?

A. Yes, sir.

Q. Now, let me ask you, how many you ever took on yourself without the intervention of the brokers, direct from New York to Worcester and Concord?

A. I never took any from New York.

Q. How many did you ever take on yourself, more or less, independent of any connection with any broker?

A. I could not say.

Q. I mean independently; yourself. How many did you ever take on?

A. I could not say. You mean where I took charge of the men?

Q. Yes, sir; when you came on with them yourself; when you paid their fare?

A. I could not say; took along a good many.

Q. How many?

A. I could not tell the number.

Q. How many do you think?

A. Two or three hundred that I would have charge of.

Q. Now, I want to ask you, when you took these men, starting from Worcester, did you buy a ticket from Worcester to Concord?

A. No, sir.

Q. Did you pay their fares from Worcester to Nashua?

A. Sometimes in the cars.

Q. Did you usually buy a ticket for them to Nashua?

A. We generally used to in Worcester, because they came early in the morning.

Q. Have you ever bought tickets from Worcester to Concord?

A. I never have, that I know of.

Q. Could you, at the Worcester office, buy a ticket from Worcester to Concord?

A. I never have.

Q. Do you know whether you could or not, and make a saving of half a dollar on the fare?

A. I could not say.

Q. Now, I want to ask you how many you have paid fares for to George Clough? How many men have you paid fares for to him?

A. Every time that I ever went over the train.

Q. How many times, will you state, that you have paid fares to George Clough?

A. I could not state.

Q. Will you state that you ever paid the fares for ten men?

A. Certainly, sir.

Q. I mean, taking George Clough alone. How many will you state that you paid fares for to George Clough alone.

A. A good many times. I always paid for myself whenever I went over the road.

Q. How many men?

A. I could not name the number of men.

Q. Give us some idea.

A. I couldn't tell you. It might be two or three or four or a dozen.

Q. Give us some sort of an idea?

A. I do not know how to give you any idea of it.

Q. Haven't you any impression whether it is one or five hundred?

A. Well, it might have been less; might have been more. I don't know. I couldn't tell anything about it.

Q. Haven't you any sort of an idea of the number of men that you paid fares for to George Clough at all?

A. I couldn't tell; a good many.

Q. Haven't you any sort of an idea as to the number?

A. No, sir.

Q. Won't you fix it within some limits?

A. I couldn't.

Q. Will you set some limits within which it was?

A. No, sir; I couldn't say.

Q. Can you give some sort of an impression or idea to the referees as to the number?

A. I could not.

Q. Who did you get these men of, at Worcester, that you took through yourself?

A. New York men.

Q. Other brokers?

A. Yes, sir.

Q. Who were the brokers?

A. I can name them in New York, some of them.

Q. I mean that you obtained the men of yourself?

A. Oh, I can name a good many of them.

Q. Were these same men in the habit of taking men through themselves?

A. Yes, sir. Sometimes they would not come through themselves.

Q. Was it your invariable practice to pay your fare rather than to buy a ticket?

A. I always paid in the cars. It may be once in a long time that I would buy a ticket.

Q. I am talking with regard to the taking of substitutes.

A. I could not say how that was.

Q. You do not mean that it was your invariable practice, when you took substitutes, to pay in the cars? Or did you sometimes buy tickets in the cars?

A. Perhaps I did.
Q. Didn't you?
A. Think likely I did.
Q. Didn't you have any trouble with Mr. Clough about passing your substitutes?
A. I don't think that I did
Q. Did you ever have any trouble with Mr. Clough about passing your substitutes?
A. I don't think I ever have.
Q. You have no recollection of any?
A. No, sir.
Q. When you took these men on yourself, how many did you ordinarily take on at a time; when you personally took them on?
A. I have had a car-load.
Q. Were they under your charge?
A. Under my charge.
Q. Without the broker?
A. Oh, no.
Q. When you came through without the broker, how many did you take yourself?
A. I never had more than four or five myself.
Q. Did you ever, Mr. Lane, take on at any one time more than four or five, when you came on and took them yourself?
A. With my own men that I had charge of myself?
Q. Yes, sir.
A. I have had charge of a car myself.
Q. I mean taking them without a broker.
A. No, I never had more than five or six men of my own.
Q. Did you sometimes come on with one or two?
A. Yes, sir.
Q. Now, sir, was there the slightest difficulty in your buying the tickets at the Nashua office for these one or two or three or four men?
A. No, sir.
Q. Not the slightest in the world?
A. No, sir. Sometimes there might be a little rush; ordinarily had plenty of time if I had wanted to.
Q. You knew the advertisement was that the ticket was ten cents less than fare in the cars?
A. Yes, sir.
Q. It was your motive to get the men on here as cheap as you could?
A. Yes, sir.
Q. Now, one question further, if you please. You didn't always when you came on yourself, on your own hook, bring these men always on Mr. Clough's train?
A. Not always.
Q. You could have come only every other day?
A. In the morning I used to.
Q. Did you, in fact, come on Mr. Clough's train from Nashua here, but one-half of the time? Would it come to that?
A. Oh, yes; more.
Q. When you came on in the afternoon, it would not be?
A. No, sir; not in the afternoon.
Q. So, for a portion of the time, you came in on somebody's else train?

A. Yes, sir.
Q. You didn't always come on his train in the morning?
A. No, sir.
Q. Give us the names of the brokers in New York of whom you had substitutes?
A. William Kelly.
Q. Where does he live?
A. Lives in New York.
Q. What is his business?
A. Keeps a —— ——.
Q. Who else?
A. Mr. Hicks.
Q. What is his given name?
A. Bill — William.
Q. You know him personally?
A. Yes, I have seen him a number of times.
Q. Who else?
A. Stephen Rowen.
Q. Who else?
A. Mr. McReady.
Q. What is his Christian name?
A. Flory — Florence, I guess they call him.
Q. Can you name any more, sir?
A. Well, I can't think of a number of the men now.
Q. Can you think of a few more?
A. There were probably five hundred in the business.
Q. Can't you think of more?
A. I can't think of many more. Lekay. Smith—James Smith.
Q. Who were the principal ones?
A. I used to have a great many of Mr. Kelly.
Q. Is he living in New York now?
A. I don't know, sir.
Q. Was he the last you knew of him?
A. Yes, sir.
Q. Give us the names of some more.
A. I can't tell the names of some I bought of.

### RE-DIRECT EXAMINATION.

Q. (*By Mr. George.*) What proportion should you judge that these substitutes that were brought on were accepted? That is, a great many were brought on that were accepted; what proportion were rejected?
A. I should think two-thirds of them.
Q. What was done with the rest?
A. Carried to different points.
Q. Where did you carry them?
A. Some to Portland and Boston and White River Junction, and all about the country.
Q. You wanted to head them off?
A. Yes, sir.
Q. Some of them started for Boston?
A. We would meet them on the way to Boston. If they were going to Boston, we would buy these tickets.
Q. How large a number could you take through to Boston in that way?

A. Oh, I have seen a great many.

Q. And you just shifted them, and then saved the tickets to be used between Boston and Worcester?

A. Yes, sir.

Q. Now was there any objection to giving the substitute tickets; that is, on the score of his escape?

A. Oh, yes, sir.

Q. A man would sit with them; and the objection to their having their tickets was that they were afraid that they would take advantage of their tickets and run away?

A. Yes, and run away.

Q. You spoke of the cars and the crowded condition of the cars.—State how that was during the whole time you were running; whether it extended over the entire period?

A. I have seen cars with none but substitutes in the cars, where it was very crowded. I have been delayed in Nashua over a train, where the cars were crowded with brokers and substitutes.

Q. I asked you with regard to the trains run during this period, how many cars were run, and how full they ran. Was your answer intended to apply to the entire period or otherwise? Whether it was spasmodic or uniform?

A. There was a great many cars, and all of them generally full—eleven to fifteen or sixteen cars, all along; I don't know exactly how many.

CROSS EXAMINATION *resumed.*

Q. (*By Mr. Mugridge.*) Do you mean that all the time you ran substitutes, that there were eleven to fifteen cars?

A. No, sir, I don't intend to state so. I have seen them so.

Q. Well, were these exceptional cases?

A. I should think it was a pretty big train, fifteen cars.

Q. Wasn't it a pretty big train when there was eleven to fifteen cars?

A. Well, fifteen cars I call a pretty big train.

Q. Wasn't eleven cars?

A. Well, it was for any time except these times.

Q. Do you mean to give these referees to understand that during most of this time from eleven to fifteen cars ran on these trains as a general thing?

A. No, I don't know as there was as many as that. There might be from eleven to fifteen cars. I have seen good long trains.

Q. When you brought these men from New York?

A. I never brought any from New York.

Q. Well, they came from New York?

A. Yes, sir.

Q. These were in a car by themselves?

A. Not always. Some men would take their men and get them away from the others. As a general thing they would take one car if they could.

Q. Ordinarily they kept them by themselves?

A. Yes, sir; occasionally there were men going who were afraid to lose them.

Q. That was an exceptional case?

A. I have seen them.

Q. (*By the Chairman.*) Won't you please to repeat that?

A. They would fetch their men—some would fetch their men and take them by themselves and get off into a car together, and get them away from the others; and sometimes twenty in one car and ten in another; and sometimes a whole car-load all together.

Q. (*By Mr. Haile.*) You are speaking of the Concord and Nashua?

A. Yes, sir, or of the other road either. When they got to Nashua they would be put together in the first car they could get.

Q. (*By Mr. Mugridge.*) How many cars have you seen put on at Nashua?

A. Three or four or five cars.

Q. In addition to the ordinary train that came through from Boston?

A. Yes, sir; I have seen a number put on in addition.

Q. Did you ever see any cars put on at any station this side of Nashua?

A. I can't recollect that I ever did.

Q. These cars put on at Nashua were put on to the train at Nashua?

A. Yes, sir; that is, all that I noticed.

Q. And there were from three to five?

A. I don't know the number exactly, but that is about the number.

Q. I understand you to say that when the substitutes came on from New York to Boston, they were headed off and turned this way?

A. That was the idea.

Q. Now who kept these coupon tickets?

A. I don't know who did.

Q. Where a ticket was purchased for Boston, and the man was turned off, going a different direction, who kept the ticket?

A. I don't know.

Q. Did the broker keep it?

A. I think likely.

Q. Did the broker keep it?

A. I shouldn't wonder if he did.

Q. Now were there many instances of these coupons, where these men were headed off?

A. There were a good many.

Q. Now let we ask you if occasionally cases did not arise where a substitute was ticketed from New York to Concord, and where he was headed off and went to Boston?

A. There might have been.

Q. Were there many cases of that kind?

A. I couldn't say; my business was this way.

Q. Do you know what became of the coupon tickets that were to be used between Worcester and Concord?

A. No, sir.

Q. Have you any reason to suppose that they were not retained by the broker?

A. I have not.

Q. Let me ask you who requested you to come here to testify about this?

A. Mr. Stanley.

Q. When did you first have a conversation with Mr. Stanley on this subject?

A. Yesterday.

Q. Did he send for you to come and see him?

A. I was telegraphed to come from Worcester to Manchester. I didn't know what it was.

Q. Then you went right to his office?
A. No, I never was in his office.
Q. Where did you see him?
A. He came into Mr. Collis's to see me.
Q. Did you have a talk there with Hod Collis and Mr. Stanley in relation to this?
A. No, sir. I had an interview with Mr. Stanley.
Q. Was anybody else present?
A. Colonel George.
Q. Colonel George?
A. Yes, sir.
Q. When was it?
A. Yesterday.
Q. Monday?
A. I think it was.
Q. What time in the day was it?
A. I can't say.
Q. Can't you tell what time it was?
A. I don't think I could tell.
Q. Was it in the morning or afternoon?
A. I think it was in the forenoon.
Q. What time of day?
A. I think it was in the afternoon, between three and five.
Q. Did you say whether it was in the forenoon or afternoon?
A. In the afternoon.
Q. You are positive of that?
A. Yes, sir.
Q. Who else was present?
A. Mr. Curtis.
Q. Sam. Curtis?
A. Yes, sir.
Q. Anybody else?
A. No, sir; not that I know of.
Q. Where does Mr. Curtis live?
A. I don't know where he lives.
Q. Does he live in Manchester?
A. I don't know that he does. I dont't see him often.
Q. How long did you stay?
A. A few minutes.
Q. Where was you?
A. I was in the place —— no; I was in the street, and they sent for me.
Q. Whereabouts?
A. On the sidewalk.
Q. Whereabouts in relation to Hod Collis's saloon:
A. O, within a few steps of Hod Collis's, I was.
Q. They sent for you to come into the saloon?
A. Yes, sir; they sent for me there.
Q. Who did they send after you?
A. Well, I don't know but what Mr. Stanley came himself.
Q. Did Mr. Stanley come himself and want you to come.
A. I think Mr. Stanley came himself.
Q. How long did you stay there?
A. Half an hour.

Q. Were you there two hours?
A. O, no.
Q. Were you there an hour?
A. Might have been.
Q. Anybody else there save Sam. Custis?
A. I don't recollect anybody else.
Q. Were you together, by yourselves.
A. Yes, sir; Sam. Curtis and I were by ourselves.
Q. You were not where people ordinarily came in?
A. No, sir; we were in a separate place.
Q. How long were you there?
A. About half an hour.
Q. (*By Mr. George.*) Did Mr. Stanley send a despatch to you at Worcester?
A. No, sir.
Q. At the time you came to Manchester did you have any idea what you were coming for?
A. No, sir.
Q. When did you first learn the purpose for which you were sent? I don't care about that. Do you know whether I came up in the afternoon train and stopped over for the purpose of seeing you?
A. I think you did.
Q. When were you summoned?
A. I was summoned this morning, by Mr. Stanley.
Q. To come here?
A. I was. I told him I wouldn't come.
Q. Who telegraphed you at Worcester?
A. Col. Cheney. He telegraphed me to come on business; and I should not have come for any other name.
Col. Jim Cheney telegraphed you to come to Manchester?
A. Yes, sir.
Q. He didn't state the purpose for which you were to come?
A. No, sir.
Q. How long had you been in town?
A. I went to Hod Collis's saloon the first thing when I got there.
Q. Did you go to Jim Cheney?
A. Yes, sir; and asked what the business was.
Q. What did he tell you?
A. That Mr. Stanley wanted to see me.
Q. Did you talk the matter over with Jim Cheney there?
A. He talked it over some.
Q. Did you talk it over with Jim?
A. Yes, sir; some. He said it was about the business.

## TESTIMONY OF SAMUEL P. CURTIS.

Q. (*By Mr. George.*) Will you state whether you were employed here in 1865 and 1866?
A. I was employed——I came here in 1864, to work for this New England Volunteer Company.
Q. For how long?
A. I stayed until the following '65. I went away the following April, from here. I came here to run on the train. I don't think I came here until the year came in.
Q. You was here how long?

A. I was here some sixteen or seventeen months. That is—I was here more than while I was employed by them; but this time I was employed so.

Q. Won't you state what your business was in connection with that?

A. My business was to run between here and Nashua; sometimes no further than Nashua, and sometimes to Boston, to engage men for this business. That was my business.

Q. Now won't you, Mr. Curtis, state how you went and what you did. State what your business was, and what train you took from here?

A. I took the afternoon train, and returned on the one in the morning.

Q. What time did the afternoon train go down?

A. Some twenty minutes past three, I think. Sometimes I would go to Manchester and wait until the express went down.

Q. Then you would come back the next morning?

A. Yes, sir; as many as four mornings in the week, unless I went on further. Sometimes I went to Lebanon.

Q. How many have you had with you?

A. I have had as many as thirty-five in Manchester, and sometimes three, or four, or five, left with me to be put there.

Q. Well, now, how were the fares of these men that you had charge of—how and where were these fares paid?

A. Oh, well, I never bought any tickets. These men always, most generally—these men pretty much all paid in the cars. There used to be men that used to keep pretty close to their men, and didn't want them to take their tickets, and paid their fare in the cars.

Q. To the conductors?

A. Yes, sir.

Q. With what conductors did you usually ride?

A. I have rode with Mr Clough a great deal—a great deal. His train, I always used to pay my fare through, during this time. I did have a three months' pass from Mr. Gilmore. Mr. Perkins got it for me. I was here some six months, that I was going every day, and I had this three months' pass.

Q. I want you to state how it was, when you came on with substitutes and brokers, how it was about paying the conductor; whether you paid the conductor in the cars

A. It was customary. They did pay in the cars. That was ruleable; they always did. I always did, and what I had for, except this three months; but I have seen a great many others, too.

Q. Then you didn't take these substitutes?

A. No, sir.

Q. How many fares did you ever see paid to Mr. Clough at one time?

A. I never paid a great many myself, although I have seen a great many. I never have paid a great many myself. I have paid more going the other way to the conductor; a great many; to Manchester to stay over Sunday, and keep them and bring them back Monday.

Q. Pay in the cars in these cases?

A. Yes, sir.

Q. To whom did you pay?

A. Well, sometimes to Mr. Clough. When he was on the train I paid to him; and when Mr. Noyes and some of the others, I paid to them. I rode a great deal on Mr. Clough's train; more than on the rest of them.

Q. How many fares did you ever see paid to Mr. Clough at any one time?

A. Oh, well, nothing that I could——I couldn't tell; couldn't give any number.

Q. Give the best number that you can.

A. Quite a number.

Q. Well, how many?

A. Well, I should judge I had seen on a train—I should judge I had seen twenty-five, as near as I can calculate. The cars have been mostly full. I have had quite a number and had them pay. They have come through by the way of Boston. We have took 'em there and put them aboard a train.

Q. In all your running over this road, if you ever bought a ticket at the ticket-office, for yourself or for your substitutes, state how that was?

A. I don't recollect that I bought a ticket to this station. I used to get on the cars and not know whether I was going through or not; if there was any men, I would go.

Q. You always paid for yourself?

A. Yes, sir; I did, and for my men.

Q. How was it about the brokers? Did they pay in the cars?

A. They did.

Q. Do you know of anybody who bought tickets in the brokers' and substitute business?

A. Oh, yes, there has been those. There has been those that have had their tickets. There was a great many who would start with their men, and when they left they didn't know whether they was to go to Boston, Portsmouth, or here.

Q. How many men did you bring to Concord?

A. There must have been a great lot.

Q. How many?

A. Twenty-five hundred; I don't know but more. Quite a lot, I know. We did most of the business here when we were in it.

Q. You spoke of paying to Mr. Clough. I want to know how it is about seeing money in his hands, large or small notes, as the case may be?

A. I have seen some of the other conductors with a hand full of bills; I don't know whether there was one or five hundred.

Q. State as near as you can.

A. Well, I don't know; quite a number. I don't know whether he took it to make change, or took it on the train.

Q. How much money in dollars have you ever seen George Clough take on the train? [Objected to.]

A. I could not tell.

Q. Now I want you to state as near as you can how many you ever saw him receive on the train? [Objected to.]

*The Chairman.* The distinction is one that has been alluded to often in the courts. If the witness has a recollection he can state.

Q. How much money have you seen Mr. Clough receive on the train? I don't mean to a cent, but according to your best recollection.

A. I couldn't tell; perhaps he might have took forty dollars; perhaps might not so much; perhaps he got more; I couldn't tell; that wasn't my business. I knew what I did myself; I couldn't count Mr. Clough's money; I couldn't tell what he had taken. Perhaps there was twenty-five paid fares, probably, from Nashua here. As regards the money he took I couldn't say.

Q. I am speaking about the money you saw him take.

A. Oh, no. I have seen him take fares; I have paid him myself; how much I couldn't say. I have seen him have money in his hands, but how much or how little I couldn't say.

Q. How many substitutes did you say you had brought up yourself in connection with a broker or alone? How many at any one time?

A. I think thirty-five was the biggest one day that the runners did for us.

Q. Where did they come from?

A. New York and Philadelphia and Baltimore.

Q. Where were their fares paid?

A. I couldn't tell you. Some of these men paid in the cars. They had laid over in the cars. Some of this party had stopped on the way. I know they had paid their fare in the cars.

Q. They did pay their fare in the cars?

A. Yes, sir.

Q. Do you recollect the conductors that they paid?

A. Yes, sir.

Q. Who was it?

A. Yes, sir; Mr. Clough was aboard that train.

Q. He was the conductor of that train?

A. Yes, sir.

CROSS EXAMINATION.

Q. (*By Mr. Mugridge.*) Where do you stay, Mr. Curtis?

A. I stop at the Riverside Park.

Q. What was your business before you went there?

A. Hack driver; have been in the liquor business.

Q. Are you, sir, and have you not been, sir, for the last ten years, a professional gambler?

A. No, sir.

Q. Hasn't gambling been your business for nine-tenths of the time?

A. No, sir.

Q. Have you ever earned an honest dollar for the last ten years?

A. Yes, sir.

Q. Where?

A. At home.

Q. Where is your home?

A. At 107 Court street.

Q. Where do you work now?

A. I tend gate at the Riverside Park.

Q. On what occasions?

A. When there are trots.

Q. What other business have you been in?

A. I have been in the horse business.

Q. Where?

A. In Boston.

Q. How long have you been in the horse business?

A. More or less for ten or fifteen years.

Q. Who have you been with?

A. Mr. Sam. Perkins.

Q. Do you mean that gambling hasn't been a portion of your business?

A. Yes, sir.
Q. Are you not under the influence of liquor at this time?
A. No, sir.
Q. How many drinks have you had since dinner?
A. Two.
Q. What did you drink?
A. Whiskey.
Q. Have you had a dinner?
A. Had a dinner?
Q. Have you had anything to eat in Concord?
A. No, sir.
Q. How many times have you drank since you came into Concord this morning?
A. Oh, two or three times.
Q. Haven't you drank a dozen times?
A. No, sir.
Q. Where did you drink the last time?
A. I think Mr. Clark and I went down and took a drink.
Q. How long before you took the stand?
A. About half an hour.
Q. How long after the other?
A. An hour.
Q. How many times have you drank since you came into Concord?
A. I couldn't tell how many times I drank; perhaps three and perhaps four.
Q. Haven't you five?
A. No, sir.
Q. Will you say that you haven't drank half-a-dozen times?
A. I couldn't state.
Q. Will you state that you haven't drank half-a-dozen times?
A. Yes, sir.
Q. How much did you take?
A. Oh, I drink light.
Q. As a general thing don't you take liquor every day?
A. As a general thing I take a drink or two.
Q. Are you at the present time under the influence of liquor?
A. No, sir.
Q. You are not?
A. No, sir.
Q. Let me ask you if this man Lane is an associate and friend of yours?
A. We don't live anywhere near each other.
Q. Isn't he a friend and acquaintance of yours?
A. He is an acquaintance—a friend, I presume.
Q. Didn't you know Mr. Lane formerly?
A. I used to work with him in a machine shop.
Q. Where do you go in Manchester?
A. To one place and other.
Q. Gamble in these saloons?
A. I never gambled three times in my life. I don't go there for that purpose, sir.
Q. Is there a week that passes over your head but what you are engaged in gambling?
A. Yes, sir.

Q. Isn't gambling your business?
A. No, sir.
Q. I want to know if you have ever earned an honest dollar?
A. I don't know what you call an honest dollar.
Q. If you have ever earned a dollar in any legitimate business, what it is?
A. I have told you, sir; keeping gate.
Q. Do you bet on races?
A. Oh yes, sometimes.
Q. Who asked you to come here and testify?
A. I came here by the request of Mr. Stanley, sir.
Q. Were you telegraphed to come to Manchester?
A. Yes, sir.
Q. Where were you?
A. I was in Boston.
Q. Who telegraphed to you?
A. Mr. Cheney. I was going to take the next train; but he said Mr. Stanley wished to see me.

RE-DIRECT EXAMINATION.

Q. (*By Mr. George.*) You were summoned here?
A. Yes, sir.
Q. Did you know before you came to Manchester for what purpose they wanted to see you?
A. No, sir; I wanted to go back this morning.
Q. When was the purpose first made known to you?
A. Saturday afternoon.
Q. Did Mr. Clark come in and ask you to take a drink?
A. I went down with Mr. Clark.
Q. The same gentleman that came and spoke to Mr. Clough?
A. Cooper Clark.

## TESTIMONY OF ELIPHAZ W. UPHAM.

Q. (*By Mr. George.*) Mr. Upham, where do you live?
A. Nashua.
Q. Will you state what connection you had with the Concord Railroad, when it commenced and when it ceased?
A. I could not tell when it commenced; I was very nearly eight years.
Q. Where?
A. At Nashua. It will be two years since last November—since my connection ceased. I think it was two years.
Q. What was your business?
A. Station agent.
Q. You were station agent? What charge did you have at Nashua?
A. Well, I occasionally was called upon to run as conductor.
Q. What charge did you have at Nashua?
A. I had the whole charge—freight, tickets and wood, everything there was there.
Q. You had the entire charge?
A. The entire charge, except the road department.
Q. Does the Nashua and Worcester road unite at Nashua with the Concord road?
A. They do.
Q. One runs up on one side of the depot and the other the other?

A. Yes, sir.

Q. You said you sold tickets at Nashua?

A. Yes, sir.

Q. To what stations?

A. I sold for all stations—about all, unless some of the little way-stations. I sold on the Concord road, on the Concord and Claremont, Contoocook Valley, Northern, Passumpsic, Vermont Central, and Ogdensburgh on the Ogdensburgh road, and the Boston, Concord and Montreal road. I sold local tickets and joint tickets on all these roads. Some stations there was no tickets, where it was too small.

Q. You sold to Ogdensburgh?

A. Yes, sir; I think Ogdensburgh was the only station that I sold to on that road.

Q. How did the cars of the Worcester and Nashua road——that road leads to New York?

A. Yes, sir.

Q. Connects with the Western road?

A. Western and boat.

Q. When the cars arrived from Boston, on the Nashua and Lowell road, coming up to Concord, with what did they connect at Nashua?

A. With the Worcester. That would be the New York travel and the western travel on that line.

Q. Where did the travel from New York—where did they take the cars on the Concord road?

A. At Nashua; except those that went to Boston.

Q. I mean those that took that line.

A. At Nashua.

Q. How did these trains run? Which got into Nashua first, usually; and how did they run?

A. At the same time; but perhaps it was more particularly confined to a portion of the year. There was a time when the Worcester timed in ahead a little. Then they used to get in in season for the train—that is, for Manchester—but when they were timed in, you see, from Boston, they would not stop for anybody, but in Worcester they would want to. When they were timed on time, as in that case, the Worcester and Nashua got in first.

Q. When did this happen, that the Worcester train found the Nashua train waiting there to go to Concord?

A. Well, I speak only from recollection, but it seems to me that it was in the summer when they were in about our time; I think in the summer time. It might be reversed. That came in consequence of the Worcester road hardly ever changing their time.

Q. Take it during the time of the war, what proportion of the time was the Worcester train in late, keeping the Concord Railroad train waiting for it?

A. I don't know as it makes any difference about the war; the time that they would be running on would be the occasion. I should think there might be a third of the time that they would be liable to be late.

Q. Now when the Worcester train was late, and the Concord train was waiting, what was the practice at your office?

A. If the Worcester train was late, they would hurry across, and many of them would not buy tickets.

Q. They would hurry across from where?

A. From the Worcester to the Concord. When they got in on time, there was no need of hurrying; but when they were late, they would be hurried.

Q. How was it about buying tickets at the ticket-office?

A. In such cases as that, there wouldn't be many that would buy. There would be some come in, but not many.

Q. How many cars, and how many passengers, as nearly as you are able to state, how many would be brought in the cars on the Worcester road to take the cars on the Concord Railroad?

A. Well, there would be from three to four; they generally ran either three or four; sometimes three and sometimes four. That road is short of cars, and they economize.

Q. Now I will suppose that a train of cars from the Worcester road of four cars loaded, came into the Nashua station—take a period during the war—loaded with passengers for the Concord road. I will suppose that the Concord and Nashua train was late, and waited for them to get aboard. How many tickets would be purchased at the ticket-office?

[Objection to hypothetical cases.]

A. Speaking of the number of cars, I could not tell the number of passengers; I should have to judge from the people in the train that went over when the cars came in.

Q. You saw the passengers go across?

A. Yes, sir; I could form some opinion, of course, if I had been paying attention to it, from the increase of people that were there.

Mr. George stated the question to which objection was made.

*The Chairman.* Now the witness has already stated that generally speaking there did not hardly anybody buy—very few; now can you come any closer than that?

*Mr. George.* Very well; I am satisfied with that.

Q. (*By Mr. George.*) Mr. Clough was one of the conductors while you were at Nashua?

A. Yes, sir.

Q. And the train which he ran has already been testified to. I want to know how many cars during the war, when the substitute business was going on—how many cars (and in what manner they were filled) started from Nashua?

A. As near as my memory serves me, the general amount of cars was about five, I think. I think it was five cars; we always kept one at Nashua.

Q. (*By the Chairman.*) About five cars?

A. This would be in the height of travel.

Q. (*By Mr. George.*) Now go on, sir, if you please.

A. I say, we generally kept one car there. If the orders was to hitch that car on, it was hitched on. Frequently there would be one or two in it. Sometimes it was all filled up. But there was no particular regularity about it.

Q. How much of the time having three or four?

A. There was times when a crowd would be brought down. Sometimes they had two—might have had three—left there. It was a very common thing for these squads of soldiers to be brought down; and in fact, I believe, invariably there was—I don't recollect now that there was a train ran extra unless there was a whole regiment. These cars would be set off; not put into our car-house, but these would be taken on to wait the order of Mr. Biddle.

Q. Supposing there was more than there could be accommodated in the cars?

A. Well, if we hadn't cars enough, we have done such a thing as to borrow a car of the Worcester road to go to Manchester, and promise to pay the break.

Q. How was it about there being brokers, and substitutes, and runners?

A. There was substitutes come through from Worcester. I said "substitutes"; I supposed they were substitutes.

Q. How many?

A. Sometimes five or six; sometimes a dozen. Perhaps as many as a dozen. They generally came in the night trains. There might have been some times when there might have been twenty, supposed to be substitutes.

Q. Now, sir, so far as you have any recollection, will you state how it was about your selling any brokers or substitutes tickets at your office, at Nashua. Have you ever sold a broker or a substitute a ticket?

A. I don't recollect of selling any going over that road. I have sold some on other roads. I do not say that I have not sold to them. All the way that I can tell a substitute would be by a man that had the management of them.

Q. You knew the brokers?

A. No, I did not. I got acquainted with some of them, running up one day and back one night. I thought I came off pretty well. I shouldn't want to know them. It would only be a matter of opinion with me who was a broker, or broker's dealer. I wasn't acquainted with any of them, with the exception of that time, as I say; I had enough then.

Q. In the height of travel, how many passengers—not cars, but how many passengers would come from the Nashua and Worcester road on to the Concord road? Will you go on and tell, as near as you can, in the height of travel, how many passengers would be brought over Nashua to Worcester to take the cars on the Concord road?

*Mr. Tappan.* Has he any direct knowledge of this?

A. I could not tell anything only a matter of opinion about it. I could tell whether there was a few or a great many. In the height of travel, I could state—I don't know as it is of any service—but in the height of travel, the most of it was New York; beyond Worcester, you know. The travel on the Worcester and Nashua increased but very little in the height of travel. It is the travel that came in from New York and Albany and in through that way—Springfield—that are going to the White Mountains and such places. It increases with that kind of travel, but not with the local travel on the Worcester road.

Q. You cannot give any better idea than you have?

A. I could not—nothing more than what I have said—with regard to the amount.

Q. Did you say that you had run as conductor for a period of time?

A. I did some.

Q. What time?

A. I could not state. The first time, I ran on Mr. Corning's train, while he was gone out west.

Q. Did you run with Mr. Clough?

A. Yes, sir.

Q. Did you return the way-bills?

A. Yes, sir.

Q. You signed your name to the way-bills?

A. Yes, sir.

Q. (*By the Chairman.*) What time?

A. It was about a week, or such a matter; only a short time.

CROSS EXAMINATION.

Q. (*By Mr. Mugridge.*) Did you run as conductor during the war, while the substitute business was being carried on.

A. I did considerably.

Q. Let me ask you if some of the substitutes sometimes went with you on the train up to Concord?

A. I should think they might.

Q. How was it about their having tickets through to Concord in those cases?

A. I think they had their tickets; I don't recollect of having any trouble with them. If I had taken money, I should remember, I think. I mean those that came in over the road.

Q. Do you recollect one that came from the Worcester road as a substitute, that didn't have a ticket?

A. No, sir; no means of knowing only when I was conductor.

Q. I mean when you was conductor.

A. You see, unless they wanted to run their faces, it was a good deal cheaper to buy through tickets.

Q. Do you remember instances, Mr. Upham, when you were conductor, when a broker came on with a batch of substitutes, when that broker paid for their passage in the cars?

A. I do not know; I shouldn't want to swear. I think they bought their tickets through. I am speaking now of the substitutes that came in from that road. I have no recollection of having taken any money of any that came in from that way.

Q. Now take the through passengers. You have been inquired of as to the through passengers that came in from over that road from New York. Didn't they buy their tickets, as far as your knowledge extends? Didn't they, if they started from New York to go to Concord, buy a through ticket at New York to Worcester and right through?

A. Of course they would; that would be the natural course for them.

Q. What was the difference in fare, say, from New York to Concord, if they bought through, or if they bought to Nashua, and then from Nashua to Concord?

A. I don't know.

Q. Was there a difference?

A. I shouldn't be surprised if there wasn't any difference. It is generally the case at competing points. I should not be at all surprised if it was just as cheap.

[Adjourned.]

[SEVENTH DAY. Thursday, July 30, 1868.]

MR. G. G. SANBORN, *re-called.*

The hearing was continued at half-past eight o'clock, A. M., the cross-examination of Mr. Sanborn being resumed by Mr. Mugridge.

Q. (*By Mr. Mugridge.*) Have you reckoned the number of cars required, reckoning fifty passengers to a car, by the returns made by the conductors?

A. From two or three to four; four is the highest number, and two the smallest number required to carry the average, from April, 1862, to March, 1866.

Q. (*By Mr. Tappan.*) Can you give it each year?

A. Suppose I give it each month.

Q. That will do.

A. April 1862, 2 cars; the number required to carry the passengers returned.

Q. Are the "dead-heads" included in that?

A. No, sir.

Q. Couldn't you include the "dead-heads?"

A. No, sir; it does not include free passes.

Q. (*By Mr. Mugridge.*) Now commence, Mr. Sanborn.

A. April, 1859, 3; May, 3; June, 3; July, 3; August, 3; September, 4; October, 3; November, 3; December, 2. January, 1860, 2; February, 2; March, 3; April, 3; May, 3; June, 3; July, 3; August, 4; September, 4; October, 4; November, 3; December, 2; January, 1861, 3; February, 3; March, 3; April, 3; May, 2; June, 3; July 3; August, 4; September, 4; October, 3; November, 2; December, 2. January, 1862, 2; February, 2; March, 2; April, 2; May, 2; June, 2; July, 3; August, 3; September, 4; October, 4: November, 3; December, 2. January, 1863, 3; February, 3; March, 3; April, 3; May, 2; June, 3; July, 3; August, 3; September, 3; October, 3; November, 3; December, 2. January, 1864, 2; February, 3; March, 3; April, 3; May, 3; June, 3; July, 3; August, 4; September, 6; October, 3; November, 2; December, 2. January, 1865, 2; February, 2; March, 2; April, 2; May, 2; June, 3; July, 3; August, 4; September, 3; October, 3; November, 3; December, 2. January, 1866, 2; February, 2; March, 3. That is all I have.

Q. (*By Mr. Mugridge.*) I want to ask you one question. Did you ever know of Mr. Gilmore's receiving money for the transportation of passengers, that he did not return, but appropriated to his own use, making no account of to the transportation office?

A. You mean in any manner?

Q. In any manner?

A. There was money paid over to him that I never knew of being returned to the road.

Q. How much?

A. At one time there was thirty-four hundred dollars.

Q. (*By Judge Bellows.*) Was that for transportation?

A. Transportation of soldiers.

Q. (*By Mr. Mugridge.*) What year was that?

A. I should think that was in 1864. I have a paper at my office from which I could tell. I think it was in 1864.

Q. (*By the Chairman.*) Was it any matter of your business?

A. No, sir.

Q. (*By Mr. Mugridge.*) Let me ask you if any other money at any other time was received by him for transportation which he did not return, and made no account of at the transportation office?

A. Yes, sir; I think that year or the year following—I am not certain which—there was an amount of five hundred and odd dollars, I think, for political tickets.

Q. Was he at that time candidate for office?

A. Yes, sir.

Q. Was that the sum of five hundred and thirty-one dollars?

A. That was the sum, I guess, in the hands of the Republican committee.

Q. (*By Judge Bellows.*) It was not accounted for that you recollect?

A. No, sir.

Q. (*By Mr. Mugridge.*) Now let me ask you if these bills for transportation were in your office, and taken from your office by Gilmore?

A. You mean vouchers?

Q. Yes. Where were they kept?

A. They were kept at my office.

Q. And you say that Gilmore called for the bill, and you gave it to him?

A. Yes, sir.

Q. You properly should have received that money?

A. Properly either the cashier or myself.

Q. It should not have been Mr. Gilmore?

A. No, sir.

Q. Now, you have spoken of those two instances where the amount was considerable. Were there instances where the sums were considerably smaller where Gilmore has appropriated money in this way?

A. I do not know as I could state with any degree of certainty, because I have not investigated the matter far. There was one instance where I paid, which I never knew whether it was paid or not. The whole facts, if I may state, were that there were tickets sold from Manchester representing to be "Squog," serving to carry at the inauguration.

Q. (*By Mr. Stanley*) To Concord?

A. Yes, sir. The tickets were actually sold in Manchester. The tickets were represented to be sold over the North Weare road.

Q. (*By Mr. Mugridge.*) You may as well state the object of that dodge.

A. I was under orders of Mr. Gilmore to arrange the matter in that way.

Q. For what purpose?

A. I know of nothing only that he wanted the funds from these particular tickets.

Q. Wasn't it the purpose to keep an account separate from the Concord Railroad account at all?

A. I do not know what the object was.

Q. Can you conceive of any other object than to keep the account away from your general ticket office?

A. It was plain enough that the tickets were sold at Manchester, and sold from "Squog."

Q. Was there any cheat in this?

A. It was not accounted for in the general form.

Q. Now you may go on and state what the order was and how many tickets were issued.

A. I do not now recollect the number of tickets sold to make that. I think it was three hundred and twenty-five dollars, or in that vicinity; that was the amount of money which was paid to him. He might have turned it in to the cashier. I never like making such a statement without being able to swear.

Q. Can you conceive of any reason why this should have been done, if his purpose had been honest?

A. No, sir.

Q. State whether there were any other occasions where Gilmore appropriated the prices of transportation over the Concord Railroad, save those you have spoken of.

A. No, sir; none other that I recollect of.

Q. (*By the Chairman.*) Were those North Weare tickets issued from your office?

A. Yes, sir.

Q. Were they returned there?

A. Yes, sir.

Q. Well, why were they not matters of record?

A. Because he requested me to keep them out of the account?

Q. Was there any reason why he could not have had the tickets sold at Manchester and kept them?

A. No; only he would have had them returned.

Q. The question I make is this: at the very time that you issued these tickets on the North Weare road, why couldn't you just as well have given him the same number of tickets from Manchester to Concord?

A. I could, only I would have had to alter the Manchester account; because I received a report from them stating the number that were sold from that particular station; and I have got to alter the account, or else there would be a discrepancy in the account.

Q. If they were sold from Manchester they would go into the Manchester account. Now where were these tickets sold?

A. They were sold, really, at Manchester, by the same man who made the other sales.

Q. Why didn't he report them?

A. He did, but he reported them separate.

Q Wouldn't it have been just as easy for him to have got a lot of real Manchester tickets and ordered him to keep them out.

A. Yes, sir.

Q. Then, if I understand the matter, if the ticket-master would obey his orders, it was just as easy for him to have done that with the Manchester tickets, as it would have been with North Weare tickets?

A. Exactly.

Q. At what price were these North Weare tickets sold for?

A. At the same price as the Manchester tickets. For instance, I issued five hundred special tickets from Manchester to Concord, and at the same time I issued tickets from other points on the same occasion. Mr. Gilmore ordered me to issue so many tickets, and he wished them to come in from the Concord road account. That money he wished me to give to him and nobody else.

Q. Was there any difference in the fare from Piscataquog or Manchester?

A. No, sir.

Q. The North Weare road was under the control of the Concord road?

A. It was operated by them.

Q. (*By Mr. Haile.*) If I understand these Piscataquog tickets, there was no record, or return, or anything else; but if it had been at Manchester, you must have changed the record of the Manchester return?

A. Yes, sir.

Q. You say that the North Weare road was at that time run by the Concord road?

A. Yes, sir.

Q. They sold the tickets and took the proceeds?

A. Yes, sir.

Q. If these tickets had been duly accounted for, or made part of the record, could it have made any difference to the receipts of the Concord road, whether they had been reported as from the North Weare road, or from the Concord road—would it have made any difference in the receipts of the Concord?

A. I am not able to state exactly the manner of settlement between the North Weare and the Concord road.

Q. There was, then, a settlement?

A. There was a settlement between the two roads.

Q. And in that settlement it would make some difference whether these were Manchester tickets or North Weare?

A. Yes, sir.

Q. (*By Judge Bellows.*) Who had the proceeds of the North Weare road?

A. As I said, there was a settlement between them, and I am not able to state the manner of settlement. My impression is they paid so much for the use of the road.

Q. That depended upon the amount of the receipts?

A. I could not pretend to say. I never saw the contract, and I could not say as to that. I never was familiar with that.

Q. (*By Mr. Mugridge.*) Well, now, I want to ask you if there have been other transactions of this kind?

A. I don't recollect; that is, if you mean of any amount.

Q. Have there been transactions involving two hundred dollars?

A. I don't recollect of any now.

Q. [Showing tickets.] Was there ever any return of these tickets made at your office at all?

A. No, sir.

Q. I mean of any like this?

A. No, sir.

Q. Now, without going into the large transactions of that kind, let me ask if Gilmore was not in the habit of receiving money in smaller amounts as the price of transportation over the Concord Railroad?

A. I don't recollect of paying anything over to him of any consequence.

Q. Have you any recollection of paying smaller amounts of money to him?

A. I do not recollect the occasion now.

Q. Did he ever come to your office after money, as the price of transportation, in other instances?

A. I cannot recollect.

Q. Have you examined these tickets with reference to their dates?

A. No, sir.

Q. Will you be kind enough to do so?

A. I will.

RE-DIRECT EXAMINATION.

Q. (*By Mr. George.*) I want to begin with this transaction of Mr. Gilmore. When was it that you first received any instruction, and who applied to you with regard to furnishing any tickets?

A. The first intimation that I had was from Judge Upham. According to my recollection, it was not far from October or November, 1865. I think, perhaps, it was the first of October.

Q. Judge Upham was then President of the road?

A. Yes, sir.

Q. Was it before or after the directors had passed a vote directing the investigation?

A. I didn't know anything about the order for the investigation. He said there had been an investigation ordered.

Q. What did he say?

A. He said Mr. Gilmore would give me instructions what to do, and he represented them about as Gilmore did afterwards.

Q. What were you required to do?

A. I was to go to the office of Col. George. [Objected to testimony as to the instructions to Mr. Sanborn.]

MR. E. W. UPHAM, *re-called.*

Q. (*By Mr. Mugridge.*) You spoke of cars being left at Nashua as the train went down. On which train of cars were these left, the 3 1-2 train or the 10 1-2 train? Which train from Concord left the cars at Nashua?

A. The first train in the morning, out of Concord.

Q. Was it the first train in the morning, or the 3 1-2 train, that usually took the extra car?

A. The one that comes up the next on that particular day leaves an extra car.

Q. Which was the later train, the 3 1-2 or the 10 1-2?

A. The 10 1-2 was the latest train.

Q. Let me ask you if you ever heard Clough instruct a brakeman to get passengers out at Nashua to buy their tickets, rather than have them pay their fare in the cars?

A. I never did see him.

Q. Did you ever hear of his brakeman coming into your office to get tickets for passengers in the cars?

A. Yes, sir; he has.

Q. How many?

A. I could state any more definitely than that it was no uncommon thing. I think there was one that came in oftener than the other. He would pay me the next day; I got my money; they were perfectly good.

Q. You say this was no uncommon thing. Won't you indicate how frequently this was done?

A. I could not tell; I could not say any further than that.

Q. Give us some sort of an idea; the best you can.

A. Well, in the height of travel it was more common than when there was less travel.

Q. Was it every other day?

A. It might be every day for a week; and then there might be days when he might not be in at all. As I said, it was no uncommon thing; it was rather common.

Q. Now, in how large numbers would these tickets be purchased, and to what stations, when they would be purchased?

A. I could not say about the stations.

Q. In how large numbers did you sell?

A. I don't think it was very often that I sold more than three at one time; not more than three at one time—at one call. There may have been more than three, but that seems to be my impression.

Q. Who were the brakemen who made these purchases?

A. Well, Thayer got them; I think Thayer got more than any other one.

Q. What is his given name?

A. I could not say. He is conductor on the Nashua, Lowell and Boston.

Q. Thayer was brakeman? Didn't he go up in the morning?

A. Yes, sir; that was his train up; that was one.

Q. Who was the other one, sir?

A. I don't know as I can call his name; he is on the Northern road. If I should hear his name, I should know; I don't know as I should be able to tell his name.

Q. I want to ask you if you knew Starkey?

A. I did.

Q. You occasionally ran up on the train when Starkey was aboard?

A. Yes, sir.

Q. What was his practice as beginning at the rear of the trains and meeting the conductor, taking the tickets up as he went along?

A. I have known him to do such a thing.

Q. How frequently have you known him to do it?

A. Well, it was very apt to be the case. The first time I knew it, it was when I was doing it myself; and when I was half or two-thirds through, I met him with some tickets. And I felt a little kind of doubtful about it, not knowing what the custom was. But still I said nothing anyway, and took the tickets. I learned, by what I thought proper authority, that it was common practice with him, and I didn't make any opposition; but I always chose to take my own tickets if it was possible to get through. There is one thing I would state: it was a very common thing for Starkey, or any other brakeman, if he thought that the conductors had not got to the car, and there was a stop where the passengers were to get out, it was his custom to take their fare.

Q. You spoke, yesterday, of five cars making up a train which left Nashua and went to Boston?

A. I spoke, I guess, of its coming up—that ran up from Nashua.

Q. Now you state that five cars made up the train which left Nashua, coming up. Which train was that?

A. That would be the morning train.

Q. Now how many cars made up the train which came up at night—the evening train?

A. Well, I couldn't say definitely; sometimes there was three, but generally a less number; because all of the Northern—north of Concord—cars would be off.

Q. Now the five cars, or the three cars that you spoke of, made up the whole train?

A. That would be two long trains. I should not want to be understood to say definitely. The morning train is more distinctly in my mind.

Q. Supposing a party was going to Concord from New York state, how was it about their baggage? Was it checked to Nashua, for instance, and could their baggage be checked without showing their tickets?

A. There was an order come to us that the baggage should not be checked until they showed their tickets. I endeavored to carry it out. I instructed my baggage master; I endeavored to carry it out as rigidly as I could, but it was not done in all cases.

Q. If a party took the cars at Worcester, and was going through to Concord, he had his baggage checked from Worcester to Nashua?

A. Yes, sir.

Q. Now, then, he could not get his baggage checked from Nashua to Concord, until he presented his ticket to the baggage master at the station?

A. That was the order given me, and what I gave to the baggage master, and enforced it as strictly as I could.

Q. Mr. Upham, as a general thing, sir, how many cars came into Nashua from the Worcester road at this time we are speaking of? What was the number of cars ordinarily constituting a train then, in 1865?

A. I answered that question, I believe, yesterday. It was from three to four.

Q. That was in the height of travel?

A. I should think it was. I am pretty sure I answered that question.

Q. Was the fourth one on very often? Wasn't it an exception?

A. Well, it was not an unfrequent thing that they brought but three in.

Q. Was it an ordinary thing that they did not bring in the fourth car?

A. I should not want to make a decision.

Q. Was one of these a baggage car?

A. I spoke only of passenger cars.

Q. Have you ever known, as a general rule, of the running of more than the three or four cars you speak of?

A. Oh no, sir; that is my impression strong. I think they as often ran three as they did four.

Q. Let me ask you if your attention was ever called by Mr. Clough to the fact that he had a load of nothing but dead heads in his train?

A. It is my impression that I heard him come in jawing about it. He would find a good deal of fault about it, and scold about, and sometimes talk pretty rough about it.

Q. Give us your best impression as to the number of times you have known that done?

A. Well, I couldn't say.

Q. Give us your best impression.

A. Well, it was quite a number of times.

Q. Whether Gilmore ever gave you personal directions to pass people over the road?

A. Yes, sir; he has.

Q. How many times has he done so?

A. Oh, I could not tell; it was no uncommon thing.

Q. Have you known of Mr. Gilmore's appropriating money that was the price of transportation?

A. Not a red cent that I have had myself.

Q. How many persons have you known Gilmore to have passed by you?

A. Oh, I don't know.

Q. (*By the Chairman.*) Was that when you were conductor?

A. Yes, sir.

Q. (*By Mr. Mugridge.*) If, on any occasion of that kind you would have anything to return to the road as a representation of the passage of that man, would you have anything to indicate it on your way-bill?

A. Not upon the bill.

Q. Did you have anything?

A. After some time there was a rule that you had to give the name of the person you passed, and where from and where to. In those cases I always got as many as I could of them.

Q. That was where they had a paper pass, was it not?

A. I should rather think it was, but I couldn't say it definitely.

Q. Now, supposing Mr. Gilmore came in and says, "Pass that man, and that man, and that man," did you do it?

A. Yes, sir.

Q. In that case, did you have anything to show on your returns?

A. No, sir.

Q. If you were, as a matter of fact, when running a train of cars, to do all the rest of your business, and then attempt to return the "dead-heads" that ran on the road, could it be done? [Objected to as involving an inquiry as to how fast Mr. Upham was, and what his capacity was; and as having no connection with Mr. Clough's train whatever.]

*Mr. Mugridge.* I will modify my question.

Q. (*By Mr. Mugridge.*) Now, Mr. Upham, will you be kind enough to state to the referees, in your own way, when you ran a train, what attempts you made at collecting the tickets and returning the "dead-heads," in accordance with the instructions that you received from the road, and how you succeeded in both directions; and what you were able to perform in this direction, and how you came out in endeavoring to discharge your obligations to the road in those particulars?

A. Well, I can state the course I took. It would depend something altogether upon the size of the train; that is, the passengers in the train, and the number of "dead-heads." My first object was to get all the money that belonged to me, to take my tickets, and at the same time to carry these "dead-heads" along on my way-bill. A blank was furnished. If I got into a large snarl, when I found it was going to cut me short of time, I had rather let them go than let the money and tickets go. My system of keeping it was, if I had time, to put down a full statement of names, where from and where to, and by whose order passed; but if I hadn't time, if I got drove up close, in order to get the most of them, I would put down "man and lady"—man and wife, I suppose; "order J. A. Dodge's pass." For instance, I take that as an instance; from Concord, Nashua or Manchester, wherever it might be. In that way I would give the particulars. In some instances I could fill my blank all up, and had no space to put on more,

Q. You mean that there would be so many dead heads that you would fill up your blank. How large was it? As long as my hand?

A. I speak from recollection. It would be about that long [indicating]; rather fine ruled.

Q. Wasn't it as broad as that?

A. I couldn't tell.

Q. How many times did you get that filled with "dead heads"?

A. Not a great many times.

Q. How many do you think it would hold?

A. It would be nothing but a mere matter of opinion.

Q. Twenty?

A. Well, I should think it might.

Q. Thirty?

A. Well, I should rather doubt it.

Q. How many times do you think it happened that you filled your way-bill full of "dead heads," and hadn't room enough to put down all the "dead heads"?

A. I couldn't say. I suppose not a great many times.

Q. How many times, as near as you can say.

A. I shouldn't want to say.

Q. Can you give the number of times?

A. I should give a safe number.

Q. Should you not say ten times?

A. I should think not.

Q. How many times did it happen that you were not able to take them all?

A. That would be nothing but a matter of opinion.

Q. Give us your best opinion.

A. Well, there have been times that I could not get them all.

Q. How many do you think?

A. Oh, well, it might happen once out of four trains, perhaps—not trips, but trains.

Q. You may state whether there was more or less "dead heads" on every train?

A. I don't know as ever I ran a train but that there was. I would say—as it has been intimated that I was slow—I would say that I am perfectly willing that my balance should be compared with other conductors.

Q. Have you known a train to pass over the road where three-fourths of the passengers were "dead heads"?

A. I think I haven't found so many up as I have down.

Q. Have you known of some trains down where three-fourths of the passengers were "dead heads"?

A. I should think not.

Q. Half of them?

A. I shouldn't want to go that.

Q. A quarter of them?

A. Well, I will tell you what, I have been into a car where about half of them were "dead heads"—"dead heads" are apt to flock together—I speak now of those that had passes.

Q. How many times has that occurred?

A. About as frequently as those leading men had relatives come to see them, and sent them away free.

Q. Was it very frequently that this happened?

A. Well, I don't know as very frequently. "Dead heads" that are acquainted with each other are apt to be together.

Q. Did you ever see a case where every single person in the smoking car was a "dead-head"?

A. I never ran much since the smoking car was put on.

Q. Can you remember a case?

A. One night I recollect, I think, that I never took a fare of a person in the smoking car. They either were ticketed or had passes.

Q. What number had tickets, so far as your rec ollection extends?

A. That is a thing that I never noticed sufficiently. I generally found out who I was authorized to give, without "pitching into them," as the term is.

Q. You have spoken of the number of cars that make up the trains going out. Now you may state how many made up the train as it passed from Nashua to go down; take the train as it left the Nashua depot.

A. Both mail trains?

Q. Yes, sir.

A. Well, I could not say; I never noticed them as much as I did the other trains; but I should say that these same cars that went through in the morning, if they went down at night, we would have a car left.

Q. Left at Nashua?

A. Yes, sir; that would be the rule to leave one, but sometimes there would be two left.

Q. Then if there were passengers in these cars?

A. They would go forward into the other car.

Q. So that the number of cars that went up in the morning train would be pretty much the same as in the night train before?

A. Pretty much. Generally the conductor would order the passengers out of the car that was going to be cut off.

### RE-DIRECT EXAMINATION.

Q. (*By Mr. George.*) On what road was Mr. Thayer conductor?

A. On the Montreal.

Q. And from what place to what place did he run?

A. I think he would go up as far as Plymouth. He would go up until he met the down train.

Q. Now when Mr. Thayer bought tickets at your office, to what road did he buy them?

A. Generally bought them over their own road. I think he has bought some local tickets in the night. He bought them for the accommodation of the passengers.

Q. What do you mean by saying his own road?

A. On the Montreal road; to some stations where they were going.

Q. You stated that he was now employed as conductor on the Boston and Lowell road?

A. Yes, sir.

Q. You spoke of the number of cars on the train. Do you mean the ordinary regular train? Or what do you mean? Won't you explain?

A. What I wish to be understood to say is, that that was the common number of cars that came in, until along in the fall.

Q. When you spoke of the five cars, did you mean to include or exclude the baggage cars? In other words, does the five include the passenger cars only, or the baggage cars?

A. The passenger cars.

Q. How many baggage cars were run on the same trains?

A. I think there were passed Nashua, two—I think.

Q. You make the ordinary train of seven cars, including two baggage cars?

A. Yes, sir; I think there was two.

Q. Now I want to ask you what was the effect of the war, or the draft, upon the Concord Railroad, upon the amount of travel?

A. You mean the rate of increase?

Q. What was the actual effect upon the amount of travel over the Concord Railroad?

A. The effect was an increase of travel, people passing, moving, on the motion.

Q. But of what did this extra travel consist? Won't you name what was the extra travel brought into existence by that condition of things during that period of time?

A. All that I could say with regard to that is, that soon after the war commenced business started up very brisk, and they began to travel. People traveled to do business, and gave a good deal of business to be done. And the military business, direct or remote, that is, everything that was moved by the military was the cause of an increase of travel, as I understood it, or understand it. Any one can see at once. A son is coming up to Concord to be examined; and the father wants to come with him. And perhaps they would get up here to camp. And of course they would want to come up and see them. Cause must produce the effect. I know of no cause of business starting up, but directly the war.

Q. Mr. Upham, do you recollect when the order not to put baggage in the cars, without showing the tickets, was made?

A. I could not tell.

Q. About what time?

A. I think I could not tell.

Q. How long was that in force, either actually or nominally in force?

A. Well, it was put in force, that is, the order was given to me, and then it seemed to rather die away for a year, or such a matter, and then was revived again. I could not state. I don't know but what it was before.

Q. Supposing the Concord Railroad train got into Nashua first.—Take the case that you have stated as occurring, and gave the number of times that it occurred, and the proportion of times of the Concord Railroad trains getting into Nashua first, and waiting for the Nashua and Worcester train to come and the passengers to get into the cars; how was it then with regard to any enforcement of this regulation?

A. Could not carry it out with any regularity then.

Q. How was it with the baggage man carrying it out?

A. The baggage master would commence the enforcement of the rule, and if it came on him too hard, and the train was likely to go off and leave somebody, they did the best they could with the time they had. [Returns authenticated.]

CROSS EXAMINATION *resumed.*

Q. (*By Mr. Mugridge.*) Now, couldn't Thayer buy a ticket to Concord at Nashua?

A. He wouldn't buy it of me, because I couldn't sell it myself. He would buy it for persons in the cars.

Q. If there was anybody in his train that hadn't a ticket?

A. If it was a person taking the cars at Nashua, he would not; but if it was somebody in the cars, he would come to me. Boston sold through, and Lowell sold through. Probably all these way-stations are prohibited from selling through. Consequently people taking the cars at these stations could be ticketed only to Nashua. Coming up, Thayer and the northern brakeman probably felt it their duty—I don't know whether they were requested or not—to accommodate them.

Q. Now, then, the tickets for the passengers that Thayer purchased were for persons who had no ticket over the road north of Nashua?

A. Further north than Nashua.

Q. He would purchase a ticket for some person in the cars who wanted to go north of Nashua, and had no ticket?

A. Yes, sir.

Q. Now I want to know if there was anything to prevent Mr. Clough from taking the fare of that man, and passing him to Concord himself, and then that person buy a ticket here at Concord for the Montreal road?

A. Yes, sir.

Q. Didn't he sometimes buy local tickets from Nashua to Manchester and to Concord?

A. He did; not so common.

Q. I understand it; but occasionally?

A. Yes.

Q. Wasn't the fare higher?—if two tickets were bought from Nashua up to Concord, we will take it?

A. If they took it from Nashua to Concord, and from Concord to Plymouth or Franklin, it would cost them more, likely, than one joined. The joint ticket would come cheaper. It ain't always the case; but in this instance, it was the case.

Q. Now was there anything to prevent Mr. Clough taking the fare of that passenger to Concord, leaving the passenger to buy his ticket from Concord to Plymouth?

A. I don't know anything.

Q. If the disposition was there, the opportunity was open to him?

A. I don't know of anything to prevent. There was one of the trains that had another brakeman on.

Q. Who was that?

A. Kelly ran there a good deal.

Q. Did he buy tickets?

A. He bought some.

Q. Did he run on Clough's train?

A. Oh, I don't know, hardly, he was on somebody's train.

## RE-DIRECT EXAMINATION.

Q. (*By Mr. George.*) Mr. Upham, did Mr. Clough run below Nashua?

A. Yes, sir.

Q. Where did you go to?

A. Nashua.

Q. Where did Thayer?

A. He ran through to Boston.

Q. Now supposing anybody got on below Nashua. Mr. Thayer was obliged to buy a ticket to Nashua, and then from Nashua on? Now how was Mr. Clough to know anything about these passengers being on that train?

A. He would not know.

Q. Now what would Mr. Thayer do?

A. He would get out of the cars and go in and say he wanted tickets.

Q. And Mr. Clough would take these tickets? What opportunity would he have to take anything?

A. He could not have any, unless it was a standing request of his with these individuals. That is all the way I can see any way.

Q. Now would Mr. Clough have anything to do with, or any knowledge of these passengers, when they go to Nashua ?

A. Of course he could not.

Q. Now, supposing a ticket from Nashua to Plymouth was bought, what was the effect of that on the Montreal road, so far as the difference of fare was concerned ?

A. They would get the percentage.

Q. How much would the Concord and Nashua road get ?

A. I could not say.

Q. Do you know how much the fare of passengers going over the road is ?

A. All the experience I had in it was that they used to get about two shillings—we used to get about two shillings on the ticket. I ought not to say. I inquired of a person, the proper authorities ; and they told me so.

Q. Do you know how it was with regard to the upper roads giving instructions to their conductors to see that passengers had joint tickets to their roads ?

A. No, sir.

Q. With what roads did the trains connect ?

A. With three trains, morning, noon and night—three trains each day.

Q. Did Mr. Clough buy tickets at your office ?

A. He did.

Q. How many ?

A. Oh, I could not say ; it was no uncommon thing for him to come in there and ask for tickets, and tell who he was.

Q. I want to know to what extent ?

A. I could not tell, for I never made any record of it. I never expected to be called upon in the world. All I can say was that it was no uncommon thing.

Q. How many do you think ?

A. I should think, as a matter of opinion, that he never bought more than two or three at once.

CROSS EXAMINATION, *continued.*

Q. (*By Mr. Mugridge.*) You have been inquired of how Mr. Clough could have known of the passengers going north. Suppose his brakeman told him, that would be one very good way for him to find out ?

A. Of course.

Q. That would be an easy way ?

A. Yes, sir.

Q. And he had an opportunity to tell him ?

A. Yes, sir.

[Adjourned.]

[EIGHTH DAY. Friday, July 31st, 1868.]

## TESTIMONY OF DAVID PERKINS.

Q. (*By Mr. George.*) Will you please state where you live ?

A. I live at Manchester.

Q. If you were engaged in the procuration of substitutes during the war, will you state where your headquarters were, with whom you were associated in business, and the extent of your business as far as you may?

A. My headquarters were principally at Concord; I was connected with what was called the New England Volunteer Company; there were six members of it.

Q. Give the names of those who formed this company or association?

A. Nathaniel Perkins, Manasseh Perkins, myself, A. M. Peck, Chas. H. Branch and Luther M. Underhill.

Q. Now give a statement of the extent of your operations?

A. Well, we procured substitutes for individuals who were drafted, and furnished town quotas. We done a very extensive business; that is, in the parts of the state where our headquarters were, Portsmouth, Lebanon, and here; we had some of the time an office in Boston which lasted some months. In 1864, I think it was—putting in recruits for Massachusetts.

Q. State how it was with regard to your company taking contracts with towns. If so, to what extent?

A. We furnished a great many towns, the town of Hopkinton, Bow, Hooksett, city of Concord, Manchester and Nashua, and other large towns—a great many of the large towns.

Q. You spoke of your headquarters being at Concord. Will you state what time—what portion of the time you spent at Concord. I mean what years?

A. I came here in the latter part of 1863 and stayed until April, 1865, with the exception of a part of the time, when I was in Boston.

Q. Will you state the *modus operandi*, very briefly, by which these substitutes were procured and brought here for examination? Just state briefly, so that the referees will understand?

A. They were mostly sailors that came from New York and Boston, and we had a great many recruits from the west, by persons promising them bounties, and a great many came from Canada. They were brought here by what were called the principals, as they called them.

Q. And then what arrangement did you make with the brokers?

A. We had contracts with certain towns to furnish recruits, and then we paid the brokers a certain amount for the men, and they made arrangements to suit themselves with the men they brought.

Q. You mean made arrangements with regard to the pay, the amount received?

A. Yes, sir; the amount received.

Q. Now how was it with regard to the men you called runners? How many were there?

A. Well, we had perhaps in all places, a dozen.

Q. What did these men do?

A. It was their business to travel on the train in which the runners were. When we wanted them here, instead of at Portsmouth, we would send our men to Boston and Worcester and stop them, and have them come up this way.

Q. Supposing a number of men were bound from New York to Boston, and you wanted them to come to Concord?

A. Sometimes the men met them at Worcester, if they could. We had a man at Worcester and a man at Boston. They used to go to Boston for the purpose of going down east, to Augusta.

Q. In the case, I suppose, of any number of men starting from New

York for Boston, and you wanted to send them up here, what did you do?

A. I went to Worcester for the purpose of sending the men this way, rather than going east.

Q. What road would they take?

A. The Worcester and Nashua road to Nashua, and then the Concord road to Concord.

Q. Now, Mr. Perkins, won't you state generally—or, during this period, where was your home, your family?

A. At Manchester—always my home.

Q. Can you give the referees a general idea of the number of the men (I mean the aggregate) who were brought to you in this period, in the way suggested?

A. I have no means of ascertaining the number over the Concord road. We have had men from all parts, and the majority of them came over the Concord road, of course. We had some considerable number of men from Canada, and in some instances they would come from Buffalo.

Q. Can you state somewhere near the number?

A. I should think it was some thousands that were offered and recruited.

Q. What proportion that you offered or that were offered were recruited, and what proportion rejected? Give an estimate.

A. I could not estimate them correctly, as I know of. The old gentleman used to have strange freaks. Sometimes a great many. I should judge he might have accepted a third, and perhaps not quite as many.

Q. Then two-thirds would be rejected, of course?

A. Yes, sir.

Q. What was done supposing a man was put in here at the office and rejected? What next was done with him?

A. Well, the men rejected by Caswell were usually taken at Lebanon, and in some instances at Portsmouth and Portland.

Q. How was it about his being accepted?

A. I never knew a man that was rejected here but what was accepted somewhere.

Q. Mr. Perkins, how often during this period of 1863–4 did you use to pass over the Concord road?

A. I used to go home two or three times a week.

Q. By your home you mean Manchester?

A. I mean Manchester—yes, sir.

Q. And over what other portions of the road did you pass, and how often?

A. I used occasionally to go to Boston. I think, during the time I was recruiting, I went to New York—twice, I think; once in a great while to Portsmouth.

Q. Mr. Perkins, we were speaking of runners—how was it with regard to the runners supplying the brokers with money to pay the fares of substitutes?

A. When they hadn't money themselves we used to advance it. Some of them had money of their own—rather an expensive class of fellows, free hearted when they had on hand an amount of money.

Q. State what the practice was, as far as your personal knowledge goes, in regard to the payment of the fares of substitutes by the runners in the cars. [Objected to.]

Q. So far as you know state the course of business.

A. I can tell you what our instructions were to our men. [Objected to.]

*Mr. George.* My proposition is to show first, that such instructions were given to the runners, and second, that these instructions were carried out to Mr. Perkins' personal knowledge.

*The Chairman.* We think, if the objection is insisted upon, that it is immaterial.

Q. (*By Mr. George.*) Well, so far as your personal knowledge goes, what was the practice in regard to the payment in the cars by substitute brokers and runners of themselves, of their own fares and their substitutes?

A. Well, sir—what came under my personal knowledge—I never used to pay a fare for the reason that I had a pass; but our runners always paid in the cars.

Q. For whom did they pay in the cars?

A. For themselves and for whoever they might have with them, of course.

Q. Mr. Perkins, on which train did most of the substitutes and runners come?

A. They came in on your up train, the train which goes up through.

Q. (*By Mr. Stanley.*) The morning train?

A. Yes, sir.

Q. (*By Mr. George.*) Arrived here when?

A. About half past ten or eleven; the ten o'clock train from Manchester, up.

Q. What train did you usually return to Concord on?

A. The half past three train, and on the late express.

Q. These are the trains you went on?

A. Yes, sir.

Q. And on what train did you return?

A. On the ten o'clock train. I used, ordinarily, to make it a practice to come on the train that had the most men on.

Q. You mean by the most men, the most substitutes?

A. I mean the most men that we could use, the most substitutes or recruits.

Q. Mr. Perkins, during the war, won't you state to the referees, as nearly as you can, the condition of the trains, as to the number of people riding, and the number of cars drawn. upon the Concord Railroad—the trains run by Mr. Clough? [Objected to.]

Q. Now you may describe the trains from Concord, over the Concord Railroad, during the war—during the draft—the number of cars, the filling of the trains, during the time you were stationed at Concord and passing over the road. Describe the three trains run by Mr. Clough or Mr. Noyes, as to the number of cars and the filling of the trains, during the period you were engaged here in Concord, engaged in the business of furnishing soldiers. Go on and state as fully as you may.

A. The trains that I usually went down on were in the afternoon, the half past three train from here, and the late train. The half past three train was generally very long. I have counted the cars when there were sixteen or seventeen cars, including the baggage cars.

Q. Now describe the other—state now with regard to the filling of the cars?

A. Generally about full; pretty well filled up.

Q. State how it was; and how often, if at all, with regard to people being obliged to stand?

A. I never noticed very particularly about that as I know of. I noticed the cars were pretty full. I have frequently stood myself, because I didn't feel I had a right to crowd in ahead of others, being a dead head.

Q. Now describe the others of the three trains, so far as your observation extends. Take the quarter past ten o'clock train, down.

A. I scarcely ever was at the train at the time of the departure of the morning train from here. When I went home some three or four times a week, I was on that train that was a quarter of an hour before the half past ten train.

Q. How was it about your going through the trains?

A. My business used to take me through the cars, but I had some difficulty until a gentleman gave me a key, so I could go through the ladies' car.

Q. What was your object in going through the train?

A. There were generally men who came up on the train.

Q. How many on the half past ten o'clock train?

A. I don't think I could tell. I never counted them in the morning train: the reason why I counted those going down, was because it was an extra long train—a very long train.

Q. How many times did you count them, according to your recollection?

A. Well, I really can't tell; I counted them very frequently.

Q. Give the best recollection which you have—take the half past ten and the half past three train—with regard to the number of cars run? [Objected to.]

Q. State the best recollection you have with regard to the number of cars, that during this period were run on the half past three and the half past ten o'clock train, for example?

A. As I told you before, I had occasion to notice the half past three train; I have noticed it very often as having sixteen or seventeen cars, because I never recollect seeing anywhere trains of that length.

Q. How many cars usually compose that train?

A. Well, I noticed only when I counted.

Q. Unable to state?

A. Unable to state, sir—except when I counted.

CROSS EXAMINATION.

Q. (*By Mr. Mugridge.*) You say you were led to count these cars because you noticed that they were extra trains?

A. Yes, sir. I didn't say that they were extra trains, but very long trains. I usually counted them going round a bend.

Q. Do you remember how many baggage cars there were upon these trains?

A. No, sir; I could not say.

Q. Could you be able to say whether there were three, or four, or five?

A. No, sir.

Q. You cannot say that on the trains you counted, there were not six baggage cars?

A. Not positively; no, sir.

Q. Mr. Perkins, let me ask you whether there were a large number of soldiers upon these trains—a great many soldiers going back and forward?

A. You mean going two and fro?

Q. I mean squads of soldiers mixed up with the other passengers on the trains.

A. I never noticed the soldiers particularly. I never noticed whilst going down on the trains, and not when I was in the trains.

Q. You mean those set apart for the soldiers, of course?

A. Yes, sir.

Q. Now let me ask you if there were not soldiers mixed up with the passengers on the regular trains?

A. Before the elections, generally, I noticed a great many soldiers in the trains.

Q. Are you able to state the number of soldiers on a train at any time you counted?

A. No, sir; I could not do it.

Q. Would you be able to say as to the number of passengers who passed free upon these trains when you went down?

A. No, sir.

Q. I suppose a train as made up at Concord separated at Manchester; one portion going over the Portsmouth road, one over the North Weare road, and one over the Nashua and Lowell road—

A. I know nothing about the North Weare road. There was a train made up at Manchester. A portion didn't go through.

Q. And the passengers had to change cars?

A. Yes, sir.

Q. Did the Portsmouth road run a through car?

A. I don't know. There was a time when they ran through here—a portion of the time when they ran through here. I don't know the time.

Q. You don't mean to state that you counted these sixteen cars all packed full?

A. I don't recollect noticing particularly on those days. I noticed generally that they were filled up.

Q. You were passed free?

A. Yes, sir.

Q. For how long a time did you have a pass?

A. I got it from Gov. Gilmore; I think it commenced early in 1864.

Q. Do you know of any other men who had free passes, connected with your business.

A. There was a short time that Mr. Curtis had one.

Q. Did he have one from Gilmore?

A. Yes, sir.

Q. Do you recollect any one else, that you came in contact with, that had one?

A. No, sir; not that I think of now.

Q. How was it about your brothers?

A. My brothers had passes over the Manchester and Lawrence and on the Portsmouth road, from Manchester.

Q. Who issued that pass to them?

A. Mr. Gilmore.

Q. Do you know of other persons that had passes from Manchester?

A. No, sir. The whole of our company was on this general pass and each had a copy.

Q. The pass was good for anybody?

A. Yes, sir. It was a season ticket, and it had the names of each written on it.

Q. Then it was good for each of you?

A. Yes, sir.

Q. You didn't pay anything yourselves?

A. No, sir.

Q. Then none of you had to pay fare?

A. I don't think Mr. Peck or Mr. Branch ever had occasion to use it.

Q. How many times did you pay fare to George Clough?

A. I don't know that I recollect of any particular instances of paying to Mr. Clough.

Q. Did you know men living in Manchester, any of your associates, that were in the habit of passing over the road without paying fare?

A. No, I think not.

Q. You never had your attention called to anything of that kind?

A. No, sir.

Q. When substitutes were brought here to Concord—brought here and rejected, where did they go to as a general thing, Mr. Perkins?

A. Sometimes to Lebanon, or to Portsmouth.

Q. I understood you to say that ordinarily brokers came through with their men from Worcester and New York; that is, if a man started from New York with a lot of men, he usually came through with them to get his pay for them here?

A. Yes, sir; ordinarily.

Q. Now you wouldn't go through with him, ordinarily?

A. No; that was the business of the runners.

Q. That wasn't a part of your business?

A. Not as a rule.

Q. I speak of it as a rule. Now of course you don't know of your own personal knowledge, whether these persons had tickets or paid through?

A. Not when they came to me here.

Q. When the brokers got the men here you paid for them, took them off their hands and put them in to supply these quotas?

A. Yes, sir.

Q. Was it your business to pay the fares of these men, or did they pay them?

A. We had nothing to do with that; the men were turned over to us.

Q. You say, Mr. Perkins, that you advanced money to your runners. Now you don't know whether they spent the money in paying fares, or whether they spent it in purchasing tickets in the cars?

A. I know from the amounts returned what it was.

Q. You haven't any personal knowledge of the fact?

A. We ordinarily have no other means of knowing, than what they stated to me.

### RE-DIRECT EXAMINATION.

Q. (*By Mr. George.*) You spoke of the extravagance of your runners, won't you explain that?

A. Expensive, I said.

Q. What did you mean?

A. That is what I was telling, our instructions as to what they were to do.

Q. Mr. Perkins, I would ask you how many substitutes have you known to go up to Concord on a single train?

A. I have known many instances, when coming from Boston, that there were very near two cars, and a greater part of the men were substitutes or volunteers, as the case might be.

Q. How were the fares of these substitutes paid in the cases you refer to?

A. They were paid to the conductor.

Q. Mr. Perkins, you spoke of——

*Mr. Tappan* (interrupting). I object to this testimony unless they shew that it is paid to Mr. Clough.

*Mr. George.* Quite a late time to object now.

Q. Mr. Perkins, during the war, at any time when you were stationed at Concord, won't you explain how it was about this being a district, or about this being a general rendezvous?

*Mr. Mugridge.* That has all been admitted once.

*Mr George.* I don't understand that it has.

A. There was a general recruiting office opened here, for volunteers for any town in the state, or any district.

Q. Who had charge of that recruiting office?

A. Col. Carr, sixth New Hampshire regiment.

Q. And for how long a period of time was that general recruiting office kept open here, according to your recollection?

A. As near as I can recollect, it was opened sometime in the early part of the fall and kept open until the following spring; I think it was the fall of 1864, sir; I would not be positive.

Q. They recruited for all the towns in the state?

A. Yes, sir; a general recruiting office.

Q. Mr. Perkins, now I want you to state, if you can, what regiments were encamped or rendezvoused at Concord during the war?

A. The first, third, fifth, ninth, eleventh, twelfth, thirteenth, fifteenth, sixteenth and seventeenth regiments, I believe; I would not be positive as to my own recollection of it.

Q. Now how was it about New Hampshire cavalry?

A. There was a company, I believe, the first New Hampshire company; that might have been stationed here; I didn't think of that.

Q. How was it with regard to recruiting for the heavy artillery, in Concord?

A. Recruits were accepted, at Concord, for any regiment; but as regards recruiting for the heavy artillery, I think the men had their choice in the branches of the service, except the navy.

Q. How was it with regard to Concord being a general rendezvous. You may describe that. Describe how it was with regard to Concord being a general rendezvous?

A. Concord was considered the general headquarters for the state, particularly after the general recruiting offices were open. That is, as the main part of the regiments were here, it was so considered.

Q. Who had charge of that camp?

A. There were different individuals, Maj. Whittlesy and others.

Q. General Hinks?

A. He was assistant provost marshal, I think.

Q. State how it was that at the time the regiments were stationed here that you have remarked, with regard to the number of persons passing backwards and forwards on the cars, particularly at these times?

A. I should think it would have a tendency to increase the travel. [Objected to.] My recollection of the actual fact is that there were more people when the recruiting was going on briskly than there were before or after.

Q. (*By Mr. Mugridge.*) I suppose you don't mean to state that these two car loads of substitutes paid their fare to George Clough ?

A. No, sir.

Q. (*By the Chairman.*) Do you know to whom they did pay ?

A. Well, sir, I don't know the gentleman by name ; but I know him.

Q. Did you, at that time, know Mr. Clough by name ?

A. Yes, sir.

Q. (*By Mr. Mugridge.*) And you knew that the conductor to whom they paid their fares, was not Mr. Clough ?

A. Yes, sir.

Q. Was it either of the trains run by him ?

A. I think it was the early train that came out of Boston ; the express train.

Q. You spoke, Mr. Perkins, of the number of regiments rendezvous'd here at Concord. I suppose these regiments were not made up of substitutes ?

A. No, sir.

Q. There were a great many volunteers ?

A. Some of the early ones were.

Q. Will you please answer my question, and tell me if many of the other regiments were not made up of volunteers, drafted men and substitutes ?

A. All the regiments were originally made up of native volunteers, but the ranks were thinned, after disease and battle took them away, and they were filled up with substitutes and volunteers, with the exception of the cavalry, which were made up of recruits from abroad.

Q. They were filled up the same way ?

A. Yes, sir.

Q. When you got a squad of fifty or one hundred men, were they not sent off to the regiments ?

A. They were sent off very early.

Q. And when they were sent off they were sent off on the ordinary trains that went out from Concord ?

A. I don't remember about that.

Q. Do you mean to say that they would charter an extra train to send them off on ?

A. I mean to say that I don't know anything about it.

Q. So far as you know they left on the regular trains ?

A. I don't know how they went.

Q. Now when a full regiment went away from Concord, an extra train was sent with them ?

A. Well, sir, I think so ; I think I have seen extra trains go out from here with a regiment.

Q. Now let me ask you how many men made up the smallest squad that you ever saw sent from Concord ?

A. I couldn't tell how many.

Q. Don't you know that where an urgent call for men was made, they were recruited and sent off as soon as they were recruited ; that is, when there was a great urgency ?

A. Yes, sir; and that is at the time when recruiting was the most brisk at Concord.

Q. Now when the call was very urgent and recruiting was brisk, they were sent off in small squads?

A. I don't know how small a number; sometimes we sent on a great many; sometimes a hundred men a day.

Q. And then sent them right off?

A. Yes, sir. I don't know as they sent them right away; they sent them to camp.

Q. Well, they did not keep them here very long?

A. It is something that I had no means of knowing, except as I saw them march through with the guard.

Q. Have not you seen a small squad going off?

A. I have no recollection about it.

CROSS EXAMINATION *resumed.*

Q. *By Mr. Tappan.*) Mr. Perkins, I understand you to say that you had your recruits and your substitutes, mostly from New York?

A. Perhaps the greater part came from New York.

Q. Well, where are the next greater part, Mr. Perkins?

A. Well, we got them from everywhere—from as far west as Detroit and Chicago, and some from Washington.

Q. Now, from New York, what were the brokers names?

A. I can tell you some of the most prominent ones.

Q. Well, I should like to have you.

A. Mr. Kelley, two men by the name of Churchill, a man by the name of Greene; those were our particular customers. And then there were a great many others that sold to us.

Q. Did these men come through with their men and deliver them?

A. Sometimes they did.

Q. Was that a general thing?

A. As a general rule, they did.

Q. And you took and delivered the men here, at Concord?

A. Yes, sir.

Q. Was it any part of your business to pay the fares of men, or purchase tickets for them, or get them along in any way? [Objected to as being a repetition.]

Q. Was it any part of your business to see to the transportation of men, and purchase their tickets, or pay their fares, or anything of that kind?

A. It was a part of our business to see that the men got here safely. That was the object of the runners.

Q. Was it part of your business to pay their fares?

A. No, sir; not when they brought their own men.

## TESTIMONY OF NATHANIEL WHITE.

Q. (*By Mr. George.*) You live at Concord?

A. Yes, sir.

Q. What is your business?

A. General express business.

Q. Where is the express-office, and where has it been for the last ten years?

A. In the north-west end of the passenger depot.

Q. During the war, during the draft, during the times when the draft was going on, I want to ask you if you had made any observations of the number of cars on the two main trains on the 10.30 and the half-past three trains?

A. I had at one time my attention called by General Peaslee, I think in regard to the railroad, and a statement was made to the superintendent and the directors. It was during the war these trains were run, the time there was recruiting of substitutes and everything going on.

Q. Now I want you to state, sir, how many cars ran on to Concord on the 10.15 train at that time?

A. Well, they were the trains that I was speaking of, that I was speaking to General Peaslee about, and he went to Governor Gilmore, and he sent for me then to come to his office. [Objected to.]

Q. I want to know what the fact was.

A. Sixteen cars is what I represented.

Q. (*By the Chairman.*) Is that including the baggage cars?

A. Yes, sir; that includes the whole train.

Q. I want you to state as to this time, and as to that being an extraordinary thing.

A. In July, August and September, it was my judgment that they ran about the same number of cars; perhaps the same number of cars. The first two weeks of September it was, perhaps, about the same.

Q. (*By Mr. Haile.*) Can you tell the year?

A. I cannot tell the year, because I never thought of it.

Q. (*By Mr. George.*) Was it while Mr. Clough was on the road?

A. Yes, sir; I recollect the conductors who were all on.

Q. You say you had an interview with Mr. Gilmore; I don't ask you the details of that interview at all, because that is not competent evidence, but I want to ask you if Mr. Biddle's record of the number of cars was sent for?

A. Yes, sir; it was sent for.

Q. Did Mr. Biddle bring his record?

A. I don't know as the record was brought, but he stated ——

[Objected to and ruled out.]

Q. Mr. White, I will simply ask you if Mr. Biddle was sent for? I will leave it right there.

A. He was, sir.

Q. I want you to describe now as well as you can, during this period that you noticed the trains, how much travel there was on them?

A. I could not certify to the filling of the cars.

Q. Give the best idea you can to the referees.

A. I can only judge from the number of the cars, that they wouldn't run them unless they had passengers for them. [Objected to.]

Q. Mr. White, during the war, during the recruiting, how was the amount of travel on the Concord Railroad upon the three main trains of the road, according to your observation? I mean as to the amount, and the filling, and all about it.

A. The two what we called main trains was the first train up, and the half-past three train.

Q. Will you state how these trains ran during the war as to the amount of passengers, according to your observation and recollection—how it was during the substitute time of the war, and when the regiments were stationed here?

A. I couldn't state that the cars were full, you know, all of them.

Q. State whatever recollection you have.

A. I should think as a general thing they ran pretty well filled, and that really of course I could not tell. I could not say whether they were crowded, or half that, but I should say they were pretty well filled. I didn't notice them to think of any such thing.

Q. Did you notice the travel on the occasion that I have spoken of, so that you can state from recollection what the travel was then compared with what it was prior to the war?

A. I should judge that it was a great deal more.

Q. (*By Mr. George.*) Mr. White, how was the travel in 1857–8–9–60 in these main trains—about how many cars?

A. I should think they would average more than two-thirds the number of cars.

Q. The number of passengers is what we want.

Q. (*By Mr. Cushing.*) What is the actual number of cars, as you remember it?

A. I should think, on the three main trains, the Northern road usually ran one baggage car and three passenger cars; and then the Montreal road about two passenger and one baggage; I should judge that would be about seven cars.

CROSS EXAMINATION.

Q. (*By Mr. Mugridge.*) Then the Concord road?

A. Unless they had something extra, I don't think there were any Concord cars.

Q. (*By Mr. Tappan.*) What period will this cover?

A. Well, he said before the war they might occasionally take on a Concord car, but I don't think it was regular.

Q. (*By Mr. Mugridge.*) There is the Portsmouth road you have not got in.

A. They didn't at that time, I guess.

Q. Now, Mr. White, before the war did not the Portsmouth road run out of Concord separate from the Concord road?

A. They did until they went to Manchester.

Q. How many cars did the Portsmouth road run out of Concord before the war?

A They ran one baggage car and one passenger car, usually.

Q. Now the Portsmouth trains are run out of Concord, in connection with the Concord trains?

A. Yes, sir; that is, the afternoon trains are.

Q. Then there would be two more cars that would enter into the computation?

A. Yes, sir; on that train.

Q. Now, Mr. White, you have spoken of the train as made up ordinarily before the war. You say three cars and a baggage car on the Northern road. Now didn't the Northern road sometimes run as many as six cars?

A. No, sir; I don't know of any such occasion, not without it was some excursion.

Q. Now there were four there. Then didn't the Montreal road sometimes run four cars, in the height of travel?

A. I should think sometimes they might.

Q. Did'nt the Portsmouth road sometimes, in the height of travel, run with three cars?

A. Oh yes, sir.

Q. Then there would be sometimes eleven cars that would run out over the road before the war. Now let me ask you, Mr. White, if sometimes, in the height of travel, the Concord road didn't put a car on from here to Concord?

A. Yes, sir; they did on some days.

Q. And didn't they sometimes, in the height of travel, put on two cars?

A. I think they have.

Q. Now, Mr. White, I understood you to say, in reply to a question put to you by Col. George, that from July to September you counted, sometimes, sixteen cars on a train?

A. Yes, sir.

Q. Let me ask you how that was?

A. I was asked that question—I could not specify the year—I should think it was some two years before Mr. Gilmore gave up.

Q. Then it would be in 1864 or 65?

A. Yes, sir.

Q. Now let me ask you if, when these trains were running out, there was a lively recruiting business being done here at Concord?

A. I should think there was.

Q. Now, sir, wasn't it also at a season of the year when the travel is ordinarily the highest?

A. That is what I stated, I believe.

Q. Then there was recruiting business in Concord, and it was also a season when the ordinary travel is the greatest, that is to say, in July, August and September?

A. Well, sir, I expect most folks travel in those two months.

Q. Mr. White, let me ask you, sir, if you did not know what the practice was here, when this recruiting was going on in the city of Concord, after the regiments had gone on to the field, and the regiments were cut up by battle, and their ranks thinned by death, if as fast as the soldiers were recruited here they were marched immediately to their regiments in the field?

A. Yes, sir. Sent off to Charlestown or down to Boston. Just about half of them got away down there.

Q. The battle of Gettysburg occurred on the 4th of July, 1864?

A. I believe that is the time.

Q. Now, then, from the time when these cars went out from the depot—or, let me ask you if there was any period of the war when men were urged forward so fast as they were just after the large fights?

A. I think they would usually send along after any battle as fast as they could.

Q. Then when the recruiting was most rapid, the men were sent off most frequently in squads from Concord. Was not that so?

A. I should think so; I know they used to send off as often as they could get fifty or a hundred, I should think.

Q. Now, Mr. White, when a squad of fifty or seventy-five men were recruited, they were marched down in squads from the camp and run on the regular trains?

A. They would not be likely to run extra trains for that number.

Q. Did they put on extra trains only when regiments were sent off?

A. I only know that they put on extra trains for regiments.

Q. Do you recollect of any cases where extra trains were put on the road to take soldiers, save when an entire regiment was to be taken away?

A. I do not recollect of any such thing; there might have been. I did not charge my memory.

Q. Now, Mr. White, supposing there was a squad of one hundred and fifty men carried from Concord upon and in the 10.15 train, how many cars would the Concord road put on naturally to accommodate them?

A. They would put on three cars.

Q. They would naturally crowd them?

A. Well, they would generally march a company into a car.

Q. And they would put on at the rate of a car for every fifty that went away?

A. They could not do much less than that. They used to stow them in a little thicker than that sometimes when they were short of cars.

Q. I want to ask you something about the dead-heads. How long have you been connected with the Concord Railroad?

A. About twenty-seven years.

Q. Let me ask you if you have occasion frequently to ride in the trains?

A. I have.

Q. Have you frequently seen Mr. Gilmore tell Mr. Clough, for instance, to pass men in the cars?

A. I have heard such things, and seen such things, I believe.

Q. Would not he direct him to pass them without their having a written pass? Would not it be his verbal order?

A. Well, occasionally, when they got ready to start, I have seen him go to all of the conductors and say, pass this man.

Q. And in that case he would get into the cars without anything in the shape of a written pass?

A. In that way he would.

Q. Let me ask you, Mr. White, if you ever went down over the road when one-half of the passengers of the particular car that you might be in were dead-heads?

A. Well, I had no knowledge of any such operation.

Q. Did you ever remark to anybody that you had been down over that road at times when in some of the cars one-half were dead-heads?

A. I do not recollect ever my making any such remark.

Q. Let me ask you if you have ever taken passage down over the road when from a quarter to a half didn't pay fare?

A. I suppose there might be a quarter; there is a train on Monday morning and there are more or less persons going down. That is the train I generally go on.

Q. Have you ever been in the car when from a quarter to a half were dead heads?

A. I think, as I say, on Monday, because I have been in when there would be but a small number, and that would be those that had dead head or special tickets; I do not know what it was.

Q. Have you ever been down in the smoking car on the road, when every single man in the car was a dead head?

A. No, sir; I do not very often ride there.

Q. Have you never been in the smoking car?

A. I have been in, but I did not stop there, because I do not smoke; I did not choose to stay there.

Q. Do you know anything about the extent to which Gilmore used to issue free passes?

A. I do not, to my own knowledge.

Q. Have you frequently seen instances where persons were passed over the road by Mr. Gilmore?

A. I have seen persons that presented his pass.

Q. Have you *frequently* seen that?

A. Well, I should think I had.

Q. Have you ever known, sir, of Gilmore's distributing tickets—I mean the regular ticket of the road—to any persons who wanted a pass?

A. No, I have never known of that.

Q. Have you ever known of his giving away free tickets, for political purposes?

A. It was passes you were speaking of.

Q. Did you ever know of his giving away passes, for political purposes?

A. I have no knowledge, except hearsay.

Q. Did you ever talk with Gilmore on the subject?

A. No, sir; I never did.

Q. Did you know of his using the Spalding passes, at the time when he was candidate for counsellor?

A. I never knew of it; I have heard of it.

Q. Have you known an entire family that has been passed down over the road, on Gilmore's pass?

A. I have no particular knowledge of such things.

Q. Have you obtained passes from him?

A. I have, occasionally. I did not use to go to him more than once or twice. I generally calculated to pay my fare one way. I might occasionally go and ask him for one for some person who was poor and had not got any money, and he would give me a pass. Mr. Clough used, sometimes, to pass me down, but I used, generally, to calculate to pay fare one way.

Q. Did you ever go to Mr. Clough and ask him to pass people down over the road?

A. I think likely I have.

Q. What has he said?

A. He has taken them, sometimes. I have told him that they were poor and needed it. I never have asked him except when they were poor.

Q. Have you frequently asked him to do this?

A. Yes, occasionally, when there was not time to go to headquarters. I have done that on all the roads, and I have been to all the roads and they have always done it.

Q. Would there be anything to show for that?

A No, sir.

Q. When your family had occasion to ride with Mr. Clough, did not you always buy tickets?

A. I calculated, as a general rule, to pay fare one way—to pay the fare up or down. My wife does not go to Boston more than once a year.

Q. How was it about your children?

A. They did not ride but very little.

Q. How was the fact?

A. It was as I say: I generally calculate to pay the fare one way. Sometimes I did not, but that is what I generally calculated.

Q. (*By Mr. George.*) In 1864 you took the quarter past ten train from Concord?

A. Yes, sir.

Q. When were you giving the number of cars?

A. That was on the three and a half train.

Q. Now, sir, during that time how many cars were there, ordinarily, that went down on the quarter past ten o'clock train—say in 1857–8–9 and 1860?

A. There was one north from here. I believe that started from here, but the passengers were all changed.

Q. How many cars ran down on that train?

A. I should think about seven or eight cars perhaps, on both roads. There were two baggage cars, and the Lawrence road at that time ran the most. I should think likely there might be five cars. As a general thing I suppose they would run about eight cars. In the height of travel, there might have been more than that.

Q. (*By Mr. Mugridge.*) Now, in the height of travel let us see if they did not run more than that. How many cars started here that went over the Manchester and Lawrence road in the height of travel? I am speaking of the ten o'clock train.

A. I should think some three cars, besides the baggage cars.

Q. Now how many cars left Concord on that train and went by Nashua and Lowell in the height of travel?

A. They might run perhaps as many as they did the other way.

Q. Now there is the Portsmouth road?

A. The Portsmouth road did not run at that time.

Q. But the Portsmouth road ran separately?

A. Not at that hour.

Q. If they did not go at that hour?

A. They went in the morning.

Q. There were two trains in the day?

A. Yes, sir; they went down at seven and at half-past three.

[Mr. Stanley objected to introducing testimony in regard to the Portsmouth train, inasmuch as the plaintiff's counsel had not been allowed to put evidence as to that road.]

*The Chairman.* It is difficult for me to see what the Portsmouth road has to do with it. The Portsmouth road was a separate affair, but passengers that went there went in connection with the other train.

Q. (*By Mr. Mugridge.*) Now let me ask you if five cars, many times in the height of travel, have not left Concord to go to Boston by either road?

A. I should think there had been days that there had.

Q. Let me ask you how many baggage cars would be required ordinarily to accommodate five full passenger cars?

A. Two, usually, but they used to occasionally put on an extra baggage car.

Q. Now, then, I want to ask you if five cars have not started from Concord to go by the way of Lawrence?

A. I should think there had.

Q. Let me ask you if five cars, in the height of travel, have not left Concord to go by the Nashua and Lowell road?

A. Oh, yes, sir.

Q. Then, to accommodate these passengers, of course there would be required sometimes two and sometimes three baggage cars, and while there were two, that would make a train where there would be fourteen cars?

A. There would be some days.

Q. You said that oftentimes in the height of travel fourteen cars would leave on the three and a half train. I want to know if you have known a time when eighteen cars have left here in a period before the war?

A. Well, I do not know that I could say.

Q. Now let us see: in the height of travel, Mr. White, how many cars have you ever known to leave Concord—in the height of travel, I mean—to leave Concord for Portsmouth?

A. Without some special occasion, two cars is all they ran.

Q. In the height of travel?

A. Yes, I understand.

Q. When they were going to the beaches, let me ask you if the Portsmouth road has not run three and sometimes four cars?

A. They have sometimes had six cars.

Q. I do not mean excursions. But take the ordinary amount of travel going toward the beach. Has not the time been frequently when the Portsmouth road has run three or four cars?

A. I should think not.

Q. Have they not run three?

A. I should think two would be the number as a general thing.

Q. Do you mean to say that they have not run three?

A. No, I do not say that, but I should say that two was the general thing.

Q. Now there would be three. How many have you known brought over the Northern line which went through to Boston?

A. Well,—I could not say I am sure about that—as a general thing three cars.

Q. Have not you frequently known four or five cars come down over the Northern road?

A. If they did they were generally shifted off.

Q. How many have you known to come down on the Montreal road?

A. Well, sir, they used to run three cars generally, besides baggage cars.

Q. How many baggage cars?

A. There would be two baggage cars.

Q. Now how many cars have the Concord road put on to accommodate the travel that took passage at Concord, including the travel from the stages at Pittsfield and Alton, and the other stages that come here?

A. Well, they would not hit the ten o'clock train.

Q. How many cars have you known the Concord road to put on, independent of the cars that came down over the other roads?

A. I could not state; I know they put on some.

Q. Have they not put on four or five?

A. It may be so.

Q. Well, we will call it four. How many baggage cars?

A. They would put on one baggage car.

Q. Would they not put on two?

A. No; not as a general thing more than one.

Q. That makes sixteen cars. Now have you not known of such a case frequently?

A. There may have been; I did not happen to notice it.

Q. What would be a fair statement of the number of cars run before the war?

A. I should think I had made a fair statement of the running of the cars before the war.

Q. Do not you think the statement you have given me a fair statement of the travel, in the height of travel, before the war?

A. I should think there were extra days when more folks got in.

Q. (*By Mr. George.*) When northern trains came down and there were more than could be accommodated——

A. They put on other cars.

Q. (*By the Chairman.*) There is one affair that perhaps this witness knows: whether it was the custom of the trains that came from the upper roads, each to take its own baggage car along?

A. Yes, sir.

Q. How would it be with the Portsmouth train, if it started from here at half past three?

A. They would furnish their own baggage car.

Q. Then the Concord road put on some cars?

A. They always run a baggage car with that train, to Portsmouth.

Q. How was it with the Concord road, did they put on one?

A. Not as a general thing—they didn't.

Q. (*By Mr. George.*) But used the northern baggage cars?

A. Yes, sir.

Q. (*By Mr. Mugridge.*) Whether it was occasionally so, that they had to put on their own baggage car?

A. I stated that they did, I believe.

Mr. George then read from the deposition of Mr. George Clough, interrogatories 126 to 136, inclusive; also, interrogatories 185 and 186.

[Adjourned.]

[NINTH DAY. Tuesday, August 4th, 1868.]

## MR. G. G. SANBORN RE-CALLED.

*Mr. George.* I desire to show now, Mr. Gilmore's endeavors to squelch this investigation.

*Mr. Mugridge.* We do not, of course, desire to shun any investigation of this subject, if the referees think it is of any particular value, and that it has any legitimate tendency to show whether there is any money in the hands of Mr. Clough. If you think it proper we don't object at all.

*The Chairman.* I believe we have ruled in regard to that, that the testimony was inadmissible unless made so by evidence that showed collusion.

Mr. George read from the deposition some of the interrogatories for the purpose of showing the particular testimony in regard to collusion.

*Mr. George.* These suits were commenced on the 10th of February, 1866, on the property of this defendant and sundry of the other conductors. The annual meeting of the corporation was held on the Tuesday preceding the last Wednesday—the 29th day of May. Mr. Gilmore was, part of that time, the superintendent and executive officer of the road. I do not think that my brother can claim that the evidence that I have read from Mr. Clough's deposition does not show that immediately after the commencement of these suits, though these suits, as it appears, were commenced on the same day of the discovery of the tickets at Whitcher's,—I think my brother cannot claim that the evidence does not show that immediately Mr. Gilmore went into a consultation with the conductors who were sued for embezzling the funds of the corporation—went immediately into consultation, and combined with Mr. Clough and the other conductors of the road to turn out the old board of directors, who commenced these proceedings. Now then, there is the first thing. It shows that they went into a combination to purchase stock. Mr. Clough, I grant, says in the course of his deposition, in reply to a question as to whether there was any understanding: "I cannot remember." In reply to another question, he says: "I cannot remember in particular." But this board of referees, I think, cannot fail to see that this testimony has a legitimate tendency to show a combination. Let me ask, if these conductors went to work and bought this stock in 1867, bought a thousand shares, and pocketed the loss between 1867 and 1861,—gave bonds to Mr. Hill and Mr. Johnson, gave their notes for the loss—let me ask if that does not show that there was some object? Now, what was that object? It was to cut off these suits. Mr. Clough admits that they sustained from two to three hundred dollars loss. Now then, if I am correct, that immediately upon the commencement of these suits, Mr. Clough went into a combination with Mr. Gilmore, then I have a right to show what Mr. Gilmore, in pursuance of that combination, did; and to show that he did everything he could to prevent the investigation of this matter.

*Mr. Tappan.* I respectfully submit that there is not a particle of evidence that has been read from this deposition, or elsewhere, that goes to show that there was any combination or collusion made by Mr. Clough and anybody else for the purpose of suppressing these suits. And there never was a moment of time when Mr. Clough or Mr. Eaton —I speak of these particularly—would have consented to have had these suits go by without a full and fair investigation. The testimony does show that there was a purpose of turning out all the old officers. If that is the case, nobody has suffered anything by the change. We have no objection that the evidence should be gone into. All the evidence the statements of the counsel on the other side show is this: that at the outset, when this excitement first began to arise in regard to the management of the old board of directors and the officers of this road —when it first began to be bruited abroad of the peculations of the superintendent to some extent all along, Mr. Gilmore, it appears, ordered these suits to be instituted, for the purpose, as we say, of diverting public attention from the real parties to those who were innocent. And it appears that unless something was done to stop the proceedings in some way, it appears that Mr. Gilmore was willing to turn about and curry the favor of the conductors, and with the design of saving himself,—failing in the outset to turn public attention from himself to the conductors. And whatever combination or collusion they made was to turn out that old board and purge this thing. And it has been purged; and nobody has suffered anything by it. What I have said has been merely for the purpose of setting things right in regard to this charge of combination or collusion for the suppression of these suits. But if, after having said this much, the referees think that this at all pertains to the case, or will facilitate the investigation of the case, and show that there is one dollar, or one cent, in the hands of Mr. Clough, then we say let it go in. But if they do not, we shall claim the privilege.

*The Chairman.* If we understand the matter correctly, the proposition is to show now some unlawful acts and some fraudulent acts of Mr. Joseph A. Gilmore. And we understand that the inference intended to be made is, that because Mr. Gilmore was doing these unlawful acts, Mr. Gilmore and Mr. Clough were combining together to smother this suit; and to infer from that, that there was something for Mr. Clough's interest to get smothered. Now, we think, if I understand the matter right, that the evidence so far simply goes to show a combination on the part of Mr. Gilmore and Mr. Clough and others, by lawful means, by the means which they have, the means to prevent the legal right to pursue the investigation. Now it may possibly be that some inference may be drawn from that, from the fact that Mr. Clough was desirous of having the board of direction changed. But we think that the fact that Mr. Clough and Mr. Gilmore and others had formed a combination, in a legal way, to buy stock and effect changes in the board of direction doesn't tend to show that Mr. Clough formed that combination with Mr. Gilmore to do fraudulent acts for the purpose of smothering this suit. And we think, therefore, that, as it now stands, it does not tend to show that kind of collusion between Mr. Clough and Mr. Gilmore that would make the evidence admissible, or that would bind Mr. Clough by Mr. Gilmore's acts.

*Mr. George.* It is clear that Mr. Gilmore and Mr. Clough, at the time these suits were brought, thought they were all wrong, a great outrage, and ought never to have been commenced, etc. Now I propose to show that at that very time Mr. Gilmore was endeavoring to prevent any possible development with regard to the conductors, as I claim, by combination with Mr. Clough. And I say that the evidence tends to show it, and that Mr. Clough was part and parcel with this arrangement, with Mr. Gilmore to prevent any development of the facts, which were the basis of this suit.

*The Chairman.* The precise ruling of the referees, as I understand it, is that we do not find here any evidence tending to show a fraudulent collusion between Mr. Clough and Mr. Gilmore; any evidence in any way tending to show a fraudulent collusion as would justify us in implicating Mr Clough in Mr. Gilmore's frauds. That is the precise view that I think the referees entertain now about the matter. And I think we are all of one mind about it. This evidence, as I have already suggested, this fact that Mr. Clough was desirous to change this board of directors, and willing to spend money for the purpose of changing the board of directors (taken in connection with some other evidence), to operate upon the suits to get them adjusted and arranged—that they may have some tendency to show that here were some suits that it was for his interest to get out of the way. But that does not show that there was any fraudulent collusion. That is the precise point that we rule. We think there is no foundation now such as would make that evidence against Mr. Clough. That is our view of the thing as a matter of fact.

Q. (*By Mr. George.*) Mr. Sanborn, you have spoken here with regard to Mr. Gilmore's directing you to collect some $3400 or $3500 of the quartermaster of the United States, and that you now hold that money. Did you so say?

A. I do not recollect what I said, but I believe that is the fact.

Q. You collected the money?

A. Yes, sir.

Q. Whose receipt did you give?
A. I did not give any receipt.
Q. (*By Mr. Tappan.*) How much was it?
A. Thirty-four hundred or thirty-five hundred dollars; I do not recollect.
Q. (*By Mr. George.*) Won't you state the facts in regard to that?
A. I made out the orders for transportation the same as I would any other for the Government, and presented them to the quartermaster, by direction of Mr. Gilmore. He looked the orders over and saw the amount of them and handed me the money, and I passed the orders out and he said, "No matter about those—we don't care for them," and I took the orders and returned them to the office.
Q. Returned the money to Mr. Gilmore?
A. Returned the money to Mr. Gilmore; I kept the orders myself.
Q. So far as you know, who first brought that matter to public notice? By whom and when was that matter first made public? [Objected to and withdrawn,]
Q. With regard to the $3400 for political tickets—those tickets to whom were they delivered and sold, and with whom was the settlement made?
A. The tickets, I believe, were delivered to Mr. Chandler, and the money was presented——
Q. Did Mr. Gilmore ever account for this $3400?
A. No, sir.
*Mr. George.* For the purposes of this trial, I will agree that it is a straight theft.
Q. (*By Mr. George.*) Who developed that fraud, sir?
A. I believe it came from Mr. Kimball—John Kimball.
Q. Who took Mr. Kimball's deposition?
A. Well, sir, I don't know; I don't know anything about those depositions at all.
Q. Do you know whether Mr. Kimball's deposition was taken?
A. I don't know at all.
Q. You spoke, sir, about the tickets which were taken by Mr. Gilmore. [Objected to,]
Q. Now, Mr. Sanborn, won't you state when, by whom, and how, as to the development of this $3500, so far as you know?
A. I don't recollect as to the date, but I think it was February, 1866, that yourself came into the office to inquire if I had paid Mr. Gilmore any money, and that I was not aware at that time that it was known beyond one individual that I had consulted in regard to some of my matters, as he was a bondsman of mine and I had told him of some of these facts.
Q. (*By Mr. Tappan.*) What did he ask you?
A. If I had paid Mr. Gilmore any money at different times, or at any time, to the amount, or very near the amount which I have given—$3400. I told him that I had paid him that amount, and gave him the exact figures.
Q. (*By Mr. George.*) What direction was then given you, if any, to examine the books and see if that was accounted for?
A. I then was desired to examine the books and see if any such amount was entered upon the cashier's books. I did so and made inquiries of the cashier and found that no such item had been entered.
Q. Who was your bondsman, the only man to whom you had mentioned this?

A. William Parker.

Q. Then superintendent of the Northern road?

A. Yes, sir.

Q. And inquiries were made of you by whom in regard to these political tickets?

A. I don't recollect the particular circumstances about that; but I think it was a short time after that that the matter of those political tickets was brought up; but I don't recollect distinctly the conversation which was had at that time, only the general facts.

Q. By whom was the inquiry made?

A. That I don't recollect. I think it was yourself; but I don't recollect the circumstances, the place nor the time.

Q. At the time the investigation was going on in regard to these suits, what instructions did Mr. Gilmore give you?

[Objected to and withdrawn.]

Q. (*By the Chairman.*) I want to inquire of you whether or not the presenting of these bills to the quartermaster was any part of your official duty?

A. No, sir.

Q. In fact, was there any person whose duty it was to take these bills and present that money?

A. Yes, sir; it was the duty of the cashier, I suppose.

Q. (*By Mr. George.*) Did Mr. Gilmore give you instructions upon the subject in regard to information to me?

A. Yes, sir.

Q. Now you may state what those instructions were?

A. He told me to give no one information in regard to the affairs of the road in any shape,—the conductors or any one else—without his permission; and particularly not to give you any information at all unless it was in writing, and that writing to be presented to himself, and then the answer to be made in writing and go through his own hands.

Q. Had you, prior to this time, been instructed in regard to affording me facilities for investigation?

A. Yes, sir.

Q. By whom?

A. Mr. Gilmore and Judge Upham, both.

Q. And how long after this direction of Mr. Gilmore and Judge Upham before it was countermanded in the manner stated?

A. I am not able to state in that respect, but it was not a great while after the investigation commenced before I was required to inform him of any inquiries which might be made of me.

Q. What did Mr. Gilmore say in regard to the investigation at that time—with regard to its being permitted to go on?

A. When he gave me the first instructions he said very little about it. I mean when he first instructed me to inform him of such inquiries. When he first gave me instructions there was nothing said in regard to the propriety of it, or anything else; no remarks made, or anything about it. But soon after it commeneed he required me to inform him of any questions that might be asked. A short time after that—I should think after the investigation had been going on three weeks—he instructed me not to give information except an order came in writing, and the answer was made in writing.

Q. What did he say then, if anything?

A. I don't know as he said anything about the investigation then going on; but be said this much: that it was evidently the design of Colonel George to rip the whole of us out of office, and get us all turned out. I can't word it just as he did, but he used that expression: "rip the whole thing out."

Q. I want to know if he discharged you?

A. I understood that he did, and I went to him to ascertain the fact, and he said that he did intend to, but was going to talk with me about it first.

Q. On what ground was he going to discharge you? Did he tell you?

A. He told me that I had given information that I was not authorized to do, and mentioned these facts in regard to the money.

Q. At what time was this that you heard he had turned you off, and you went down to see?

A. I could not tell the date, but it was the last day of the month. I could not tell whether it was February or March.

Q. It was after the investigation commenced?

A. Yes, sir; I should think it was as late as March. I saw him at his house.

Q. Whether or not you informed me of this—what the result was so far as my connection was concerned, and so far as you were concerned?

A. When you came to me for information, I told you I could not give it to you without you presented an order in writing, and requested you to do so, or else I could not give the information.

Q. I will ask you how it was with regard to my sending, as soon as this information was communicated to me about the $3400 worth of tickets—how it was with regard to sending for Judge Upham and Mr. Peaslee, and having an interview with them?

A. I recollect your sending for me once when Mr. Peaslee was present. My impression is that Judge Upham was not present; but I was sent for and desired to state in regard to this $3400; and I told you I could not make a statement because I was required not to, and did not, but I believe you stated to Gen. Peaslee yourself the facts, and I was present and assented to it as far as ——

Q. Did not deny it?

A. I did not deny it.

Q. Do you recollect whether Judge Upham was sent for to come in?

A. I do not recollect it; it has gone from my mind.

Q. Mr. Sanborn, since the adjournment of the court on Friday, won't you state whether you have gone over carefully with the tables of the number of passengers carried over the Concord Railroad, including the soldiers carried in squads and represented in any way by tickets; including all persons returned by the conductors as having paid in the cars, and all persons having tickets either from the upper road or from the Concord road?

A. Yes, sir.

Q. Now I want you to state what is the total, what it includes and what is not included.

A. It includes everything, military and civil, except regiments as they went away.

Q. Full regiments?

A. Full regiments went away in special trains. The special trains only came into Concord, as for instance, at an inauguration. Where they ran large trains, those trains are not in; but everything that came in on the regular trains, including all these small fares. Only it possibly may be some passengers on the Portsmouth road—it has occurred to me since coming that some of these might have been left out.

Q. Everybody carried on the regular trains without exception?

A. Yes, sir; except this exception that I have spoken of.

Q. I want to see whether it includes soldiers' transportation by the regular trains on the road?

A. Yes, sir; it does.

Q. Won't you state how carefully you have been over it?

A. I have been over with it entirely from 1859 to 1866.

Q. In the other table which was submitted to the referees and copied by them, what were the numbers, trains or trips?

A. Those were trains. But perhaps it is due to say that these tables were not designed to show every passenger, but to show the per cent. between one time and another. This has been got up with a view of including everything that passed over the regular trains.

Q. And the other was trips, while these are trains?

A. Yes, sir.

Now I desire to put this in.

| MONTH. | 1859–60 | | | | 1860–61 | | | | 1861–62 | | | | 1862–63 | | | |
|---|---|---|---|---|---|---|---|---|---|---|---|---|---|---|---|---|
| | Whole No. of Passengers. | No. of Trains. | Average No. of Passengers per train. | No. of Cars, at 50 per Car. | Whole No. of Passengers. | No. of Trains. | Average No. of Passengers per train. | No. of Cars, at 50 per Car. | Whole No. of Passengers. | No. of Trains. | Average No. of Passengers per train. | No. of Cars, at 50 per Car. | Whole No. of Passengers. | No. of Trains. | Average No. of Passengers per train. | No of Cars, at 50 per Car. |
| APRIL | 11.983 | 156 | 77 | 2 | 12.946 | 156 | 83 | 2 | 11.195 | 156 | 71 | 2 | 10.870 | 156 | 70 | 2 |
| MAY | 11.418 | 156 | 73 | 2 | 12.103 | 156 | 77 | 2 | 10.211 | 156 | 66 | 2 | 10.065 | 156 | 65 | 2 |
| JUNE | 10.163 | 156 | 65 | 2 | 10.921 | 156 | 70 | 2 | 10.691 | 156 | 69 | 2 | 12.421 | 156 | 80 | 2 |
| JULY | 14.992 | 156 | 96 | 2 | 16.180 | 156 | 104 | 2 | 12.547 | 156 | 81 | 2 | 14.184 | 156 | 91 | 2 |
| AUGUST | 19.887 | 156 | 128 | 3 | 20.377 | 156 | 130 | 3 | 15.074 | 156 | 97 | 2 | 18.816 | 156 | 121 | 3 |
| SEPTEMBER | 18.453 | 156 | 119 | 3 | 12.138 | 156 | 77 | 2 | 13.423 | 156 | 86 | 2 | 14.593 | 156 | 94 | 2 |
| OCTOBER | 14.834 | 156 | 96 | 2 | 17.484 | 156 | 112 | 3 | 13.744 | 156 | 88 | 2 | 14.199 | 156 | 91 | 2 |
| NOVEMBER | 12.575 | 156 | 81 | 2 | 13.741 | 156 | 88 | 2 | 9.427 | 156 | 61 | 2 | 11.402 | 156 | 73 | 2 |
| DECEMBER | 8.832 | 156 | 56 | 2 | 10.953 | 156 | 70 | 2 | 8.298 | 156 | 56 | 2 | 9.875 | 156 | 64 | 2 |
| JANUARY | 7.497 | 156 | 48 | 1 | 8.320 | 156 | 56 | 2 | 7.069 | 156 | 46 | 1 | 9.701 | 156 | 63 | 2 |
| FEBRUARY | 8.765 | 150 | 59 | 2 | 9.839 | 150 | 66 | 2 | 7.114 | 150 | 47 | 1 | 9.985 | 150 | 67 | 2 |
| MARCH | 11.783 | 156 | 76 | 2 | 12.374 | 156 | 79 | 2 | 9.042 | 156 | 58 | 2 | 12.211 | 156 | 78 | 2 |

| MONTH. | 1863–64 | | | | 1864–65 | | | | 1865–66 | | | |
|---|---|---|---|---|---|---|---|---|---|---|---|---|
| | Whole No. of Passengers. | No. of Trains. | Average No. of Passengers per train. | No. of Cars, at 50 per Car. | Whole No. of Passengers. | No. of Trains. | Average No. of Passengers per train. | No. of Cars, at 50 per Car. | Whole No. of Passengers. | No. of Trains. | Average No. of Passengers per train. | No. of Cars, at 50 per Car. |
| APRIL | 13.205 | 156 | 85 | 2 | 18.797 | 208 | 90 | 2 | 17.987 | 260 | 69 | 2 |
| MAY | 12.965 | 156 | 83 | 2 | 19.876 | 208 | 102 | 2 | 16.818 | 260 | 61 | 2 |
| JUNE | 15.821 | 156 | 102 | 2 | 24.091 | 208 | 116 | 3 | 20.381 | 260 | 79 | 2 |
| JULY | 19.101 | 156 | 123 | 3 | 27.931 | 208 | 134 | 3 | 26.907 | 260 | 104 | 2 |
| AUGUST | 20.618 | 208 | 99 | 2 | 33.519 | 208 | 161 | 4 | 28 510 | 260 | 109 | 3 |
| SEPTEMBER | 19.227 | 208 | 93 | 2 | 27.708 | 208 | 133 | 3 | 31.563 | 260 | 122 | 3 |
| OCTOBER | 21.970 | 208 | 106 | 2 | 18.815 | 208 | 90 | 2 | 26.377 | 260 | 102 | 2 |
| NOVEMBER | 18.744 | 208 | 90 | 2 | 16 977 | 208 | 82 | 2 | 20.033 | 260 | 77 | 2 |
| DECEMBER | 16.321 | 208 | 79 | 2 | 14.512 | 208 | 70 | 2 | 18.819 | 260 | 68 | 2 |
| JANUARY | 17.801 | 208 | 86 | 2 | 15.392 | 208 | 74 | 2 | 19.991 | 260 | 77 | 2 |
| FEBRUARY | 16.276 | 200 | 82 | 2 | 14.305 | 250 | 57 | 2 | 17.827 | 250 | 72 | 2 |
| MARCH | 17.415 | 208 | 84 | 2 | 18.218 | 260 | 70 | 2 | 20.546 | 260 | 79 | 2 |

Q. (*By Mr. George.*) Mr. Sanborn, I want you to state to the referees how many seats there are in the ordinary passenger cars of the Concord Railroad; how many they will carry? You have reckoned these at fifty. What is the ordinary number of seats on a side?

A. They have made some cars that were sixty, and I don't know but there are some made so now. As they were originally counted they were fifty to the car, but the cars they make are larger now.

Q. Will you be kind enough to go and look at the latest cars that run on the Concord road?

A. Yes, sir.

Q. Now, sir, there has been some testimony in regard to Mr, Upham's running trains over the Concord road. Have you way-bills or statistics of way-bills of such trains run by Mr. Upham, so that you can show the returns that he made? [Objection to.]

*Mr George.* It appeared that Mr. Upham ran trains. On the cross-examination Mr. Upham was asked whether substitutes ever paid him in the cars. Now I propose to show just what Mr. Upham did return.

*Mr. Mugridge.* The only objection to that would be that the return was made in such a way that Mr. Upham running both trains, the trains of Mr. Noyes and Mr. Clough, makes his return in such a way that the returns of one train cannot be distinguished from the returns of another train. Now, then, we suppose it is improper here for him to show returns of other conductors and attempt to judge us by those returns.

*Mr. George.* I propose to introduce simply the trains upon which Mr. Upham testified, on cross-examination, as to receiving money from substitutes and the details of the way that trains were run. [Handing a paper to witness.] [Objected to.]

*The Chairman.* It seems to us that without touching this question of per cent. at all, that in this particular instance the inquiries that have been made in cross-examination on one side entitle the other party to complete the examination, as it were, by showing the actual returns that were made by that conductor. We do not, by that, mean to say that the conclusions can be drawn from that for per cent., only it seems to us to be the ordinary course of practice to allow one party to complete the cross-examination commenced by the other.

*Mr. Mugridge.* Will the referees be kind enough to indicate the part of the cross-examination?

*The Chairman,* We think the witness was examined in reference to the number of cars, the number in the cars, and also as to the returns which he made, and the number of free passes which he returned, or endeavored to return. And this he went into, in point of fact, pretty fully. And as to the witness' recollections of what he states in these tables. [Defendant's counsel excepted to the ruling of the referees.]

*Mr. Mugridge.* We make the farther exception that this paper cannot be introduced by this witness.

Mr. George stated that he proposed to show the amount of money returned by Mr. Upham, as returned upon the train upon which he has testified in the cross-examination by the other side.

*Mr. Mugridge.* We object to any compilations that have been made from any way-bills that they assume to have been used. We desire them to prove them.

Q. (*By Mr. George.*) Have you the original way-bills of '65?

A. I think I have.

*The Chairman.* Now, if I understand right, the objection is that the cross-examination did not open this inquiry?

*Mr. Mugridge.* Yes, sir.

*The Chairman.* And secondly, that there are no such way-bills from which he can prove?

*Mr. Mugridge.* Yes, sir.

*Mr. Tappan.* We object that it is a question of per cent.

*The Chairman.* The purpose for which the testimony is offered, perhaps, does not effect the question of its accuracy. The point that seems addressed to the referees is, whatever may be the effect of evidence tending to show the facts being put in, whether the other side have not the right to go forward and extend the inquiries in the same direction, to limit it, or test it, or put in evidence which tends to show how far it is completed. That is the only view which the referees have taken, and that without reference to what particular train it was on. The referees cannot, of course, anticipate all the possible uses that can be made of testimony. We can only consider whether for any use it is competent.

[Testimony ruled in on the ground of its being an extension of the inquiry as to the same matters that were opened on the cross-examination. All the documents shown objected to until proved.]

Q. *By Mr. Tappan.*) Mr. Upham returned these papers, did he?

A. Yes, sir; that is, I don't recollect of Mr. Upham's coming there at that particular time—coming there and leaving way-bills.

Q. (*By Mr. George.*) Do you know Mr. Upham's hand-writing?

A. I don't think I could swear to it, but I should think that was his hand-writing.

Q. How was it about Mr Upham's returning you way-bills when he returned on the trains?

A. Yes, sir.

Q. (*By Mr. Mugridge.*) Do you recollect of these particular way-bills being handed in to you?

A. No, I do not.

Q. Do you know that that is his hand-writing?

A. I should say it was, but could not swear.

Q. Did you ever see him write? Will you state here, unqualifiedly, that he wrote those way-bills?

A. No, sir.

Q. (*By Mr. George.*) You have seen him write?

A. Yes, sir.

Q. Have you any doubt that that is his handwriting?

A. No, sir; I have none at all.

*The Chairman.* Is there any other objection to these way-bills except that they are not sufficiently identified?

Q. (*By Mr. Mugridge.*) Is there a word or a figure that you will state is Mr. Upham's handwriting?

A. I should say that that is his signature, I could not swear to anybody.

Q. Will you undertake to say that you have any knowledge at all as to its being Mr. Upham's way-bill, or of either one of them being his?

A. I have no doubt it is his, but as I have said before I wouldn't swear.

Q. How came they in your office?

A. I don't recollect of his rendering them at that particular time. I don't recollect the date that he ran.

Q. And you cannot swear that a single figure contained in those way-bills was made by him?

A. No, sir.
Q. Now, sir, take the word October?
A. That is written in my own hand, I think.
Q. Now take October 24th; in whose handwriting is that?
A. I should say that it is his.
Q. Would you swear that it is?
A. No, sir, I would not.
Q. Now, is your knowledge derived from the fact that Mr. Upham made returns to your office, and that you saw these there and concluded that they were his?
A. I have a memorandum of every conductor; at that time I entered against every one E. W. U., showing that he ran the train.
Q. You do not know that he returned them to the office?
A. Only from that indication.
Q. And you would not swear to a mark or scratch on them as his?
A. No, sir. [Objection renewed.]
Q. (*By Mr. Haile.*) When did you write the word October on that?
A. On the day that it was brought in. Sometimes conductors neglected to put on the month, and when they did I put it on myself.

*The Chairman.* I suppose there is evidence here competent to be considered, tending to show that these are the way-bills that were returned by Mr. Upham. If that is all the objection that is made, that they are not sufficiently identified as Mr. Upham's way-bills ——

Q. (*By Mr. Mugridge.*) Won't you look on this way-bill and see what the words George Noyes mean?
A. It means that these were printed for his particular use.
Q. Does it indicate that these were used by Mr. Upham?
A. Mr. Upham's indicate that.
Q. Then it indicates that the returns were all made on Noyes's train?
A. Yes, sir.

*Mr. Mugridge.* Now then we object that if evidence at all it is evidence only as to what was returned on Noyes's train, and that not a single one of Clough's trains were being run by him at that time. The question as it arose before the referees covered the way-bills returned on both trains—the trains run by Noyes or Clough—and I understood the referees to rule that to be incompetent. They now offer way-bills that indicate on their face that they are way-bills of Mr. Upham, who ran Mr. Noyes's train alone, and do not relate to a train which was ever run by Mr. Clough at all.

*Mr. George.* Here are the other way-bills about which inquiries were made, and I propose to have all these returns put in. The evidence already before this tribunal is that one conductor ran two trains on one day and another the next.

*The Chairman.* I understand that it is put a little different from that. I understand that they propose to offer all the way-bills.

*Mr. George.* All that the road has in possession.

*Mr. Mugridge.* We now object to these bills because they are not returns on Mr. Clough's trains.

*The Chairman.* It does not appear to us, from the view we have taken in the case, that this makes any difference, because we understand that the evidence tends to show that these are some of the way-bills that Mr. Upham testified about, the purpose for which we have deemed this admissible being simply to further the examination in regard to the same matters cross-examined upon, and it appearing to the referees that these way-bills do relate to this, and that they are admissible simply on that ground. [Defendant's counsel objects.]

Mr. George then read the first three way-bills, from Oct. 24, '1865.

October 24, 1865.—First train, by E. W. Upham. Amount returned as taken in the cars: Three from Concord to Suncook at 45 cents each; three from Concord to Manchester at 85 cents; carried out, $3.90. One from Suncook to Manchester, 65 cents; carried out, 65 cents; making the whole of that way bill $4.55. The next is dated October 24. Third train, by E. W. Upham. From Concord to Suncook, 45 cents; three from Concord to Manchester, 85 cents; one from Concord to Nashua, $1.60; one from Concord to Boston, $2.85. These are all carried out $8.00. Then there is one from Suncook to Hooksett, 25 cents; three from Suncook to Manchester, 65 cents; two from Suncook to Boston, $2.75. These are all carried out $7.70. Then there were two from Manchester to Nashua, 85 cents—$1.70; making the entire way-bill $17.40. The next, October 24, second train—Mr. Sanborn, how should that be?

A. That is a mistake in making out the date.

Q. What should it be?

A. It should be the 25th.

Then the second train by E. W. Upham, October 25th: Three from Concord to Manchester, 85 cents; one from Concord to Nashua, 60 cents—$5.15. Two from Hooksett to Nashua, $1.35—$2.70; one from Manchester to Nashua, 85 cents; making the entire way-bill $8.70.

The next is October 26—the 10.15 train—by E. W. Upham: One from Concord to Suncook, 45 cents; three from Concord to Martins', 70 cents—$2.10; two from Concord to Manchester, 85 cents—$1.70; three from Concord to Nashua, $1.60—$4.80; one from Concord to Boston, $2.85; carried out $11.90. One from Suncook to Nashua, $1.40; one from Manchester to Reed's, 50 cents; five from Manchester to Nashua, 85 cents—$4.25. Whole amount of way bill, $18.05.

The next is first train, October 27. Two from Concord to Manchester, 85 cents—$1.70; one from Concord to Nashua, $1.60; one from Concord to Boston, -2.85; carried out $6.15. One from Suncook to Manchester, 65 cents; one from Manchester to Nashua, 85 cents. Making the entire way-bill, $7.65.

Q. (*By Mr. George.*) Mr. Sanborn, will you please state whether or not there are not other way-bills of Mr. Upham's in your possession?

A. Not that I have a memorandum of.

Q. (*By the Chairman.*) Mr. Sanborn, these are marked trains, but I see that the heading of these tables is "local, down and up."

A. That is what we generally term a trip.

Q. Then each of these papers indicates a trip?

A. Yes, sir.

Q. (*By Mr. Haile.*) The conductor then uses it down and back?

A. Yes, sir.

Q. (*By the Chairman.*) That is understood as being what technically you call a trip?

A. Yes, sir.

Q. Well, now, the first of these bills, Mr. Sanborn—the first one that was offered here—was October 24th, the first train, and October 24th the third train. Would those be the alternate trains?

A. Yes, sir.

Q. For instance, if the first train there was Mr. Noyes's train in the regular course of business, what would the third train be?

A. It would be Mr. Noyes's.

Q. Then on the 25th, which would be the second train the next day up?

A. Mr. Noyes's.

Q. Then October 26th, the next day, whose was the second train?

A. Mr. Clough's.

Q. Is the 10:15 train the second train?

A. Yes, sir.

Q. And whose train would that be?

A. That should be Mr. Clough's.

Q. I see that Mr. Clough's name has been erased there. Does that indicate anything?

A. That is supposed to indicate that Mr. Upham did that when he used the way-bill.

Q. Whose would the first train down be on the 27th, if the second train was Mr. Clough's on the 26th?

A. It should be Mr. Clough's.

Q. I understood you, or somebody, to say that in the regular course of business the conductors alternated in that way?

A. Yes, sir; that is the regular course.

Q. I was looking to see what the indications would be here—for instance, on the first that was put in I see October 24th; there was the first train which was run by him. Now if the first train was his the third would be?

A. Yes, sir.

Q. And the second on the 25th would be Noyes'?

A. Yes, sir.

Q. Are you able to tell, from the course of business, whose those were?

A. I cannot say whether Mr. Clough changed off with Mr. Noyes and ran the second train, as they sometimes do.

Q. I wish, then, to inquire how these way-bills would be in the regular course of business?

A. I go by the lists as to the ordinary time they went. My recollection of these way-bills I should not go by, except as they agreed with my memorandum.

Q. (*By the Chairman.*) I understand that conductors do change?

A. Yes, sir.

Q. Would your list enable you to determine?

A. It would enable me to determine more accurately.

Q. (*By Mr. George.*) Mr. Sanborn, I want to ask you how these way-bills are printed. For instance, the fare printed in here is 85—what was the ticket-office fare?

A. Seventy-five.

Q. After the rule was adopted requiring an extra charge of ten cents, where tickets were purchased in the cars, how were the way-bills printed in regard to that additional ten cents?

A. They were always printed in.

Q. What was the fare from Concord to Suncook?

A. Twenty-five cents.

Q. How was it about their being printed in on the way-bills at that time?

A. That was always the case from the time that the rule was adopted.

Q. I will ask you how it was in regard to Mr. Clough? Whenever Mr. Clough returned money for tickets purchased in the cars what return did he make, and was it variable or otherwise?

A. He made a return by the rates which are printed there, generally.

Q. You mean including ten cents paid in the cars?

A. Yes, sir.

Q. You were asked, when you were on the stand before, when the system of ticketing to the Worcester road and beyond—when that system was adopted, and when it was renewed. I want you to state, if you can, accurately, when that ticket system to New York commenced and when it ceased, and when the interregnum was, if any?

A. It commenced in October, 1857.

Q. (*By Mr. Stanley.*) In 1857 you ticketed from where?

A. From Concord to Manchester.

Q. (*By Mr. George.*) To what points?

A. Groton Junction, Keene, Bellows Falls, Clinton, Worcester, Springfield and Albany, New York (by boat), New York (by rail), and Providence.

Q. (*By Judge Bellows.*) What time did you say that was?

A. October, 1857.

Q. (*By Mr. Stanley.*) Providence, via Worcester?

A. Yes, sir. The sales from Manchester to Worcester were discontinued in November, 1862—resumed June, 1864. The sales to New York, by railroad, were discontinued December, 1863—resumed December, 1867. Sales to New York, by boat, discontinued March, 1864—resumed June, 1864. Sales to Groton Junction discontinued January, 1862; to Springfield, October, 1862; to Albany, October, 1862; Clinton, May, 1863; Providence, July, 1860.

Q. (*By Judge Bellows.*) Not resumed?

A No, sir.

Q. (*By Mr. George.*) Mr. Sanborn, you were asked the other day to give the number of passengers and the number of cars required to carry the passengers up to and including March, 1866. I want you to give the amount returned, as money taken in the cars, on Mr. Clough's train, in January, February and March, the same three months.

*Mr. George.* Now I propose to offer to show the amounts returned by Mr. Clough during the three months of which you have put in statistics.

The proposal of Mr. George was objected to, and discussed, pending which the hearing was adjourned.

[TENTH DAY. Wednesday, August 5th, 1868.]

## TESTIMONY OF MR. SANBORN (*Continued*).

Q. (*By Mr. George.*) You stated, yesterday, the number of passengers carried. Will you state clearly what that embraces? What does the paper that I put in yesterday—what does that embrace? [Objected to.]

*Mr. Mugridge.* It seems to me that this evidence is admitted upon premises that are not exactly correct, so far as the evidence is concerned. The referees seemed to proceed upon the admission that we have made inquiries as to all the passengers that passed over this road. Now we have made no such inquiry at all. We have not asked for anything of that kind; but the referees seem to have assumed it.

*The Chairman.* I thought I had stated just as distinctly as a man could, that this was to be confined to just the matter which you had inquired about.

At the request of Mr. Tappan, the ruling of the referees was put in writing, as follows:

"The defendants having introduced in cross-examination, evidence of the number of cars required to carry passengers on the Concord road, as shown by the conductors returns of tickets and fares paid in the cars, and not including free passes or soldiers who did not have tickets, the ruling is that the plaintiff may inquire about the particulars and data from which those results were derived."

To this ruling the counsel for the defendant excepted.

*The Chairman.* Now here is a paper which was used yesterday, and which after being used was placed in the hands of the referees. I believe that the witness testifying from this paper which he had drawn up himself, gave certain evidence; this being used as a sort of memorandum from which he testified. And if Judge Bellows will be good enough to look and see what he said when he testified here in regard to it—what he said it was.

*Judge Bellows* [reading]. "Includes everything save military, in the regular trains, and in special trains. It includes soldiers' transportation in regular trains."

That was the way in which this was put in; and my impression is that it was not objected to. Now if the witness has any explanation to offer; that is, if he stated that any different from what it was, I suppose he has a right to correct it.

Q. (*By Mr. George,*) What result did you mean to give yesterday as contained in that paper?

A. I meant to give that it contained everything which is upon the ticket book of the Concord road, and I suppose—the reason that I did not say anything about free passes is that the free passes are not on my books. I hadn't thought of it before.

Q. (*By Mr. Tappan.*) Does it embrace small squads of soldiers?

A. Yes, sir; such as went off on the regular trains.

Q. (*By Mr. George.*) Supposing a squad went on a regular train, without a government order and without tickets, does it embrace those?

A. Yes, sir; those were put in; that is according to the memorandum on my book, which I made at the time.

Q. (*By Mr. Mugridge.*) What is the difference between that paper and that one there?

A. That was got up for the purpose of showing the per cent. between the years. It was made two years ago and I do not recollect what it embraced, but my impression is that it was to embrace everything that went over the Concord road from the lower roads. The business done here don't embrace a good deal of the business that is done by the Northern roads.

Q. Now the question that I asked did not cover any of the travel over the upper roads?

A. I think it did not.

Q. You know what paper you had before you when you were inquired of. Do you know what it contains?

A. No, sir. I said it was made two years ago, and I do not know except according to my recollection.

Q. You did not know what you were saying as to the number of cars used here?

A. Well, sir, I supposed I knew.

Q. I ask you, sir, if you are able to state what that paper contains?

A. The testimony from my books. I could figure it up.

Q. Do you know?

A. No.

Q. In reply to my question—did you know what basis you were figuring upon?

A. I supposed that I knew. I have found since then that that basis was not what I supposed it was.

Q. Have you not found out since that the figures that you gave were not upon the basis that I assumed?

A. I think it is so.

Q. Now I want to ask you, sir, if you gave me a reply to the question that I put to you when you gave off these figures to me?

A. Well, just as I said before, I supposed that the figures that I was giving you were correct, but I found that they did not contain what I supposed they did.

Q. Did you give in answer to the question I put to you any accurate, truthful account?

A. If that word truthful means ——

Q. I do not mean to imply any falsehood on your part. I ask you, did you in fact give me any correct account as to the question I asked you?

A. I think they were not, now.

Mr. George objected that the further continuation of the examination was a little irregular.

Mr. Mugridge stated that he desired to show exactly what was before the referees upon which the counsel asked to extend this inquiry.

*The Chairman.* Well, the witness, as I understand when he was testifying was testifying for the purpose of correcting a misapprehension which he thought might have been derived from what he said yesterday. He wanted to have the court understand exactly what was embraced in his testimony yesterday. And he has stated that exactly, so that we now know; and it appears that there was a misapprehension about it. Then you desired to make some inquiries as to the other paper. The counsel assented; and you went on with your inquiries, and, as I understood the matter, were really making inquiries into the very matter that you objected to having an inquiry made into; that is into the figures to which you objected.

[This question was the subject of some further discussion between counsel, after which a question was proposed to the referees by Mr. Mugridge.]

*Mr. Mugridge.* Supposing I ask the witness a question as to an answer that he made yesterday, and he says that the table is made up of returns from the Concord and Montreal road; that the passengers returned by them enter into that computation. Now if I push the inquiry in that way, do the referees rule that it permits Mr. George to put in the number of passengers that passed over that road.

*The Chairman.* What the referees have ruled is this: that the plaintiff may inquire as to the particulars and data from which these results are derived. Mr. Sanborn having stated that at a certain time, for instance, it would take two cars to carry the passengers, we rule that the other side may go on and ascertain the particulars and data from which he reached these results. If that should lead to the showing of the amount of money that was paid in, in a legitimate way, then that will appear.

*Mr. Mugridge.* That is just the point that I wanted to understand. I do not propose to ask any more questions at present.

*Mr. Tappan.* I was going to inquire, in a word, about the question of fare paid to the conductors on the Concord and Portsmouth road, if it should lead to the amount paid to the conductors on the Concord and Montreal road. I do not know how far this goes.

*The Chairman.* We suppose it does follow from this ruling that the plaintiff may inquire into the number of passengers and the points between which they have traveled. And I believe the evidence of the price is already before the court. It is mere calculation after that, to show how much they paid.

Q. (*By Mr. George.*) I want to have you state further how it is about that paper's embracing every passenger, or otherwise.

*The Chairman.* I think, Colonel George, you should confine your inquiry to what he has testified.

Q. I want to know how it is about that paper's embracing the passengers.

A. After the question was brought up in regard to the Concord and Portsmouth road, it occurred to me that for that road we had an account separate; so far as books were concerned, and the other road at the time was changed over; that it would still be the Concord and Portsmouth road, and there was still a separate book. In taking off these figures I used that book, and everything that came into the Concord road, as is commonly done. That is the impression that I had, and I looked over the books this morning to see.

Q. Suppose a man bought a ticket from Concord to Portsmouth. Does that appear?

A. No, sir; I think it does not. As I understood, I think when that was brought up, the Portsmouth book was not referred to. If it would be proper for me to make an explanation, I should like to, in order to clear the matter up a little. The explanation I wish to make is this: that the paper referred to was originally made for another purpose, as I understood—for the purpose of per cent. made some two years ago, and when the tickets were called for I found on looking for the record of that report—I found some report, and I supposed that to be just what I wanted and I copied it and brought it here. Now when the question was brought up as to including the whole number of passengers and soldiers, I went through it and found that it was not a full report.

Q. (*By Mr. George.*) You found you were in error and corrected yourself in this?

A. Yes, sir; that was my design, but I did not include the Concord and Portsmouth road.

Q. (*By Mr. Rolfe.*) What was it you found you were mistaken in?

A. I found they did not contain all the books which should have come in.

Q. (*By Mr. George.*) And this does contain them?

A. Yes, sir.

Q. And this does not contain passengers over the Portsmouth road, nor free passes, nor soldiers' transportation on the special trains?

A. No, sir.

Q. Does it include tickets sold from any points on the Concord road to the Manchester and Lawrence road?

A. Yes, sir.

Q. (*By Judge Bellows.*) That is your last one?

A. Yes, sir.

Q. (*By Mr. George.*) Now, Mr. Sanborn, will you state the amounts returned for passengers paying in the cars by the conductors from 1859 to January, 1866, and take the trains embraced in your reply to the question proposed by the counsel for the defendant which has been the subject of the preceding ruling?

*The Chairman.* We have not ruled in that question yet. We have ruled that the counsel might inquire as to the particulars and data from which Mr. Sanborn gave his answers in cross-examination. But he has expressly stated, as I understand, these were not the particulars and data. It is proposed now, that the witness having yesterday given us some testimony as to the number of passengers that went on certain trains — I do not know as I could state exactly what you said about it — won't you be good enough to give it from your memory?

*Witness.* The number of cars required to carry the passengers on the Concord Railroad?

*The Chairman.* What you gave yesterday. What you said was contained in that paper you was inquired of yesterday.

*Witness.* The paper put in contained everything that was on the books of the Concord Railroad.

Q. (*By Mr. George.*) Will you state the number of passengers who paid in the cars to the conductors on the Concord Railroad every month during the years embraced in your answer to the counsel for the defendant, as to the number of cars required to carry said passengers?—("As appears from the same source from which your answers were derived," added by the chairman.)

Mr. Mugridge took exception to the question, on the ground (1) that it was a matter of record; (2) that the defendants had made no inquiries as assumed in the questions put by plaintiff's counsel; and (3) that the evidence would be generally incompetent.

Q. Now, Mr. Sanborn, can you state the number of free passes?—Have you the statistics that enabled you to state the number of passengers that paid their fares to the conductors during this period of time?

A. The only statement, I think, is the record which I have on my books. I kept accounts up to 1864, I think.

Q. What is the source from which you answered the questions as to the number of cars.

A. The record of the books.

Q. Now, from that same source, can you state the number of passengers who paid in the cars, embraced in the same period?

A. The books will show what is returned on the way-bills. [The books were procured at the request of the examining counsel.]

*The Chairman.* We are trying to limit the inquiry to the particulars and data from which the witness derived his answer. Well, now, if it should so happen that the particulars and data from which the witness derived his answer, connected with what has already been put in, should lead to the amounts, why, they would be in there. But it would be one of those things which results from the position the evidence is in.

Q. Now, sir, will you state from what source did you derive the information on which your answer to the inquiry, as to the number of cars that would be required, propounded to you by the counsel for the defendant?

A. From those books.

Q. Now, sir, will you state the number of passengers who paid in the cars from the same source of information that you made your reply from, as to the number of cars that would be required. Just begin with January, 1859.

A. They are scattered all through the book. [Books explained.]

Q. Then, Mr. Sanborn, I want you to prepare a schedule from 1859 to January, 1866, of the persons and of the stations between which sales were made in the cars. I want to ask you a question or two, however. I have put in some printed instructions which I have understood were admitted. I want to know whether these printed instructions were given to the conductors?

A. They were.

Q. (*By the Chairman.*) Are these the ones that he referred to?

A. They are.

Q. (*By Mr. George.*) They were furnished to Mr. Clough?

A. Yes, sir; this one with his signature to it shows that it is one. They were left in my office; that is, so far as my recollection goes. Every one of that kind. I recollect distinctly about the instructions being left there for being left with the conductors, except this one that his signature is on.

Q. After this investigation was ordered, you may state anything that Mr. Gilmore did,—any marking upon the instructions, and what instructions were given to the conductors.

A. The old issue of instructions was sent to them again, and underscored with red ink—certain instructions; I do not recollect the exact items which were underscored, to be given to the conductors at that time.

Q, Have you a copy that was underscored in this way?

A. I have one that was underscored, some of it; I could not identify it as being the one underscored at that particular time.

Q. Who gave you those particular instructions?

A. Mr. Nutter.

Q. Who was Mr. Nutter?

A. Mr. Gilmore's clerk.

Q. Were they underscored by you?

A. No, sir.

Q. Mr. Sanborn, you have spoken about tickets which Mr. Gilmore used to get through you. and make you throw out of the account. I want you to state to the referees, just as near as you are able to, how many of such tickets Mr. Gilmore got in the course of the year—just as near as you are able to.

Q. (*By the Chairman.*) What years?

Q. (*By Mr. George.*) In all the years right straight along. I want him to state how many such tickets there were, and what was the reason Mr. Gilmore asked for these tickets.

A. The tickets were given in small numbers; likely one or two at a time. He would sometimes remark that he did not wish to give a pass and he would ticket the individual the ticket was designed for. And I should think, perhaps, an average number for each year might be one hundred and fifty; possibly more, but I should say as much as that.

Q. Is that as near an estimate as you are able to make of the number he had taken?

A. Yes, sir.

Q. Now take these political tickets and all the other tickets he ever got. I want you to make the best estimate you are able to of the number he got. These political tickets were when Mr. Gilmore was Governor, in 1863 and 1864; I suppose that is a public fact.

A. There was but very few tickets given out for that purpose, until he was candidate. I do not recollect the year he was candidate. After he was, it would be almost impossible for me to give the number of tickets given by his order. I think they would go up into thousands. I do not mean regular tickets; I mean all kinds, political tickets, used for political purposes.

Q. The political tickets, used for political purposes by both parties, were paid for, were they not? I want to know the fact of these being accounted for.

*Mr. Rolfe* Will it appear that this witness knows?

Q. (*By Mr. George.*) I want to know about the tickets furnished legitimately, if I may so call it, to the two parties.

A. I think that previous to 1866 there was but a few of the political tickets which really brought any funds in to the road. That is my impression, so far as my knowledge went.

Q. Did you do the settling, for instance? You had more or less accounts with the democratic state committee and the republican state committee.

A. Very little with the democratic; mostly with the republican.

Q. Did the democratic state committee pay their bills?

A. I think they did not.

Q. Now at what time were these political tickets obtained?

A. What time of the year?

Q. Yes, sir.

A. The winter season: January, February and March, mostly.

Q. Mr. Sanborn, how long have you been acquainted with Mr. Clough?

A. I think it is nearly twenty years; about nineteen.

Q. How far have you lived from him?

A. Lived up here in the same ward.

Q. And you were on the road together until since he was removed?

A. Yes, sir.

Q. I want to know whether he has had any ostensible business; whether he has had any business apart from his conductorship on the Concord road; and if so, what?

[Objected to on the ground that the plaintiff had not brought them under the rule in the case of the *Worcester Railroad* v. *Dana*; and that they had not shown any loss by the railroad, so far as Mr. Clough was concerned; that they had not shown that Mr. Clough had any of the company's money; that they had not laid the foundation, therefore, for making this inquiry.]

*The Chairman.* My impression was that that was not the exact point ruled in the case of *Dean* v. *Farmer*. But without undertaking now to rule this in, it does occur to us that it would be well to postpone it at this time until pretty much all the evidence that the plaintiff is to put in shall be in.

*Mr. Tappan.* That we do not object to.

*The Chairman.* When we have before us the whole of the evidence that is offered, or likely to be offered, we shall be very glad to rule.

CROSS EXAMINATION RESUMED.

(*By Mr. Mugridge.*) You were inquired of in regard to the returns made by Mr. Upham, and you have put in the returns in October, 1868?

A. I think so.

Q. Let me ask you if at the time Mr. Upham was running this train Carney was upon the road, paying out, as it was understood, large sums of money—fourteen, or seventeen, or eighteen hundred dollars? Did you know of that?

A. Will you repeat the question?

Q. Did you know that at this time, covered by these way-bills, Carney was running on the trains as a detective, and paying out large sums of money?

*The Chairman.* You mean way-bills of Upham?

*Mr. Mugridge.* Yes, sir.

A. I think it was the first of October that Mr. Carney commenced.

Q. How long did he run?

A. I could not say as to that.

Q Wasn't he there at the time covered by these way-bills?

A. I should think he was.

Q. Will you take the way-bills returned by George Clough during the months of October and November, and put them in here for each day, and the amount returned by him. You have these way-bills?

A. Yes, sir.

Q. Will you give them?

A. The first day of October, I think, was Sunday. October 2d, second train, $9.75.

| | 1st Train. | 2d Train. | 3d Train. | | 1st Train. | 2d Train. | 3d Train. |
|---|---|---|---|---|---|---|---|
| Oct. 2, | | 9 75 | | Nov. 1, | | 5 95 | |
| 3, | 5 30 | | 6 15 | 2, | 6 15 | | 7 45 |
| 4, | | 10 35 | | 3, | | 9 00 | |
| 5, | 4 70 | | 4 50 | 4, | 3 40 | | 5 45 |
| 6, | | 5 70 | | 6, | | 12 40 | |
| 7. | 4 75 | | 6 50 | 7, | 5 85 | | 8 30 |
| 9, | | 9 30 | | 8, | | 10 60 | |
| 10, | 4 80 | | 5 80 | 9, | 19 35 | | 10 40 |
| 11, | | 7 40 | | 10, | | 10 30 | |
| 12, | 2 80 | | 4 30 | 11, | 5 55 | | 12 95 |
| 13. | | 12 10 | | 13, | | 12 50 | |
| 14, | 3 95 | | 6 00 | 14, | 2 95 | | 28 25 |
| 16, | | 8 05 | | 15, | | 11 55 | |
| 17, | 5 50 | | 5 15 | 16, | 2 55 | | 5 75 |
| 18, | | 7 75 | | 17, | | 10 00 | |
| 19, | 6 95 | | 4 80 | 18, | 5 10 | | 20 25 |
| 20, | | 7 10 | | 20, | | 11 25 | |
| 21, | 4 30 | | 5 35 | 21, | 14 10 | | 15 70 |
| 23, | | 8 45 | | 22, | | 15 95 | |
| 24, | 4 70 | | 13 20 | 23, | 11 35 | | 30 40 |
| 25, | | 17 40 | | 24, | | 15 90 | |
| 26, | 4 65 | | 17 50 | 25, | 5 60 | | 15 10 |
| 27, | | 4 45 | | 27, | | 8 65 | |
| 28, | 85 | | 9 80 | 28, | 7 05 | | 7 68 |
| 30, | | 13 15 | | 29, | | 11 20 | |
| 31, | 5 15 | | 5 95 | 30, | 3 80 | | 8 65 |

Q. (*By Mr. Mugridge.*) Mr. Sanborn, you have produced here five way-bills returned to your office by him. These would show five trips over the road?

A. Yes, sir.

Q. Let me ask you if, shortly before and after that time, he made other trips on the road save these five?

A. Yes, sir.

Q. And there should be way-bills indicating these returns at your office?

A. Yes, sir.

Q. Now if you will search and see if you have these way-bills?

A. There is one of them here.

Q. Have you read that?

A. I read that, not as one of Clough's train; November 30th, first train.

Q. Are there any other way-bills which were made by Mr. Upham, which you have been able to find?

A. I have not looked through the way-bills for that purpose. I can only say that the memorandum that I have kept don't indicate that there are.

Q. Have you any doubt that he did?

A. Oh, yes; he ran at other times.

Q. There was an inquiry made by Mr. George as to the number of seats in a car. Won't the way-bills themselves show the trains that Mr. Upham run, signed with his name on the back of the bill?

A. That was the usual way they did.

Q. I would like to have you search your office, and find the way-bills returned by Mr. Upham. How would the amount of travel on the road from 1847 up to 1850 compare with the amount of travel from 1850 up to 1860?

A. I know very little of it previous to 1857, or in that vicinity. I knew but very little about the travel.

Q. Then you have no information antecedent to 1857?

A. No, I do not know.

Q. How did the travel up to 1860 compare with the travel afterwards?

A. I should think it was not as large as afterwards.

Q. Wasn't there a very material increase of travel on the road after 1860, compared with what existed before?

A. I should think in 1860, compared with 1861, perhaps there was not so much travel as there was in the subsequent years. That is, I mean by that that the amount of travel seemed to fall off a little in 1860.

Q. Let me ask you if there was, at one time, such an arrangement on the road that Mr. Clough and Mr. Corning remained in Boston over night?

A. There was an arrangement, when there were three conductors, when one of them stayed in Boston—when Mr. Wright was running.

Q. You do not remember the arrangement made by Mr. Gilmore, by which either Mr. Corning or Mr. Clough remained in Boston over night?

A. I do not remember that.

Q. Take it from 1860 up, what proportion of the travel upon the road should you think went on the two main trains of the road—that is, the ten o'clock and the three o'clock trains—up to the time they left the road?

A. I should think a large portion.

Q. Will you be kind enough to indicate what portion went over the road in the two main trains from 1860 down to the present time? You say a large proportion; now, what proportion do you think?

A. It would vary in different times. Perhaps before this half-past seven train was put on I should state it a little different from what I should after.

Q. When was that put on?

A. It was put on permanently, I think, in '63.

Q. Then I will confine my inquiry to from '60 to '63. What proportion of the entire travel on the road went in the two main trains?

A. The down travel I could state more clearly. The down travel I should think twenty-nine thirtieths of it went on these two trains.

Q. From '60 to '63?

A. I should think so.

Q. Now, after the half-past seven train was put on—take the trains as they run: the ten and a half, the three and a half, and the seven and a half train—from '63 up to January, '66, will you be kind enough to state what proportion of the travel went on these two trains?

A. I should say fifteen-sixteenths.

Q. We want you to make a little statement in regard to that, and hand in this afternoon. You were inquired of by Col. George as to when the Concord Railroad began to ticket toward New York, and toward stations in that direction, and you replied. Let me ask you, sir, if they haven't always ticketed from New York, and from other stations, to Concord, over the Concord road?

A. They have to that time.

Q. You designated certain times when the road used to ticket to New York and stations lying in that direction, and you said there were times when they did not so ticket. Now, let me ask you if, during all the time, they have not ticketed from New York to Concord, right through?

A. Yes, sir.

Q. Before I go on to another branch of the examination, I want to ask you about these coupon tickets. Has it been the habit of the officers of the road, and persons connected with the corporation, to deal in coupon tickets, with the full knowledge of the officers—to buy and sell them, and take to themselves the profits of such sales?

A. I don't know to what extent that has been carried on.

Q. Have you known of any cases?

A. I don't know as I know exactly what you mean.

Q. Have you ever known of any of the officers buying these coupon tickets, and turning them over and taking the money for them?

A. I do not recollect.

Q. Have you done that thing?

A. I don't think I have; but I do on some tickets from the Vermont Central and Passumpsic road. They were tickets which were sold by way of Lowell and went by way of Lawrence. These tickets were taken up.

Q. These were coupon tickets, were they?

A. Yes, sir.

Q. What happened with regard to these tickets?

A. These tickets were returned by the conductor.

Q. Where?

A. And put into the record.

Q. Where were they returned?
A. To the general ticket office.
Q. To you?
A. Yes, sir. I think some of them were returned by the Nashua and Lowell road, or by the Manchester and Lawrence.
Q. They came to you?
A. Yes, sir. These tickets I put into the office at Manchester to be sold.
Q. What became of the money?
A. I could tell you if I could find my memorandum. My impression is that he gave me an equal number of tickets to Boston the way that they should go.
Q. Mr. Henry Hurlburt?
A. Yes, sir.
Q. Now did you personally realize any advantage from that transaction?
A. No, sir.
Q. You mean to say that you never put into other ticket offices, tickets that have been sold and you be benefitted by that transaction?
A. I don't recollect anything of that kind; not on selling tickets; tickets were exchanged. For instance, that was when the Concord and Portsmouth road was in existence from Concord, that once in a while there would be a person who happened to wish to change his ticket, and would give up his ticket to Boston for a ticket to Portsmouth?
Q. What did you do with that ticket?
A. I sold it.
Q. What did you do with the money?
A. I kept it myself.
Q. Won't you just be kind enough to repeat that?
A. Once in a while, in the summer season, (while there is through travel, of course) the passengers would come from the west with tickets to Boston and wish to go to Portsmouth. I made that exchange.
Q. You took their ticket and sold it?
A. Yes, sir.
Q. And put the money into your pocket?
A. Of course I had to pay for a ticket to Newmarket Junction, whatever the difference was.
Q. How much was it?
A. I think it was half a dollar at that time, or something of that kind.
Q. You at that time were acting in the capacity of ticket seller at the Concord station?
A. Yes, sir.
Q. How many such instances occurred where you have done this thing, and to what extent did you traffic in that way in these tickets?
A. I should think there was some half dozen such cases.
Q. Won't you refresh your recollection between this and the afternoon?

Q. Mr. Sanborn, I was inquiring of you at the time of the adjournment as to your transactions in these coupon tickets. Let me ask you now, having refreshed your recollection, if you are able to say whether you have dealt in these coupon tickets from western roads, to any extent?

A. I do not think that I have; I do not know that I have. I would like to make one further statement in regard to those which I spoke of. I went and took my books at the depot, and I found a record of what I wanted to show: three instances where I carried tickets to Mr. Hurlburt. That was by understanding.

Q. You say there were three instances where you took tickets to Mr. Hurlburt at Manchester, and you took the money?

A. The money—I don't recollect about that, but my impression is that I took tickets. My first impression was that I took tickets in return.

Q. Will you say, Mr. Sanborn, that Mr. Hurlburt didn't pay you money?

A. No, sir; I couldn't swear now.

Q. Did you ever put these tickets into the Concord office to be sold for yourself?

A. I don't recollect that I ever did.

Q. Didn't Mr. Whipple sell any of them?

A. I don't think he did.

Q. While you were ticket-master how many did you put in?

A. I know it was all done in one season. I think it may have been half-a-dozen times. It was very seldom. It was when A. C. Pierce was one of the directors of the road. People would come and want to change their tickets, and I would decline to do it. Perhaps in half-a-dozen instances I did so. And then there was a doubt in my mind about it; and I asked Mr. Pierce. My impression now is that Mr. Upham was away; and my recollection is that I asked Mr. Pierce whether I had better do so, and talked the matter over; and I don't know that I ever did so afterwards.

Q. You are quite confident you have not done it since?

A. I do not recollect.

Q. You are quite confident that you haven't done done it since Mr. Pierce went out?

A. I am not positive, because I may have forgotten.

Q. Did these coupon tickets come into your possession very often?

A. I don't recollect except these three instances that I mentioned.

Q. Never any others?

A. Well, I won't be positive about that. These are transactions that occurred a number of years ago. There was so many and so great a variety of transactions that I cannot now remember. I remember of receiving from the other road. The business was changed from the Boston & Lowell to the Manchester & Lawrence, and when this happened it caused the tickets to go a contrary way from which they were sold. I could not recollect with any degree of certainty about it.

Q. What became of this accumulation of tickets?

A. I sent them to the roads where they were.

Q. Did you ever sell these tickets?

A. No, not except those instances that I have spoken of.

Q. You spoke of instances where you sent to Manchester. How many did you send the first time?

A. The first time I sent down one hundred and seventy-one tickets.

Q. How much did you receive net yourself from the sale of that one hundred and seventy-one?

A. I did not receive anything.

Q. Nothing at all?

A. No, sir; that was an open transaction which was a matter that was talked up and known by the superintendent.

Q. Where is your return? Have you any account of it?

A. Nothing only what is on the book. This is a memorandum taken off from my book, ticket-book for 1865.

Q. What is the next instance?

A. One hundred and thirty-nine and a half.

Q. Did you realize anything from that?

A. No, sir.

Q. Did you from any?

A. No, sir; that is, I don't know why I should; I have no recollection of it.

Q. Didn't you receive the difference between the cost of the tickets?

A. I don't know any further than this: it was a person who came from the west. In the first place they wanted to go to Portsmouth; and they came here and exchanged with me. The fare, I think, was $1.50 to Portsmouth, and I think it was $1.75 to Boston. That is my recollection.

Q. And whatever it was you kept the difference?

A. I did.

Q. How many instances of the kind you have last described have happened since your connection with the Concord Railroad?

A. I should think there was some half a dozen; it might be more and it might be less; there was very few instances.

Q. (*By the Chairman.*) Did you say that was all under the directorship of one man?

A. Yes, sir. My recollection was that it was all in one season—two or three months, I think it was.

Q. (*By Mr. Mugridge.*) Did Mr. Gilmore know of that transaction?

A Oh, that was before he was superintendent of the road, I think.

Q. Who was superintendent?

A. Judge Upham.

Q. Did Judge Upham know of it?

A. My impression is that it was talked with him after that.

Q. Before you made the sale?

A. No, afterwards. I know I made the sales and the question came up in my mind, and I asked him.

Q. Now give me the third instance.

A. One hundred and eighty-eight.

Q. Now these tickets—the 171, 139 1-2 and 188—were put into Mr. Henry Hurlburt's hands?

A. Yes, sir; I either sent them or carried them down.

Q. They were tickets to different stations on the Concord road?

A. They were either to Lawrence or Boston; I think now there might have been some to both.

Q. Let me ask if these tickets, the 171, the 139 1-2 and 188, were put into your general ticket account? That is, if you charged these tickets to Henry Hurlburt the same as you charged to others?

A. No; an exchange was made, I think, with a separate account.

Q. Did you charge those tickets to Henry Hurlburt's office in the same way as you charged the tickets to his office, ordinarily, that he had?

A. I did in one sense; that is, I charged them to him. I don't think I entered them on with the regular memorandum. That is, at that time he was furnished with a regular number of tickets. As I recollect this, he either paid me money or tickets. My impression is that it was tickets, but I am not certain. Whenever those tickets were given up to me, they were offset by those.

Q. Why did you carry them to Manchester?

A. I don't know why it was, unless because some tickets were sold from that station, and they would get sooner sold from there.

Q. Who directed that way?

A. Mr. Gilmore.

Q. He knew you had those tickets?

A. Yes, sir.

Q. And he directed you to carry them to Hurlburt to be sold?

A. Yes, sir.

Q. Did Hurlburt make a return of those tickets in money or in tickets?

A. I could not swear, but my impression is that it was in tickets, but it might have been part of it in money.

Q. Are you able to state, positively, whether he made a return of these tickets to you in tickets to Boston, or in other tickets, or in money?

A. My impression is that it was in tickets.

Q. Have you got any sort of account anywhere in your office to show the returns that he made to you of these tickets?

A. Nothing but what I have got written off on my books.

Q. Did you, whenever a return was made by Mr. Hurlburt, put it down as so much money received?

A. I should do so.

Q. Can you tell any reason why you should not have made the same return in regard to this that you should make of the tickets together?

A. I should, I suppose; that is, so far as the money came into my hands.

Q. If you have got any return on your books that shows that it was returned in money, won't you produce it?

A. I know I have not.

Q. Now, if he made that return in tickets, should you have made a return showing that it was returned?

A. No, sir: only among collections.

Q. Not returned by the conductors?

A. No, sir. If he gave me tickets the conductors didn't have them at all. The account balanced itself when he handed these tickets to me.

Q. You made charges to him when you put these tickets into his hands?

A. I charged them unless he gave me the tickets at the time.

Q. Do you mean to say that you made no record of it?

A. There would be no need of it.

Q. Supposing you put the tickets into his hands, and then received something in compensation for them. Nothing appears upon your book to indicate that transaction anyway?

A. The system of selling tickets—the way that we did often—was to charge a man for a thousand tickets. Those that he failed to produce he was to account for and pay for. He was to report them at the end of the month. If he put them in correctly, his sales and my collections would agree then.

Q. Supposing you put a dozen into Mr. Hurlburt's hands, you charged him for them?

A. Yes, sir.

Q. Then he is chargeable for the balance?

A. Yes, sir.

Q. I understand you to say that these tickets did not go into the general account?

A. No, sir.

Q. I want to ask you if there is any record of these New York tickets? I mean where you charge them to him, and where you credit him with anything you received from him?

A. As I said before, there is no necessity for any charge or credit.

Q. What did you do with the tickets you received from him?

A. All the tickets were returned, I suppose. If the tickets were from Manchester to Lawrence or to Boston—if they were tickets from Manchester to Lawrence, I will say—if he gave me 171, I took these tickets and put them into the drawer among the tickets that the conductors returned to me daily. At the end of the month he returned all the other numbers. Now, 171 of that 500 was of that, and the conductors should make up the difference.

Q. Where did you get these?

A. Got them from the Nashua and Lowell road.

Q. Who gave them to you?

A. They were sent up by the general ticket agent.

Q. What kind of tickets were they?

A. I couldn't describe them any further than that they were upper road tickets.

Q. What places did they run to or from?

A. From all points.

Q. Did he send them up to pay you in any way?

A. He sent them up for us to give him credit for that number of tickets from Lowell to Boston.

Q. (*By the Chairman.*) How came they into his hands?

A. The tickets were sold by the way of Lawrence. When the business was changed over, or previous, or perhaps during that time, everything was sold that could be sold by way of Lawrence; and for that reason who would have gone by way of Lowell would get tickets the other way; and when they came to go down, they went the other way. The conductors collected these tickets on the Lowell road, and returned them to me to collect.

Q. The passengers who kept these coupon tickets were permitted to ride either way, and the conductors took these tickets?

A. Yes, sir.

Q. And they were returned by the conductors to the ticket agent at Nashua or somewhere?

A. Yes, sir.

Q. And that ticket agent returned them to you?

A. Yes, sir. He requested us to make a transfer of our tickets to Lawrence or Boston, as the case might be. They came from the Nashua and Lowell road.

Q. That is, they are tickets sold over the Lawrence road?

A. Yes, sir.

Q. And were taken up over the Nashua and Lowell road?

A. Yes, sir.

Q. And you gave the Nashua and Lowell people credit for the carrying of the amount of these tickets?

A. Yes, sir.

Q. Is that the way of it?

A. Yes, sir.

Q. (*By Mr. Haile.*) And this was done for the accommodation of passengers, in the first instance?

A. Yes, sir.

Q. What object was there to send them to you?

A. Merely to dispose of them, and have them go where they belong. The coupons belonged to the Boston & Maine road. There were a great many more tickets sold there from Manchester. I don't know who made that suggestion; but either Mr. Gilmore or some one else.

Q. (*By the Chairman.*) That is, you had paid them for carrying the passengers; and, meanwhile, the other people had not carried the passengers?

A. Yes, sir; but had got the money.

Q. So the object was to sell the tickets again to somebody that would ride over the road, and realize the money again?

A. Yes, sir.

Q. Now the thing of it was that you let the Manchester agent have a quantity of these tickets, and he gave you some of the tickets that you had already let him have instead of them; was that the way you did business?

A. Yes, sir; that would balance his account then.

Q. You took back part of the tickets you let him have to sell; and gave him these, and considered him as having so many tickets, no more and no less?

A. Yes, sir.

Q. And these went back into your general amount of tickets to be used again?

A. No, sir; these tickets as they came back I put in with the collections.

Q. What was the object of that?

A. So as to have the calculations agree. We have always made a comparison of sales and collections.

Q. Did you in that way make him pay you money for these tickets and also for the ones that he had?

A. No, sir; he didn't pay for but one.

Q. Well, then, in point of fact, who got the money that was realized in the sale of these tickets?

A. There was no more realized by anybody, any more than what should be.

Q. Well, but who got it?

A. Each road got their proportion. For instance, if one of these tickets rode from Manchester to Boston by way of Lawrence and was collected by way of Lowell; and that it was collected at Manchester, Mr. Hurlburt takes that and sells it in place of one of our own tickets. We have reported it, and have got our proportion of it. It merely settles the affair of the exchange of tickets.

Q. (*By Mr. Mugridge.*) Mr. Sanborn, you were inquired of a little in regard to Colonel George's connection with this investigation. Let me ask you how long Colonel George has been counsel for the Concord Railroad corporation?

A. I could not tell you; I think he was when I came on the road.

Q. Has he been consulting counsel?

A. I could not say.

Q. Has he been confidential counsel and adviser of Gilmore and Upham for the last ten or fifteen years?

A. I know he has been very intimate with them.

Q. From what you have understood, and from what you have seen, since you have been on the road, has he been the confidential adviser of Gilmore for the last ten or fifteen years, up to the time a new clerk was put on the road?

A. I think that my first acquaintance with the management of that direction, Gen. Pierce was a member.

Q. After Mr. Pierce left?

A. I have no means of knowing except as I saw that intimated.

Q. From what you have seen, I make the inquiry, has he been the confidential adviser in the law business, and a considerable portion of the time the clerk?

A. I should not want to make that statement with any degree of certainty, because I do not know.

Q. So far as you know, has there been any other person who has been the confidential adviser of Gilmore, up to 1866, except Colonel George?

A. I cannot state that.

Q. Let me ask you how it has been about his being called frequently to the office, from day to day and from week to week?

A. It was a very common thing for him to be sent for.

Q. Did he ever have anybody else?

A. I think Judge Fuller was.

Q. Were these instances exceptional instances?

A. Yes, sir.

Q. So far as you know, has he attended all the directors' meetings?

A. I never was at any of the directors' meetings, and I could not say that he was ever at one of them.

Q. Do you know whether he has taken an active part in the management of the Concord Railroad for the last ten years?

A. I could not say as to that.

Q. Do you know what part he has taken?

A. No, sir.

Q. Do you know whether suggestions in regard to the management have come from him?

A. I could not say.

Q. Did you ever hear Gilmore say anything of that kind?

A. No; I don't think I have.

Q. Did you ever hear Gilmore say anything tending in that direction?

A. He might have, in some particular cases, perhaps, spoken of things of which Mr. George spoke.

Q. What were such instances?

A. I could not recollect.

Q. Has he frequently made such statements to you?

A. No, sir; I don't know as he has.

Q. Hasn't he ever spoken of pieces of policy as emanating from Col. George?

A. I could not state that.

Q. You say that Mr. Gilmore at one time told you to furnish information to Col. George for the discovery of the alleged frauds committed by the conductors? He at one time told you to do that, did he?

A. To give information?
Q. Yes, sir.
A. Yes, sir.
Q. To furnish him all he wanted?
A. Yes, sir.
Q. Now you say that after that he came to you and told you not to do it, unless he put it in writing; and not to answer him unless you put it in writing and furnished it to Gilmore?
A. Yes, sir.
Q. Did he tell you that the reason was that Col. George would get you into trouble, and get them all into trouble?
A. I don't recollect as that last was said.
Q. Have you heard him say to anybody that he wanted these things down in writing, or he [George] would get you all into trouble?
A. I don't recollect just that conversation. But when he told me, he opened on Col. George and said he would get us all turned out there —"rip it through."
Q. Did he say that this was his scheme for getting himself in as counsel for the road with Stark as manager?
A. I don't recollect.
Q. Anything like that?
A. He said it was Col. George's scheme to get him turned off in the first place, and then the rest of us afterwards.
Q. Did he state that Col. George had this in view?
A. I dont know as he has; but somebody has.
Q. Did he say that it was in order to root out the old superintendent for the purpose of getting himself some particular place of power on the road?
A. He may have stated that first part, but I don' recollect as he did the last.
Q. Will you say that Gilmore did not make that statement?
A. No, I wouldn't say that he didn't.
Q. Will you say that at the time he asked you about having this put in writing whether he stated anything else, and what it was?
A. It was to that effect, that this was a case that Col. George instituted for the purpose of getting him out, and that we should all follow in the same category.
Q. I want you should give that conversation—the substance of it.
A. The substance of it was that he wanted we should follow him; that if we wouldn't, we should all go.
Q. Did he always say to you, after that, that this was started through Col. George's influence?
A. Yes, sir; after the first instance he intimated to me that that was the case.
Q. Did he ever direct you to withhold a single fact, but only that the fact should be in writing?
A. No, sir; only in that way.
Q. Only that they should be put in writing?
A. Yes, sir.
Q. Did he say you ought to do that for your own benefit and your own safety?
A. I should think very likely he might. It would be very natural for him to.

Q. Now, sir, did ever you see any inclination on the part of Mr. Gilmore to withhold from Col. George any information that he desired with reference to this prosecution against the conductors?

A. I think his general tenor indicated that he did not wish to have him have information; that is to say, to judge from his manner.

Q. I want to know if he ever told you that he wanted any information withheld from Col. George that he desired for the prosecution of these conductors?

A. No, sir.

Q. I want to ask if you heard Col. George say, at any time, that for the fifteen years last past he had been engaged in covering up the frauds and iniquities of the old board of directors?

A. There was a statement made here the other day, during the recess——

Q. No, sir—I do not mean here. Did you attend a committee hearing before the Legislature when Col. George addressed the committee?

A. No, sir.

Q. Did you attend a meeting of a legislative committee in one of the committee rooms, two or three years ago—where Col. George addressed the committee?

A. I don't think I did now; I don't recollect now.

Q. You did not hear any statement that he made at that time?

A. No, sir; I think not.

Q. You say there were some tickets furnished republicans and some to democrats—did the republicans pay for the tickets they had?

A. A part of them I suppose they paid for.

Q. Were there not a large number of tickets used in the campaign that were never paid for?

A. They never were paid for until sometime afterwards.

Q. How many?

A. I said I could not tell; I should think that it would run up to thousands.

Q. Were there not five thousand?

A. There might have been that.

Q. Will you say that there was not more than five thousand?

A. No, sir; it would not surprise me at all.

Q. How much would these amount to in dollars?

A. I don't know; I know I figured it up at the time.

Q. How much should you think?

A. Between two and three thousand dollars.

Q. (*By Mr. Rolfe.*) Besides the five hundred and thirty-one?

A. I don't recollect whether that came in or not.

Q. (*By Mr. Mugridge.*) You said you had no record of the number of passes during the last years under the reign of the old board; you could not tell the number?

A. No, sir, I could not tell that.

Q. You have been through with these tickets, Mr. Sanborn?

A. Yes, sir.

Q. Now, won't you be kind enough to give us the result. I inquire simply as to those which could have come into Mr. Clough's hands.

A. I gave you a minute of that the other day; and that I haven't got now.

Q. Let me ask you to say as to those tickets dated after January, 1866; I mean those tickets which could have come into his hands in the ordinary course of business.

A. Thirty-six, I make them, since Mr. Clough left the road.
Q. Let me see some that were dated before that.
A. There was five that were dated previous to that.
Q. Dated previous to January 14?
A. Yes, sir.

*The Chairman.* There is one thing I do not understand about this inquiry. Are you seeking to ascertain this before and after Mr. Clough left the road?

*Mr. Mugridge.* Yes, sir.

*The Chairman.* What day was this?

*Mr. Mugridge.* January 14th, 1866.

A. [Reading]:

| | | | |
|---|---|---|---|
| 1866. | February | 3. | Nashua to Manchester. |
| | " | 6. | " " |
| | " | 6. | " " |
| | " | 3. | " " |
| | " | 2. | Manchester to Nashua. |
| | " | 2. | " " |
| | " | 8. | Nashua to Manchester. |
| | " | 3. | " " |
| | " | 6. | " " |
| | " | 3. | " " |
| | " | 6. | " " |
| | " | 3. | " " |
| | " | 6. | Manchester to Nashua. |
| | " | 6. | " " |
| | " | 5. | " " |
| | " | 5. | Lawrence to Concord. |
| | " | 2. | Concord to Manchester. |
| | " | 6. | " " |
| | " | 5. | " " |
| | " | 3. | " " |
| | " | 2. | " " |
| | " | 9. | Concord to Hooksett. |
| | " | 6. | Hooksett to Manchester. |
| | " | 5. | " " |
| | " | 5. | " " |
| | " | 5. | " " |
| | " | 5. | " " |
| | " | 6. | " " |
| | " | 5. | Suncook to Manchester. |
| | " | 5. | Manchester to Concord. |
| | " | 5. | " " |
| | " | 5. | " " |
| | " | 3. | Lawrence to Concord. |
| | " | 3. | " " |
| | " | 9. | Concord to Hooksett. |
| | " | 8. | " " |

There has two got in here by mistake; December, 1865. They should have gone into the other account. There are six of this date previous. I think I said five before.

Q. (*By Judge Bellows.*) Does that take off one from the other lot?
A. Yes, sir.

Q. (*By Mr. Mugridge.*) Now, do you say that all the other tickets are dated before January 14th?

A. All the others that are dated plain enough so that I can see.

Q. How many of them are there?

A. Seven.

Q. (*By the Chairman.*) Are those all the tickets that you say in the course of business would have come into Mr. Clough's hands?

A. I don't recollect the number that I said the other day.

Q. (*By Mr. Mugridge.*) There are thirty-four of these tickets that you have found, that you swear absolutely bear date since January, 1866?

A. Yes, sir.

Q. Now, there are seven of them that you swear absolutely bear date prior to that time?

A. Yes, sir.

Q. Now, take the other tickets and go through them, and state how many there are that were dated before that time, and how many there are that you can't give the date of?

A. That is what I want to state here. I have looked them all over. I found fifty-four tickets with full dates. I find thirty-five with the month on them, without stating any year.

Q. Are you able to say what year those thirty-five tickets bear date in?

A. No, sir; I could not. There is nothing to indicate at all.

Q. (*By Mr. Cushing.*) Does it appear that those where the year does not appear—is that from a different stamp, without the year upon it?

A. Yes, sir. My impression is that most of them came from the Montreal road, and they don't put on the year.

Q. (*By Mr. Mugridge.*) Whether these are precisely the same kind of tickets that were being used in 1866?

A. I should think they were.

Q. Let me see one of them. That bears no date as to the year?

A. I think not.

Q. Let me ask you if all these have the appearance of being new tickets?

A. I should think so.

Q. (*By Mr. Cushing.*) Are they different from those in use in 1865?

A. I don't know that they have changed them at all for ten years.

Q. (*By Mr. Mugridge.*) All of these came over the Boston, Concord and Montreal road, did they not?

A. Yes, sir; I think so; those in that package?

Q. (*By Mr. George.*) Just cards or coupons?

A. Yes, sir.

Q. (*By Mr. Mugridge.*) Will you produce a package of tickets that bear date since January 14th, 1866, for the purpose of comparison?

A. They are not like those. These are some of the Boston, Concord and Montreal road. [Package shown.]

Q. Have you got coupons from the Concord and Montreal road that bear date up to that time?

A. No, sir.

Q. You have spoken of thirty-five tickets and of fifty-four tickets. What do you say with regard to the others?

A. There was twenty that had only the year, I think, without regard to the day or the month either. That is, I was not able to discover it.

Q. What was the year?

A. Various years.

Q. What were they?

A. I should have to look them all over.

Well, look them all over.

A. I think that none of those that had only the year could have come into his hands. Most of those are entirely separate from the road.

Q. Mind you, I do not make any inquiry of any ticket that could not come into his hands in the regular course of business.

A. These were all of foreign roads.

Q. And could not have come into his hands? What was the number that you arrived at?

A. I made them two hundred and thirty-five. The rest had no date on them.

Q. Had no date to indicate whether they were before or after January 14th, 1866?

A. No, sir.

Q. (*By the Chairman.*) Were these tickets coupons that you say could not have come into his hands?

A. I do not recollect how many of them could have come into his hands.

Q. (*By Mr. Mugridge.*) Now pick out how many there are without dates that could possibly have got into his hands. How many of these do you find?

A. One hundred and three and a half.

Q. (*By Judge Bellows.*) That might have come into Mr. Clough's hands?

A. Yes, sir.

Q. (*By the Chairman.*) What is the half ticket?

A. A child's ticket.

Q. (*By Mr. Mugridge.*) What is the appearance of these tickets? Do they appear like new or like old tickets?

A. Most of them appear like new. There is a few that are old, I should think; that is, they look very much worn.

Q. Is there anything upon these tickets which indicates that they might not have been dated after January 14, 1866?

A. I think not.

Q. *By the Chairman.* Is there anything on these numbers that take the place of the date?

A. No sir; that is the number of the station from where they were sold.

Q. (*By Mr. Mugridge.*) The only number of tickets that you are able to testify to as bearing date prior to that is seven?

A. Yes, sir.

Q. In making your examination of those tickets, did you undertake to ascertain how many came down from the northern roads?

A. No, sir.

Q. How many of these one hundred and forty-three—I don't inquire of any other—are there that might have come into Mr. Clough's hands from the northern roads and branches?

Q. (*By the Chairman*—interrupting.) Mr. Sanborn, what do you say that means? [Indicating upon the ticket.]

A. That is the year that it was printed.

Q. (*By Judge Bellows.*) What is the whole number you say that could have gone into his hands?

A. One hundred and forty-three, I think.

Q. (*By Mr. Mugridge.*) You may count them now.

A. I made them fifty-two.

Q. (*By Judge Bellows.*) You say fifty-two came over the northern roads?

A. The Northern and Claremont and the Montreal.

Q. (*By Mr. Mugridge.*) The rest were over the lower roads?

A. Local and lower roads.

Q. What do you mean by the lower roads?

A. The Lowell, and the Boston and Maine. Here is one from the Worcester road.

Q. Whether there are any from Lawrence to Boston?

A. I have not reckoned those in at all, because you said not to put in those.

Q. There is one question more that occurs to me. In what proportion did the soldiers increase the travel over the Concord Railroad, as nearly as you can state it, sir?

A. I could not tell.

Q. What would be your idea of it?

A. You mean the passage of the soldiers themselves, or the friends they brought?

Q. I will take the soldiers alone, first.

*Mr. George.* You mean substitutes and soldiers?

*Mr. Mugridge.* Yes, sir.

A. Well, I couldn't tell. I couldn't give any sort of an idea; because they came from all directions. A good many came from the north as well as the south. I could not tell you what proportion.

Q. Can't you give some sort of an estimate, Mr. Sanborn?

A. I don't know how I could, I am sure.

Q. Was it largely increased?

A. Certainly it was.

Q. Mr. Sanborn, were you familiar with the tickets that the western roads used to sell, where the passengers came down over the Concord road, in 1860–1–2–3–4–5 and along there?

A. Yes, sir.

Q. Let me ask you whether they were marked upon them, "good for thirty days."

A. Yes, sir; some of them were thirty, and then they reduced it to twenty days, and fifteen, and sometimes to ten days. I guess ten days was the lowest.

Q. Did the time range all the way from ten to thirty days that the tickets were good for?

A. Yes, sir.

Q. You say you think mostly they were for fifteen days?

A. No, sir; mostly above that. I think there were some of them good for ten days only, but the most from fifteen above. I should say that twenty days would be what my recollection would be.

Q. The average?

A. Yes, sir.

Q. Did any of them extend higher than thirty days?

A. I don't recollect.

Q. In regard to the tickets sold over the Grand Trunk road. Was there any limitation?

A. There was some of these tickets that there was no limitation of time upon them.

Q. Good for any time, so far as anything on the ticket shows?

A. Yes, sir.

Q. (*By Judge Bellows.*) You say some of these, or all, were on the Grand Trunk road?

A. Yes, sir; I think on the Grand Trunk road there was some that had no time at all.

Q. (*By Mr. Mugridge.*) Were not a great proportion?

A. No, sir. I should think most of them were limited.

Q. I am speaking about the Grand Trunk,

A. I am not able to separate them.

Q. Do you know a portion of them were unlimited as to time?

A. Yes, sir.

Q. Was there anything upon the tickets to prevent their transfer from one person to another?

A. I don't recollect of ever seeing that.

Q. Was there anything like this: "not transferable"?

A. I have seen tickets of that kind; but they were special tickets.

Q. Was there anything that limited their use to the person who bought them?

A. No, sir; I think not.

Q. What is that, sir [showing a ticket]? Is that a coupon ticket sold over the Grand Trunk road?

A. Yes, sir; it purports to be that.

Q. Now, so far as you have observed, is that the ticket that was usually sold?

A. I think it was—on the Grand Trunk road.

Q. Now take that particular ticket and say when the date indicates that that was sold?

A. November 15th, 1865. It is not perfectly clear—the year. It is a half ticket.

Q. Is there anything upon that ticket which limits its use to the particular person who purchased it? Or is there anything which limits its use as to time?

A. No, sir; I don't see anything.

### RE-DIRECT EXAMINATION.

Q. (*By Mr. George.*) You have been asked what Mr. Gilmore has said about the matter, whether I have been the confidential and legal adviser. If you have any knowledge of either good or bad advice yourself, I want you to state it. [Objected to and question withdrawn.]

### CROSS EXAMINATION.

Q. (*By Mr. Mugridge.*) There is one matter that I want you to explain to us. You have got here, by a paper put in by Mr. George, the number of cars that it took to accommodate the passengers that passed over the road, from 1859 up to 1866. You have testified, before, that the travel in 1857–8–9 was light upon the road; and that it was very much increased in 1865.

A. In 1860 and '61, I think it was, according to my recollection.

Q. Was it light in 1859?

A. I don't recollect about that; I recollect it being a very common remark in 1860.

Q. Now, sir, the travel was very essentially increased after that, wasn't it?

A. Yes, sir.

Q. Now will you be kind enough to explain to the referees how it is, where the travel is so very much increased as you say it was during the war—that it takes only two cars still to accommodate the passengers?

Q. The estimate I make of these cars—for instance, if a car holds fifty, if it was fifty-six passengers you want two cars, or, if it was ninety-nine, you call it two cars.

Q. And if it is a hundred and four you call it three cars?

A. Yes, sir; I have in some instances.

Q. (*By the Chairman.*) What passengers does that show? Those accounted for or those not accounted for?

A. Those accounted for. That is what I designed to have. So that ninety-nine passengers would take no more cars than fifty-five.

Q. (*By Mr. Mugridge.*) Do you mean to say, Mr. Sanborn, that there was not increase of travel enough to bring the cars up beyond the general average of two, that it had required for years before to accommodate the passengers?

A. Well, I thought my statement was correct here, according to these figures.

Q. I put this question to you: Up to the time of the war the number of passenger cars required was two, upon the average?

A. Yes, sir.

Q. Now I mean to ask you if you can say that the increase of travel on the road was not sufficient to raise that above that general average, as it existed before the war?

A. I suppose it did not.

Q. Look and see if it did, in a single instance, save one.

A. This is per train, as I understand it. Before the war, in 1859, there was three trains; while in 1864 there were five, I think—four or five trains. That would make a difference, wouldn't it, in the number of passengers?

Q. You have put in here a statement that it required two cars, upon the average, to accommodate the passengers over the road, up to the war—the passengers returned each train. Now I want you to explain why it is, with the large amount of increased travel returned, that you have not been able during the war to get these cars up beyond that general average of two?

*The Chairman,* Do you understand that there is a large increase in the amount of travel returned?

*Mr. Mugridge.* Yes, sir. [Reads from the statement to show the increased number of passengers returned.]

Q. Now we will take it here: in October, 1863, there were 21,970 returned, and only two cars to accommodate these passengers when it was so many; in June, 1859, 10,163, and there were only two cars.

A. Well, I should explain it in the way I did before. In April, 1859, we ran three trains. In 1865——

Q. What time in the day?

A. Half past five in the morning; quarter past ten; and three-thirty P. M.

Q. Now take 1865.

A. In 1865 we ran five trains: half past five in the morning; half past seven; quarter past ten; three-thirty, and eight P. M.

Q. Now, Mr. Sanborn, take the cars which made up the three trains, as run in 1859, and the cars that made up the five trains; how many more cars did you run in 1865 than you ran in 1859?

A. You want me to figure out the calculation?

Q. Yes, sir. Take the entire number of cars run on the three trains, in 1859, and the entire number of cars run on the five trains, in 1865, and be prepared to tell me how many more cars you ran in a day in 1865 than you ran in 1859. Do you see the point?

*Mr. George.* According to that calculation?

*Mr. Mugridge.* Yes, sir; according to that calculation.

A. Do you want me to figure it out now?

Q. Yes, sir; and we will put that in by another one.

RE-DIRECT EXAMINATION *resumed.*

Q. (*By Mr. George.*) Mr. Sanborn, if you ran five trains in 1865, and three trains in 1859, you would run three-fifths as many in one case as the other? That is palpable?

A. Yes, sir.

Q. Now, if you ran two cars on an average in 1859, and three trains down and three up, that would make twelve cars a day? And if you ran five trains down and five up, it would be twenty, wouldn't it?

A. Yes, sir.

Q. (*By Judge Bellows.*) When did you commence this change from three to five trains?

A. It was gradual, sir; up to August, 1863, I think is the first one.

Q. That is 208?

A. Yes, sir.

Q. It goes from 150 to 208?

Q. Yes, sir.

Mr. Stanley offered in evidence the income returns of Mr. Clough for the year 1868. [Objected to.]

*Mr. Tappan.* I claim that the basis is what his property cost him, and not its worth now. The railroad has nothing to do with what his property is worth to-day. It may have something to do, perhaps, with what his property cost him. And therefore his income returns do not have any bearing here.

*The Chairman.* In that point of view, perhaps as much as could be expected would be to show how much is in his hands, and leave him to show how much it cost him. The referees are of the opinion that this evidence is subsidiary, and would more properly come in as subsidiary evidence, after the other.

*Mr. George.* I put it in now for the reason that we have testimony that will take up the whole of the forenoon, and we wanted to put in documentary evidence now.

Q. (*By Mr. George.*) Mr. Sanborn, you put in the average paid Mr. Clough per trip, and the various changes of tariff, and the returns as paid in the cars, in 1858–62; average amount per train, $2.09 (I call it from the commencement of the year wherever a change is made); from 1862 to 1864, $2.84; from 1864 to 1865, $4.13; from 1865 to 1866, $3.74. Let me ask you how you made that up?

A. I took the number of trains, and who ran the train, for the day.

Q. You mean trip for the day.

A. Yes, sir. And then the days in the year. Then I took the whole amount which he returned in the year, and divided that by it.

Q. (*By the Chairman..*) Then you got the whole number of trains run in the year, and then who took each train?

A. Yes, sir.

Q. (*By Mr. Mugridge.*) And you called it a trip for a day?

Q. Yes, sir.

Q. (*By Mr. George.*) And three trains a day?

A. Yes, sir.

Q. (*By Mr, Mugridge.*) And that is the basis that runs through the entire calculation?

A. Yes, sir.

[Adjourned.]

[ELEVENTH DAY. Thursday, August 6th, 1868.]

*Judge Bellows.* This deposition of Mr. Colby has been examined; and we have come to the conclusion that it is admissible, and upon this ground: that it furnishes evidence competent for a jury to consider on the question whether Mr. Clough had not received pay for these tickets. It, of course, is not conclusive in the matter, but it is evidence tending that way. Unless the referees should find, in point of fact, that Mr. Colby had received pay for these, of course this could not be connected. We think it is proper evidence to submit to a jury as tending to show that it is not necessary that the evidence offered, in order to be competent, should show an issue, but have a tendency to show. It does not appear from this deposition that he did or did not receive compensation for these tickets. The question is not upon that. We understand that it is upon the ground that he had workmen who, while this work was going on, received tickets to their homes. It is not negatived in any form; and therefore it is not in conflict with the rulings that have been made before. We bave ruled heretofore that where it appears that they were furnished as goods this was not evidence to be received in this count for money had and received.

Mr. Stanley then read from the deposition of Mr. Colby, omitting interrogatories 4, 6, 8, and 9, and cross-interrogatory 13.

Mr. Stanley then offered the deposition of Herman Straus as evidence, admissible upon two grounds: (1,) as upon the ground already ruled upon, and (2,) as showing Mr. Clough's dealing in the coupons of the other road.

Defendant's counsel objected to that part of the deposition relating to free passes.

The chairman stated that as far as the deposition tended to show that it was the practice to use the coupons referred to, it would be admissible.

Defendant's counsel stated that there was no objection upon their part to that portion of the deposition.

Mr. Stanley thereupon read a portion of the deposition of Mr. Straus.

The deposition of Jacob F. Smith was offered in evidence by plaintiff's counsel and read.*

* The Depositions will be found on the closing pages of this work.

## TESTIMONY OF GEORGE J. CARNEY.

Q. (*By Mr. George.*) Won't you please state your residence, your age, and your present occupation, Mr. Carney?

A. Lowell, Mass.; age, 33; my present occupation is treasurer's clerk of the Merrimack Manufacturing Company, in Lowell.

Q. Who is the treasurer?

A. Francis D. Crowninshield, of Boston.

Q. Where does the treasurer have his office?

A. The treasurer's office proper is at Lowell.

Q. Will you state whether you were employed at any time to make an examination as to the matters of the Concord Railroad? If so, by whom, and the circumstances under which you were employed?

A. I was employed to make an examination by yourself and Gen. Butler, associate counsel of the Concord road, in, I think, October, 1865.

Q. State who first spoke to you on the subject?

A. Gen. Butler.

Q. State how it was. After Gen. Butler spoke to you, when did you see me first; and whether, prior to that time, you had any acquaintance or communication with me whatever?

A. Gen. Butler spoke to me, I think, one afternoon, about two o'clock, and gave me a letter to you, and I came up in the train which left, I think, at six o'clock, and I came directly to see you. I think I saw you the same evening, and delivered my introductory letter to you. I don't remember exactly the conversation, but I think you engaged me to undertake an investigation, and I commenced at once.

Q. State how it was prior to that time about your having any acquaintance either with the conductors of the road or myself.

A. I think I had never seen you before.

Q. The conductors?

A. I didn't even know them by name.

Q. Now you may state how it was with regard to your keeping a journal of your transactions?

A. I did keep a memorandum of everything that I did or caused to be done, I think.

Q. Will you turn to your journal, or memorandum, of November 10, 1865—or, first, I will ask you whether you have any recollection, apart from your memorandum, so that you can testify apart from it; and whether that memorandum was made at or about the time, and whether it was true? [Memorandum shown.] You may read the whole or a portion, just as you desire.

A. [Reads.] "Friday, November 10, '65. Saw Gen. Butler this morning, and, after talking with him about my business, I took the one o'clock to Nashua; paid 60 cents for ticket to Nashua. At Nashua took cars for Concord; paid Clough $1.50 for ticket to Concord, and he gave me no ticket."

Q. Now, November 11?

[Defendant's counsel objected to the reading of anything from the memorandum except as to what the witness did.]

Q. Please read what you did.

A. [Reads.] "Saturday, Nov. 11, 1865. Took the 3.30 train from Concord. Clough, conductor. Draper came down with me as far as Nashua. I paid Clough $4.25 for my fare to Boston and Draper's to Nashua."

Q. (*By Mr. Mugridge.*) You mean to say that the $4.25 covered both?

A. Yes, sir.

*Mr. George.* The next relates to his instructions to Draper.

*Mr. Mugridge.* That we object to.

[Witness continues to read, omitting the portion objected to.] "Clough gave me a ticket to Boston which is marked No. 6, issued by stamp in ink on the back, at, I believe, East Concord, on Oct. 2, -86-."

Q. Is that the ticket?

A. That is the ticket.

Q. What is the endorsement on that ticket?

A. "Nov. 11, 1865. No. 6. G. J. C."

Q. Is that the only ticket you were given from East Concord to Boston?

A. Yes, sir.

Q. Now, sir, you may turn to November 14th.

[Objected to on the ground that the conversation in the memorandum was between the parties.]

Q. Go on and read the last part of November 11th. I thought it was November 14th.

A. [Reads.] "At Manchester, or below it, rather, Clough came around for the tickets. Two men behind me had tickets which Clough told them were good only on the freight trains, and therefore he wanted fare from them."

Q. Do you know whether the next remark was made in presence of Mr. Clough?

A. I think it was not.

Q. Do you recollect the fare paid between them—between what points?

A. I have no recollection at all.

Q. Do you know how far they rode?

A. No, I don't know at all, unless there is something here about it. I should trust entirely to this, and not to my memory.

Q. Won't you see if there is anything there that shows how far they rode? Read the whole.

A. I don't know how far they rode; but I have a minute here which says. I can state that I overheard one of the men state what he paid; and I made a minute of it at the time.

Q. Won't you go on and say what you were going to say, without developing the fact you were going to allude to?

A. Shall I be allowed to state the facts as they are here?

Q. Any conversation not in the presence of Mr. Clough you will not state.

A. [Reads.] "There were men behind me of whom Clough wanted fare, and one of the men stated, after Clough had passed, what he had asked. Whether they paid it or not, of course I could not tell."

*The Chairman.* Have you any recollection of noticing when they left the cars?

A. I have no notice, I merely heard the remark. I did not see the fare paid.

Q. Now you may take the next.

A. [Reads.] "Clough also collected the fare from a gentleman with a small reticule or leather bag slung over his shoulder. This person was on the train from Manchester to Nashua, and here I left the train to give Draper some directions. and then entered another car.—Gave Draper some directions. The two men above mentioned were aboard in the cars at Concord when I got aboard. Mr. Clough gave me no check for Draper to Nashua."

Q. Now you may turn to the 14th—Tuesday, November 14th.

[Defendant's counsel objected to testimony in regard to any collusion between Mr. Kendrick and Mr. Clough.]

*The Chairman.* We think that the point is well taken here; that this evidence as we understand it does not tend to show any money in the hands of Mr. Clough. Even if it did tend to show irregularity, it does not tend to show that he got any money by it; and therefore that testimony is, we think, incompetent on that ground.

[Plaintiff's counsel submitted in writing what the evidence was which was offered.]

*The Chairman.* We consider that the testimony proposed here is inadmissible, as it does not show that any of the money was received by Mr. Clough. The proposition is to show that Mr. Kendrick, after seven dollars issued, issued to Major Carney four tickets used by Mr. Clough, and thereby the Concord road was defrauded to that extent. But there is no evidence to show that any of that money went into the hands of Mr. Clough. We are inclined, now, without absolutely ruling out the testimony, to hold that there shall be some evidence tending to show that Mr. Clough and Mr. Kendrick divided the proceeds in this matter in some way—some evidence tending to show that Mr, Clough had some pecuniary benefit from it—before putting in the other evidence. It is a mere matter, I suppose, in the discretion of the court, the order in which the evidence should go in.

*Mr. George.* There is no other evidence except as to the circumstances under which the transaction occurred. We propose to show that Mr. Carney came up on Mr. Kendrick's trains; he ran trains from Portsmouth; that he purchased of him four fares, and paid him seven dollars, which was a dollar short of the regular fare—made a "knock down" of a dollar; (that is, instead of taking the regular amount, he charged him seven dollars instead of eight;) that he then gave him these irregular tickets; that then Mr. Carney was in the ears of the Portsmouth road, separated by a baggage car; and that Mr, Clough came in at Manchester and took up these irregular tickets which he had no right to do, from the circumstances of the case; and that the road was defrauded out of seven dollars.

*The Chairman.* If I understand you right that is the substance of the evidence?

*Mr. George.* That is the substance of the evidence.

*The Chairman.* I believe we are all of the opinion that the testimony, being substantially what you state, is not competent. It does not tend to show money into the hands of Mr. Clough.

Q. (*By Mr. George.*) Well, now; the 14th, the same day; you may go on.

A. This afternoon, the 14th, I took the 3.30 train for Nashua; and had only three men with me; but intended at first to give to Clough—— [Objected to; and objection withdrawn.]

Q. You may go on and read the whole.

A. I intended, at first, to give to Clough just seven dollars and a half, the fare of us to Nashua 3, and to Boston 1, thus: 3 at $1.50—$4.50; 1 at $3.00—making $7.50. But as there were a number of soldiers aboard, I determined to try a little game of my own; and accordingly, when Clough came around, I told him I wanted to pay for six to Nashua and one to Boston, thinking, that if required to identify my men, I could do so for three of them, but hoped to allay suspicion by stating that very possibly they went down to Lawrence by mistake. I gave into George Clough's hand, thirteen dollars—one ten dollar, one two and one one dollar bill; thirteen dollars total. After we left Manchester, he came along and asked how many I paid for. I answered six. He then said he would be back to see me soon, and I became very nervous. I took a book, to appear unconcerned; and Clough passed me twice, whether for scrutiny or not, I know not. Then he approached me from behind and thrust something into my hand, and immediately went back. On examination it proved to be fifty-five cents. He made no remark, and I supposed he intended it as a present. Not being a multiple of six or seven, I am not able to state how he arrived at this result. Clough gave me a ticket for Boston which is marked "No. 7." The fifty-five cents I have retained, and may be found in envelope marked 55 cents, in figures. [Ticket shown.]

*Mr. George.* It is marked, "No. 7, November 14, 1865, from Concord, Manchester and Lawrence road check; tickets are sold ten cents cheaper at the office than in the cars; passengers are requested to keep their tickets in sight. G. G. Sanborn, T. M." It is punched through the word Manchester.

*The Chairman.* Is there any objection to the reception of evidence that Major Carney delivered to Mr. Clough any of these tickets that we talked about this forenoon. You remember, Mr. Mugridge, that Major Carney has testified to having obtained some tickets from Mr. Kendrick; and the question was as to the admissibility of that testimony; and after Col. George stated here what he proposed to offer, he has finally reduced it to writing. It is a simple statement of what he proposes to prove by Major Carney; and one part of it is what you, I suppose, understood, that Major Carney had four tickets on which he rode from Manchester to Concord?

*Mr. Mugridge.* I did not know of that fact before.

*The Chairman.* I want to know if that is not true, that he had these tickets and rode on them to Concord, and gave them to Mr. Clough?

*Mr. Mugridge.* No.

*The Chairman.* The other I don't think can be admitted unless by showing that here was a fraudulent irregularity of which Mr. Clough was knowing.

*Mr. George.* We can only show it by the circumstances of that case. We have no further testimony.

*The Chairman.* Simply offered by itself, I think that must be the ruling.

The ruling was excepted to by plaintiff's counsel.

Q. (*By Mr. George.*) Describe the tickets.

A. First, a ticket with a pentagonal, five sided punched hole in it, purporting to be to Sanbornton Bridge from Lowell, signed in printing, "B. F. Kendrick"; large figure 2 on the left end. Second, a yellow ticket, C. M. & L. R. R., Lowell to Concord, 1864; punch stamp like a horse shoe, thus [indicating]; inked stamp on back, May 20, 1865, C. M. & L. R. R. Third, green ticket on one side, white on the other; to Concord; horse shoe stamp. Fourth, white on one side, red on the other; issued by Boston and Maine; no punch and no ink stamp.

Q. Have you any recollection or record with regard to between what stations that was?

A. No. These tickets I gave personally into the hands of George Clough, conductor from Manchester to Concord, and he made no objection nor remark.

Q. I want to ask you, simply, when you rode from Portsmouth to Concord, what train was it?

A. The first train from Portsmouth; 5.30 A. M. train from Portsmouth. Saturday, November 25th, took 3.30 train for Lowell; paid Clough $1.60; Batchelder on the same train; see his report of to-day. Draper paid to——

[Objected to unless witness saw Mr. Draper pay.]

Q. Have you any recollection apart from what is on your minutes at all?

A. I have not.

Q. Won't you state what you wanted to state in regard to that; what the fact is?

A. The fact is, I cannot remember whether I saw Draper pay or not.

Q. (*By the Chairman.*) Are you able to state that you did or did not make a record from your personal knowledge of the fact?

A. I cannot say, because I do not remember. I may have seen him pay, but I don't remember.

Certainly that is not within the rule; not that part of it. [Ruled out.]

CROSS EXAMINATION.

Q. (*By Mr. Mugridge.*) I want you to take your report of November 24, 1865.

A. "Took 10.15 train for Boston. Paid Clough $1.60 to Nashua."

Q. Now, sir, I want to commence right here, and take Mr. Clough's way-bills. On the 10 15 train you paid George Clough a fare to Nashua?

A. Yes, sir.

Q. He has returned three, has he not?

A. I suppose that is what he means.

*The Chairman.* That is November 24?

*Mr. Mugrigde.* The same day All my inquiries will relate to this now.

Q. I will ask you next as to November 25th, on Saturday. You paid Mr. Clough one fare to Nashua, $1.60, on that day. Now will you be kind enough to take the way-bill for November 25th, on the 3.30 train, and see if there is a fare returned to Nashua?

A. I think there are five returned.

Q. The last one you paid was under date of Nov. 14. You say you paid to him six fares to Nashua and one to Boston that day, on the 3.30 train?

A. Yes, sir.

Q. Now, take the way-bill of Nov. 14, the 3.30 train, and see if he has returned any fares ro Nashua and Boston; and, if so, how many?

A. Nashua, there appears to be ten; Boston, via Lowell, two.

Q. The next paid before that was on November 11th, Saturday, on the 3.30 train. You say you paid Mr. Clough $4.25 for your fare to Boston and Draper to Nashua; that is, one ticket to Boston and one to Nashua.

A. Yes, sir.

Q. Now, sir, I will pass you the way-bill of George Clough, on the 3.30 train, and ask you if he has made any return of fares to Nashua and to Boston?

A. There appears to be three to Nashua and one to Boston, via Lowell.

Q. The first one that you presented was dated November 10, on Friday. "Saw Gen. Butler this morning, and after talking with him about my business, took one o'clock train for Nashua; paid sixty cents for ticket to Nashua." You paid that on the Nashua and Lowell road, didn't you? That wouldn't be returned to Concord?

A. No, sir.

Q. "At Nashua I took the cars to Concord; paid Clough $1.50 for a ticket to Concord, and he gave me no ticket." That is under date of November 10th. Now, sir, look at the 10.15 train of Mr. Clough's returns, and see whether he has returned any fare from Nashua or not?

A. There appears to be two between Nashua and Concord.

Q. Now, sir, I want to ask you how many men were in your employ to conduct this system of espionage that you carried on against the conductors of the railroad?

A. I had Mr. Draper, a man named King, a man named Vally, Dr. Carney, and others; although they are not under pay.

Q. Is Dr. Carney your brother?

A. Yes, sir.

Q. He lives in Boston?

A. In Boston.

Q. How many physicians did he have with him?

A. I can tell by looking at his report.

Q. Won't you be kind enough to tell how many he had, who were not under pay as you say? Haven't you the facts in your mind without looking your documents all over? Do you know whether he had one or fifty?

A. I know he had one; I think he had two or three more.

*The Chairman.* I understood the witness to state that by looking at the report he can refresh his memory.

Q. Have you any idea, sir, outside of your report?

A. I have not. I purchased tickets for five persons — [looking at report.] [Objected to.]

Q. What does it state?

A. It does not say.

Q. Then your report comes to nothing?

A. I don't know, sir.

Q. Do you see anything there that indicates the names of the men?

A. It shows the purchase of tickets for the five persons; that is all.

Q. You are the man that this matter was entrusted to?

A. Yes, sir.

Q. Who employed your services in this direction?

A. Colonel George and General Butler.

Q. Was it one or both?

A. I suppose they were together.

Q. Were they together at the time they employed your services?

A. No, sir.

Q. Now who was the man who contracted with you? Col. George or Gen. Butler?

A. I should think Col. George more than Gen. Butler.

Q. Did General Butler employ you a little?
A. I don't know what you mean.
Q. What do you mean, that Col. George employed you rather more?
A. I think the matter was placed in Gen. Butler's hands to determine; and after that he gave me his note to Col. George.
Q. Who was the man who employed you to carry on these operations? Col. George or Gen. Butler?
A. Col. George, I think.
Q. Altogether, didn't he?
A. I couldn't say certain.
Q. I want to ask you what compensation you were to receive and were promised for these operations that you were to carry on?
A. I was promised no definite sum.
Q. Will you state what inducements, in the line of compensation, were held out to you by the person who employed you, as consideration for your services?
A. I think no inducement was held out at at all. I was told that if I came up here——I think the conversation was with Gen. Butler that afternoon before I came; I asked him what would be a fair price to charge, if I came up and was engaged by the road, and was not successful at all, basing it upon the fact that I might be engaged in the investigation two months; and I think he said it would be about five hundred dollars.
Q. I want to know if you made up your mind, before commencing, as to what your pay ought to be?
A. No, sir.
Q. Have you stated that you had made up your mind to charge twenty-five thousand dollars?
A. Not that I remember.
Q. I want to know if you have stated anywhere, that since you got through with the operation you have concluded what your compensation would be; and that it would be a sum from twenty to twenty-five thousand dollars?
A. I have not said it in those words.
Q. What have you stated?
A. I have stated I was told, I think, by Col. George, that the revenues of the road were increased fifty thousand dollars a year by my services. I took the first ten years of the benefit that would accrue to the road, and cast upon that five per cent., which would be a proper commission.
Q. How much would that be?
A. That would be twenty-five thousand dollars. I think I stated that I did not think that it would be fair to claim that I had a life lien.
Q. You say you reckoned five per cent. on that as a fair commission?
A. Yes, sir.
Q. That would amount to twenty-five thousand dollars?
A. Yes, sir.
Q. Now you say you made up your mind that you ought to receive a commission on the revenue how long?
A. I didn't make up my mind that I should; I think I was told so by Gen. Butler.
Q. For how long a time did he tell you?
A. He didn't limit it at all.

Q. Have you limited it at all, as being your own idea of propriety in the matter?

A. No more than what I have told you.

Q. You have given a deposition in this case?

A. Part of two.

Q. I want to ask you if you have stated anywhere as to the length of time that you ought to have twenty-five thousand dollars a year, for services rendered by you in your system of espionage?

A. No, sir; I think I never have made any such statement.

Q. Are you confident?

A. I don't remember any such statement.

Q. Have you stated what, in the aggregate, would be a fair sum to compensate you?

A. I think I have expressed an opinion that twenty-five thousand dollars would be satisfactory.

Q. Let me ask you if you regard twenty-five thousand dollars, to-day, to be a fair compensation for your services?

A. I should decline to answer the question, because that would bring up what my services were.

Q. What is your idea on that subject?

A. My idea is that if I was offered twenty-four thousand nine hundred and ninety-nine dollars, I should not take it.

Q. Do you make a claim against the Concord Railroad for the services rendered by you and for your operations against the conductors?

A. I have not decided what to do. I should take the advice of counsel for the road—Col. George and Gen. Butler.

Q. I want to ask you if you do not mean to say here that, in your judgment, a fair compensation for the services rendered by you to the Concord Railroad, would be twenty.five thousand dollars; and if it is not your present intention to claim that?

A. In my opinion, twenty-five thousand dollars would be a fair compensation.

Q. And isn't it your present intention to claim that sum?

A. Yes, sir. I may increase it, or may decrease it.

Q. It is your present intention to present a claim for twenty-five thousand dollars?

A. Yes, sir; at this minute.

Q. How long were you employed in this system of espionage?

A. About six weeks.

Q. Let me ask you if you have already presented a bill for services; and if so, what is the amount of that bill?

A. I think they have a bill from me for two thousand dollars?

Q. When was that bill presented?

A. That I don't know.

Q. Shortly after you rendered your services?

A. I think so; I don't know.

Q. Has that bill been paid, sir?

A. No.

Q. Now will you be kind enough to state any reason that you have assigned why that bill has not been paid by the Concord Railroad? If you have assigned a reason, what reason have you assigned?

A. The only reason I have assigned, was one given me by Mr. Weld, who said that the directors would decline to do anything with any claim that I might have, until after I had given my testimony.

Q. Did you ever say that a bill had been presented to the Concord Railroad; that they never objected to it; but that they thought it better for you to appear here unrewarded; that your testimony would then carry more weight?

A. I think, as a matter of recollection, that your words express the idea of the matter. I do not recollect exactly that I ever made such a statement; that is, exactly as you express it.

Q. How did you express it?

A. That I don't remember. I think it is somewhere in one of my depositions.

Q. Have you testified substantially to this? Will you look at your deposition and see?

A. Here it is; 117.

Q. Now let me ask you one thing. When this deposition was taken, did you take every single question down, and then did you write your answer down, and read it off, before the magistrate wrote it down?

A. I think I did.

Q. And then, after having done that, the magistrate wrote it down?

A. I don't know; my impression is that I did do so; although probably some of them wrote while I wrote.

INT. Do you know that a committee has been appointed by the board of directors to consider the justice of your claim, and have reported, upon advice of counsel, that it is very much too large a sum for the services rendered? ANS. No, sir; I do not.

INT. Have you rendered an account for your services as detective? ANS. Yes.

INT. Has it been paid? ANS. It has not.

INT. Has any part of it been paid? ANS. No.

INT. Do you know why? ANS. My impression is that the directors think my evidence would go further with a jury if I went into court unrewarded.

*Mr. Mugridge.* I want it distinctly understood that I do not put in this deposition at all.

Q. And the reason, as you understand it, that your bill is not paid, is that the directors have an impression that you would stand better unrewarded?

A. That is the impression that I have. I don't recollect where I obtained it.

Q. Do you recollect whether Gen. Butler ever made that suggestion to you?

A. That I could not say.

Q. Have you had any conversation with him as to whether you would stand better?

A. That I do not remember.

Q. Have you and the general had frequent conversations as to your compensation, and what he thought would be proper and just in regard to the matter?

A. No.

Q. Let me ask you if you have got a pass on the Concord road?

A. Yes, sir.

Q. How long have you rode on a free pass over the Concord road?

A. Whenever I came to Concord, since early in 1866. I won't say all the time, because when I had that deposition I think I paid fare to come up.

14

Q. You stated that it was your present intention to charge the Concord Railroad twenty-five thousand dollars for services rendered. When you went into this matter, was your compensation regulated or fixed in any way by the amount that you should discover as the peculations of the conductors?

A. My idea was that the amount of compensation that I should receive would be regulated to a certain extent by the benefit received, as it might be shown by a tabular statement. I think Col. George has stated to me, in general terms, I should say, at least half-a-dozen times, that he had purposely abstained from mentioning any amount.

Q. Did you call it that you were entitled to a good round sum?

A. I should say so. He never indicated any sum.

Q. He has heard you state before this, at the time you gave your deposition, that twenty-five thousand dollars was your price?

A. I never stated that it was my price.

Q. Did he ever indicate to you, after that, that it was too much?

A. He never indicated any idea about any specified sum.

Q. Now, sir, your idea is this: if you were an instrument in discovering half a million dollars of frauds, you would receive more money than if you had not discovered more than ten thousand dollars' worth?

A. Yes, sir; that is my impression.

Q. That is the way you went into it?

A. I think I have stated in my deposition somewhere that I should receive a given sum whatever was the result of the suit.

Q. But you expected to receive more if you discovered large frauds?

A. The greater the result, the greater the reward.

Q. If you detected half a million, and the road saved that amount, your idea was that you would receive more than if you detected only ten thousand dollars' worth?

A. Perhaps I don't make myself fully understood. I have stated in one of my depositions that the amount of compensation did not depend upon whatever might be recovered from the conductors, but it would depend in a great measure upon what might be deemed to be the benefit coming to the road.

Q. Have you any doubt, sir, but what you will receive the sum of twenty-five thousand dollars, if this suit is decided in favor of the road?

A. I do not know; I wish I could tell you.

Q. Have you ever made a statement on that subject? Has that question been put to you, and have you made a reply to it?

A. I think not; not a definite question and answer.

Q. Let me ask you if this question was put—"Have you any doubt but that you will receive said amount—twenty-five thousand dollars—provided said suits are successfully prosecuted by the Concord Railroad?" And was your reply—"I think I shall obtain this amount whether it is decided in favor of the road or not?"

A. Yes, sir; I think I stated somewhere that I did not mean to charge any specific sum; because I did not know what to charge.

Q. Did you make that specific answer?

A. I did; yes, sir. But I would not be understood as confining myself to twenty-five thousand or two hundred and fifty thousand dollars; because that is a matter that you will have to settle for me at some other time.

Q. Let me ask you if you have a direct pecuniary interest in the result of the suit of the Concord Railroad against George Clough?

A. I think I have a pecuniary interest.

Q. A direct pecuniary interest in the result of the suit?

A. Yes, sir; that is, George Clough's suit and all the suits; his proportion of the whole of them.

Q. Have you not a direct pecuniary interest that the verdict in the suit against George Clough shall be swelled to as large an amount as possible?

A. I think I am disinterested in that.

Q. Has that question been asked you?

A. I don't recollect.

Q. Will you turn to your deposition, interrogatory 62, and read the question and answer?

A. [Reading.] "Have you not a direct pecuniary interest that the verdict in the suit against him shall be swelled as large as possible? A. I believe I have." That was merely my opinion then; and that has changed.

Q. But your opinion has changed?

A As the question comes to me now; as my thoughts come to me.

Q. Was it your opinion then, when the question was asked? Was that your opinion?

A. That was my answer then.

*Witness.* I should like to ask the referees whether I have not the right to amend my answers so that the deposition shall be a whole one. Neither of the depositions have been signed or finished.

*The Chairman.* I should understand this: that in this cross-examination the counsel have a right to ask you if at any other time you have made certain statements; and I suppose that you are to answer whether you made these statements or not; and I suppose, also, that you have a right to make any explanation that you see fit to make.

Q. I put you this question. Have you not the motive of a direct personal pecuniary interest to do all you can toward the procurement of a large verdict against George Clough, in this suit? Will you answer that question without referring to your minutes?

A. I don't think I have. I think the amount to be recovered from Mr. Clough I am disinterested about.

Q. Now, sir, when you gave your deposition was that same question put to you? You may look now at your deposition, interrogatory 63, and read the question and the answer to it.

A. [Reading.] "Have you not the motive of a direct personal pecuniary interest to do all you can towards the procurement of a large verdict against George Clough, in this suit? A. I believe I have."

Q. That was the answer given by you at that time?

A. Yes, sir. I have the word "additional" added on my notes. I read this deposition after I got back to Lowell; and I put on the word "additional."

Q. Have you not, from the moment you set out as a detective, performed your duty and rendered the services you have rendered, under the stimulus of a direct personal interest in all you have done?

A. My opinion now would be that I had done so. I don't know how I answered then.

Q. Have you done any act, or made any movement, for the discovery of the alleged frauds upon the part of the conductors, that has not been done and made by you under the stimulus of a direct personal and pecuniary interest?

A. I think everything I have done has been done with the hope of reward.

Q. Have you done any act for the discovery of the alleged frauds that has not been done by you under the stimulus of a direct personal, pecuniary interest?

A. No, sir; I think every act that I have done has been done for the purpose of obtaining the pay for it.

Q. And under that stimulus?

A. Yes, sir; daily bread.

Q. Do you know, or have you understood by conversation with the officers of the road, that you were the principal witness relied upon in the suit against George Clough? [Objected to.]

Q. Do you know, or have you been informed by the officers of the road, that without your testimony they had no suit against George Clough? [Objected to.]

*Mr. Mugridge.* I propose to show the inducements that were made, and the understanding he has had as to the importance of his testimony, and the inducements under which he testifies in giving that testimony.

*The Chairman.* It does not appear to me that, at this time, the plaintiff can be bound here by anything so general as a statement of the officers of the road.

*Mr. Mugridge.* It is showing the motive; it is not to bind the officers of the road.

*The Chairman.* We think that, in a general way, he may be asked if he understood that he was a material and important witness.

Q. (*By Mr. Mugridge.*) I desire to put this question: Have you ever been informed by the officers of the road, or any of them, and if so, by whom, that your testimony would be very material to them in the trial of this case, and that without it they have no case against George Clough? And is that your present understanding? [Objected to.]

*Mr. George.* The objection is, that supposing the officers of the road did or did not say so, it is an entirely immaterial matter. It is put in here for a purpose; and I suggest that it is not competent for any purpose. All the purpose with which it is offered is to get a confession from the officers of the road, who cannot make a confession for the road.

*Mr. Mugridge.* All I have to say is, that may be true or it may not be true. There is a strong inducement for him to come into court. And I have to say that there is a very strong inducement, as is apparent from his own testimony, for a man to make up a record. But supposing it was made three years ago, he gives his testimony as of to-day; and it is to be considered a record as of to-day. And we say it is proper for us to show all the interests which surround the witness at the time he gives his testimony.

*The Chairman.* I do not think the question is admissible quite in this general form. We suppose that if any of the managing agents of the road have made any statements from which the witness could infer the great importance of his testimony, he might presume upon the stand to know about the situation of the case that might be put in. We do not think that generally the question can be put as to whether any of the officers—people who might or might not have anything to do with a competent cause—stated to him as to the importance of his testimony.

Q. (*By Mr. Mugridge.*) Is it your present understanding, and has it been, that without your testimony the Concord road have got no case here against Mr. Clough?

A. I think now it would be my impression that they have a case against him.

Q. How did you testify as to that when you gave your deposition? Look at interrogatory 67.

A. Because I supposed when I made my report that that was all the evidence. I am no lawyer.

Q. Your understanding is that now they have a case? Now I want you to read that interrogatory 67.

A. [Reading.] "Have you been informed by them that without your testimony they have no case against the conductors or either of them? And is that your understanding?"

*Mr. George.* I understand that the question of the deposition has now been ruled incompetent. Now they propose to show what answer he gave to an incompetent question; and put that in.

*Mr. Mugridge.* The question is made up of two parts; first, as to what the officers of the road informed him, and second, as to his understanding. I do not propose to put in the first part.

Q. Read the interrogatory and answer.

A. [Reading.] "Have you been informed by them that without your testimony they have no case against the conductors, or either of them? A. I have not been so informed; and I believe, myself, that without my testimony and evidence there will be no case."

Q. Have you ever been informed by the officers of the road, its managing agents, in relation to these suits, that your testimony was of a very material character in relation to them?

A. I cannot say that they have informed me so. I have believed so, I believe that no one has ever expressed the opinion.

Q. Any of the managing agents of the road? Have the attorneys of the road?

A. I think not.

Q. You say they have not, or either of them?

A. I think not.

Q. Has not Gen. Butler told you so?

A. I think not.

Q. But it has been your impression? Did you testify, when you gave your deposition, that you believed yourself that, without your testimony in evidence, there would be no case against the conductors?

A. Yes, sir.

Q. Did you testify to that?

A. Yes, sir; I supposed then that my record was evidence.

Q. Do you understand and know that upon your testimony made in this case, depends the fact as to whether you receive or do not receive a considerable sum of money?

A. I think it does not depend on my testimony. That is, you are asking my present opinion?

Q. Yes, sir. Do you understand and know that upon your testimony given in this case, depends the fact as to whether you receive or do not receive a considerable sum of money?

A. Well, I understand that what I believed to be testimony and evidence cannot be used by me; and that the result of the case depends upon whether witnesses will be brought in to substantiate what I have here; and that upon them depends the result of this suit. I do not believe that what I shall get from the road depends upon the result of it.

Q. Will you answer my question now? Do you understand and know that upon your testimony in this case depends the fact as to whether you receive or do not receive a considerable sum of money?

*Mr. George.* I think the witness' first answer was a proper and direct one. And I submit to the referees that the counsel has no right to direct the witness how he shall answer; and if he shall answer properly, he shall not dictate any particular manner.

*The Chairman.* I think the court generally holds that if a question is put in a form such as to require yes or no for an answer, he must answer yes or no; but he may put in an explanation.

Q. Now I mean to put this question: Do you understand and know that upon your testimony given in this case, depends the fact as to whether you receive or do not receive a considerable amount of money?

A. No, sir; I do not believe it does.

Q. Now, sir, was this question put to you in your deposition, and did you make this answer to it: "Do you understand that upon your testimony given in this case depends the fact as to whether you receive or do not receive a considerable sum of money? A. Yes, sir."

A. If this case should fall through, I should say the evidence in my possession, in any suit, that there might a case be brought against the Concord Railroad for services rendered.

Q. So far as you know or understand, in all your operations as detective, and, further, in this case, are you not subject to the same feelings and interests and biases of a party who has a direct personal, pecuniary interest in the suit?

A. I am.

Q. Have you not a strong personal feeling that the plaintiff should prevail in this suit?

A. Yes, sir.

Q. Are not your feelings and your actions influenced by your desire, to a certain extent?

A. Undoubtedly.

Q. Have you any doubt that your conduct and operations heretofore, in connection with this case, may have been influenced to a certain extent by such a desire?

A. I have no doubt of it.

Q. But what they may have been?

A. No. sir.

Q. Do you regard yourself, in your actions in this case, as in the attitude of a disinterested person?

A. No, sir.

Q. (*By Mr. Mugridge.*) In putting down your account here of the transactions of Tuesday, November 14th, you got something like this as I have taken it down: "After we left Manchester, he came along and asked how many I had paid for." Then you go on and tell him how many you had paid for. "Clough soon came back to me, and I became very nervous, and took up a book to appear unconcerned; and Clough passed me twice; whether for scrutiny or not, I know not; then he thrust something into my hand from the back; on examination it proved to be fifty-five cents; he made no remark, and I supposed he intended it as a present, not being a multiple of six or seven." Now, sir, you say here, and swear that you supposed Mr. Clough intended the fifty-five cents as a present to you?

A. Yes, sir; I was under a misapprehension then as to the amount. My examination of the latter part of this testimony shows that this is exactly the right amount that he should have returned.

Q. What do you mean, sir, by recording there in that report and swearing to the fact that at that time you thought that Mr. Clough gave you fifty-five cents as a present?

A. Because, as I recollect it, my impression is that the price he should have given me was not the correct amount, until Mr. Sanborn had given me some information. I found after writing that down that Mr. Clough had made the right amount.

Q. Why didn't you, when you put in your testimony, make this explanation?

A. I was willing to put in the whole book.

Q. Didn't you put in that testimony this forenoon, swearing to it exactly as you recorded it?

A. The testimony that I have put in I have sworn to.

Q. Then didn't you know that that testimony was false when you were putting it in?

A. I knew just this: I had to read according to your instructions.

Q. Did you know you were testifying to what was not true when you were putting that in?

A. Yes, sir. I don't know as it was hardly to be called false testimony; because the whole testimony was not put in.

Q. Have you, until the inquiry was put to you, suggested any explanation of that testimony that you unqualifiedly gave this morning?

A. Nothing except what I have suggested.

Q. Nothing except in reply to my question?

A. Nothing before you said you were coming to it.

Q. Did you offer to explain until I commenced to cross-examine as to what you put in this morning?

A. No, sir; I think not.

Q. Now, sir, didn't you know and understand, as well as you know and understand anything, when you put in that testimony this morning that you were giving the referees a false impression?

A. It was not my intention.

Q. If it wasn't your intention, why didn't you accompany the fact with an explanation?

A. I really cannot tell. I read whatever the other side consented that I should read. Whatever was read then was the recollection I had in the matter.

Q. Is that your reason for not accompanying that fact with an explanation?

A. Yes, sir.

Q. Now, sir, did you, this morning, in giving your testimony, knowingly give to these referees an erroneous and false impression in regard to what you regarded as a material fact?

A. No, sir; I did not.

Q. Didn't you know of this explanation at the time you gave this testimony?

A. I wanted to put in the whole testimony.

Q. Didn't you know of this explanation at the time you gave this testimony? Do you mean that you didn't have it?

A. I do.

Q. Let me ask you, sir, if you didn't mean this morning, when you gave this testimony to these referees, to give them to understand that George Clough had given you a present of fifty-five cents?

A. I knew that he didn't. I didn't mean that they should understand anything about it.

Q. Didn't you mean to have these referees understand just what you testified to, as being true, this morning?

A. Well, there comes up the matter that I believe I have stated two or three times. All I know about this case consists in these records that were made by me at the time. I am willing to put the whole in. This morning I thought that the whole that were excepted were those that the other side objected to. I have no desire to tell any untruth here, and I believe I have not.

Q. I want to ask if you meant to give these referees to understand, when you testified, that you understood that George Clough gave you fifty-five cents?

A. No, sir.

Q. Why didn't you accompany that with an explanation?

A. I cannot give any other season than that it escaped my mind at the time.

Q. Now, sir, you mean to intimate here, that one of the reasons why u did not give that explanation is because you were not allowed to put in your entire records?

A. Yes, sir.

Q. You do?

A. Yes, sir.

Q. Now let me ask you if there is a single word in that record that explains this transaction?

A. I think there is.

Q. Will you find it?

A. Shall I read it?

Q. You may read.

A. I think it is that entire page. It is merely accounting for the thirteen dollars.

Q. Do you mean to say that there is a word of explanation on that page with regard to this testimony?

A. I think there is.

Q. Were you not conscious this morning of the existence of this?

A. Not that I know of.

Q. You didn't?

A. I knew that it existed,

Q. Then why didn't you accompany it with an explanation?

A. You have asked me that three or four times.

Q. Well, can you explain it, sir?

A. I cannot.

Q. Now I want to ask you, sir, if George Clough, when he returned you this fifty-five cents, didn't return the exact change that he should have returned you from the thirteen dollars?

A. Yes, sir.

Q. He did?

A. Yes, sir.

Q. Do you see, sir, with all the facts that are before you to-day—— do you see or know of the slightest irregularity connected with the act of George Clough in giving back that fifty-five cents?

A. Not the slightest. [Adjourned.]

[TWELFTH DAY. Friday, August 7th, 1868.]

The hearing was resumed, the cross-examination of Mr. George J. Carney being continued by Mr. Mugridge.

TESTIMONY OF GEORGE J. CARNEY—*continued.*

Q. (*By Mr. Mugridge.*) I understand you to say when Mr. Clough paid you the fifty-five cents he approached you from behind and thrust something into your hand, and immediately went back. Do you mean by that that he acted slyly and stealthily?
A. I meant so at the time. I supposed that he did. A that time I supposed that he approached me stealthily.
Q. And slyly? You meant to indicate that in your testimony?
A. Yes, sir, at the time.
Q. Now, as it turned out, doesn't it appear that that was a good deal imagination?
A. Well, in the ordinary way, I should say not.
Q. Do you see any occasion, as the fact turned out, for Mr. Clough to have approached you stealthily or slyly?
A. Well, I should have to explain myself on that matter. I think the ordinary way of business would be to approach a passenger in front instead of passing the money over his shoulder.
Q. Then you think there was something extraordinary, out of the usual course, in this?
A. My impression about it is that his manner was not that of an ordinary conductor; that is, the manner of an ordinary conductor doing his business.
Q. Won't you please to describe what there was so extraordinary in his conduct?
A. Nothing except the mere fact that he approached me from behind, and passed me the money over my shoulder.
Q. And that was the important fact in your mind?
A. Yes, sir; it is now.
Q. You attach a good deal of significance to that fact?
A. No more than the fact warrants.
Q. Well, do you attach importance to that fact that he approached you from behind?
A. I do.
Q. From your view of the case, this is a matter that you think is entitled to especial consideration and note?
A. If I was called upon to judge of Mr. Clough's action, and had only that, it would have an influence upon my decision.
Q. Now, sir, I want to ask you how much money the Concord Railroad put into your hands to carry on this operation?
A. I think I have received — I see by my memorandum — received from the Concord Railroad thirteen or fourteen hundred dollars.
Q. That amount of money was paid into your hands to conduct this operation that you were carrying on?
A. Yes, sir.
Q. And you have expended it, sir, almost entirely?
A. Yes, sir; I think there is nothing due the Concord Railroad.
Q. That is, you have paid it all out?
A. Yes, sir. But let me explain that matter. I think when I finished — this report was dated, I think, in December, 1865 — I think it

was all expended except somewhere between forty and fifty dollars.

Q. That you have given the road credit for?

A. That I have given the road credit for; and I think there are charges since that.

Q. What are those charges?

A. I have had to pay hack hire and postage; I don't recollect what now.

Q. Have you attended the annual meetings of the Concord Railroad corporation?

A. All of the last ones. In '66 was the first one; '66, '67 and '68.

Q. At whose request did you appear?

A. General Butler's; and, I think, Colonel George's; I am sure the first one was by Gen. Butler.

Q. Did you make any charge against the Concord Railroad for this?

A. I think I charged hack hire from the station in Lowell to the meeting, in 1867; I think that is the only one that I remember.

Q. Make a charge for your *per diem?*

A. I have not.

Q. Have you made a minute of it?

A. I have made a minute of it.

Q. Have you intimated what you proposed to charge?

A. I think I have stated—I don't remember exactly what I have stated—I think it was something in this fashion: If the question came to be settled upon per diem charges, I should charge twenty-five dollars a day. I had decided to charge in that. But I should like to state this, as I believe I have stated before: that I do not wish to be understood, in making my testimony, anything binding myself to any price.

Q. Don't wish to be bound to any price?

A. I don't wish to put any price.

Q. Let me ask you if you do not think, in a certain event, your fees might amount to a great deal more than twenty-five thousand dollars?

A. That I cannot say.

Q. Haven't you the impression that in a certain event your fees might amount to very much more than twenty-five thousand dollars?

A. My impression is that it may amount to more than twenty-five thousand dollars. And that event, of course, would be a decision favorable by the board of referees.

Q. Without saying anything about any decision that the board of referees may be called upon to make, I ask you if it is not your own impression to-day that in a certain event your fees may amount to a great deal more than twenty-five thousand dollars?

A. If you will give what the event is, I can tell you.

Q. Supposing the result of your developments is that the conductors should be charged with peculations and defalcations, we will say, to the amount of five or six hundred thousand dollars, don't you think, in that event, you might get a little more than twenty-five thousand dollars?

A. Well, sir, I suppose if the road was benefitted by my services two millions, that I should charge them somewhere between two and five per cent. on the amount.

Q. You think that would be a fair charge?

A. Fair and ordinary brokerage.

Q. Has Gen. Butler ever told you that that would be a fair charge?

A. Yes, sir; that is my impression, that I was entitled to a per centage on whatever the road was benefitted.

Q. Did you and Gen. Butler ever talk that over?

A. My impression is that I did. I would not say certainly that I did; but my impression is that there was something said.

Q. Let me ask you if there was any conversation between you that there was to be a divi in any event?

A. Not one particle.

Q. Who was the man from whose brain the plan that you subsequently followed in making your discoveries originated?

The mode of operation?

Yes, sir.

A. I did, myself.

Q. Now, did not Butler help you?

A. Incidentally.

Q. Do you remember from whom the original proposition sprung?

A. I think it was from myself.

Q. Do you remember whether you made the original suggestion of this plan, or whether Butler did?

A. I did, I think.

Q. Did you suggest it to Butler?

A. No; I thought of it myself.

Q. You afterwards communicated it to Gen. Butler?

Am I to answer generally?

Yes, sir.

A. I think that when I first came to Concord we were talking about my living here; and I said that if I lived here I must have some ostensible business; and it was suggested that I had better take some general business—buy some kerosene lamps.

Q. Then Gen. Butler proposed to put you on "your kerosene" at once, did he not?

A. No, sir; not exactly.

Q. I should like to know now whether Butler suggested the kerosene business, or whether you suggested it?

A. There was no suggestion made at all.

Q. How is that, Mr. Carney?

A. I said that I should have to have some ostensible purpose—an apparent purpose for being here in Concord. As I remember now, there was something said about a patent right to some oil lamp.

Q. Was it kerosene?

A. I have forgotten.

Q. Didn't you say a moment ago that it was in connection with kerosene?

A. I supposed, from the term, that is was in connection with kerosene.

Q. Did the kerosene oil business commend itself to your judgment?

A. No, sir; it did not.

Q. Very well, sir; you took umbrage at that?

A. I believe the general was not offended at my manner.

Q. Let me ask you how much time you and Gen. Butler consumed in this?

A. I think Friday afternoon I went to his office at two o'clock. I waited there until about 2 1-2 o'clock. It was a stormy day, and the general was in his front office; and I remember I had my cavalry boots on, already to march. Sometime, I think, after half past two.

Q. (*By Mr. Tappan.*) What Friday was that?

A. I think it was the last Friday in October, 1865. I would not be positive.

Q. (*By Mr. Mugridge.*) You had on your cavalry boots?

A. Yes, sir.

Q. Did you have your sword on at that time?

A. Not much.

Q. Was the general in uniform?

A. He had his clothes on.

Q. How long did you talk this thing over?

A. I think I was there less than half an hour.

Q. Were you closeted by yourselves?

A. The key was in the door; the door was not locked. He wrote a note to Col. George, and I brought it up; and before I saw the general again I had decided upon the business.

Q. Then he submitted the matter to you untrammelled?

A. Yes, sir.

Q. And you decided to appear as a license agent?

A. As a claim agent.

Q. Well, sir, what did you do next? Did you tell Butler what you had concluded upon?

A. Yes, sir; I suppose I did.

Q. How did it seem to meet his views?

A. I don't recollect any expression of disapproval.

Q. Did he still insist on the kerosene business?

A. No, sir; I think not.

Q. What did you do after you came up to Concord?

A. Well, I don't know whether I carried the letter that night to Col. George; but I think it was that night.

Q. What did you do when you came here?

A. I came here as a claim agent.

Q. Did you advertise as a claim agent?

A. Yes, sir.

Q. With whom did you do business, ostensibly?

A. L. D. Stevens.

Q. Did you advertise pretty extensively?

A. As extensively as the business would warrant.

Q. What did you pay for a license?

A. I don't know; it is marked on the license, I think.

Q. Five dollars and eighty-three cents?

A. Whatever is paid is there.

Q. Did you advertise in the newspapers of the state?

A. Some, I think.

Q. But you did not do that business at all, did you?

A. I wrote one or two letters to friends of mine, who wrote and sent me claims to collect, and told them they had better send to others.

Q. Then you put yourself forward in the cars as a claim agent?

A. That is what was announced.

Q. How old are you?

A. Thirty-three years old, I think I said.

Q. What business have you been engaged in since you were twenty-one years old, sir?

A. I have been engaged in merchandizing,
Q. Where?
A. In the city of Boston.
Q. What else?
A. Banking.
Q. What position did you hold in the banking business?
A. Assistant treasurer for the "Lowell Institution for Savings." I am still holding that position.
Q. Are you on the retired list now?
A. No, sir; I am under bonds; and am officially on duty, though I am not under pay.
Q. How long has it been since you held that office, and have not received any pay?
A. Since December, 1862.
Q. Then it is practically nothing but a title?
A. It is nothing but a title. If my father should be sick, I should step in and take it.
Q. You hold that title under bonds? Have you any other titles?
A. I was going on.
Q. Go on in your own way.
A. I have been in the service of the United States, in the army; then as a civil officer. Then as an employe of the American Manufacting Company.
Q. What capacity?
A. As treasurer's clerk.
Q. Who is the treasurer?
A. Mr. Crowninshield.
Q. Do you get any pay from this office?
A. Yes, sir.
Q. You do now?
A. Yes, sir.
Q. How long have you held that office?
A. Since the third of June, this last year. I am temporarily filling a place there as clerk.
Q. Any other office?
A. And I am, I suppose, managing director of the Haverhill Gas Light Company. That office I have held since July—since last month.
Q. Any other titles of office that you have?
A. Yes, sir; officer in the Massachusetts militia—lieut. colonel.
Q. Do you get any pay for your connection with the Gas Light Company?
A. I won't be certain; but I think the directors passed a vote giving them a dollar apiece.
Q. Any other offices that you hold?
A. Not that I remember of——yes, sir, I am justice of the peace.
Q. And quorum?
A. That I don't remember—whether the quorum is hitched on or not.
Q. Were you connected with the Army of the James?
A. I was.
Q. Major general Butler was, at that same time?
A. Yes, most of the time; he succeeded Gen. Ord and Terry.
Q. Let me ask you what titles you held there?

A. Captain, assistant quartermaster, and afterwards major and quartermaster. I believe the general appended A. D. C; but I don't know whether he had the right to or not.

Q. You say you were captain?

A. Captain and assistant quartermaster, and then major and quartermaster.

Q. And the general appointed you A. D. C. did he?

A. A. D. C.

Q. What does A. D. C. mean?

A. Aide-de-camp.

Q. Now let me ask you, sir, were you in the Army of the James during all its campaigns?

A. I was attached to it; I was not in the field.

Q. You were attached to it—was it an honorary position?

A. It was an honorary commission.

Q. You were with the Army of the James in all its campaigns and sieges?

A. Yes, sir; although not in the field.

Q. Were you at the Dutch Gap canal?

A. I have had the honor of being in the bottom of the Dutch Gap canal, and was very glad when I got out of it.

Q. Were you at Fort Fisher?

A. No, sir; I was at Newbern at that time.

Q. What title did you hold at the time of the Fort Fisher affair?

A. I was major.

Q. After you got out of the army what did you do?

A. The first thing was to make this office up here; I was treasurer of the Revere Woollen Mill, Ipswich, Mass.

Q. Were you active in the discharge of the duties of this office?

A. Yes, sir; I run it into the ground.

Q. While you were holding the office, was it run into the ground?

A. Yes, sir.

Q. While you held office as treasurer of this woollen mill, did the thing run into the ground?

A. Well, I am unable to tell you, unless I tell you the full particulars.

Q. Did you say a few moments ago that you were treasurer of that mill, and run the thing into the ground?

A. I may have said so, but I did not mean to say so.

Q. Now you may explain.

A. Four persons were associated. John W. Hill, Mr. Wright, Dwight Foster and myself purchased the mill property and appointed me treasurer. Mr. Hill was thrown from his wagon and killed. We had a large stock of goods on hand, the market falling; but as we had enough money to carry it, we did not sell. And then on the death of Mr. Hill, Mr. Wright retired, and Judge Foster agreed to purchase the property with me. Then the property was sold at auction. Foster held half the stock, and I a quarter; it was knocked off to us for five hundred dollars.

Q. In fact did the thing run out?

A. The thing has not run out. The thing of it was Judge Foster would not sign the deed. I am still treasurer; and the mill has been taken by Francis Skinner & Co.

Q. Have you any funds in your hands?

A. Yes, sir; I have.
Q. How much?
A. I wouldn't say certain; I think between one and two hundred dollars.
Q. Have you been in any way connected with the United States service?
A. I have been revenue inspector.
Q. When did you hold that office?
A. I think I held it from sometime in 1866 until, I think, November, 1867, when I was dismissed.
Q. Who dismissed you?
A. Hon. E. A. Rollins, I think, Commissioner of Internal Revenue.
Q. For what reason?
A. The letter dismissing me said, "very improper conduct."
Q. Very improper conduct on your part?
A. Yes, sir.
Q. Were you charged, sir, with receiving a bribe of a thousand dollars in the discharge of the duties of your office?
A. Yes, sir.
Q. Do you know a man by the name of John Leighton, of Boston?
A. Yes, sir.
Q. You do? While you were in the employ of the United States as inspector of revenue, were you instructed to investigate charges against Leighton for making a false return of his income?
A. No, sir.
Q. Were you instructed by C. C. Estey to make some investigation?
A. Yes, sir.
Q. Did you make that investigation?
A. I did, sir.
Q. How long were you engaged in it?
A. I think somewhere about thirty days.
Q. How much were you paid a day for your time?
A. My impression is that it was four dollars.
Q. During this investigation, did you make an appointment with Estey to have an interview with Leighton and Estey, for the purpose of examining the books.
A. Yes, sir.
Q. For the purpose indicated? Did you both keep that engagement, sir?
A. Yes, sir; I think we did, finally; all three, in fact. I think Mr. Leighton appointed one time and was taken sick, and could not come.
Q. Then the first time you did not keep it?
A. My impression is that the first time we did not. This appointment was to be at Framingham, where Mr. Estey's office is.
Q. Did you write a letter to Mr. Estey, saying that Mr. Leighton was unwell, and could not keep the engagement, and that you yourself were obliged to go in another direction?
A. My impression is that I did; it is merely a matter of impression.
Q. Did you and Estey together ever examine Leighton's books?
Do you mean whether we both examined separately, or jointly?
Whether you examined jointly?
A. No, sir; I examined them, and Mr. Estey did too. That is my impression.

Q. Will you state here, sir, that Mr. Estey ever examined these books?

A. I will swear this: that Mr. Leighton carried his books out to his office, and I saw Mr. Estey looking them over. So far as I noticed, he was examining them.

Q. Did you make a report, sir, in the John Leighton case, so called?

A. I did, sir.

Q Was it approved by Estey?

A. I don't know whether he approved it or no. I should like to state to the referees that this case of John Leighton's was a matter which came up before (I think) Mr. Chase; and I declined to answer the questions after they had gone to a certain extent; and for this reason: that after I was dismissed from my official position, I learned that District Attorney Hyde proposed to proceed against me criminally.

*Mr. Mugridge.* Is there any occasion for this?

*Mr. George.* I think the witness has a right to make an explanation.

*The Chairman.* We all think that the witness cannot be denied the right to make any explanation, which he desires to make, of the testimony which he gives. The witness is asked to answer a question. If it is insisted that he shall give a direct answer, he has a right to give any explanation which he sees fit to give to it. I do not know as the rule of practice goes any further than that; and if the witness feels, after the answer which he has already given, that he should make an explanation, I think the referees ought to hear it. If it turns out that the explanation does not bear upon the preceding examination, then the referees may decline to hear it. But we think the witness has a right to any explanation in regard to the testimony he has given.

A. Well, then, for the reasons given, I declined to answer the interrogatory about the John Leighton matter, after they had gone to a certain extent, on the ground that if they proceeded against me—Mr. Chase advised me that I should not furnish any evidence that might be used against me. But this morning I am in a different position, altogether. I propose to tell all I know about the John Leighton matter.

Q. Will you allow me to take my own course in your examination?

A. Yes, sir. I ought to add that I did not, at the time, and do not now believe that I had done anything wrong.

Q. Did you make two reports, sir, as the result of your investigations; one to Mr. Estey and one to the department?

A. Yes, sir; those were my instructions in all cases.

Q. Did they both bear date of Oct. 5th, 1866?

A. Well, I could not say.

Q. Have you any doubt but what they did?

A. I do not know that they did.

Q. Were they in the month of October?

A. I could not say.

Q. I want to ask you, sir, if they were mailed from Lowell—these reports? And dated in Lowell?

A. Yes, sir; I think so. That is, I may have dropped them in the Boston office; because I was visiting in Boston. I think they were mailed in Lowell; it was my usual custom to mail most of my letters there.

Q. Did the report that you made to the department exonerate Leighton from all intentional frauds?

A. I think my report covered the year —— [Objected to.]

Q. Now, I want to ask you if after you made the report, Leighton paid you any sum of money?

A. I think — my impression is, that Mr. Leighton, Mr. Estey and myself were at Framingham; I won't say what day, but I will suppose it to be Wednesday; I think on Friday Mr. Leighton gave me a check for a thousand dollars, which I kept. That was before the letters were sent from Lowell, although the report was all finished, and submitted to Mr. Estey, and he had dismissed the case. That which Mr. Rolfe obtained in Boston I think a little was done after Mr. Leighton sent the thousand dollar check.

Q. Did you receive a thousand dollars from Mr. Leighton before your report was sent away?

A. The report was received.

Q. Will you answer the question directly?

A. Yes, sir. The examination was made, I think in June, 1866 — June or July — but it was made from Mr. Leighton's books. I had better give you the whole case. [Objected to.]

A. You told me I might make any explanation I please.

Q. Mr. Carney, had you as lief let me ask you one more question before you give your explanation?

A. Yes, sir.

Q. Was the thousand dollar check enclosed in this letter?

A. I think it was.

Q. [Reading letter.] "26 Kilby st., Boston, Sept. 27, 1866.

My Dear Major: The gentlemanly manner in which you have performed what I know must have been a disagreeable duty, entitles you to my respect and esteem. The time employed in the investigation necessary to perform that duty, I wish to compensate you for. And I feel that I can do so now, without the imputation of a desire to influence your action. Please accept enclosed check, and believe me, with a desire to know you better. Truly your friend,

John Leighton.

To Maj. George J. Carney, present."

Q, Now, did you receive a thousand dollar check in that letter?

A. Yes, sir.

Q. And the date that the letter bears is September 27, 1866?

A. That I could not say. It is my impression it was that day.

Q. Let me ask you if you know that letter?

A. I think that letter was brought to me in the office of Lawrence, Reed & Co., Federal street, Boston, where my desk was when I was treasurer of the manufacturing company, by one of Mr. Leighton's boys.

Q. Was that check on Elliot & Co., for a thousand dollars? And did you endorse it and get the money?

A. I don't know what company it was. I did endorse it, and did get the thousand dollars.

Q. Were you dismissed from the treasurer's department on account of this transaction?

A. I was.

Q. Let me ask you, sir, if, before you were dismissed, your conduct was investigated by the department, fully, by Mr. Legrow, of Rochester?

A. Mr. Legrow came to Lowell and I gave him the facts in my possession. I am not sure that I gave a copy of the report; I gave him a copy of the original letter.

Q. Is Mr. Legrow a detective?

A. That is not his title.

Q. Is that his business?

A. His title is "revenue agent."

Q. Let me ask you if he was on your track for any length of time?

A. No, sir; I was not aware of it.

Q. Do you mean to say that he was not on your track for some time?

A. Mr. Legrow had instructions from Washington, asking him to come and see me and ascertain whether I had received a thousand dollars from Mr. Leighton; and he came directly to me.

Q. Do you not think Legrow was on your track for a long time, as a detective, before he came to you?

A. No, sir.

Q. Did he not trace the check of Leighton's back to you?

A. If he was employed weeks and weeks, he ought to be dismissed.

Q. Do you mean to assert that you did not discover that he had been on your track weeks and weeks?

A. Yes, sir; I do say so. I didn't care anything about Mr. Legrow.

Q. Were you arrested by him?

A. No, sir.

Q. Were you ever in his custody?

A. No, sir; and never got out a writ to Legrow to be released.

Q. Do you mean to say that you were not arrested, and that Gen. Butler was not retained as counsel? Was the general retained as counsel?

A. No, sir.

Q. Was he consulted by you?

A. He was.

Q. How long after Legrow met you at Lowell before this was done?

A. I think I wrote to Gen. Butler as soon as Mr. Legrow left. I think it was immediately.

Q. Did you retain him?

A. Yes, sir.

Q. Have you had frequent consultations?

A. Yes, sir.

Q. Has Gen. Butler interfered to protect you, and succeeded in adjusting the prosecution with which you were arrested?

A. Not that I am aware of.

Q. Do you mean to say that Gen. Butler has not operated to stifle it?

A. No, sir; Gen. Butler told me at the beginning there was nothing criminal whatever; and it was by his advice that I declined to answer the questions of the deposition in regard to this matter.

Q. Do you mean to say that it was under advice of Gen. Butler that you acted?

A. Yes, sir.

Q. Where was he at the time this deposition was given?

A. Gen. Butler was in Washington, I think.

Q. Do you know that Gen. Butler went to the secretary of treasury in Washington, and told him that if George J. Carney was indicted for taking a bribe, that Mr. Jordan was also to be indicted with him?

A. I never knew that he said it. It would be very proper for him to say.

Q. Did you ever hear that Gen. Butler held such intimations over the department?

A. No, sir; I never did.
Q. Was any such thing ever alluded to?
A. Gen. Butler told me.
Q. Did you ever hear of that?
A. I heard of Jordan receiving a present in the case of J. D. Williams's suit. Allow me to say that Secretary McCulloch said I had done perfectly right.
Q Did you ever hear such a suggestion by anybody else, that if George J. Carney was indicted Jordan was to be?
A. No, sir; I never heard the names of George J. Carney and Jordan mentioned together until you have mentioned them.
Q. Let me ask you if you knew of Hyde, the assistant district attorney, sending to Washington for instructions in regard to prosecuting you, sir?
A. I think I did. I think I have copies of everything—all the documents in the matter. I think there is one to Judge Hallett.
Q. You say Gen. Butler told you not to answer these questions. Was he present at the time the deposition was taken?
A. No, sir.
Q. When did you consult him?
A. I think Gen. Butler wrote a note, saying: "Say nothing until I see you."
Q. On what ground did Gen. Butler instruct you to decline answering such questions?
A. He gave me no grounds at all. I will give you a copy of the letter if you desire it. He said: "Say nothing about it until I see you." I cannot recollect the conversation, but when I saw him next—when he came north the next time, I met him, and told him as my honor and reputation was impugned, I was desirous to have the matter investigated. The general said that in the event of a case being made out against me, it would be for the purpose of tormenting and making trouble and perplexing. He said that no judge would entertain it for a minute. He said there was really no corruption whatever; but he said that I had better, for prudential reasons, not talk about it.
Q. Did you decline on the ground that it was criminating yourself?
A. I declined to answer on the ground of instructions of Mr. Chase.
Q. State whether you declined to answer on the ground that it might criminate yourself?
A. That would be used against me. I think I stated in the answer before that I was not bound to furnish anything whatever.
Q. Did you ever hear it assigned as a reason why you were not prosecuted, that Mr. Hyde sent to the commissioner of internal revenue that a gentleman had been there and stated that if you were prosecuted Gen. Jordan would also be prosecuted?
A. No, sir; I never did.
Q. Since you gave your deposition, have any of your friends been to intercede with Mr. Hyde?
A. No, sir.
Q. Do you mean to say that?
A. Yes, sir; I do.
Q. Did you employ anybody to do that thing?
A. No, sir.
Q. Have any of your friends employed anybody?
A. Not that I am aware of. There was no occasion for them to.

Q. You say there have been no movements made to stifle this threatened prosecution?

A. I believe nothing has been done whatever.

Q. And that you swear to?

A. That I cannot say; but I do not know of any.

Q. Do you mean to say that no efforts have been made by your friends in that direction?

A. I don't remember that any have been made.

Q. And you swear positively that there have not?

A. I cannot swear positively; but if they have been made, I have not known of them. I know that if Gen. Butler was in power I should be reinstated at once.

Q. Now you may go on with your explanation.

A. Mr. Leighton had a clerk named Nickerson. Leighton's business was that of a stock broker, a man who had got up corners, and who had built himself quite a name in the city, I had heard of him; never had seen him, personally, until one day Mr. Estey said that he had a broker who had made very incorrect returns of his income; and that he should like to have me examine the case. I had taken the office of revenue inspector, after this investigation here, because Mr. John Nesmith thought it would be well to have some competent person; but I found, afterwards, that Mr. Nesmith proposed to use me for his own interest, and I declined. After I undertook the Leighton matter, I found that in Worcester and New York and Boston he had been in the habit of buying these gambling stocks and then going into the market and selling the papers, and making transfer of certificates; and I found in his trunk, I don't know how many certificates of these. This clerk, Nickerson, when he looked over this income return, says: "I don't believe that Mr. Leighton has made correct returns." And upon examination, and refreshing his memory of the sales, he found that Mr. Leighton had made false returns; probably, I think, as high as four hundred thousand dollars. And upon that, charges were preferred against him. And being in my district, I examined the case. I went through Mr. Leighton's books, throughout; and wrote to Washington for instructions as to how I should settle the case. I suggested that the fairest way to settle would be that if he bought these stocks and held them for a year and then sold them for fifteen, he should pay on that fifteen. I wrote, I don't know how many times, to Rollins; and I got a letter to this effect, that profits could not be estimated until the stocks were sold. I think it was a wise decision. At the time Leighton was carrying on this corner, I found that Mr. Nickerson, and the companions of Leighton, tried to obtain information from me in regard to Leighton's matter, and I declined to communicate with them. After an examination of the books, he said that he believed he had made fifty thousand dollars; and he was willing to pay on fifty. The commissioner asked if he would swear that that was exact. He said that he would. He finally made about thirteen hundred dollars more. After the books had been examined, for thirty days, Mr. Leighton took his accounts out to Mr. Estey; and they were examined over, and Mr. Estey said that under the instructions I had received in regard to the matter, Mr. Leighton was not liable for taxation upon these sums, as asserted in the charges. Two days after that had been done, Mr. Leighton sent me a check for a thousand dollars; and I kept it. Then Mr. Legrow came to see me; and I supposed the matter had passed by. I was doing no business as revenue inspector. I kept away. I told Mr. Nesmith that I would take it with this understanding, that I was not to take such a specialty. After they supposed they had dismissed me from office, I received instructions in regard to the distillers' business; and I wrote to Mr. Nesmith. They wrote to Washington and I received this letter; and I think they wrote to Boston, to Judge Hallet.

Q. Let me ask you if just as soon as you were set back, Gen. Butler was retained as your counsel?

A. Yes, sir; I think Gen. Butler was retained as counsel; and then I think John Andrew was retained.

Q. The question was whether, just as soon as you were back, Gen. Butler was retained as counsel?

A. Oh, he was retained; but whether it was just as soon as I got back I don't know.

Q. Was it before your report was made?

A. Yes, sir; I think it was. I ought to add, here, that there was an attempt to bribe me in that case; but it was on the other side.

Q. Well, you may state that, if you please.

A. Mr. Nickerson came to me. He came to Lowell, and we seated at a table in the Savings Bank; and he says: "Major Carney, I have found out that I am entitled to a reward as an informer in this case; and whatever I get I will make right with you. That was all the money matter on the other side.

Q. That you understood to be an attempt to bribe you?

A. Yes, sir.

Q. Have you ever been approached by other men to bribe you?

A. In this matter?

Q. In any other transaction?

A. Yes, sir.

Q. How many times have men attempted to bribe you?

A. I don't know how many; I guess they have tried it a good many.

Q. Were they men who were acquainted with you, who knew your character?

A. No; I think they were not.

Q. Were you ever approached for a bribe by a man who did not know you intimately?

A. Yes, sir.

Q. Who and when?

A. After Gen. Butler left Virginia, in 1865, after he was relieved from duty, the collector of the port came to me with Mr. Jones, I think it was; and at that time I was commissioner of negro affairs. I had the states of Virginia and North Carolina.

Q. What was your title then? Commissioner of negro affairs?

A. Yes, sir; commissioner of negro affairs.

Q. (*By Mr. Haile.*) Freedman's bureau?

A. That was the beginning of it. There was no head to the freedman's bureau at that time? I had on the Taylor farm, opposite Fortress Monroe—I had a nice saw-mill there; and there was a large quantity of fine lumber. These two men came in and said they saw some fine timber down there, and if I would sell them some sticks of timber they would share. I said if they would reduce their proposition to writing, I would consider it. They never did.

Q. How many times can you recollect now, where attempts have been made to bribe you?

A. Do you want me to swear to a definite number?

Q. Would you say that you have been approached a hundred times?

A. I would not say that I haven't been approached five hundred times?

Q. Will you say that you haven't been approached a hundred times?

A. I couldn't say anything about it.

Q. Let me ask you whether in many of the instances where you were approached with bribes, the men who approached you were acquainted with you?

A. I think they were not in most of the cases.

Q. In any of the cases?

A. In some of the cases I think they claimed acquaintance.

Q. Did they live in Lowell, where you live?

A. No, sir.

Q. Did they live in Boston, where you lived?

A. No, sir.

Q. Did they live anywhere where you lived?

A. Not that I think of now.

Q. Have you been approached with propositions involving bribes by men in the neighborhood of where you lived?

A. No, sir. I think not. I never have been approached by Gen. Butler.

Q. You never received one?

A. I never received a bribe of any description.

Q. In these two or three cases that you have been approached, have you always sternly refused?

A. There was one case where it turned out to be a bribe that I accepted.

Q. Money did you get?

A. No, sir.

Q. What was it?

A. The Sisters of Charity in Norfolk, Virginia, wanted their hospital repaired. I had charge of pretty much everything in that line; so I repaired their buildings, and in came their basket—a basket of fresh fish and flowers. As they were ladies, I did not send it back. In a short time there came an application to me for the hay off the farm.—And I declined to send it. Afterwards they said my reward would not be in heaven.

Q. (*By Mr. Mugridge.*) Have you had any interview with General Butler in regard to this matter?

A. I have had an interview, but not in regard to this. I had an interview for the purpose of having him present here.

Q. What did you want him here for?

A. Col. George met me one day in Lowell. He said, the other side will attempt to show that you were very corrupt in your army life, and I want you, if possible, to get Gen. Butler to come to Concord and stay while you are giving your testimony; so that he, having been on the spot, and having given you orders, may be able by his personal presence and personal words to perhaps convince the referees that Major Carney is not as bad as he is painted. And I went to Gloucester for that purpose. I told him that he could come here and take no part.

Q. Did you think his presence here would intimidate anybody?

A. I think I told Mr. Mugridge that I would give him notice so that he could run if he wanted to. I think that after the conversation which you had with me, you said you would see whether General Butler ruled New Hampshire.

RE-DIRECT EXAMINATION.

Q. (*By Mr. George.*) Won't you look at your journal of November 10th, and the way-bill of November 10th. What train did you take and how much did you pay to Mr. Clough on the 9th?

A. November 10th I paid $1.60 for two tickets to Nashua, and $1.50 for a ticket to Concord; and he gave me no ticket.

Q. Won't you examine that way-bill, and see if he returned two tickets to Nashua?

A. There are two between Nashua and Concord, $1.60, as it appears on this way-bill.

Q. Is there any return there of a fare between Concord and Nashua?

A. No, sir.

Q. Now will you take the 11th?

A. Took the 3.30 train to Concord, Clough conductor; I paid Clough $4.25 for my fare to Boston and Draper's to Nashua.

Q. How much was that to Nashua, and how much to Boston?

A. I don't know.

Q. Well, sir, won't you figure it up and see?

A. I don't know what the fares are at all.

Q. You paid $4.25 for the two?

A. Yes, sir.

Q. That would be $1.50 and $2.75? Won't you look at the way-bill for that day, and see whether any fare is returned at $1.50 between Concord and Nashua, on that way-bill, or any fares between Concord and Boston at $2.75?

A. No, sir; there appear to be none.

Q. You have given testimony all the time covering Mr. Clough's being on the road?

A. Yes, sir.

Q. Whether this included the time that you paid fare to Mr. Clough?

A. Yes, sir; I believe it did.

Q. Now, there was something said yesterday with regard to your giving an untruthful statement with regard to reading from your journal. Will you state what you read from, and what you did read?

A. I read the whole account, I believe, of Tuesday, November 14th, in regard to having paid Mr. Clough $12.45, and his giving me back fifty-five cents. On November 16th, the next day, I find an entry explaining that. The first portion of the record of November 14th was admitted, I believe, by the counsel; and if I had turned the page over —I knew that it was here, and if I had turned over ——

Q. Did you read just what the counsel told you?

A. I read just what the counsel agreed to jointly.

Q. I will ask you if you knew what was legal evidence and what not?

A. No, sir; I did not.

Q. Did you suppose ——

A. I supposed that my whole report was evidence. My report was made up at the time, and I think the conductors were turned off on the strength of my report.

Q. Did you know what was legal evidence, as to whether you could put in what your subordinates did?

A. No, sir; I supposed it would be something like martial law.

Q. Won't you read that explanation?

A. "Thursday evening, Nov. 16th, 1865. Since writing what I have in the previous pages, that the fare to Boston is $2.85, and to make up the amount paid Clough, I can account for it only by supposing that his suspicions were aroused, and that he charged me ten cents more than the regular fare, as required by rule of the road; but if he charged ten cents more on the fares to Nashua, why did he not charge ten cents more on the fare to Boston, thus: one to Boston, $2.85; six to Nashua, at $1.60, $9.60; making $12.45; 55 cents added makes $13." This does not account for the money, and I jotted these notes down merely to show that I noticed them at the time.

CROSS EXAMINATION RESUMED.

Q. (*By Mr. Mugridge.*) You say you paid Mr. Clough six fares in the cars to Nashua?
A. Yes, sir.
Q. They should have been $1.60?
A. Yes, sir.
Q. And he charged you that, did he?
A. Yes, sir.
Q. That makes $9.60 in all?
A. Yes, sir.
Q. He charged you just that, and you paid him just that?
A. Yes, sir.
Q. He should have charged you $2.85?
A. Yes, sir.
Q. Didn't he do it?
A. Yes, sir.
Q. Now, $2.85 and $9.60 makes $12.45?
A. Yes, sir.
Q. And he gave you 55 cents, and he should have given you 55 cents?
A. Yes, sir.
Q. Is there anything wrong about that?
A. No, sir; the fare from Concord to Boston is $2.85, and thus I can account for all the money I gave Mr. Clough.
Q. You paid Clough $1.50 on November 10th?
A. Yes, sir.
Q. And he returned $1.60?
A. Yes, sir.
Q. You don't know whether he made a mistake or not in this?
A. No, sir.
Q. You don't know whether he returned your tickets or somebody else's?
A. No, sir; I don't know anything about it.

TESTIMONY OF WESLEY R. BATCHELDER.

Q. (*By Mr. George.*) Will you please state your age, where you live, what your business is?
A. My age is 24; residence, Lowell; business, manufacturing hardware.
Q. Will you please state, sir, whether in 1865—and, if so, at what time—you were in the employment of Major Carney in the investigation on the Concord Railroad?
A. I was.
Q. What time did you go on? Can you remember?

A. I cannot tell.

Q. Have you a journal?

A. I have minutes that I gave to him at that time.

Q. Won't you look at your minutes and see when you were employed? I mean when you first went on to the road? How long did you continue? Can you state from your recollection?

A. Nothing, only from this.

Q. Was it five years, or three months, or three weeks?

A. It was less than three months, I should judge.

Q. Now, sir, won't you turn to your minutes. Have you any recollection apart from your minutes?

A. Nothing connectedly; although there might be some little items that I might remember at the time.

Q. Were these minutes made up at the time?

A. Yes, sir; they were made up every evening.

Q. Were they true?

A. Yes, sir.

Q. Would you have sworn to them at the time?

A. Yes, sir.

Q. Will you look at your minutes of November 21st?

A. Tuesday.

Q. Now read your minutes so far as they relate to Mr. Clough.

A. This starts from Lowell. "Took the 8 o'clock train from Lowell for Nashua. Above Nashua paid Conductor Clough fare to Concord; no ticket." Nothing more that day that relates to Mr. Clough.

Q. You may turn to Nov. 23d. Just read what relates to Mr. Clough.

A. "I went to Concord at 10.30 in the morning. At 3.30 P. M. I sent him down with two others; bought tickets there to Boston of Conductor Clough, $8.55; tickets numbered 7 and 8. Also, paid conductor."

Q. (*By Mr. Mugridge.*) Is that Clough?

A. Yes, sir; it must be. One fare to Nashua. Man left train at Manchester.

Q. Did you say how much that ticket cost?

A. Yes, sir; $1.50; ticket No. 9.

Q. (*By Mr. George*) Won't you produce those tickets?

A. Yes, sir; 7, 8 and 9.

Q. That ticket was purchased from Concord to Nashua [showing ticket]?

A. That was ticket No. 9.

Q. These are the two Boston tickets just referred to?

A. Tickets 7 and 8.

Q. (*By the Chairman.*) I want to know whether these tickets were punched when you took them, or whether afterwards?

A. My impression is that the tickets were not punched when I bought them, though I am not willing to swear.

Q. How far did you ride on them?

A. I rode on them to Lowell. I came through on that ticket—No. 6. [Showing.]

Q. And all these punches you think were made while they were in your possession?

A. While they were in my possession—while I was in the cars—yes, sir.

Q. (*By Mr. George.*) Did you make your minutes on the books at that time?

A. I made them after my arrival home; when I made this report and wrote it out.

Q. Now you may go on and read the balance, sir.

A. At Manchester, met one of yesterday's companions with a ticket to Boston, bought of the agent at Concord. Not having sufficient funds with me to procure another, came through on that ticket to Lowell. No. 6. [Ticket No. 6 produced.] November 25th, Saturday, 3.30 P. M., left Concord for Lowell with two companions; paid three fares to Nashua, $4.80; no tickets. For change from $5, Conductor Clough gave me $1.20; and it was not until I called his attention to this—seemingly a mistake—that he took the surplus change with a "thank you."

Q. That when you paid him in the cars he charged you $2.80 instead of $3.80?

A. I paid for three fares to Nashua, $4.80, and for change from a $5.00 bill he gave me $1.20, a dollar too much, and I called his attention to it, and he took it and thanked me for it.

Q. Now, sir, will you turn and see if you went on the 27th, or the 28th, or the 29th?

A. My record does not show whether it was Clough's train, or whose it was. I left Manchester at two o'clock, and returned at eleven o'clock; this is dated the 27th. Left train at Manchester, and returned at eleven o'clock to Lowell; paid two fares to conductor to Nashua, $1.50; no tickets.

Q. Now read right on.

A. The 28th, sir, one-half ticket, Nashua to Concord.

Q. You started from Lowell at what time?

A. At 3.30 P. M. Left Concord, paying conductor one fare, $1.50.

Q. (*By Mr. Mugridge.*) Fare to where?

A. To Nashua. No ticket; see memorandum.

Q. (*By Mr. George.*) Now, sir, look at the 29th. You stopped on the 30th going on Mr. Clough's train?

A. I don't remember.

Q. Well, just look and see whether you ever went on Mr. Clough's train after that?

A. Wednesday, the 13th day of December, Left Manchester for Concord at 2.20 P. M., paying Conductor Clough two fares to Concord, 85 cents each, $1.70; no tickets.

Q. Is that the last time you went on Clough's train?

A. Friday, December 15th, at 11 o'clock, left Manchester for Nashua; paid fares for self and lady and girl to Nashua, $1.70.

Q. How old was the girl?

A. Well, I don't remember of ever having any child on the train with me, except a niece of mine at Manchester; I think she must have been about eight years old.

Q. 1865, Saturday, December 16th?

A. Left Concord at 3.30 P. M.; paid fares to Nashua, $3.20; no tickets. There I have a pencil mark—I have written: "see memorandum." Memorandum says to procure no tickets to Worcester. Monday, December 25th, paid Conductor Clough fare to Concord.

Q. (*By Mr. Mugridge.*) Where from?

A. This must have been from Nashua. Paid Conductor Clough fare to Concord, $1.60; no tickets.

Q. (*By Mr. George.*) Mr. Batchelder, will you state whether you at any time were appointed conductor on the Concord road?

A. I was.

Q. Do you remember when?

A. I don't remember; it was in the spring of 1866, or the last of the winter; I cannot tell.

Q. Following this investigation?

A. Yes, sir; as my recollection serves me. Now I remember, it was in February.

Q. Where did you run? What route?

A. Well, I had charge of the train from Manchester to Lawrence. Occasionally, I think, between here and Manchester. That was very seldom.

Q. You ran all the train and part of the Manchester train?

A. Three trains down and three trains up, and came to Concord every night.

Q. (*By Mr. Mugridge.*) Well, sir, when were you discharged as conductor, and by whom? [Objected to on the ground that it was proposed to show that the witness was discharged by Mr. Gilmore on account of having conducted this investigation.]

*Mr. George.* There has been evidence put in here in regard to Mr. Gilmore's action in stifling this investigation. I propose to show that on the very night preceding the annual meeting, without preceding notice, Mr. Gilmore discharged Mr. Batchelder.

*Mr. Mugridge.* We don't make any objection to that fact as you state it.

Q. You may state when and under what circumstances you were discharged, and by whom.

A. I was discharged by a letter from Mr. Gilmore, as superintendent, on the evening (immediately upon the arrival of my train in) of the annual meeting of the stockholders, for the year 1866. After my discharge, I wrote Mr. Gilmore, asking him the cause of my dismissal, whether from direliction of duty, or because of a change in the direction of the road. Either he or his clerk—he signed it, so it must have been he—he wrote me a letter.

*Mr. Mugridge.* Produce the letter.

Q. Have you the letter?

A. I have it; but I neglected to bring it, because it slipped my mind.

CROSS EXAMINATION.

Q. (*By Mr. Mugridge.*) Let me ask you who employed you to carry on this investigation that you did carry on against Mr. Clough upon the cars?

A. Mr. Carney.

Q. Were you an acquaintance or friend of his before that time?

A. Yes, sir.

Q. Were you acquainted with Gen Butler before that?

A. If you will allow me, I will state the circumstances.

Q. All I want is, did you or did you not know him?

A. I did know him.

Q. Were you in his employ?

A. I was one of his confidential clerks, during his connection with the Army of the James.

Q. Were you with him in New Orleans?

A. No, sir.
Q. Were you with him at Fort Fisher?
A. No, sir.
Q. How long a time did you remain in the capacity of confidential clerk of Gen. Butler, sir, while he was commander of the Army of the James?
A. From February up to the time of his removal.
Q. How many months?
A. I should judge it was about eleven months, to the best of my recollection.
Q. Were you a chum of Major Carney's?
A. Major Carney was at North Carolina, and I was at Fortress Monroe.
Q. Did you have communication with Gen. Butler before you went on the cars as a spy?
A. I did
Q. Before you went on to the Concord Railroad, as a spy, did you talk with Butler as regards your going on?
A. I talked with him—yes, sir.
Q. Was it the understanding that, if you went on the road as a spy, and succeeded in discerning peculations and frauds on the part of Mr. Clough, the old conductors would be turned off from the road, and then you would be put on?
A. There was no such understanding at the time, at all.
Q. Was it the talk between you and Gen. Butler, or anybody else, that if you went on the Concord road and conducted operations which resulted in the discovery of frauds and peculations on the part of the conductors, they would be removed?
A. No, sir; not until after I got all through my investigation, then I proposed the subject myself.
Q. Did you have the thing in mind before you got through?
A. Not in the least.
Q. Then it occurred to you suddenly?
A. Yes, sir; it occured to me that, if the matter turned out as I supposed it would, if I could have the appointment, I would ask for it and have it.
Q. Will you please state, sir, how long it was after you concluded your operations, before the idea occurred to you that it would be desirable for you to have a position on the road as conductor?
A. I think it was in the next two weeks afterwards.
Q. And the idea never crept into your mind before?
A. No, sir.
Q. Butler never suggested it to you?
A. Not until afterwards.
Q. And Carney never intimated it to you?
A. No, sir.
Q. You never thought of it until two weeks afterwards?
A. Until within two weeks.
Q. You state that, do you?
A. Yes, sir; to the best of my recollection and belief.
Q. When that idea came into your mind, who did you communicate to? Or, first, let me ask you who procured you the appointment?
A. I cannot tell. I made application to Mr. Upham. I made application for Mr. Clough's train.

Q. You had your eye on Mr. Clough's train, had you?
A. Yes, sir.
Q. That is the train you had your eye on?
A. Yes, sir.
Q. You didn't ask to go on Noyes's train, or Jones's train?
A. I asked for Clough's train.
Q. Now, sir, let me ask you if you went to Mr. Butler to get him to engineer your desires through?
A. After I had thought of it, and come to the conclusion myself that it would be desirable, I made known my wishes to the general.
Q. General Butler?
A. Yes, sir. And asked him if he would aid me to get the position. He said he would as far as he could.
Q. Did he aid you, sir?
A. I think he did, although I am not able to state at the present time. It does not occur to me in what manner; but I know I always alleged it to his influence or through the kindness of Mr. Upham.
Q. Did you make an application to Colonel George for aid in that direction?
A. I cannot state whether I did or not. I know I saw Col. George. I was just thinking—I could not tell whether I saw Col. George until after my apppointment. It does not occur to me clearly whether I saw him or not.
Q. Wasn't it put upon the ground of the services you had rendered the road?
A. Not in the least, that I know of.
Q. It was upon grounds entirely independent of that?
A. As I understand it.
Q And you don't connect your position as conductor with your operations as spy at all?
A. I never have made any connection in that way. I have often thought that whether I had any connection with the road prior to that, it might have been the same result by my applying.
Q. Did you ever have any agreement of this kind with them?
A. No, sir; not until after I got through.
Q. Now, sir, what was your business before you were connected with the Army of the James?
A. I was clerk of the register of deeds in Lowell, about nine months, I think, before I went to Fortress Monroe.
Q. How long were you employed in this business of spying out the doings of the conductors?
A. I don't know that I could tell how long I was, directly and indirectly. It was less than three months.
Q. What compensation did you receive for the services you rendered?
A. That I am unable to state positively.
Q. Won't you be kind enough to state approximately if you cannot positively?
A. I don't think it was over a hundred dollars?
Q. For the entire three months?
A. For the whole time.
Q. Have you got anything in expectation?
A. Not in the least.
Q. Have you got any arrangement that you are to have part of the twenty-five thousand dollars?

A. No sir; I have no claim against the Concord road whatever.

Q. And that hundred dollars liquidates your claim?

A. Yes, so far as at present.

Q. Are you here swearing by the day?

A. No, sir.

Q. Have you been summoned here?

A. No, sir.

Q. Have you understood that you are to be paid for it?

A. Not in the least.

Q. What do you come here for, then?

A. Well, I suppose that the laws of New Hampshire provide for necessary expenses.

Q. Do you come here on a pass?

A. I come here on a pass of the Boston and Lowell Railroad.

Q. Whose pass is it?

A. John D. Winslow's.

Q. Did you ask for it?

A. I believe Col. George gave it to me.

Q. Been riding back and forth on that pass?

A. Yes, sir.

Q. Have you had any money advanced to you to pay your fees?

A. Not a dollar; merely came up at the request of the parties.

Q. Now, sir, will you be kind enough to turn to your first record of November 21st. You say on that day you paid fare to George Clough from Nashua, in the cars. Will you be kind enough to take his way-bill and see if he has not returned a fare as paid in the cars from Nashua to Concord on that day?

A. He has returned two fares from Concord to Nashua. May I be allowed to state further?

Yes, sir; anything you please.

As the regulations were when I was on the road, as I came from Nashua to Concord, these marks should have been at the top instead of the bottom. Fares up should be marked from the top down, and fares down from the bottom up. These were my instructions.

Q. Did Clough return fares from Concord to Nashua?

A. He has returned two fares between Concord and Nashua.

Q. Now, sir, the next day is the 23d. I understand you to say that upon that occasion you bought three tickets to Boston, and you paid one fare to Nashua. Will you be kind enough to take the return of Mr. Clough on that day, and see if he hasn't returned fares to Boston and also to Nashua? See how many he has returned to Boston and how many he has returned to Nashua?

A. He returned eight fares to Boston by way of Lowell, and two fares over the road between Nashua and Concord.

Q. That is where you paid your fare?

A. It may be.

Q. (*By the Chairman.*) Is that a case where you used a ticket?

A. Yes, sir.

Q. A ticket over the Vermont Central road?

A. Yes, sir.

Q. (*By Mr. Mugridge.*) The next time you make a return is Nov. 25th; paid three fares to Nashua. Will you be pleased to take the way-bill of the 3.30 train on the 25th, and see if Mr. Clough has returned any fares between Concord and Nashua on that day, and if so, how many?

A. Clough has returned, as I read, five fares between Nashua and Concord. There are four down and one across it. I don't know whether it is to erase it or to make it five; it is one of the two.

Q. (*By the Chairman.*) What did you say you paid for these three tickets?

A. Bought three tickets to Boston, and paid $8.55.

Q. No; these tickets from Concord to Nashua?

A. November 25th, paid fares to Nashua, $4.80.

Q. (*By Mr. Mugridge.*) That would be $1.60 a fare?

A. Yes, sir.

Q. Well, now, you have seen fit to note the fact, and put it in here, that you gave him a five dollar bill, and he gave you $1.20; and that you called his attention to it, and that he thanked you. That is all there is of the transaction?

A. Yes, sir.

Q. You saw the mistake and handed it back. That is all there is to that?

A. That is all there is, so far as I know.

Q. Let us take the next record that you have, which is under date of November 27th. On the 10.15 train you say you left Manchester; paid two fares to Nashua, $1.70; no tickets. How would that be?

A. From Manchester to Nashua.

Q. Now, sir, will you be kind enough to take Mr. Clough's way-bill, and see if he does not return two fares?

A. There are two fares returned over the road from Manchester to Nashua.

Q. (*By Mr. George.*) What price?

A. Eighty-five cents.

Q. (*By Mr. Mugridge.*) Now the next one is under date of November 28th, on the 3.30 train. You say you left Concord, paying the conductor one fare to Nashua; no ticket. Will you be kind enough to take the way-bill of November 28th, and see if Mr. Clough has returned a fare, and if so, how many?

A. Between Concord and Nashua?

Q. Yes, sir.

A Between Nashua and Concord one fare.

Q. Just one fare?

A. Yes, sir.

Q. Now, sir, the next is December 13th, Wednesday. You left Manchester, you say, at 2.20 P. M.; paying Mr. Clough two fares between Manchester and Concord, 85 cents each; and that is the fare, as returned upon the way-bill—85 cents. Now, see if Mr. Clough has returned it?

[Mr. George objected, stating that he did not like to have statements made by counsel as to what the witness had said, when it was stated incorrectly.]

Q. You have returned, under date of December 13th, two fares from Concord to Manchester, $1.60?

A. Yes, sir.

Q. Eighty-five cents each?

A. I have put in pencil; "Eighty-five cents each," and after that "$1.60."

Q. Now, will you look at the way-bill between Manchester and Concord?

A. Either six or seven.

Q. Now, sir, the next record you notice is under date of December 15th, Friday. At 11 o'clock left Manchester for Nashua; paid for self, lady and girl. That is just two fares; and he didn't take fare for the little girl?

A. My memorandum merely shows that there was a little girl.

Q. Was it the custom then to take fare for a little girl?

A. I don't know.

Q. Will you be kind enough to take the way-bill and see if Mr. Clough has made any return of fares between Manchester and Nashua?

A. Between Manchester and Nashua, four fares.

Q. Now, sir, the next is December 16, Saturday. Left Concord in the 3.30. How many fares did you pay, making up $3.20?

A. I will state to the best of my recollection and belief; it is six. I came up with one companion to Concord. When I went back I paid him $3.20. It must have been that the companion went back with me.

Q. Then that should have been how many fares?

A. Two fares.

Q. Each way?

A. No; only one way.

Q. Now, sir, will you be kind enough to look at his way-bills of December 16th, and see if he has returned any fares between Nashua and Concord?

A. Between Nashua and Concord, four fares.

Q. Now, the next is under date of December 25th, Monday. Paid Clough a fare from Concord to Nashua, $1.60. That would be exactly right, would it not, sir?

A. I don't know, sir.

Q. Just look at that and see what the fare was from Nashua to Concord?

A. Fare from Nashua to Concord, $1.60 That is just what I paid him.

Q. See if it is returned in the fares?

A. Two fares, $1.60.

Q. That is the entire record you have got against Mr. Clough?

A. That is the entire record from which I have made the memorandum against Mr. Clough, so far as I know. I have not seen these records for about three years.

RE-DIRECT EXAMINATION.

Q. (*By Mr. George.*) When you had completed your record, what did you do with it?

A. I made them out day after day, and sent them to Major Carney.

Q. And you haven't seen them from that day to this?

A. From that day until day before yesterday.

Q. Was a man by the name of Draper on the cars?

A. I think I did see a person who was pointed out to me by the name of Draper. There was a man by the name of Draper—the major told me after I got through—who was running at the same time. He was pointed out to me as being Mr. Draper.

Q. Do you know where he is?

A. No, sir.

Q. Have you made any inquiry?

A. I went yesterday where he used to be, and asked the young man there if Draper was there. He said he was not; he had been gone about three weeks.

Q. (*By Mr. Mugridge.*) Did he tell you the circumstances under which Draper left town?

A. He said the last he saw of him he was going down street, and he said something afterwards about two men after him.

Q. Did he tell you whether they had badges on their right shoulder?

A. I think he said afterwards that two policemen were after him.

Q. And Draper hasn't been seen since?

A. No, sir. [Adjourned.]

[THIRTEENTH DAY. Thursday, December 17th, 1868.]

This day was entirely devoted to the discussion, by the counsel, of the exception taken to the evidence offered by the plaintiff.

[FOURTEENTH DAY. Friday, December 18th, 1868.]

The hearing was resumed at 10 o'clock, A. M.

RULINGS.

The chairman announced the decision of the referees on the questions submitted:

*First.* As to the comparison between the returns of trains on the Concord road: *Held,* that such comparisons may be made by proper testimony, reserving the question as to comparisons between other roads, though the opinion of the board is not against receiving such testimony.

*Second.* On the motion for a non-suit, the referees rule that there is evidence for them to consider.

## INCOME RETURNS.

Mr. George read to the referees, and offered in evidence, the returns of income made by George Clough to the assessor of internal revenue during the years 1864, 1865, 1866 and 1867, as follows:—For the year commencing January 1, and ending December 31, 1864. (Statement sworn to, May 16, 1865, before W. Odlin, assistant assessor of 7th division, 2d district, New Hampshire.)

RECEIPTS.

| | | |
|---|---|---|
| From rent of buildings, | | $2522.30 |
| From interest on notes, bonds, mortgages, or other personal securities, | | 585.44 |
| From interest or dividends on stock, capital or deposits in any bank, trust company, savings institution, insurance, railroad, canal, turnpike or slack water company. | | 1600.00 |
| From dividends of any incorporated company, other than those above mentioned, | | 13.00 |
| From salary, other than as officer or employee of the United States, | | 840.00 |
| Gross income, | | $5560.74 |
| DEDUCTIONS. | | |
| Interest paid out or falling due within the year, | $75.00 | |
| National, state and local taxes paid within the year, | 882.08 | |
| Amount actually paid for insurance on homestead, | 100.00 | $1057.08 |
| Net income, | | $4503.66 |

TAX.

| | | |
|---|---|---|
| Exempt by law, | $600.00 | |
| Amount in excess of $600, and not exceeding $5000, subject to 5 per cent., | 3904.00 | |
| Total tax, | | $195.20 |
| Deduct amount of tax withheld by institutions under sec. 120 and 122, | | 64.00 |
| Amount of tax due, | | $131.20 |

TAXABLE ARTICLES.

| | |
|---|---|
| One carriage valued at $50, and not exceeding $100, including harness used thereon, | $1.00 |
| One carriage valued at $100, and not exceeding $200, | 2.00 |
| One carriage valued at $200, and not exceeding $300, | 3.00 |
| Two gold watches valued at $100, or less, | 2.00 |
| One gold watch valued at above $100, | 2.00 |
| One piano forte, valued at not less than $100, and not above $200, | 4.00 |
| Sixty-two ozs. plate of silver, kept for use, per oz. troy exceeding 40 ozs. used by one family, | 3.10 |
| | $17.10 |

For the year commencing Jan. 1, and ending Dec. 31, 1865. (Statement sworn to May 17, 1866, before W Odlin, assistant assessor.)

RECEIPTS.

| | | | |
|---|---|---|---|
| From rents of lands, | | | $250.00 |
| From rents of buildings, | | | $2769.17 |

PROPER DEDUCTIONS.

| | | | |
|---|---|---|---|
| Repairs, | $577.66 | | |
| Insurance, | 80.99 | 658.65 | $2110.52 |
| From interest or dividends on stock, capital or deposits, &c. | | | 1684.21 |
| From dividends of any incorporated company, other than those above mentioned, | | | 12.50 |
| From interest on United States bonds and treasury notes, | | | 483,00 |
| From interest on notes, bonds, mortgages or securities, other than those above enumerated, | | | 562.58 |
| From salary, other than as officer or employee of the United States, | | | 900.00 |
| Gross income, | | | $6002.81 |

DEDUCTIONS.

| | | |
|---|---|---|
| Interest paid out or falling due within the year, | $512.27 | |
| National, state and local taxes paid within the year, | 1016.25 | 1528.52 |
| Net income, | | $4474.29 |

TAX.

| | | |
|---|---|---|
| Exempt by law, | $600.00 | |
| Amount in excess of $600, and not exceeding $5000, subject to 5 per cent., | 3874.70 | 193.70 |
| Deduct amount of tax withheld by institutions enumerated in paragraphs 6 and 7, | | 84.21 |
| | | $109.49 |

TAXABLE ARTICLES.

| | |
|---|---|
| One carriage valued at above $100, and not above $200, | $2.00 |
| Two carriages valued at above $200, and not above $300, | 6.00 |
| Two gold watches valued at $100 or less, | 2.00 |
| One gold watch valued at above $100, | 2.00 |
| One piano forte valued at above $200, and not above $400, | 4.00 |
| Sixty-two ounces silver plate, exceeding 40 ounces, | 3.10 |

For the year commencing January 1, and ending December 31, 1866. (Statement sworn to May 14, 1867, before John Kimball, collector.)

RECEIPTS.

| | | |
|---|---|---|
| From rents, | | $2619.10 |
| From profits realized by sales of real estate purchased since December 31, 1863, | | 120.00 |
| From interest on any bonds, or other evidence of indebtedness, &c., | | 1800.00 |
| From interest on notes, bonds, mortgages, or other securities other than those enumerated above, | | 957.53 |
| Gross income, | | $5496.63 |

DEDUCTIONS.

| | | |
|---|---|---|
| Exempt by law. | $1000.00 | |
| National, state, county, or municipal taxes paid within the year, | 769.14 | |
| Amount actually paid for water and land rents, | 101.00 | |
| Amount paid for repairs, | 475.94 | |
| Interest paid out or falling due within the year, | 761.71 | |
| Interest on dividends from corporations, | 1800.00 | $4907.79 |
| Taxable income, | | $588.84 |
| Amount of tax at 5 per cent., | | $29.45 |

TAXABLE ARTICLES.

| | |
|---|---|
| 1 gold watch valued at $100, or less, | $1.00 |
| 3 gold watches valued at above $100, | 6.00 |
| 62 oz. silver plate, exceeding 40 oz., | 3.10 |

For the year commencing January 1, and ending December 31, 1867. (Statement sworn to April 1, 1868, before W. Odlin, assistant assessor.)

RECEIPTS.

| | | |
|---|---|---|
| From rents, | | $2832.00 |
| From interest on bonds, or other evidences of indebtedness, &c., | | 4810.00 |
| From interest on notes, bonds, mortgages or securities, other than those enumerated above, | | 92.68 |
| Gross income, | | $7734.68 |

DEDUCTIONS.

| | | |
|---|---|---|
| Exempt by law, | $1000.00 | |
| National, state, county and municipal taxes within the year, | 815.11 | |
| Amount actually paid for water and land rent, | 101.00 | |
| Amount paid for repairs, | 291.14 | |
| Interest paid out or falling due within the year, | 1698.32 | |
| Interest on dividends from corporations enumerated above in paragraph 6, | 4810.00 | $8715.57 |
| Taxable income, | | none, |

TAXABLE ARTICLES.

| | |
|---|---|
| 1 gold watch valued at $100, or less, | $1.00 |
| 3 gold watches valued at above $100, | 6.00 |
| 62 oz. silver plate, exceeding 40 oz., | 3.10 |

Mr. George called the attention of the referees to the fact that in neither of the four statements was there any return made of receipts from farming operations; or value of live stock; or profits on sale of gold or stocks.

## TESTIMONY OF PETER DUDLEY.

Q. (*By Mr. George.*) State where you live, and your occupation.

A. I live in Concord; I keep a livery stable.

Q. When did you first become acquainted with Mr. George Clough, and for how long have you known him?

A. I have known him for thirty years, or more.

Q. What was Mr. Clough's occupation when you first knew him, and where was he engaged in business?

A. When I first knew him he was driving stage into Lowell, for a man by the name of Osgood. He was driving on what I think was called the Dover and Lowell route. I know that he went into Lowell, but what was his stopping point at the other end, I cannot say.

Q. Was it the route from Dover through Pittsfield?

A. No; from Dover through Windham and Derry.

Q. In what year was that?

A. I should think it was four or five years before the railroad was opened, which was in 1842.

Q. How long did he continue to drive stage for Osgood?

A. He drove there until he came on to the Mammoth road, and bought some property of me.

Q. What was the "Mammoth road?"

A. It is the road leading from here to Lowell by way of Hooksett, Londonderry, Windham, &c.

Q. At that time, who owned the staging on the Mammoth road from here to Lowell?

A. There were different interests. I owned half of one line to Lowell, and a man by the name of Newell owned the other half. We owned the line down on Tuesdays, Thursdays and Saturdays. On the opposite days, Mr. Walker and Mr. Wood owned a stage line on the same road.

Q. When you speak of "line," you mean one day down and the next day back?

A. Yes.

Q. How many horses were required?

A. I worked twenty-four horses from here to Lowell; twelve were mine, and twelve were Newell's. Twenty-four horses constituted the line.

Q. You say that you sold some property to Mr. Clough. What did you sell him; and when, and for how much did you sell him?

A. I had twelve horses on the road, I think; I am not certain whether it was ten or twelve; I sold him two teams, but whether it was ten or twelve horses I cannot say.

Q. You sold him your interest in one-half of the line?

A. Yes.

Q. What did he pay you for that interest?

A. He paid me two thousand dollars.

Q. What did the purchase embrace besides the two teams?

A. It embraced the coach, harness, blankets, &c.

Q. State as nearly as you can the time when you sold this property to him?

A. I do not think that I could state the year. As near as my recollection serves me, it might have been three years before the railroad cars ran into Concord, which was in the fall of 1842. It was three years before that, which would probably make it sometime in the year 1839.

Q. How and when did he pay you that two thousand dollars?

A. He paid one thousand dollars at the time that we made the trade; and he gave me a note for one thousand more. I do not recollect when he paid the note; at any rate he paid it before I asked him for it; it might have been five or six months.

Q. What did he do after you sold him one-half of that line?

A. He took the property and drove it, and took care of it.

Q. How long did he continue to run that property?

A. He ran it up to the time that he was appointed conductor on the Concord road, and still longer than that, for he hired a man to drive.

Q. He ran it after the cars ran to Manchester, and before they ran to Concord?

A. Yes.

Q. Which end of the line did he run?

A. He ran from Manchester here until the cars came here; and I think that he ran for a while from Manchester to Lowell, after the cars ran up as far as Manchester.

Q. How long was it after the cars ran to Concord before he disposed of the stage property? Did he run any stages after the cars came into Concord?

A. I am not certain, but I think that he did.

Q. Where?

A. It occurs to me that he might have run from Manchester to Hopkinton, but I would not say that he did or did not.

Q. At the time that Mr. Clough was appointed conductor on the Concord Railroad, what property had he, as far as you know?

A. I do not know what he had. I know that he bought this property of me, and paid me for it, and that he owned a house.

Q. Where was the house?

A. On South street.

Q. How large a house was it?

A. It was a two-story house.

Q. Did he have any other property?

A. I do not know of any other property.

Q. Did you own any interest in the stage line after you sold to Mr. Clough?

A. Not in that line; I owned an interest on the same road on opposite days.

Q. You ran one day, and Mr. Clough the next?

A. Yes; on the day I went down, he came up.

*Mr. Mugridge.* Mr. Dudley is a witness that we propose to call hereafter, and we do not care to examine him until then.

Mr. George offered in evidence and read the following deposition of Ira Foster, sworn to on the 24th day of November, 1868.

## DEPOSITION OF IRA FOSTER.

1st Int. Where do you reside, and how long have you known Geo. Clough? Ans. I reside in Concord, and knew Mr. Clough as much as three years before the Concord Railroad started.

2d Int. In what business were you engaged, and was he engaged, from the time you first knew him, up to the time of the opening of the Concord Railroad to Concord? Ans. We were both engaged in staging. I run from Concord to Nashua, and he from Concord to Lowell, over the Mammoth road, according to the best of my knowledge. I think he run opposite days from me, generally.

3d Int. Who were the first conductors on the Concord Railroad from Nashua to Concord? Ans. Myself and Mr. Clough.

4th Int. State what property Mr. Clough owned when he first commenced to run on the Concord road, and your means of knowledge, whether from what he said to you or otherwise. State fully. Ans. I have no means of knowing whether he had property or not, or how much, otherwise than that he had a line of stages from Concord to Lowell, that I suppose he owned, but I do not know how much of it. I recollect one time of being in the cars with him, and he told me that he had subscribed, or took, in the Northern road, I think, fifty shares. This talk, I think, was about the time they commenced on the Northern road.

5th Int. Of what did the line of stages from Concord to Lowell consist? Ans. I suppose it would consist of about twenty horses, and harnesses sufficient for them, and one or two coaches—I don't know whether he had one or two—and stage-sleighs, blankets, &c.

6th Int. Have you any knowledge or information that George Clough had any other property at the time when he commenced to run as conductor, except the line of stages, and property, as specified in the preceding answer? Ans. I have not.

7th Int. State whether or not Mr. Clough ever said anything to you about the amount of property he was worth at the time that he went on to the road as conductor. Ans. Never.

8th Int. Have you or not any knowledge whether Mr. Clough owned that line of stages? Ans. No positive knowledge.

### CROSS EXAMINATION, BY H. P. ROLFE, ESQ.

1st Int. Did not Mr. Clough, at the time he went on to the Concord road as conductor, own a house and land on South street, in Concord, and don't you recollect of his putting his team up there, and taking care of it himself, and being his own hostler, for a year or two before he went on to the Concord road? Ans. I suppose he did own a house there. He lived there, and I suppose he owned it. I have no doubt in my mind that he owned it, but I had forgotten it. I do not recollect about his putting his team up there. As he ran on opposite days to me I suppose that I should not know it.

2d Int. Has not Mr. Clough, while you have known him, been a remarkably energetic and industrious man? Ans. I do not know as there is anything remarkable about it; but he has always been an energetic and industrious man, the same as all good business men are.

### MAIN EXAMINATION, RESUMED.

9th Int. Will you describe the house that Mr. Clough owned, on South street, referred to by you? Ans. It was a two-story, white house, about 24x32 feet, and an L part, and barn. It was on a large lot; if my memory serves me, I think it was nearly an acre, maybe not over three quarters.

10TH INT. Do you or not know whether he owned it free and clear of debt, or whether it was mortgaged or encumbered at the time he went on to the road? [Objected to.] ANS. I could not tell.

11TH INT. Do you or not know how long Mr. Clough had owned that house and lot at the time he went on to the road, and what he paid for it? ANS. I should think he lived there for a number of years; I couldn't tell for how long. I do not know what he paid for it.

12TH INT. Do you or not know of any other property which Mr. Clough owned, or which he had in his possession, at the time he went on said railroad as conductor? If so, what was it? ANS. I do not recollect any.

[Signed] IRA FOSTER.

*Mr. Stanley.* We propose to put in evidence the first paragraph of the answer to the 14th interrogatory in Mr. Clough's deposition, which is as follows:—"The first eight or ten years that I ran through to Boston I used to carry produce down—buying here and selling in Boston. I used to bring up stuff at the same time."

*The Chairman.* Is there anything that shows in what capacity he was then acting?

*Mr. Stanley.* He was running as conductor; he never run in any other capacity.

*Mr. Mugridge.* What is the object of this testimony?

*Mr. George.* Mr. Clough claims in his deposition that he made a sum of money carrying produce on the cars. If he did, we say that he made it without right, and that the money that he received belongs to the railroad.

*Mr. Mugridge.* We object to the testimony.

*Mr. George.* We propose to show that he made money in this way just as he made money on the coupons; we propose to show further that he made money by putting a boy on the cars to sell peanuts; and we claim that all the money he made in these several ways, he made contrary to law, and that it belongs to the corporation.

*Mr. Tappan.* We suppose that this evidence stands upon the same footing as the evidence with regard to free passes.

*Mr. George.* Not at all. We put it precisely upon the same ground as the ruling that has been already made with regard to the two thousand dollars profit on the coupon tickets.

*Mr. Mugridge.* That ruling does not cover this point; that is a double ruling.

*Mr. George.* What do you mean by "double ruling"?

*Mr. Mugridge.* We will show you by and by. There is no claim made for this in the specification.

*Mr. Stanley.* Yes there is.

*Mr. Mugrigde.* Where?

*Mr. Stanley.* The specification will show it.

*Mr. Mugridge.* If this evidence is offered with a view of showing Mr. Clough's property, or the method and manner in which he accumulated that property, then we do not object to it; but if it is put in for the purpose of showing that Mr. Clough is specifically chargeable, then we do object to it.

*The Chairman.* This evidence is perhaps competent enough in certain respects; but whether the referees will hold that money made in such a way as is just indicated belongs to the corporation, is a matter that we do not propose to settle at this time. In other respects the evidence may be competent enough, and we do not think therefore that we can exclude it.

*Mr. Mugridge.* We do not wish to be understood as objecting to it as showing the manner in which Mr. Clough accumulated his property; but if it is put in for the purpose of charging us in this action, we do object to it.

*The Chairman.* Is that in the specification?

*Mr. Mugridge.* We claim that the specification does not cover it; the specification is "for moneys the defendant received for the plaintiff, and has not returned the same in pursuance of his duties as conductor."

*The Chairman.* Whether the money that he might make in that way—by carrying on a business which at present it appears is not a part of the business of a railroad corporation—could be recovered in this form of action, and under this specification, is a question that will have to be considered hereafter, I suppose.

*Mr. Mugridge.* If it is more agreeable to have the evidence go in, with our exception, which may be considered at some time agreeable to the referees, it may be received in that way.

*The Chairman.* The evidence may go in if competent for any purpose; whether it is competent for the purpose it is claimed, is yet to be shown.

*Mr. Stanley.* (Continuing the deposition.) "The first eight or ten years that I ran through to Boston I used to carry produce down, buying here and selling in Boston. I used to bring up stuff at the same time. That is what I used to do when I ran on the cars. On the stuff that I used to carry down and bring up I made more than my salary, considerable, I think."

Now turn to the 306th cross interrogatory:

"Q. You have heretofore stated, in your direct examination, that when you were first on the Concord Railroad you carried produce from here to Boston, and speculated in it. Will you state the articles in which you dealt from here to Boston and from Boston back, and also from Franklin to Boston and from Boston back? ANS. I used to carry chickens and turkeys and geese, butter and eggs, apples, cranberries, chestnuts and berries. I used to send potatoes down by the freight train. I used to carry carcasses of veal and mutton from Franklin to Lowell and Boston, and the same articles back, that I brought to Concord."

The 307th cross interrogatory:

"Q. State, if you are able, the financial results of said operations, the length of time you were engaged in them, when you stopped, and the occasion of your stopping the business. ANS. I kept an account for a year or two of my speculations of that kind, and I think I made from five to eight hundred dollars per year. I stopped, I think, in 1849 or 1850. Isaac Spaulding was the president of the road at that time, and told me that he did not want me to carry any more. He said that if I carried stuff, the others might want to. He said that he did not care anything about what I had carried, but he did not want that I should carry any more; and I have carried nothing since."

The next question enquires whether he has any memorandum; and the answer is: "I have not; I cannot find it; it must have been burned when my house was burned."

I read now from the 407th direct interrogatory:

"Please state the amount you claim to have made in the business transactions you have spoken of. Ans. On the chicken and produce trade, $5200."

I will next read the 372d cross interrogatory:

"Q. Did you, during the time that you run on the cars, employ a boy, or boys, to peddle sweetmeats, &c., in the cars? If so, with what pecuniary result to you, giving the length of time that you were so engaged? Ans. I did employ a boy some seven or eight years, to peddle fruit and nuts. I used to pay him from fifty to seventy-five cents per day; and he used to take from a dollar to five and six dollars per day. I think the profits would amount to something over a dollar and a half per day.

## THE DEFENDANTS' CASE.

Mr. Rolfe, in opening the case for the defendants, said:

*May it please you, gentlemen of this Honorable Board of Referees:*

This action, it appears, was brought against Mr. Clough by the Concord Railroad Company, with the intention, undoubtedly, when the action was brought, (as there are various counts in the declaration) to charge Mr. Clough not only with money, or money's worth, that he collected for the plaintiff corporation, but also for injuries done to the Concord Railroad by reason of his allowing persons to pass free over the road.

The referees, as we understand, have very properly excluded any proof, except as to the money that he has had and received in behalf of the Concord Railroad, and for money's worth. So that this action, as it is now before you, is substantially for money which George Clough, as a conductor upon the Concord Railroad, has received, and has dishonestly and frudulently withheld from the railroad, which money belonged to the Concord Railroad.

This case has a very extented history at the present time. Allow me to say, in behalf of Mr. Clough, that upon the day when the directors voted that he should be dismissed from the road, he supposed that no man had a better reputation as an honest man, or had better deserved it, or had more faithfully earned it than he. He knew full well of the customs, and of the corruptions of the Concord Railroad, and of its officers; but in the midst of these corruptions he had tried to steer clear of any infamy that might rest upon them, and to save his good name and reputation. For twenty-four years he had served the road, as he supposed, faithfully, promptly and efficiently; having never, during twenty-four years, missed a train; and being at the time that he was discharged, and for years previous, a large stockholder in the road, he had endeavored to do the duties of a conductor for the Concord Railroad as he would do them for himself.

The counsel for the plaintiff, in opening his case, stated, somewhat at length, what he expected to prove, and what he hoped to show. But alas! how frail and uncertain are all human hopes, and all human expectations. During the time that we have been engaged in this hearing (as I wish to suggest to you now, and as my brother, when he makes his closing argument, will undoubtedly argue to you,) more than nine-tenths of all that has been put in here as evidence have been suggestions and inuendos and incompetent testimony, put in with the idea that they were going to connect it in some way with something else, and make it competent. In this they have signally and entirely failed.

I ought to say that I am obliged to Col. George for the memorandum he has furnished me, of the positions taken. I propose, in the outset, to analyze these positions; and wherein they are false I propose to show that they are so; and wherein they are true I propose to admit the truth.

*First*, he says that tickets which had been once used were re-sold at stores in Manchester to the extent of several hundred dollars. Now, gentlemen, that may be, or it may not be. Tickets that may have been used and re-sold, but with which George Clough had no connection, nor any knowledge of whatever, have nothing to do with this case. This position, I admit, must have referred to the tickets of Mr. White and the tickets of Mr. Weston, neither of whom testify that they ever had a ticket of George Clough, or that they had any knowledge, or idea, that any of the tickets that they had ever came from Mr. Clough. There is no *inuendo*, even, in their testimony, that they did come from Mr. Clough. When this testimony was put in, it was proposed to connect it, in some way, with Mr. Clough. It was proposed to show that George Clough let James Whitcher have the tickets, and that James Whitcher let the two witnesses, White and Weston, have them; or that he let Mr. Starkey have them, and that Mr. Starkey let the two witnesses have them. There might be considerable pertinency in my asking at this time, in respect to this testimony, *where is James Whitcher?* Is he living, or is he dead? I simply ask that question; and if James Whitcher is not produced before the referees before the close of this trial, my brother, when he comes to place this matter before you, will, I think, do ample justice to James Whitcher, and to the two gentlemen who went down to arrest James Whitcher, and who brought him here and kept him here over Sunday under arrest.

The second position taken by the plaintiff is, that Mr. Clough sold tickets and passages, for boots and shoes, to John Grier; and that he improperly passed his tenants, workmen, and others, for his own private benefit. As to the charge that he passed his tenants, workmen, or others, for his private benefit, I submit that there is not one particle of proof that we are called to answer, except it is in the single instance where Mr. Colby testifies that he went down to Boston on Mr. Clough's pass, to look at a heating apparatus; and when Mr. Clough comes upon the stand I think he will explain that to your entire satisfaction. As to the charge that Mr. Clough sold tickets or passages for boots and shoes, to John Grier, I assert that it is not true; and Mr. Clough will come before you and state that it is not true; he will state to you that Mr. Grier came to him and said that he was so poor that he was not able to earn bread for his wife and children during the six days of the week, but had to work on Sundays; he will state that because of Mr. Grier's poverty, he never in his life took of him one cent's consideration for riding over the railroad; that he paid him for all the work that he ever did (and he charged very high prices), to the last dollar, and to the last cent; and that he has receipted bills to show for it.

I have heard one of the counsel on the other side, before public meetings, and on other occasions, before Mr. Clough ever had an opportunity to reply to this charge, state that one of the conductors of the Concord Railroad had been furnished, for two or three years, with boots and shoes in exchange for free rides over the railroad. I have to say, in reply to this charge, that it is a part and parcel of a continual stream of tirade and slander on the part of that counsel and his friends, of

George Clough, which has lasted for more than three years. In the cars, in his office, on the street, and at the hotels, his mouth has been an open sepulchre, to prejudice everybody against the defendant. I was about to say, although it may not be proper for me to say it, that because of the froth and foam that the newspapers have stirred up against the conductors of this road, it was hardly to be expected that the three honorable gentlemen who compose this board of referees could come here and sit down to the fair and candid trial of this cause, without being, in some degree, prejudiced against Mr. Clough. It has been extremely hard for Mr. Clough to set quietly by for the space of three years and maintain any sort of good nature, knowing how outrageously he was being wronged. But I have said to him, "Be quiet, and wait. Time will at last bring all things right. If Mr. Carney, nor anybody else, has found you retaining money belonging to the Concord Railroad, you are all right; and no matter what powerful array of counsel there may be against you, time, and the judgment of fair and disinterested men, which is the tribunal before which we hope sometime to have this case heard, will set this matter all right." And now, gentlemen, we are here before *you*,—three disinterested gentlemen of learning, of candor, and of intelligence; and we are satisfied that there can be nothing, so far as the law is concerned, that can escape your scrutiny in relation to George Clough. We can now hope that you will see substantial justice done between George Clough and the Concord Railroad. If Mr. Clough has retained but the ten thousandth part of a cent of the funds which belong to the Concord Railroad, he will answer for it in your judgment; and to such a tribunal as you constitute, we have confided our whole case; and whatever that judgment may be, everybody in this community, the Concord Railroad and Mr. Clough, the counsel upon the other side and the counsel on this side, is bound to be satisfied with, because it will be the judgment of the tribunal before which, by mutual agreement, our case has been presented.

We shall show to you that this Mr. Grier was a poor man, a shoemaker by trade; that he did business for George Clough; that he occasionally passed over the road, and that Mr. Clough passed him free,—taking no money from him, and receiving no favors of any kind in return. During all the time that Mr. Clough drove a stage, when he was the proprietor, and when he was not, he never left a man by the road-side because he had no money with which to pay for a ride; he never put a man out of the cars because he had no money; and he never accepted a pecuniary favor of any kind, or in any way. from a man so poor as John Grier; and no man can truthfully say that he has. It is an imputation upon Mr. Clough's character, by anyone who has known him, to say that he ever accepted anything from so poor a man as John Grier, supposing that he had offered it to him. We wish to suggest to you this: here was a great "hullabaloo" about Mr. Clough; the counsel upon the other side were "blowreating" (if I may be permitted to use an expression more expressive than elegant,) everywhere about the conductors, and the immense amount of money that was stolen. *I* have not heard the counsel upon the other side say that George Clough was worth two hundred and fifty thousand dollars, and that he had stolen all but fifty thousand dollars from the railroad, but I have heard a great many persons say that the counsel has said so. What I wish to suggest is, that Mr. Grier, hearing these charges, might have supposed that he, too, was coming to grief, for having ever ridden free over the road, and thought perhaps that if he could show that he had paid for every passage, he could go free of blame, and so gave the testimoney that he did. Mr. Grier was not perhaps sane at the time; at any rate, a few months afterward he went down to the river in this city and committed suicide, leaving, I believe, an account against Mr. Clough, which he has since paid. Mr. Clough will state to you all about his transactions with John Grier, and you will be able to form a correct judgment whether or not his deposition is true.

As to the charge that Mr. Clough did not punch the tickets as they were taken up, thus rendering them susceptible to a second use, I have to say that there is no evidence here as to the truth of that position. I understand what the charge refers to. Mr. Roby testifies that he and his wife were riding in the cars, that he handed his tickets to Mr. Clough, and that Mr. Clough slipped his punch so far past the ticket that it was not punched. The counsel upon the other side charge that Mr. Clough did so designedly, and for the purpose of having the tickets used over again. This is the only instance proved. Mr. Clough will tell you, in the first place, that he never did any such thing; in the second place, that he never had any occasion to do any such thing; in the third place, you will see at once that he never had any occasion so to do; and in the fourth place, a man who knows anything about the duties of a conductor will know at once that there is no necessity for any such thing. There is not one passenger in a thousand would ever notice or care whether the conductor did or did not punch his ticket. The counsel upon the other side propose to argue, because they have proved in one single instance that Mr. Clough's punch slipped by, and did not punch the ticket, that it was by design, and done for the purpose of using the tickets over again. This shows the strait to which he is driven. I shall have occasion to suggest that if such was his purpose, there was no occasion at all for such a deception. Supposing that everything that is charged upon the other side is true, there is still no occasion for using a local ticket over the road for the purpose of obtaining money out of its use. If a passenger comes into the cars, having a local ticket, the conductor takes up the ticket and puts it into his pocket; if he has no ticket, but pays his fare on the cars, all the conductor has to do, if he wishes to retain the fare, is to put the money received into his pocket. There is no need, in such a case, that the conductor should give the passenger a ticket, and thus afford a means of detection. All he has to do, if dishonestly disposed, is to take the money and put it in his pocket, and that is the end of it. Not only, in the first place, was there no necessity for punching a local ticket, but, in the second place, unless a passenger stopped over at a station for another train, the practice never obtained of giving a passenger a ticket in the cars. If, in a single instance, the conductor gave a passenger a local ticket, on his paying his fare, while a thousand people rode in the cars without ever receiving a ticket, there might be some very keen-eyed, inquisitive gentleman in the cars, who should enquire "What does the conductor give that man a ticket for? All the rest of us have no tickets." If the punch, in this particular instance, did slip by the ticket without punching it, as the witness says, it proves nothing; it might occur very frequently; and instead of proving anything, it shows simply that that accident, or circumstance, occurred once on the Concord Railroad during the twenty-four years that Mr. Clough ran as conductor, and proves nothing at all. It simply shows the ridiculousness of the suspicions that the counsel for the plaintiff entertains as to the way in which the conductors discharge their duties.

The third charge is "the finding of three hundred and forty-three tickets, most of which had been once used, on the person and at the house of James Whitcher, at substantially one time." The counsel might have omitted "substantially"; it *was* at one time. Let me say here, in my view (and I think it will be yours,) that had not Mr. Gilmore and Mr. George gone down there at that time, and found those tickets on the person of James Whitcher, and in the possession of Mr. Weston and Mr. White, you never would have been called as referees to hear and decide this controversy; this suit would never have been brought; the distinguished gentleman who appears as counsel for the plaintiff would never have earned an immortality of fame in this case; and George Clough would have been freed from the expense, the annoyance, and the mortification that he has been subjected to in this suit. Let us see how frail is this evidence in support of this charge. If these tickets came to James Whitcher from George Clough, why not *prove* it. In the entire absence of proof, they say that George Clough admits that he was in the habit of supplying tickets to James Whitcher, to a greater or less extent, for more than fifteen years. That position is not true; it is false. Mr. Clough admits, in his deposition, that he has let Mr. Whitcher have tickets, one or two at a time, to give to poor people, as he said. When he comes upon the stand he will testify further, that during times of high political excitement, he was requested, as Mr. Sanborn was, to furnish Mr. Whitcher with a few tickets. But Mr. Clough will testify that never in his life, of his own motion, did he let Mr. Whitcher have as many as twenty tickets, (not including those that Mr. Gilmore directed him to let him have,) and also all that he let him have for other persons. Mr. Whitcher wanted them to give to poor people. It is said that Mr. Whitcher was once a Shaker; and his home has been a rendezvous for people who have once lived with the Shakers. Mr. Whitcher used to go to Mr. Clough and say, "There are two ladies who wish to come up from Nashua, or some other place, and visit me; and I want you to give me one or two tickets for them"; and Mr. Clough will tell you that he let Mr. Whitcher have some tickets in mere charity. The question may be asked, "Why did not these poor people ride on his train, and so be passed without a ticket? The reply is that they could not know, nor could he know, whether they would ride on his train. There is therefore no admission on the part of Mr. Clough, as is claimed by the plaintiff. The statement made here, that there were three hundred and forty-three tickets found upon the person of Mr. Whitcher, proves too much; because, in the first place, Mr. Sanborn has testified that two hundred of these tickets never could have come into the hands of George Clough in any of the usual transactions or arrangements of the road; and of the remaining one hundred and forty-three tickets that were found on James Whitcher, thirty-four were dated January 23, 1866, eleven days after George Clough left the employ of the road, and while he was sick in bed; there were three conductor's checks, and three only, from Concord to Boston, which checks, or tickets, if George Clough ever had them, he must have been charged with by the road, and whatever tickets of that kind he ever had in his possession, he must have obtained from the ticket agent, and they were therefore as much his property as any property that you, gentlemen, possess, is yours.

*Mr. George.* [Referring to one of the thirty-four tickets alluded to by Mr. Rolfe as dated January 23, 1866.] This ticket is dated January 23, 1865 instead of 1866.

*Mr Rolfe.* I may have made a mistake in the date of one ticket.

[Adjourned.]

AFTERNOON SESSION.

*Mr. Rolfe.* When the chairman suggested that it was a proper time to suspend my opening remarks, I had just said that I might have been mistaken in the date of a single ticket. It is a mistake of my sight, if it is a mistake at all; and if it will be of any benefit to the other side, I will waive my statement in respect to that one ticket, and say that it *may* have been dated in 1865. But in relation to the thirty-three other tickets, of which this was the thirty-fourth, I am not mistaken. I hold in my hand a ticket which purports to have been found upon the person of James Whitcher, which is dated on the back, "February," and on the front, "1866." I hold in my hand, also, six other tickets found on the person of James Whitcher, dated February 9, 1866,—more than twenty-two days after Mr. Clough left the road. I also hold in my hand another ticket, which purports to have been found upon the person of James Whitcher, where the date is so plain that it must have come under the inspection of Mr. Gilmore and Col. George, at the time they obtained them.

Gentlemen, let me say that if I had been situated as Mr. Gilmore and Col. George were, and had found these tickets, which their dates show must have been issued from the ticket office after Mr. Clough left the road, and must have come into the hands of Whitcher after Mr. Clough ceased to have any connection with the road, and should conceal those facts favorable to Mr. Clough from the public, and should continue to circulate everywhere, as Mr. Gilmore and more particularly Col. George did, the charge that these tickets were nearly all obtained by Whitcher of Mr. Clough. If, I say, I should do all this, when I knew Mr. Clough had left the road and was at home on a sick bed at the time these tickets were sold from the ticket offices; and should, *finally*, have the effrontery to bring these tickets in here as proof before the referees of Mr. Clough's connection with them, I am sure I should be set down by every honest, disinterested person as either a great knave or a great fool. It would be a conclusive argument that I was guilty of great dishonesty, or was possessed of immense stupidity. These gentlemen both knew then, as Col. George cannot fail to see now, that Mr. Clough could have had nothing to do with any tickets dated after the 14th day of January, 1866, the day on which he ceased to run on the road.

This ticket, I say, found on the person of James Whitcher, and dated in February, 1866, and all their testimony taken in connection with it was taken and hid away and locked up in Gen. Butler's safe, in Lowell, or some other place where nobody could see it, until after the commencement of this trial. I hold in my hand, also, eight other tickets, dated February 6, 1866. I desire the referees to look at the date on these tickets. A man can scarcely have his eyes so dimmed by age as not to see it without glasses. To hold these tickets, to conceal them from us until the day of the trial, and then bring them in here and presume to put them in as evidence against Mr. Clough, when they could have, in no possible way, any connection with Mr. Clough, was a sort of refining of that cruelty with which Mr. Gilmore and the clerk of the road commenced and have continued these proceedings against him. I have, also, in my hand, eleven other tickets over the Concord road, dated February 5th, 1866; the dates upon all of them are very distinct; and if those gentlemen who discovered them upon the person of James Whitcher had been possessed of the slightest disposition to treat this defendant fairly and honestly, or even decently, they would never have brought any such testimony in here against him. I also hold in my hand ten tickets dated February 3d, 1866, found on the person of James Whitcher. I also hold in my hand two tickets dated February 2d, 1866. These tickets number in all, not counting the one as to the date of which I may be mistaken, thirty-three. I ask you, gentlemen, if it is not perfectly apparent to any one desiring to be just and fair that with these tickets which were found upon the person of James Whitcher, and dated after Mr. Clough left the road, he could have had no connection; but that the mischief, if any mischief was done, must be laid to the charge of somebody else than Mr. Clough. I do not know but that the introduction of such evidence might have been an oversight on the part of the counsel for the plaintiff. If so, and if, as they claim, they desire to do justice, and to be no more than fair, I can only say that such an oversight argues immense stupidity. In the package found upon the person of James Whitcher, are sixteen other tickets. The date of many of them is apparent, and is "February." On many of them the day of the month and the year is not so apparent. In relation to those tickets I have to say that when you come to examine them you will find that on the end of every one of them there is the date "February," and on a large number of them is the date "February 5"; and we propose to show that they were not dated in 1865, as is claimed. If you will take the trouble to look at an almanac you will find that February 5, 1865, came on a Sunday. It is well known that the ticket offices on the Concord and other railroads are not open on Sundays. Those tickets could not, therefore, have been issued by the company in that year, as is claimed, but must have been issued February 5, 1866. We shall undertake to show you, in relation to the tickets issued by the railroad, that there is a certain publication of tickets, and that the tickets are dated when they are published; they are then used until the number issued is exhausted, and a new supply, bearing a new date, is published. My suggestion is that these tickets, bearing the date "February 5," must have been issued February 5, 1866, and are therefore in the same category with the others, with which, as I have shown, Mr. Clough could have had no connection.

I have still another package of tickets; it is the package of coupons over the Boston, Concord and Montreal Railroad, found upon the person of James Whitcher, and upon which, as you will recollect, there was the date "February," and the day of the month, but without the year. The dates of the tickets vary from February 2d to February 9th; three of them are dated February 5, without the year, in relation to which I repeat the suggestion made in reference to the other package,—that they could not have been published February 5, 1865, for that would have been on a Sunday, and they must therefore have been issued Monday, February 5, 1866, which was after Mr. Clough left the road; he could have had nothing to do with them; and the inference is that the same hand that purloined those other tickets, purloined these.

There were also found upon the person of James Whitcher, and at his house, various other tickets; they are somewhat mixed, but here they are; and I wish you to examine them. For instance, here are twelve tickets: one over the Rockport and Boston road, one from Exeter to Dover, one from Exeter to New Market, one from Great Falls to Dover, one from Danville Junction to Lewiston, one from South New Market Junction to New Market, &c., with all of which tickets we submit that Mr. Clough could have nothing more to do than the clerk of the road, the counsel upon the other side, or the referees. These tickets, being found upon the person or in the possession of James Whitcher, go with the others.

We have here another package of tickets from Great Falls to Salmon Falls, one from Great Falls to Union Village, one over the Great Falls and Conway road, and others, numbering twelve in all, with none of which Mr. Clough could have anything to do.

Here is another package of tickets over the Boston, Concord and Montreal road, most of which are from Lawrence to Manchester, quite a large number of them being clean tickets. They are tickets with which Mr. Clough could have nothing to do, as they are not tickets over his road. There are in this package nineteen regular tickets and eight package tickets, all of them below Manchester. I call them "regular" tickets in contradistinction from the package tickets that are with them. I have here a Concord, Manchester and Lawrence ticket, marked "White," which I suppose came from Mr. White, the witness; here is a package ticket from Boston to Lowell; here is a Boston and Maine ticket to Manchester, on which are the initials of John H. George and Joseph A. Gilmore; and another of the same kind and having the same initials, making two tickets that have the distinguished autographs of those distinguished gentlemen, one of whom has been said to have covered up more frauds on the Concord Railroad than any other man in New Hampshire, and the other boasted before a committee of the Senate that he had done more than any other live man to help him do it.

I have here another package of sixteen tickets, from Boston to Gloucester, Boston to Salem, Salem to Newburyport, Gloucester to Boston, Boston to Newburyport, and Rockport to Boston,—all of them being over the Eastern road; and it is very doubtful if George Clough was ever on the Eastern road in his life, or ever had anything to do with it, or ever held in his hand a ticket over that road.

I have another package of tickets from Salem to Lawrence. I understand that those that have "G" on them were found on the person of James Whitcher, and that those that have "G 2" on them were found at his house.

I held another package of eight tickets, from Portsmouth to Suncook, Portsmouth to Concord, Portsmouth to New Market Junction, Concord to Portsmouth, New Market Junction to Dover; all of which are numbered in the one hundred and forty-three that might have come into Mr. Clough's possession.

*The Chairman.* When you say they "might" have come into his possession, you mean in the usual course of business?

*Mr. Rolfe.* They might have come into his hands in the usual course of business. The others, if he came by them at all, he must have come by them dishonestly.

I hold in my hand eight clean tickets over the Eastern Railroad. They appear to be clean tickets, and they appear also to be coupon tickets. In the regular course of business, George Clough could not have had anything to do with them.

Here are one hundred and eighteen tickets that Mr. Clough might have had to do with. They might have come into his hands in the regular course of business. The largest share of them are the Concord local tickets and tickets over the Northern and the Vermont Central roads. Here is a ticket over the Northern Railroad, but no tickets on that road were dated at the time that Mr. Clough left the Concord Railroad, nor until some time after. The tickets, prior to their being dated, simply had the number of the station on it. A large number of such tickets were found to be in the possession of James Whitcher. These go to make up the one hundred and forty-three tickets that were testified to, that could in the regular course of business have come into the hands of Mr. Clough.

Having shown you thus much in respect to these tickets, we shall not stop there. Before we conclude this hearing we shall show you, by the admissions of Mr. Starkey himself, who furnished those tickets to James Whitcher, just where they came from, and how they came there, and thus remove every particle of doubt in respect to them. In respect to the tickets which, as we have suggested, could not have come into Mr. Clough's possession in the ordinary course of business, I submit that there is no need of evidence ; and in respect to those which could have come into his hands, that were found upon the person or in the possession of James Whitcher, we shall show, to your entire satisfaction, how they came there. Having shown that, we might perhaps be relieved from going any further. Having shown that, I think we might safely argue that there is no evidence, and no probability, of Mr. Clough's having done anything irregular on this road. But we will not stop there; we will proceed ; and we will proceed with Col. George's analysis. He states, in the sixth place, that it was perfectly practicable for a conductor, if dishonest, to sell and use tickets "over and over again," without detection, so far as any return was concerned. I have already suggested, in relation to that, that the tickets were nothing but an obstruction. It could be of no benefit to the conductor to retain tickets and sell them over again, except in this way; and if he desired to retain tickets and sell them to periods who should ride on some other train than his, for the purpose of putting money into his own pocket, he might do that; and the suggestion that the counsel makes, that it was perfectly practicable for him to do that, is true to a certain extent, but, as has been testified, to no considerable extent. There has been no proof yet offered that George Clough has ever sold a ticket a second time to anybody. There is proof that he has retained tickets ; that he has not punched them ; and he will testify that he has retained tickets and given them away to poor people to ride upon ; and that be has given them to Mr. Gilmore, at his request, and otherwise. He will state all that there is to be known in connection with these transactions.

In the seventh place, he says that "Mr. Clough did use such tickets, according to his own admissions," which is true. The referees have ruled that that was of no account if he did not use them improperly.

He says, in the eighth place, that it is perfectly practicable, so far as detection is concerned, for a conductor, if dishonest, to appropriate money taken in the cars. *That is true.* It was not only true of Mr. Clough, but it is true of every other railroad conductor upon earth, and it always will be true of every other railroad conductor. You can never institute a set of rules—you never can hedge the conductors around so but that they can, to some extent, put money into their own pockets and appropriate it to their own use; and I submit that it was of no use for the directors of the Concord Railroad, and Mr. Gilmore and Mr. George, to take a journey down to Boston, and have a sort of *revelry* at the Revere House, or at the American House, in order that some scheme might be devised to detect these conductors; because, I submit to you, gentlemen, that if the officers of the road, during the twenty-four years that Mr. Clough had been acting as conductor, had suspected that he did not return what moneys he took in the cars, there were ten thousand good opportunities whereby the president of the road, or the clerk, or the counsel of the road, with his immense acuteness, could have discovered it to the most perfect satisfaction of himself and Mr. Gilmore; and there was no sort of necessity for their employing this distinguished detective down at Lowell, who on one 28th day of September, took a thousand dollars, as a reward, to make out a report in favor of a criminal, eight days after.

Have the counsel for the plaintiff shown that Mr. Clough has received any money due the company that he did not pay to them? That he could, to some extent, we admit; but have they shown that he did? Mr. Clough will state upon the stand that, to his knowledge, he never retained or withheld a single dollar, nor a single cent, of the money that belonged to the Concord Railroad. And up to the day when this charge against him was published in the Boston Journal, I submit that there was no man in this community that had, nor any man in this community that may hope to have, a better reputation with all that he has ever dealt with—(save the Concord Railroad)—for honesty, integrity, uprightness and truthfulness, than had George Clough; a reputation, gentlemen, that he has been nearly forty years in earning, and that is richly worth to him forty years of honest service He will come before you and state that to his knowledge and intention he never withheld one dollar, nor one cent of the funds belonging to the Concord Railroad. He will come before you and state that, in the midst of the corruptions of the times, from the day when Joseph A. Gilmore was a candidate for senator from the fourth senatorial district, and Col. John H. George was a candidate for congress in the second congressional district, when the prerogatives and the power and the money of the Concord Railroad began to be used for political purposes, up to the time when he was discharged from that road, he has endeavored, fairly and honestly, at all times and under all circumstances, to do his duty.

In the ninth article of this analysis, Col. George says that "during the war the amount of travel was very largely increased; for months together the number of cars on the lighter of the two heavy trains from Concord, numbering from fifteen to eighteen cars, were filled with passengers." I do not propose to say anything about how that was. You know whatever testimony has been offered to show it; but if there was any such thing as trains from fifteen to eighteen cars filled with passengers, running on the Concord Railroad during the war, and Mr. Clough and the other conductors appropriated what was earned by the road over and above its former earnings, they must have appropriated an enormous amount. But I think we shall be able to show that no such state of facts could have existed. I have this suggestion to make: that Mr. Clough was a large stockholder in the road, and he noticed from time to time the way in which the dead-heads were passed over the road; and he has stated in his deposition that he thought that the number of dead-heads that were passed over the road was equal to two per cent. of the capital stock of the Concord Railroad; he does not know but that he may be mistaken, but it seemed to him that it was equal to that. If there is any argument to be drawn from that ninth proposition, this is the suggestion that we have to offer in defence: Mr. Gilmore, from 1859 to 1864, was actively engaged in political matters; he was a candidate for state senator in 1859 and 1860, and was also a candidate in 1860 and 1861, and was elected both years; that during that time it was perfectly notorious that he used the patronage of the Concord Railroad not only for his own political purposes, but to secure the friendship and the control of the legislature, with a daring and recklessness that was never equalled by any man in New England; that he gave out his free passes by hundreds, to secure political influence; and it is in evidence here that on a certain day, when Mr. Harriman was nominated for governor, there were more than four hundred of these irregular tickets, without the ticket agent's stamp upon them, issued, that went back into the collections, and of which no account was taken. I might say that Gilmore contributed more than four hundred dollars, in tickets, for political purposes, during the year 1863–4. I might say that Gilmore went to the conductors and, because he did not want so many free passes issued, *ordered* them to reserve these tickets and return them to him, in order that he might give them away to produce political influence in his favor. I might say to you that all the ministers rode free; that all the doctors rode free; and I was about to say that all the lawyers rode free, but I believe they did not; I believe that Judge Bell, and Herman Foster, and George W. Morrison invariably paid their full fare; I believe that there was no other member of the bar that did; I believe that Mr. Stanley and Mr. Clark paid full fare for a time, and that afterwards they were allowed to pay fare one way, receiving a free ticket back. I might say that all the bank directors, north and south, in Boston and about Boston, who had any pecuniary favors to bestow upon Mr. Gilmore, passed free. I might say that all the hotel keepers and their families passed free. I might say that all the wholesale dealers who had a certain amount of freight over the road passed free. I might say that there were season tickets given out to persons, upon which all the members of their firm could ride free at all times. I might say that all the politicians in the state, who had any influence, had free passes. I might say that every member of the legislature that Mr. Gilmore thought that he could influence had a free pass. I might say that public meeting after public meeting was gotten up, not only at the time that Mr. Harriman was nominated, but at other times, and that free tickets were thrown out with a profuseness that Mr. Sanborn has forgotten. I might say that Boston men came up in the trains, and Mr. Clough and the other conductors made them pay their fares, but that the fares were paid back to them again; and Mr. Clough will tell you when he paid back the fare how he had to get it again. I will suggest to you (what was doubtless observable to you all) that during the war all the southern, and a large share of the northern mountain travel was cut off, so that during the war, in the season when the road was reaping its richest harvest, there was not so much travel to the mountains as there was afterwards. I might suggest, what it is not necessary to put in proof, that since the war the business of the road has increased. There has been an increase in the population of Manchester, Hookset, Suncook, Concord, and all along the line of the road, and there has been a corresponding increase in the travel and business over the railroad. The increase has been so great as to require, within the last year, an increase to double the former capacity of the road, in order to carry freight over the lakes and over the roads below. The simple suggestion of these things is, we think, a sufficient answer to his ninth position.

He says next, that "the fares during the war were twenty-six and a half per cent. higher than they are now. Suppose they were. I happened to be a member of the railroad committee in 1864, and many petitions were presented asking the legislature to interfere and cut down the fare on the Concord Railroad. The Concord Railroad has been the only one that the legislature has attempted to put under discipline. What was the answer of the road? They came into the legislature and showed that in the single article of cotton waste the expense was many, many times higher than before the war. They showed that the wood bill was higher than before the war. They showed great increase in the expense of running to prove that there was an immense increase in the outlay of the road. During the war there were a great many soldiers that passed for soldiers' transportation, which was two cents per mile, while the regular fare was three and a half cents per mile; and a large share of the extra travel was the travel of soldiers at this reduced rate of transportation. The Concord Railroad would not have received near so much income from this as from the ordinary travel, if they had really got it, but Mr. Sanborn testifies that there were about thirty-four hundred dollars worth of soldiers' transportations that the Concord Railroad never received. I have still another suggestion to make, and it is indicative of the perfect looseness with which Mr. Gilmore used the patronage of the road to accomplish anything and everything that he desired; it is this: During the time that Mr. Clough was on the road, there was to be a councillors' convention held at Manchester, and men in the interest of the railroad distributed to all the delegates such tickets as these [handing a package of tickets to the referees]. Mr. Spalding, a director of the road, was then a candidate. I suggest the absurdity of a suit in which such a man as Mr. Spalding, a man who has been for many years the president of this road, one of the directors of the road, and a candidate for office, should attempt to have George Clough refund money belonging, as alleged, to the Concord Railroad, when, in all probability, the fares of all the poor and destitute, and of every one else that George Clough ever passed free over the road, would not amount to so much as it cost the Concord Railroad to carry the delegates to that convention and back again on that one day. The tickets that I have shown you are not half of the free passages that Mr. Clough took notice of on that single day. He showed them to me, and I took them out of his hands and put them into my pocket, and when he wanted to return them to the road I told him that he could not have them. That is a sample of the way in which the Concord Railroad was used during the time that Joseph A. Gilmore and Isaac Spalding were candidates for political honors; and now the suggestion is made to the referees that because there were no more returns made of fares taken from passengers in the cars, that Mr. Clough and the other conductors have stolen the money that accrued from these fifteen to eighteen cars-full of passengers that it is claimed passed over the road. There must have many things transpired in relation to this free pass system, and in relation to people riding free over the road, and from whom no return was made, that neither you nor I can know. It would require perhaps the diligence that is exercised by the United States in collecting the entire census of the country, to collect the statistics of this free travel; yet George Clough is now sued for stealing what *might* have been received had there been no free traveling It is suggested that the number of cars required to carry the full travel during the war, would average about two passenger cars per train, containing fifty passengers each. It would have been as well to have stated that just as it was. Let us see how that was. The train that runs from Concord to Nashua in the morning (which, it will be admitted by everybody, runs light,) must carry cars enough to bring the passengers back; and the train that goes down in the afternoon must carry passenger cars enough, if there was not enough taken down in the morning, to bring the passengers up at night. Thus it is that so many cars go down over the road and so many cars come back over the road, year after year. The cars may be full; one car may be full, or one car may be only half full. I have known an instance of a train starting out from Concord in the morning with fourteen passengers in one car and none in the other, and every single one of those passengers was a dead-head; there was not a paying one among them. We shall introduce the testimony of men who have passed over the road on the early morning train, in repeated instances, when there was hardly a paying passenger in the cars. Still, all that train of cars had to move down to Nashua. Testimony in relation to the number of passenger cars on a train has already been given by Mr. Upham, one of their own witnesses. We shall introduce other witnesses to show, and shall show clearly, the actual amount of free passes issued, to account for the small return of income received from so many cars.

In his twelfth position, Col. George says that the amount now received from passengers is much in excess of what it was while Mr. Clough was on the road, with the travel largely diminished.

That "*largely* diminished" is an assumption contrary to the fact; and I do not understand that the referees have allowed any testimony to illustrate that position to be put in, unless by agreement; and all that we proposed, by putting in that agreement, was to show what Mr. Clough returned during two years of the time that he was conductor, as compared with the returns of two years since.

The eighteenth proposition is, that Mr. Clough, according to his own admission, bought and sold coupon tickets to such an extent that his profits amounted to two thousand dollars. That is substantially the truth. He admits it, and testifies, according to his best belief and understanding, that he sold that amount of coupon tickets. The position is taken that the Concord Railroad was defrauded thereby. I do not understand, if the Concord Railroad lives up to its contracts, and does what it agrees to do, and carries out its stipulations with other roads, that it can thereby be defrauded, either in law or in fact. I understand that this position is good in law: that if the Concord Railroad does business with the Grand Trunk road, or with any other western road, entering into a contract with it with the express understanding that that contract shall bring passengers over their road, and agreeing to carry them for less than local fare, nothing that is done by anybody within the compass of that contract, can be a fraud upon the road.— The Concord Railroad adopted an arrangement with the upper roads, to receive coupon tickets over the Northern road, without reference to their dates.

In the case of Johnson *v.* The Concord Railroad, Mr. Gilmore testifies that he bought some of these coupon tickets from some persons going over the road, to show the directors how the road was defrauded. It was not necessary for him to buy those tickets to show the directors how the road was defrauded. The Concord Railroad was doing business; it had business relations with the upper road, and it must sustain these business relations. The directors wanted to bring the travel over this road, and to compete with other railroads. Tickets were issued without any limitations; they were to be received at any time; there was no restriction in relation to their use by third parties; they were not marked "not transferable," nor "not good after a certain time." These tickets were issued by the roads above; they were used without limitation; they were used in the hands of anybody and of everybody, without any objection from the road. I now present to the referees a couple of the tickets such as were then issued. As you will see, they are without limitation. The man who buys the ticket has a right to a passage over the road. As they are not marked "not transferable," they are good in the hands of anybody; and if a passenger, after riding a part of the way, hands or sells his ticket to another party, it was the invariable custom and practice of the road to receive them from that party. Had the directors made a rule against it, it would have been of no avail. Here are two other tickets, issued in the regular order. I submit that no conductor of a railroad train can tell when they were issued. They are dated, but it is impossible to tell what the date is. Suppose the ticket is transferred, and the person who holds it offers it for fare on the road; what can the conductor do? He cannot stop to inquire whether the passenger presenting it was the original buyer of the ticket. The *holder* of the ticket is entitled to a ride. After a while there was a limitation made. A rule was passed that the tickets should be good only for a certain time—perhaps thirty days. The Concord Railroad was instructed to receive them, without any quarrel or row. That was the business arrangement with the upper road, and the rules that were published here did not annul or over-ride the contracts with other roads, or the relations that were sustained to the other roads. Mr. Clough saw these tickets in the hands of other parties; he knew that other men were buying them and speculating in them; he knew that although he did not buy them, others would; he knew that the road would not be defrauded any more if he bought them than it would be if others bought them; in fact he did not understand that the road would be defrauded at all. He went to Mr. Gilmore, the superintendent, and stated to him that other parties were speculating in those tickets, and asked if he could not buy them. Gilmore told him that he had no objection to his buying any of the tickets that came over the road from this side of Detroit; that he might as well make a little money out of them as anybody else, and that he might buy them and sell them; and Mr. Clough thereafter did buy and sell those tickets, and made by the operation, as he testifies, two thousand dollars. Mr. George, on behalf of the railroad, now claims that two thousand dollars as the property of the road. Now, admitting for the sake of the argument, that everything is as the plaintiff claims, and that the law is such that if Mr. Clough bought those tickets he is bound to respond to the Concord road for them—even admitting all this, he certainly cannot be called upon to respond for any more than the Concord Railroad's share of those tickets. The Concord Railroad has no right to the tickets over other roads. The counsel for the plaintiff alleges that the Concord road was defrauded out of one dollar and seventy-five cents on each ticket. If there can be any such close calculation as that, in this transaction, my stupidity prevents me from seeing it. The fare on the Concord road is only about a dollar and a half. I think it was never more than that from Concord to Nashua. The Concord Railroad permitted the Northern Railroad to issue those tickets. The person who buys one of them is entitled to a passage down over the Concord road. They are not marked "Not transferable," nor are they limited as to time. The Concord road has contracted to receive these tickets. They contracted to do it for what purpose? for the purpose of inducing more travel over this road than passes over the competing road in your section of the country, Mr. Chairman. No matter who buys one of these tickets, if the Concord road is simply called upon to give one passage over the road in exchange for it, nobody is defrauded in law or in fact. The Concord Railroad does only what it has bound itself to do, and cannot be defrauded thereby; and if Mr. Clough simply buys that ticket by the direction, or by leave of the superintendent of the road, he is guilty of no fraud, intentionally or otherwise, in law or in fact.

*Mr. Stanley.* In 1864 the Concord Railroad received, on local tickets, \$2.50 on each ticket, and on the western tickets 52 cents, so that they lost on each ticket, and received, in fact, only 28 cents per ticket.

*Mr. Rolfe.* The Concord Railroad were entitled to only \$1.50 when the fare was the highest, and to \$1.10 as the fare is now. If they lose by the transaction \$1.84 or \$1.75, as they would have it, or if they lose more than their fare, I can simply say that I can't see it.

*Mr. George.* The testimony of Mr. Sanborn was that on a local ticket from Concord to Boston, the net profit to the Concord road for the passage, and for the use of the cars, was \$2.12, which was a very different thing from the local fare to Nashua. On a local ticket from Concord to Boston they received \$2.12; on the western tickets the highest rate was 52 cents; and the average rate, as Mr. Sanborn testifies, on western tickets, was 28 cents, which would be a loss of \$1.24; the average rate, he thought, might be 45 cents, which would make \$1.67 loss.

*Mr. Rolfe.* No matter how much they lose. They stipulated and arranged to do business with the upper roads in that way. They not only arranged to do that, and if some one buys a ticket and does not use that ticket to go over the road, another person will. Therefore it was that Mr. Gilmore said to Mr. Clough, "If you do not buy the tickets, and save what is to be saved, somebody else will. The road will neither gain nor lose by your buying them."

The twentieth charge is that Mr. Clough received, and was cognizant of bogus tickets being sold and used by other conductors. There is no evidence of that kind. They may intend to put such evidence in, but there is no such evidence in now.

*Mr. George.* The testimony of Major Carney was that he received from ——

*Mr. Mugridge.* That is not in. It was ruled out distinctly.

*Mr. George.* The testimony of Major Carney was ruled in for the purpose of showing that the tickets came into Clough's hands. The testimony was put in for that purpose, and confined to the purpose of showing that these bogus tickets went into Mr. Clough's hands.

*Mr. Mugridge.* It was ruled out upon the ground that any tickets that Mr. Carney might have received from Mr. Kendrick was not competent unless there was collusion shown between him and George Clough, and that in this form of action he could not be held for collusion.

*Mr. Rolfe.* I understand what the gentleman alludes to. There were some irregular tickets given to Mr. Carney, and Mr. Clough took them up. Let me suggest that if a conductor below has occasion to use several tickets,—for instance, if Mr. Clough, starting from Concord here, has six of these conductors' tickets, and there are seven men who enter the cars here in Concord, who want to go to Boston, Mr. Clough takes a ticket, or a piece of paper, and writes on it: "Good for this passage to Boston," or, "The conductor will please recognize this as a ticket," or, "The conductor will take this ticket up and return it to me." It is only given as a matter of convenience, and because the conductor does not happen at the time to have tickets enough. Now, in this case, where Mr. Carney came up over the road, the conductor gave him a certain ticket which he happened to have. When he got up to Manchester, Mr. Clough would take charge of the train from there, and would take those tickets up, and so Mr. Kendrick goes to him and says, "If you will recognize those tickets now, when I get up home I will redeem them of you." And thereupon Mr. Clough takes up those tickets, and when the conductor comes up to Concord he gets his new supply of tickets, and redeems those tickets of Mr. Clough.

[Mr. Mugridge read from the stenographer's report of the proceedings of August 6, 1868 (page 203), the discussion and ruling upon the disputed point.]

*Mr. Rolfe.* I have now made an explanation of the manner in which such a transaction might occur ordinarily. That Mr. Clough could receive any benefit or gain from it, I cannot conceive. The reason Mr. Kendrick did not give him the proper tickets was because he did not have them; he gave such as he had. They were not bogus tickets, by any manner of means. They were real tickets, and they were good over the Concord road, and Mr. Clough could not help taking them up, no matter in whose hands they were. The probability was that some of these tickets gave the right to a passage beyond Concord, and as the passengers who held them were only coming to Concord, Mr. Kendrick, the conductor, informed Mr. Clough of the fact, and Mr. Clough took the tickets up, and Mr. Kendrick redeemed them. I might say in relation to this, not perhaps in the language of another, but a little stronger, that "Trifles lighter than air, are, to the suspicious, confirmation stronger than Holy Writ." No transaction of George Clough's that savors the least of irregularity, that the counsel can lay his finger upon, can escape being brought in here with an air and a significance that, as against George Clough, something is wrong. Coming from the source it does, I will say that during a long series of years that the counsel was connected with the Concord Railroad, in the language of the Arab, Joseph A. Gilmore was God, and John H. George was his prophet; and if there was anything in the shape of iniquity—anything in the shape of corruption on the part of Gilmore during all that time, that Colonel George did not help him to cover up and conceal, and was not ten times more the counsel of Gilmore than he ever was of the railroad, then the largest share of this community, that I am acquainted with, are greatly mistaken. He might well say before a committee of the senate, that he had done more to cover up and suppress the frauds and iniquities of the railroad, in connection with Gilmore, than any other man. I do not say this upon my own personal recollection alone, but upon the personal recollection of my friend and associate counsel, who was present at the time. It is very proper that if these suspicions against Mr. Clough should come in at all, they should come from the source in which they do, for who on God's earth that has had so much to do with Gilmore for eight or nine years would fail to suspect that something was irregular and dishonest about everybody. I have only to suggest further, that the simple fact that Mr. Clough took some tickets that were irregular, is but something that happens every day; and Mr. Sanborn and Mr. Kendrick testified here the other day that if the conductors should recommend anything in that way, when they were out of tickets, that they would respect the recommendation on another road, and take the tickets up, and that this was the case here. I can conceive of no way in the world in which George Clough could be benefitted by taking up those tickets; and there is no evidence that he did receive any benefit from it.

The twenty-first proposition is that the buying, selling or using of any tickets not furnished by the ticket agent, was prohibited by the rules of the road, signed by the clerk. It was not any more an imperative rule because it was signed by the clerk. The rules of the road he was bound to obey. The name of the superintendent of the road is very properly given. The superintendent is the executive officer of the road. In other words, he is the commander-in-chief of the army represented by the employees of the road. He is the man to tell when trains shall run and when they shall not run. He is the man to say to this employee, "Go," and to that one, "Come;" and they must go, and they must come. No railroad can be operated safely in any other way. Now, understand me, gentlemen: I do not mean that if Mr. Gilmore, the superintendent, commands Mr. Clough, or any other conductor, to pick another man's pocket, that he is bound to do it; but I understand that the superintendent of the railroad has the entire regulation of the affairs of the road. If the superintendent directs, for instance, that a certain person shall have a free pass, that is the end of the law. If the superintendent directs that such and such persons shall ride free, that is the end of the law. If the superintendent directs that the fare shall be so much, that is the end of the law. If the superintendent orders that a certain train shall start at a certain time, the board of directors cannot come in and say that the train shall start at a different time. The superintendent must be the executive officer of the road, and command and direct every employee of the road, and no one else can. It is not proper that the directors should come in and interfere with the superintendent of the road, any more than it is proper for the commander-in-chief of a naval vessel to tell a man under the executive officer of the ship when to touch off a gun, and when not to. The directors can remove a superintendent and appoint another; they can advise him what to do; they can order him what to do; but they cannot go to Mr. Clough, or to any other conductor, and tell him what to do, in contradistinction to what Mr. Gilmore has told him to do, because some single head must have the direction of the affairs of the railroad; and the superintendent is the man to have the direction of the affairs of the railroad. If an employee of the road goes to Mr. Gilmore, and says to him, "I want you to draw me a car-load of wood over the road," and Mr.

Gilmore draws it over the road, and says, "There is no freight to pay upon that; I give you the freight," that is the end of it. If Mr. Minot, one of the directors of the Northern road, receives a car-load of wood over that road, and Mr. Stearns gives him the freight, that is the end of it. No action for the recovery of freight can be sustained. That is the custom of the road. The employees have certain favors and advantages; and this custom is universally recognized and acknowledged. Now then, whatever this superintendent permits one of the employees to do, the railroad cannot turn around and sue the employee for doing. For instance: on the Concord road the wood is brought down over the Montreal Railroad. In the charge that is made is a charge for car expense, because the Montreal road uses a car of the Concord road, and the employee of the road simply pays for the drawing of it. Now, if the doctrine is true, that the conductor is to be governed by what the directors direct, and not by what the superintendent directs and consents to, then the Concord Railroad, if they see fit, can turn around and sue any employee, during the last six years, for the use of a car to draw their wood on over the Boston, Concord and Montreal Railroad, because he has no right to give such privileges. That is what is said.

Now, in relation to these tickets, Mr. Clough wanted to trade in them; he wanted to receive the benefit from the sale of these tickets that somebody else necessarily would receive if he did not, and which Mr. Gilmore understood and knew that somebody else would receive. Mr. Clough asked Mr. Gilmore if he might trade in them; Mr. Gilmore consented, and he did trade in them, and made money in that way. Before I was interrupted, I was suggesting that the Concord Railroad would receive so much out of these coupon tickets, and that the rest would belong to the roads below. I think that Mr. Clough will testify that on one occasion a family of seven came to Manchester from the west, and when they got to Manchester they gave their tickets to Mr. Clough, and he sold them. He will testify that it was frequently the case when men were ticketed through to New York, and did not make the connection at Nashua, and did not want to stop there, and could go on in the Boston train and take the boat and reach New York by the next morning, that they would do so; and so they would offer their tickets to Mr. Clough and he would buy them, and generally at a large discount; and he would then sell the ticket to the passenger for New York who had no ticket. Frequently it would happen that persons coming from the north and going to Boston would have a ticket for New York by Boston, and being unfamiliar with the line of travel, and not knowing that they could go directly on to Norwich, and the conductor ascertaining that they wanted to go by the short route, would say to them, "You can keep right on, and take the boat at 11 o'clock to-night, and wake up in the morning in New York, and not have to go around by way of Boston." Thereupon they would frequently say that they had a coupon ticket to Boston, and would offer it to Mr. Clough, and he would buy such tickets of them. And in these various ways Mr. Clough has made the sum of two thousand dollars; and, as affecting the Concord road, he did it with a fair understanding and with the perfect knowledge of the superintendent of the railroad; and the superintendent understood, at the time, as anybody else would, that if Mr. Clough did not buy these tickets somebody else would; and there is evidence to prove that other persons did buy them, and sold them, and that other persons rode upon them.

His twenty-second item is, that large amounts were paid Mr. Clough personally, by Lane, Perkins and others, which sums were not accounted for. Now, I am not a particular friend of Mr. Perkins, but if I was, I should object to his name being mentioned in connection with Henry P. Lane's and Samuel Curtis'. I did expect and hope that if George Clough was to be found to have appropriated the funds of the Concord Railroad, that the plaintiff would undertake to bring before the referees some decent kind of testimony. Mr. Perkins, I think, testifies in relation to paying fare on the cars. He does not come up to the mark of Col. George by any manner of means. And in relation to Lane and Curtis, I say to you, gentlemen, that if you had dragged the whole length and breadth of the purlieus of hell you could not have found two viler and more despicable men. There is not a decent man on God's earth that would trust them with money; and their testimony, as my friend will suggest ——

*Mr. Stanley.* He can go and borrow a hundred dollars at the bank in Manchester quicker than you can.

*Mr. Rolfe.* He may if he applies to the bank where the gentleman is president, and if he goes on the credit of the house of ill fame that he runs in the city of Worcester, or on the credit of the other establishment in Boston that the other man, Sam. Curtis, runs.

Now, this Henry Lane rode back and forth in the cars, and Mr. Clough collected the extra ten cents of him when he had not a ticket. He wanted him to pass him for nothing, and he told him that he could not, that he must pay his fare; and he made him pay his fare; and he made him pay the extra ten cents; and after that he didn't pay his fare in the cars, but bought his ticket at the office, and had to pay as other men have to pay; and that is the reason that Henry P. Lane is brought here; and of all that vast multitude of substitute brokers, he was not worthy to be dignified by the name of broker. He was nothing but a runner; he only went down to Worcester to "turn the men this way." We shall show you that the substitutes who came on from New York bought their tickets there, and the substitute brokers came through with them. It was the custom, all through the war, with the men here that took substitutes, to contract for them to be delivered safely in Concord; the substitutes did not pay their fares on the way, but the fares were paid by the agent in New York, and the parties here took them out of the agent's hands when they were delivered here. It escaped, in Lane's testimony, that he "had charge of them," and had to see that they did not go to Portsmouth or Boston.

Now, in relation to Mr. Curtis: He wanted to ride free, and Mr. Clough would not let him. Mr. Perkins wanted to get a free pass for him from Mr. Gilmore. On one occasion he was drunk, noisy and unruly, and Mr. Clough took him out of the passenger car and put him in the baggage car, and put men over him, and kept him there so that he would not disturb the peace of the other passengers in the car; and it was for that reason that he appeared here against Mr. Clough. I think, when my brother Tappan comes to argue this matter, that he will convince the referees, if they are not already entirely satisfied of it, that neither one of those men can tell the truth, unless accidentally.

We now come to that part of the case which is perhaps more significant than any other. The argument of the counsel upon the other side is, that at the time that Mr. Clough went on to the Concord road he was a man of very moderate means; that he was possessed of a very few thousand dollars; and that his expenses must have been quite large—must have been in the direction of the expenses of the counsel, of the referees, and of others. Now, gentlemen, in that direction, I do not know what the early expenses of the referees were. I know something about the early expenses of the counsel; and this I have to say, that George Clough, at no time within my knowledge, or within the knowledge of any man, ever run his establishment with two barrels of whisky in his cellar. The difference between Mr. Clough and the counsel is, that when Mr. Clough was fifteen years of age he was industrious, and commenced to earn, and commenced to save. I do not say that the counsel are not industrious; but I say that when George Clough, being industrious, commenced to earn, at that instant he commenced to save, and not to spend. George Clough's life has been one of energy, industry and accumulation. When Mr. Clough went upon the road, as we shall show to you, he was worth somewhere in the vicinity of eleven thousand dollars; and we shall show you how he got that property. The counsel says that the first we know of Mr. Clough he was a boy tending a livery stable; that then he drove stage for a time, and then was the proprietor of a stage line until the time when the railroad was opened, and that at that time he was worth but a few thousand dollars.

George Clough, at the age of fifteen, left his father's home, with all the worldly goods that he then considered his own done up in a cotton handkerchief. He went away and worked until he earned some clothes, with a farmer; and from there he went to Newburyport, and at Newburyport he stayed and worked in a hotel as a waiter and boot-black, and afterwards rose to the position of hostler in the stable. For two years and upwards he remained there, attending to his business, saving all the time, spending nothing, wearing old shoes and boots, and old clothes, until he had accumulated the sum of three hundred dollars, and was sixteen or seventeen years of age. Two hundred dollars he put in the Newburyport Savings Bank, and with the balance visited his home, gave fifty dollars to his mother, and retained the rest in his pocket. We shall show in the outset, by the secretary of that institution, and from the records, the deposits that were made, when they were made and when they were withdrawn. Mr. Clough stayed at home but a few days. During the time that he had been absent from his father's house, a little over two years, he never lost a day, he never spent a cent for useless things, nor was he sick. After staying at home a few days he went up to Raymond, and went into a stable and worked for a man by the name of Osgood, with a promise from him that he should have a chance to drive stage as soon as an opportunity offered. He worked there for a while at twelve dollars per month. In the course of three or four months he went on to the foot-board of the stage coach at the pay of fifteen dollars per month, which was soon increased to twenty dollars. He drove stage for Mr. Osgood for three or four years, receiving, most of the time, twenty dollars per month and found. He was not off the foot-board but eleven days of that time, and then he was sick with the measles, and Mr. Osgood paid him for that time. During that time he drove stage from Raymond to Pittsfield, and then from Pittsfield to Lowell; and the proprietor of the stage route gave him all the perquisites that he made. When he was passing back and forth he carried money; he carried chickens, turkeys, chestnuts, cranberries, and sometimes mutton and veal. This became so notorious that the line of stages that he drove received the name of the "chicken line." He spent no money uselessly; and, by the way, gentlemen, he used no ardent spirits, and has used none from the day that he was born to this time. He never spent money for ardent spirits for himself; he never drank; he never spent one cent for any ardent spirits for another; and when he drove stage he was known by the name of the "cold water driver." And this, gentlemen, is a matter of more significance in this case than it would have been to have said, when George Clough was twenty-one years of age, that he had inherited twenty-five thousand dollars. These are his characteristics: sobriety, industry, energy. These characteristics have been his all the way through life. To these he has added a keen, ripe judgment; and thus he has been working and accumulating all his life,—*honestly* working and accumulating. He accumulated so much money while he was driving stage from Pittsfield to Lowell that he was enabled to purchase one part of the line from Lowell to Concord, over the Mammoth road. He made a bargain with Mr. Dudley, and paid him two thousand dollars for one half of the line, one thousand dollars of which he paid down. He had the other thousand loaned out to different individuals, and in a very short time he collected it and paid the balance,—perhaps in the course of three or four months. So that within two or three months after he bought out that line of stages he was able to own it, and still have considerable money at interest. He drove that line of stages for a little time, and then Mr. Parker, who then kept a public house in the city of Lowell, because the stages could not put up at his house, proposed to sell out his interest in the stage line to Mr. Clough, and Mr. Clough bought out his half of the line and paid him for it,—not paying him all down at the time, but paying him in a very short time, from the earnings of the stage and from the collections that he made from those who were owing him. He then owned the whole line of stages,—that is to say, the whole line up over the road one day and back the next. He then owned, I think, twenty horses, stage coach, sleigh, extra coach, harnesses, blankets, &c., suited to the business. This was about the time that the Concord road was being constructed, and the travel increased immensely. Mr. Clough continued, in his quiet way, to drive his stage and attend to his business. He bought a house here in South street, and put his team in his own stable, thus not only driving his team from Lowell to Concord, but taking care of it after it was here. From the time that he bought the line of stages, and during a period of three years, Mr. Clough lost not a trip over the road, nor was he absent from his foot-board a single day, except when he went to Pittsfield and married the lady who is now his wife. Mr. Clough will testify, what perhaps is known to many here, that many and many a time has he driven into Concord upon the walk, and driven into Lowell upon the walk, with so great a load of passengers that he could not drive faster, and sitting upon the foot-board with his feet hanging down, because the passengers occupied the driver's seat. In that way Mr. Clough saved up about the sum of eleven thousand dollars. Then he went on the Concord Railroad. He sold a portion of his staging to a man by the name of Stone for about two thousand dollars, retaining that portion of the line which ran from Manchester to Concord. He continued to own that line, and during the first year that he owned it he made a certain amount; the travel increased so that during the next year he made a considerable larger amount; and during the last year before the Concord Railroad was opened to this place, he made out of his staging nearly four thousand dollars net. That matter will be substantiated by the fact that when Mr. Clough went on the road—when he could no longer drive the stage himself—he let a certain part of the staging out to Mr. Charles H. Clifford, who paid all the expenses of running, and paid him four dollars per day for the use of his team besides. He had in the meantime bought a house in South street for six hundred dollars. That was the visible property that he had when he went on to the Concord road. While he was driving stage he had taken out his money from the Newburyport Savings Bank, and had loaned it out in other localities. While he was driving to Lowell, a man by the name of Daniel R. Kimball noticed Mr. Clough as a man of sobriety, industry and integrity, and assisted him in any way that he could, by his advice, in such a way that Mr. Clough looked up to him as to a father; and whenever Mr. Clough accumulated any small sum of money, he would loan it to Mr. Kimball, and take his note for it; and by and by, when a large sum had accumulated, he would take up the small notes and give a larger one. Mr. Clough thus made Mr. Kimball a sort of banker; and Mr. Kimball kept his money for him until some considerable time after Mr. Clough went upon the Concord Railroad. Mr. Clough had the means, and was able at that time, to subscribe for fifty shares, or any larger amount, of the stock of the Northern Railroad; and I believe the fact is he did subscribe, and that he was advised to get rid of his subscription, and so he never took the stock. After a time, Mr. Elkins, who was then in some way connected with the railroad, said to Mr. Clough that if he had some money he could make something in stocks, and Mr. Clough loaned him a certain amount of money, and Mr. Elkins operated in stocks for a time in Boston, sharing the profits with Mr. Clough. During the early days of the Boston, Concord and Montreal Railroad, Mr. Clough took several shares of that stock, and made something out of it. He also bought some stocks, at the instigation of Mr. Spalding, in the Old Colony road. Mr. Spalding told him when to sell, and he also made something by that transaction.

Time would fail me to go through with the biography of Mr. Clough's life. Suffice it to say that inasmuch as it is required, in the testimony that we shall give you, we shall show that from the time that Mr. Clough left his father's house, with all his worldly goods on his body and in his handkerchief, up to the time when he was sued by the Concord Railroad for money which, it was alleged, he had appropriated, Mr. Clough has been a man of remarkable industry, of remarkable energy, of great sagacity. He has been one of those wonderful men that we occasionally meet. I asked Mr. Foster, in his deposition, if he was not a wonderful man; he said, "No more so than any *good business man.*" It is something wonderful to find a smart, good business man; especially is it wonderful to find a smart, good business man like Mr. Clough, who has had such limited opportunities for an education. The habits that Mr. Clough acquired during the first few years of his life, while he was at work in the hotel as a porter, boot-black and hostler, have followed him all along through his life; and it has been one of constant industry, constant sobriety, and constant accumulation, and accumulation, too, in an honest way. And I will say to you, gentlemen, that I will give you, in behalf of Mr. Clough, and to the Concord Railroad, a *carte blanche*, and I will not object to any testimony you may put in here; and you may summon all the men that Mr. Clough ever did business with, and they will testify here that they never knew him to be dishonest, or ever saw in him any disposition to wrong any mortal man out of the first cent that was honestly due. I have known Mr. Clough, in that particular, as well as I have known anybody. For twenty years Mr. Clough has been my tried and cherished friend; and I have here to say in his behalf, that in all his dealings with men; in all his goings in and comings out among men,—aside from what has been said of him as a Concord Railroad conductor,—I never yet knew any man to charge him with the first particle of dishonesty. He has been strictly and scrupulously honest to the last cent. He has earned a reputation in that particular which no man can hope to surpass. And yet, gentlemen, it is your duty to try him upon the evidence as it is here; and if there is anything that he has withheld from the Concord road—if he has intentionally and fraudulently done anything by which he has taken the funds of the Concord Railroad,—he ought to return them, and it is not necessary for me to say that you ought to say that he should return them. If he has done nothing more than to be generous with the means of the Concord Railroad, in passing poor people, and in passing his friends,—if he has done nothing more than, according to his habits of life, to make money as an agent, by carrying this thing and that thing back and forth on the Concord Railroad, with the consent of the higher officers of the road,—then he has done nothing more than you or I, or anybody would do, and what we should be justified in doing. He has accumulated what property he has. It is his. He does not want it taken away from him; he would not object so much to any reasonable amount of money taken from him, but he does not want, nor can he afford to have that good name, which has cost him more than twenty-five years of constant and steady effort, taken from him, and taken from him, as I say, in a manner most cruel and unrelenting. I believe that this whole scheme against Mr. Clough was conceived in sin and pursued in iniquity; and if George Clough had only continued to sign Joseph A. Gilmore's notes, with all the other efforts he had made to exalt Mr. Gilmore, I do not think that Joseph A. Gilmore, either in his drunken revelries or in his sober moments, would ever have told that Rip Van Winkle board of directors that during the twenty-four years that they had slept, George Clough and his associates had annually been stealing fifty thousand dollars of the revenue of the Concord Railroad. We have been dragged over the whole course of Mr. Clough's life, during twenty-two days spent in taking his deposition; we have been summoned to attend the taking of depositions that included the doings of all the conductors of the Nashua and Worcester road, and of the Lowell, Nashua and Boston road; we have been required to be present at all these proceedings before you; we have submitted to it with as good a grace, under the circumstances, I think, as any man could do. We intend to submit with equal grace to whatever other things we may be called upon to submit to in the course of this trial. But, gentlemen, if there is one thing that we have reason to thank God for, it is that he has spared your lives, and given you health and strength to come back here, and give us a reasonable prospect and hope that, at last, this case shall be decided.

And let me say, in conclusion, that an intimation has been thrown out here that George Clough has combined with Joseph A. Gilmore to stifle an investigation into these charges. No man lives who can say that George Clough has ever intimated, or asked or desired, that this matter should be stifled. When Mr Clough was sued, and his real estate and other property all attached—when, while he was on a sick bed, the sheriff came into his house, and attached all his personal property, (which has never been returned upon the writ), feeble as he was, he said then and there, that as between him and Joseph H. Gilmore and John H. George, and everybody else who wanted to rob him of his property and his fair name: "This thing will be pursued to a clean victory —or to a clean defeat." Such is our position now; and at no time, and under no circumstances, could George Clough afford to have this matter squelched. He sent letters to the board of directors, asking permission to appear before them, and to hear them make what charges they had against him; but they never gave them; we never knew them; they were locked up in the safe of Gen. Butler; and when we wanted to get them, they were spirited away to the city of Washington; and when at last we got them here, what were they? All this noise and confusion, and all this slander and detraction that has gone abroad in this community against George Clough, resulted, when the evidence came to be brought before these referees—when it came to be unearthed from the vault of Gen. Butler—to this: that five times—*five times only*—did George Clough fail to take the ten cents extra. That, gentlemen, is all the evidence there is. Things are sometimes done for effect, and I do this for effect. Not only did George J. Carney act as a detective, but he employed his brother, and his brother took a large number of witnesses with him, and he came up here and paid a visit to the asylum for the insane in this city, and rode back over the Concord road, paid fare up and paid fare back, and he came up over the Portsmouth road, paying fare up and paying fare back — paying fare to George Clough; and Mr. Tappan, when he comes to argue the case, will ask, with some degree of significance, where is Dr. Sidney H. Carney, the physician and the brother of this immaculate disciple of Major Gen. Butler "late of the Army of the James?" He has not run away, as Draper has; he is not *non est*, as others are, but he is in the city of Boston, alive and well. He has paid large sums of money into the hands of Mr. Clough. If Mr. Clough is dishonest, and has not refunded them, why does he not appear? Mr. Tappan will ask where is Dr. Carney, the brother of this model detective who took a thousand dollars for making a report four days after he drew the thousand dollar check?

This, gentlemen, concludes my opening.

## TESTIMONY OF GEORGE CLOUGH.

Q. (*By Mr. Rolfe.*) Are you the defendant in this suit?

A. I am.

Q. What is your age?

A. Fifty-two.

Q. Were your formerly employed by the Concord Railroad? If so, in what capacity?

A. I was employed, as a conductor, from July, 1842, to January, 1866.

Q. Do you recollect the date, in January, 1866, when you left the employ of the road?

A. It was January 14th or 15th, I think.

Q. Do you recollect of selling to D. B. Egerly, tickets for half price, or of passing him free over the road, or of giving him a ticket back on his paying fare one way?

A. I recollect that.

Q. State to the referees all the facts connected with that; and state why and at whose direction you did it.

A. I was introduced to Mr. Egerly by a man by the name of Garland, a stage-driver who drove from Pittsfield here. He came to me and wanted that I should pass him. [Mr. George objected to any statement of what transpired between the witness and Mr. Garland.]

*Mr. Tappan.* This is simply introductory.

*The Chairman.* Is it a necessary introduction?

*Mr. Tappan.* I think it will be more satisfactory to the referees to hear this whole transaction.

*Mr. George.* This has been ruled out in two instances: once when we offered it, and once when you did.

Q. (*By Mr. Rolfe.*) Did you receive any instructions from Mr. Gilmore in relation to him? If so, what?

A. I did. Mr. Gilmore instructed me to carry him for half fare.

Q. Go on and state the whole conversation between yourself and Mr. Gilmore.

A. Mr. Gilmore told me to let him have some tickets at half price.

*Mr. George.* The referees will understand that our exception to this is for any other purpose than the purpose presented. It might be understood as tending to rebut our proposition.

*The Referees.* I suppose that if the testimony is admissible on one ground, we cannot anticipate any other purpose that may be submitted.

*Mr. George.* I do not want it to go in without that suggestion. Perhaps we had better save an exception to the evidence for any purpose. We except to it for the purpose of saving any right we may have in the matter. We except to the ruling of the court in admitting it for any purpose.

Q. (*By Mr. Rolfe.*) Was anything said to you by Mr. Gilmore as to how you could manage it? If so, what?

A. Yes. He told me to let Mr. Egerly have tickets at half fare and pay the money into the ticket office.

Q. State whether you paid the money into the ticket office.

A. I did.

Q. What, if any thing, did you receive when you paid the money into the ticket office.

A. I received tickets.

Q. What did you do with the tickets?

A. I purchased them, and put them into my collections, and delivered them up to the special ticket agent.

Q. (*By the Chairman.*) You were to take this money, and pay it into the ticket office, and then they gave you the tickets?

A. Yes; that is, into the ticket office of the station, and not into the general ticket office.

Q. That is, under the direction of the superintendent, the ticket seller let you have a ticket at half price, which you let Egerly have?

A. No; I gave the ticket seller the full price; that is, I bought less tickets; just half the number of tickets that I let Egerly have. I let Mr. Egerly have tickets at half price; I let him have two tickets for the price of one, and then I went to the office and bought one ticket, and that made the money all right.

Q. (*By Mr. Rolfe.*) Did you do the same with the others?

A. Yes.

Q. And by Mr. Gilmore's directions?

A. Yes.

Q. How was it with Mr. Clark and Mr. Stanley?

A. It was the same.

Q. Who called your attention to their cases?

A. Mr. Garland.

Q. At the time that Mr. Gilmore directed you to let those gentlemen go for half fare, did he, or not, remark to you that Mr. Garland had made application on their behalf?

[Objected to as leading. Objections sustained.]

Q. What did Mr. Gilmore say?

A. Mr. Gilmore told me to pass Mr. Clark and Mr. Stanley at half price.

Q. Did he say anything about Mr. Garland to you?

A. Yes. He said that Garland came to him and wanted him to pass him. He came to me first and I told him that I could not do it; that he must see Mr. Gilmore.

Q. State how it was in relation to John R. French.

A. I passed John R. French at the request of Mr. Garland; it was a Mr. French, but I do not recollect his given name.

Q. State whether Mr. Clark, Mr. French and Mr. Egerly all formerly lived at Pittsfield.

A. I understood that they did.

Q. State whether, when Mr. French paid his fare in the cars, and rode only over the Concord road, you gave him any ticket.

A. No local ticket, I do not think. If he went below Nashua I gave him a ticket.

Q. Persons who rode only over the Concord road on your route from Concord to Nashua, as I understand you, never received a local ticket.

A. I never gave them any unless they wanted to stop over at Manchester.

Q. What would you do in case they wanted to stop anywhere?

A. I should either give them a ticket, or write on a card and give them.

Q. What kind of tickets do you mean? You had no local tickets.

A. No. I sometimes managed in this way: If a man paid fare to Nashua, and wanted to stop at Manchester, I would take a ticket from another passenger that was going to Nashua and give it to the man that wanted to stop over. I used to manage in that way; but it was very seldom that anybody wanted to stop over.

Q. In relation to the ticket testified to by Mr Bacheller, that he received from you at Manchester, when he left the train. If he paid you the fare to Nashua, and wanted to stop over, what would you do in that case.

A. If he paid his fare from Manchester to Nashua and wanted to stop over I would give him something to pass him.

Q. Would you be particular about what you gave him?

A. I would give him anything that the other conductors would take, and sign my name to it, as "good for one seat to Nashua," or something of that kind.

Q. If he had paid his fare and stopped over, in order to ride on he must necessarily have something to show that he had a right to pass?

[Objected to as leading, Objection sustained.]

Q. Will you please to state now whether you had business with John Grier, who formerly lived in this place?
A. I had.
Q. What business relations did you have with him?
A. He used to make boots for myself and boys.
Q. Have you passed him over the Concord road?
A. I have.
Q. Do you recollect how many times?
A. Very few.
Q. Why did you pass him?
A. He said that he was poor, and was not able to pay full fare; that he had a large family and found it difficult to support them.
Q. State whether you ever passed him for any consideration,—any pay.
A. Never. I never received a cent from him, myself.
Q. Have you bills receipted by him?
A. I have.
Q. State whether you have the bills.

*Mr. George.* What is your object in introducing the bills?

*Mr. Tappan.* To negative your charge. We offer them to show that the charges made against Mr. Clough are false. We say that he paid them in cash. We offer them first. to show his dealings with Mr. Grier, and then we offer them to show how the bills were paid. I do not know how much time they cover, but they are his yearly bills.

*Mr. George.* My objection is that they do not prove that fact.

*The Chairman.* We think those bills are admissable in evidence, the purpose being, as we suppose, to contradict Mr. Grier. [Witness produced the bills.]

Q. (*By Mr. Rolfe.*) Are those the bills?
A. Those are the bills; I paid the money on them.
Q. State whether or not Mr. Grier ever did any business for you, or furnished you with anything that you did not pay for.
A. Never,—not a cent's worth.

The following are copies of the bills referred to.

"Concord, N. H., Sept. 1864.

MR. GEORGE CLOUGH,

Bought of JOHN GRIER,

Boot and Shoe Manufacturer, No, 1, Exchange Building, Concord, N. H.

1 pair boots for son Geo., $9 00

Received payment, JOHN GRIER.

"Received from George Clough, Esq., the sum of sixteen dollars, being the amount of his bill in full up to this date,

Concord, May the 1, 1852. JOHN GRIER.

"Concord, February 9, 1863.

MR. GEORGE CLOUGH,

To JOHN GRIER, DR.

To 1 pair of boots, $8 00

Received payment, JOHN GRIER.

"GEORGE CLOUGH, ESQ.

To JOHN GRIER, DR.

1 pair boots for George, $7 00

Received payment,

Concord, July 1, 1864. JOHN GRIER.

"Concord, N. H., Dec. 1, 1865,

"MR. GEO. CLOUGH,

Bought of JOHN GRIER,

Boot and Shoe Manufacturer, No. 1, Exchange Building, Concord, N. H.

| | |
|---|---|
| 1 pair of boots tapped, heeled and foxed, | $2 50 |
| 1 pair of boots tapped and heeled, | 1 00 |
| 1 pair of boots tapped, | 0 75 |
| 1 pair of boots to son George, | 10 00 |
| 1 pair of boots tapped, heeled and foxed, | 2 50 |
| 1 pair of boots tapped and heeled, | 1 00 |
| 1 pair of shoes tapped, | 0 75 |
| | $18 50 |

Received payment.

JOHN GRIER.

Concord, N. H., July 23, 1865.

"MR. CLOUGH,

Bought of JOHN GRIER.

Boot and Shoe Manufacturer, No. 1, Exchange Building, Concord, N. H.

1 pair of French calf boots, $10 00

Received Payment.

JOHN GRIER."

Q. Do I understand you to say that these bills were paid by you in money?

A. Yes; I paid every one of them in cash.

Q. State whether Mr. Grier ever did any other work for you that you did not pay him the cash for.

A. No, sir; he never did.

Q. "Concord————1867. Received of George Clough, twenty dollars, the same being the amount due the estate of John Grier. B. F. Prescott, Administrator."—State about that receipt.

A. I lost the original bill, and went to Mr. Prescott and got a duplicate of it. Mr. Prescott came to me one day and wanted to know if I was owing Mr. Grier. I told him that I supposed that I was owing him something. He said that he could not find any books, but that his wife told him that I got my work done there; and I asked my boys (I knew that I had got nothing done there for myself) if they had had anything done there, and they said they had had a pair of boots each, and some mending done. I called it twenty dollars and gave the money to Mr. Prescott.

Q. I want to ask you, once for all, whether you ever received any consideration from anybody—Mr. Grier or from anybody else, during the twenty odd years that you were on the railroad, for passing people over the road?

A. No, sir; I have no recollection of ever receiving a dollar.

Q. How was it with reference to Mr. Starkey? Did you ever know of his selling any ticket to anybody, until after you left the road?

A. No, sir; I do not know that I did.

Q. State whether you ever let him have any tickets.

A. He used to ask me for tickets occasionally, and I let him have them once in a while, and sometimes I denied him. He would frequently have somebody in the baggage car with him, that he would want me to pass, and a majority of the time I would pass him, but I have refused, and made him pay. He used to have a good many friends.

Q. During the last year or two that Mr. Starkey ran on the road, was he ever in the habit of taking up a part of the tickets?

A. Yes; he was.

Q. What was the course of proceeding?

A. He used to run up nights from Boston, over Long's road, and we came up altogether from Manchester,—both trains came together; and he used quite often to take up the tickets in the large cars. He used to come along and offer to do it, and I used to allow him to. Sometimes when I had a large train, I let him begin at the rear end, and meet me, so that I could get my tickets all up before I came to the first station.

Q. State whether you ever let James Whitcher have any tickets, or passed him free. And if you did let him have any tickets, state when, and to what amounts, and under what circumstances.

A. I have let Mr. Whitcher have some tickets. He wanted them for poor people; and I let him have some tickets by the order of the superintendent, Mr. Gilmore. I never let him have but a few tickets myself. I never let him have one except when he represented that he wanted it for a poor person who was not able to pay.

Q. State who these poor persons were, and what were the circumstances. [Objected to by Mr. George.]

*Mr. Mugridge.* We desire to show the circumstances under which these tickets were given, and the representations made at the time, to Mr. Clough, and upon which he acted, to rebut the idea that is advanced that they were improperly given by Mr. Clough to Mr. Whitcher.

*Mr. Tappan.* We want the referees to see whether the tickets were given to him under such circumstances as would justify the giving.

*The Chairman.* There was something in the regulations to that tenor, was there not?

*Mr. George.* Yes, sir; but that has been ruled out We cannot hold him to the regulations. We propose to charge him for all the tickets that went out of his possession, and for which he received a consideration.

*Mr. Tappan.* Then this is clearly competent. If he is to be charged for these tickets, we have a right to show that they were given as a matter of charity.

*Mr. George.* There is a question of veracity between Mr. Whitcher and Mr. Clough. Mr. Clough says he did not give Mr. Whitcher any tickets for a consideration. If he did not, we cannot recover under this form of action.

*The Chairman.* Mr. Clough says that under the circumstances in which he furnished these tickets, he did not get anything for them. I cannot say that I see any materiality.

*Mr. Tappan.* It might have a tendency to show the number of tickets that Mr. Clough let him have for these parties.

*The Chairman.* Is the number material?

*Mr. Tappan.* That is all this case was founded upon. If anything is material, I suppose that is.

*The Chairman.* Taking the ruling of the court which has been already made, and which we do not at present feel inclined to depart from, that Mr. Clough is not to be charged here for tickets for which he received no consideration, it does not seem to us to be very material what the number was.

*Mr. Mugridge.* The gentleman has said that he proposes to argue that these tickets were received from Mr. Clough for a consideration; that that was the fact with reference to these tickets. If that is the theory asserted by the plaintiffs, is it not competent for us to proceed as far as we may, and introduce the circumstances attending the giving of the tickets, to combat this theory?

*Mr. George.* I remember that I argued at very considerable length, that when Mr. Clough passed people free, that is, without receiving a fare, when it was the duty of the conductor to take fares, and when the rules required him to take fares, and expressly prohibited him from passing people without taking fares,—I ought to be permitted to introduce that evidence upon the presumption that he did receive money, or money's worth for the passage. The referees took a different view and ruled it out, and my subsequent impression is that the referees were right in that ruling. The converse of the proposition is now presented. Mr. Clough says that none of these tickets were given for money, or for money's worth. If the referees find that to be the fact, we cannot recover for the value of those tickets. He swears that he did not; therefore the question rises, whether Mr. Clough tells the truth or not.

*Mr. Mugridge.* The evidence is material only in the view that these tickets were *given* by Mr. Clough to Mr. Whitcher; and it is only in that view that we seek to attach any sort of consequence to the testimony as it is now before the referees. That being so, have we not the right to inquire into all the circumstances connected with the transfer of all tickets which Mr. Clough did actually give to Mr. Whitcher?

*Judge Bellows.* Mr. Clough has already testified that they were given to him for poor people. Now I think it would not make any difference in our opinion if he went one step further, and said that Mr. Whitcher told him so. It is a pretty sharp point, and whether or not it would be admissible, might be open to some little doubt; perhaps it would be.—It strikes me, however, as very sharp, and is altogether, in my judgment, immaterial, and could have no influence on our minds one way or another.

*Mr. George.* I will withdraw the objection. It is immaterial whether they were for poor or rich people, but it may be material in another view of the question.

*Mr. Mugridge.* To Mr. Clough the suggestion might have been quite important, whether they were rich or poor.

*Mr. Tappan.* It may strike me differently from what it does the rest of the counsel. That would seem to be the very reason why it would be proper for Mr. Clough to state whether he believed what was represented to him; and he may himself have had some knowledge that the persons for whom he gave Mr. Whitcher tickets, were poor.

Q. (*By Mr. Rolfe.*) State who these people were, and why Mr. Whitcher wanted the tickets.

A. I recollect that he once spoke of some Shakers that were at his house, and said that he would like me to pass one of these. I do not recollect of anybody else.

Q. Was Mr. Whitcher formerly a Shaker?

A. He was.

Q. Could you not have had them ride on your train?

A. I think he said that they were going on another train.

Q. Those tickets that you gave under the direction of Mr. Gilmore: please state the circumstances under which you gave them.

A. Those were for political purposes, I believe. They were used to get home voters, and to get them back again.

Q. Do you recollect the number, or near the number?

A. No; there might have been six or eight at a time.

Q. How many tickets in all, did you give Mr. Whitcher, by the order of Gilmore, or in any other way?

A. I used to give him some at state elections, for three or four years.

Q. How many at a time?

A. I should think somewhere from six to ten.

Q. Besides these, how many tickets did you ever give to Mr. Whitcher,—besides the three or four years that you gave him from six to ten?

A. I do not think that I ever gave him, myself, more than a dozen or fifteen.

Q. Over what time did this extend,—for how many years?

A. I should think perhaps fifteen or sixteen years; I do not recollect exactly.

Q. State, as near as you can recollect, when you gave him the last one.

A. I did not give him any tickets for more than six months before I left the road; I do not think I had given him any for a year; I should think that it was nearer a year than six months.

Q. What kind of tickets were they?

A. They were local tickets. I could not give any others away without paying for them. Joint tickets were charged to me.

Q. Joint tickets, when in your pocket, represented so much money?

A. Exactly.

Q. How was it with tickets over the Portsmouth road, and over the Manchester and Lawrence road?

A. It was the same over tbe Portsmouth road, and over the Manchester and Lawrence road; and all the tickets from Nashua, to Boston, and Lowell, we had charged to us.

Q. (*By the Chairman.*) Were any local tickets also counted out to you and charged to you?

A. All the tickets that went over the Portsmouth road, and the tickets over the Manchester & Lawrence road, and all the tickets from Nashua to Boston and Lowell, were counted out and charged to us. The local tickets were not charged to us.

*The Chairman.* I understand Mr. Sanborn that all the tickets were counted out and charged.

*The Witness.* The general ticket agent counts out the different tickets to the different sellers at the different stations, and they are charged to them.

Q. (*By Mr. Rolfe.*) But you have no local tickets charged to you?

A. No, sir; the joint tickets were charged to us.

Q. (*By Judge Bellows.*) What were the tickets that you delivered to Mr. Whitcher?

A. Those were local tickets between Manchester and Nashua.

Q. How did you get possession of those tickets?

A. Those I kept out in my collections.

Q. (*By Mr. Rolfe.*) That is, you took them up and did not punch them?

A. Yes, sir.

Q. So that when they went into the hands of another conductor, he recognized them as tickets that had not been rode upon.

A. Yes, sir.

Q. State whether or not Mr. Gilmore ever gave you any instructions in relation to saving out tickets for any purpose.

A. He did. He frequently told me to save out some tickets for himself.

Q. Would he state the number?

A. No, sir.

*Mr. George.* What is the point of this testimony? I would like to see the bearing of it.

*The Chairman.* Mr. Rolfe stated in his opening argument, that he takes the position that the superintendent was the supreme arbitrer of this matter; and that, notwithstanding any rules or regulations,—he had the right to make exceptions. Having taken that position, I suppose that he offers now to show that Mr. Clough was obeying the order of the superintendent.

*Mr. George.* Then we object to any testimony upon that point.

*The Chairman.* It is now about the usual time for adjournment. Perhaps before we adjourn, it would be well for the gentlemen to state the positions they take.

*Mr. Rolfe.* My question is: What instructions did Mr. Gilmore give you in relation to saving out tickets for Mr. Whitcher, that had been once used? That is the point we want to make.

*Mr. George.* And to that we object.

*Judge Bellows.* Another question may arise: What is the bearing of that? Does it tend to rebut the inference that there was any money derived from the transaction?

*Mr. Rolfe.* The answer might tend to show that Mr. Clough acted from no motive of his own, but was only governed by the superintendent.

*Mr. George.* This question raises the whole doctrine, of what might be denominated in this case, the doctrine of estoppel. We have very decided views upon this subject, which we would be glad to state at reasonable length before the referees. We suppose that the whole of this question might practically be considered under the general doctrine of estoppel. [Adjourned.]

[FIFTEENTH DAY. Saturday, December 19th, 1868.]

The hearing was resumed at 10 o'clock, A. M.

The examination of Mr. Clough was suspended by consent, that the plaintiffs might examine H. G. Keyes.

## TESTIMONY OF H. G. KEYES.

Q. (*By Mr. George.*) State where you live, and what relation you have had to the New England Fire Insurance Company, of Hartford, Connecticut?

A. I reside in Concord, and have been the agent of that company since its formation.

Q. [Handing papers to witness.] Look at that policy and these papers, and state whether these are the original papers of insurance of property, in that company, and the statement of loss?

A. These are the original papers.

Q. From whom, and to whom, is the policy issued?

A. The policy is issued from the New England Company, to Eliza R. Clough.

Q. Who is Eliza R. Clough?

A. The wife of George Clough.

Q. The property insured in that company was burned, was it not?

A. Yes, sir.

Q. State whether or not this paper is the statement of loss of the property burned?

A. It is.

Q. In whose handwriting is it?

A. In mine.

Q. Who prepared that statement of loss? Who did the business with you as regards making out the claim, and with whom did you negotiate?

A. With George Clough,—all except signature.

Q. How was the signature obtained?

A. It is my impression that I went to his house and obtained her signature.

Q. At what place was the negotiation with regard to the value of the property, &c., as to the figures there, conducted?

A. A part of it was arrived at at the house of Lewis Young. I have an impression that there was something that concerned the South Church, but will not be certain.

Q. Did you ever see Mrs. Clough at all before that?

A. I think she was present at the appraisal of some damaged goods.

Q. Who else was present?

A. Gilbert Wolfe.

Q. Was Mr. Clough present?

A. I could not say whether he was or not; I do not recollect.

Q. Was the amount of the insurance received from the company, and paid.

A. I obtained a check from the New England Company, of Hartford, and delivered it to George Clough.

*Mr. Mugridge.* We except to this testimony. There is no evidence to show that Mr. Clough had anything to do with making out the statement of loss. The witness says that he does not know whether Mr. Clough was present or not.

Q. (*By Mr. Mugridge.*) Let me ask you if Mr. Rolfe did not prepare that statement of loss, and if you did not meet him at his house, with Mrs. Clough, Mr. Clough not being there?

A. Mr. Rolfe had something to do with it.

Q. Will you be kind enough to look at that schedule of loss, and state if it was not prepared by Mr. Rolfe, and then copied by yourself from the schedule prepared by him.

A. I have no knowledge of its being prepared by Mr. Rolfe; I never saw him at work upon it, and cannot say that it was.

Q. What did you get that from?

A. From a statement.

Q. Have you the original statement?

A. I have not.

Q. In whose handwriting was the original statement?

A. I have no recollection.

Q. Was it in the handwriting of Mr. Clough?

A. I have no recollection.

Q. Are you able to state that Mr. Clough himself had anything to do with preparing that schedule of loss? Was it not prepared by Mrs. Clough and Mr. Rolfe, as far as you remember?

A. My impression is that Mr. Clough has as much to do with it as anyone.

Q. Where did you see him?

A. In my office.

Q. In regard to the preparation of that schedule, where did you see him or have any conversation with him?

A. I had repeated conversations with him.

Q. Where?

A. I could not state at what places I saw him, because the thing was a long process, and there were a great many parties to bring together, and during that time I saw him at various times and at different places.

Q. Was this schedule prepared by Mr. Clough or by Mrs. Clough?

A. Mrs. Clough, to my knowledge, had nothing to do with it, with the exception that she was present at the appraisal of the damaged goods, and when they obtained her signature.

Q. Was Mr. Clough present at that time?

A. I do not recollect.

Q. Did you have any communications with Mrs. Clough in regard to the preparation of the schedule?

A. Only at that place.

Q. Did you ever meet her at Mr. Rolfe's house?

A. I do not recollect; I may have met her and I may not. I have no recollection.

Q. Will you say that you did not?

A. I cannot say; I cannot say where Mr. Rolfe lived at that time.

Q. Will you say that you did not meet Mrs. Clough at Mr. Rolfe's house with reference to the preparation of this schedule of loss?

A. I cannot say; I have no recollection either way.

Q. Who put the valuations to the goods lost?

A. I am unable to tell you.

Q. Do you remember that Mr. Clough did it?

A. I cannot say whether Mr. Clough did or not. I have no recollection as to who made these figures.

Q. You have no recollection that Mr. Clough himself had anything to do with affixing the value of goods lost?

A. I would not say that he had or had not anything to do with making the figures.

*Mr. Mugridge.* We submit that there is no evidence here sufficient to show that Mr. Clough had any such knowledge of this schedule, or any connection with it, so far as the value of the goods lost is concerned, as to render this testimony competent.

Q. (*By the Chairman.*) Who gave that schedule to you.

A. I am under the impression that George Clough brought it to my office.

*The Chairman.* The objection that is taken, as far as I understand the gentleman, is, that there is nothing now before the board which makes this testimony evidence against Mr. Clough as to the value of the property.

Q. (*By Mr. Mugridge.*) When Mr. Clough's house was burned, was not some of the property taken to the South Church and put in there?

A. Some of it was.

Q. And some to dwelling-houses?

A. That is my impression.

Q. The damage upon that property was estimated by a committee selected by you, as the agent of the company, was it not?

A. The damage to the injured property was.

Q Did not that committee assess the damages to the property injured?

A. I will not say that they did it conclusively; they did it in the main.

Q. Was it not their duty to affix the damage to the property injured?

A. Yes, sir.

Q. Did they not do it?

A. Yes; but after having done that, those figures have a right to be altered.

Q. The question I put to you is, whether they were altered after they were fixed by this committee?

A. My impression is that they were.

Q. Who altered them?

A. They were altered by agreement of the parties.

Q. Who made the agreement?

A. Mr. Rolfe and Mr. Clough.

Q. Do you mean to swear that Mr. Clough and Mr. Rolfe, after the committee had affixed the value to this property, altered it? Do you mean to swear to that?

A. I swear to this, that as far as we were agreed, some of these figures were changed after the appraisal.

Q. The point I put to you is this: whether you mean to state that after the committee affixed the damage they altered the figures?

A. I do.

Q. And do you state that you gave the check to Mr. Clough here, and that Mr. Clough carried it up to his house?

A. I do not know where he carried it.

Q. You gave the check to Mr. Clough?

A. My impression is that I did.

Q. Did he receipt for it to you?

A. My impression is that he did.

Q. Who do you think receipted for it?

A. That paper shows.

Q. Who do you *think* receipted for it?

A. I should say that Mr. Rolfe receipted, as counsel.

A. But you gave the check to Mr. Clough, and he carried it to Mrs. Clough?

A. I cannot say where he carried it?

Q. Did you give it to Mr. Rolfe?

A. I cannot say; my memory serves me that I gave it to Mr. Clough.

Q. To whom was the check payable?

A. I would not swear.

Q. Did you not give that check to Mr. Rolfe? and did he not receipt for it, and carry it to Mrs. Clough?

A. By this receipt I cannot tell whether or not it was given at the time I received the draft,—whether I took the receipt at the time I delivered the draft, or not, I cannot say; but I went to Hartford and brought this check back. In some instances I have taken the receipts and sent them on, and had the check sent back; and in others, I take the receipts when the check is delivered.

Q. Do you mean to swear that Mr. Clough, personally, had to do with fixing the value of that property? Didn't you swear a moment ago that you could not swear whether he did or not?

A. My memory has been refreshed, and I recollect that he was with me in the South Church.

Q. Your memory has been refreshed since you first testified?

A. It has.

*Mr. Mugridge.* We take the exception still, that there is nothing here to bind Mr. Clough to the value of that property.

*The Chairman.* I do not know that any of the referees think that there is anything here to bind Mr. Clough conclusively, but they think that they may consider, as against Mr. Clough, evidence of his belief that that property was in existence, and was of the value stated in that schedule. We do not mean to say that he is estopped from bringing evidence to the contrary. We think that like any other evidence, that may be considered hereafter. We consider that Mr. Keyes' testimony is evidence which may be established or contradicted. The ruling is that the evidence of Mr. Keyes now makes a *prima facia* case for the admission of this testimony for the purpose of showing Mr. Clough's belief that that property existed, and was of the value stated in the schedule. [Exception to the ruling, by the defendants.]

*The Chairman.* What objection is there to the introduction of the policy of insurance?

*Mr. Mugridge.* We have an idea that the obtaining of the policy of insurance was an act between Mrs. Clough and the company, with which Mr. Clough is shown to have had no connection whatever. Mr. Keyes' testimony, thus far, has not connected Mr. Clough with the policies at all; and more than that, the policies themselves show their connection with Mrs. Clough.

*The Chairman.* The objection appears to be well taken. I do not remember that there has been any evidence connecting Mr. Clough with the policies.

Q. (*By Mr. George.*) Who procured that insurance, and who paid the premium?

A. George Clough.

Q. Had you any negotiation whatever with Mrs. Clough, with regard to it, at the time of the negotiation?

A. No, sir.

Q. (*By Mr. Mugridge.*) Who signed the application?

A. I am not certain whether or not we issued an application then.

*The Chairman.* I presume this testimony can be held back until you have laid the proper foundation for it.

*Mr. George.* The testimony that Mr. Keyes has already given can be considered *de bene esse* until we have laid the proper foundation.

## GEORGE CLOUGH RE-CALLED.

*Mr. Rolfe.* In order that we may make a little further examination into the proper relation which the superintendent sustains to the directors of the road, and to the conductors, we will, if the referees please, waive, for the present, the question as to what the superintendent directed Mr. Clough to do in relation to taking out those tickets, but will call his attention to it at another time.

Q. (*By Mr. Rolfe.*) I do not ask you now as to the tickets which you gave by direction of the superintendent, but as to the tickets that you gave Mr. Whitcher when he requested them of you. State, as near as you can recollect, the number that you gave. I want to make a distinction between the two.

*Mr. Stanley.* That distinction has been already made.

*Mr. George.* I do not see the object of this testimony. and I will object to it, for it may bring us to a point as to which we are very well prepared to submit our views; we want to see what the object of this is.

*Mr. Rolfe.* We have been making a little further examination into the authorities as to the relations existing between the conductors and the superintendent of the road, and as to the scope and extent of the superintendent's power and authority in the matter; but we desire to examine further, and therefore propose to waive any questions which may have a tendency to raise that discussion, until some time in the future.

*Mr. George.* The question was in relation to certain transactions between this conductor and Mr. Gilmore.

*Mr. Mugridge.* We now waive that question.

*The Chairman.* There was evidence offered tending to show a very large number of tickets found in the possession of Mr. James Whitcher. This question appears to me to be directed to showing how those tickets were procured by him.

*Mr. George.* That was inquired about yesterday, and answered.

*The Chairman.* If the counsel consider that sufficiently answered, we do not desire to go any further.

*Mr. Rolfe.* As far as concerns the instructions from Mr. Gilmore to Mr. Clough, we waive the testimony for the present; and if the question is sufficiently answered in relation to the Whitcher tickets, that is enough.

Q. State the number of tickets that you gave to Mr. Starkey, the time when you gave them, and the circumstances under which you gave them.

A. I never gave Mr. Starkey any tickets for a year and a half before I left the road. Mr. Starkey used to come to me, just before the train started, and tell me that he had a friend in the car that he wanted me to give him a ticket for, and I occasionally gave him a ticket in that way. The tickets generally were given for the train that I ran. Sometimes I gave him a ticket, and sometimes a card, so that I should know the person. Sometimes Mr. Starkey would have a person in the baggage car with him, that he would want me to pass. I have passed some that way, and some I refused to pass, and made them pay their fare.

Q. State whether you ever received any consideration for any of those free passes, either from Mr. Starkey, or from any of the individuals that were passed.

A. No, sir; never a cent.

Q. State as to whether you ever gave Mr. Starkey any tickets for any purpose except to be used by persons about going down on the train.

A. No, sir; never.

*Mr. Mugridge.* For the purpose of contradicting Mr. Starkey we have a right to put leading questions.

*Witness.* I do not think that I ever gave Starkey a ticket except it was to be used on the train that I ran myself. I often gave him a card, or something of that kind, so that I should know the person when I went through the cars to take up the tickets.

Q. State how it was to substitute brokers, and substitutes paying fare in the cars over the Concord road, on the trains on which you run.

A. On the trains that I ran, it was very seldom that a substitute broker paid in the cars. I will not say but that there have been instances of that kind, but they were very rare; I refer to substitute brokers and their men.

Q. What was the practice.

A. They had tickets, These that came from New York and that way, had tickets through to Concord. The fare was less by paying through, than it would have been by paying twice, or by paying for a part of the way, and then paying again.

Q. Are you acquainted with Henry P. Lane?

A. I cannot say that I am acquainted with him. I know him by sight, and think I have known him by sight for as many as eight or ten years.

Q. Did he frequently ride with you?

A. He did. He used to go down occasionally.

Q. State the fact as to his paying fare in the cars.

A. Mr. Lane paid me a few times in the cars—two or three times, and then he wanted that I should pass him, and said that as there were other brokers that rode free, he wanted the same privilege, He wanted to know if I could not pass him. I told him that I had no authority to do so, and that he must see Mr. Gilmore. I recollect that on one occasion I had considerable trouble in getting his fare; he was not willing to pay; and hindered me in collecting my tickets on the train. Finally I got his fare, and after that he got his ticket at the office. He has never spoken to me since, that I recollect of.

Q. State the fact as to his paying fares for substitutes in the cars.

A. He never paid me a fare for a substitute in his life. Those substitutes had tickets through to Concord.

Q. What did you do with the fares that he did pay you in the cars? Did you take the extra ten cents of him?

A. I did; and put it into the way bill.

Q. Did you hear Mr. Lane testify?

A. Yes.

Q. You saw him illustrate the manner in which you took money?

A. Yes.

Q. State whether you ever put the money in between your fingers in the way he stated?

A. No, sir.

Q. Or in your hand in any way?

A. No, sir.

Q. What did you do with it?

A. When I took money I doubled it up and put it in my vest pocket. The only man on the road that I ever knew to carry money in that way was Levi P. Wright.

Q. State whether Mr. Samuel P. Curtis rode with you frequently on the cars.
A. Yes, sir; he did.
Q. State your business transactions on the road with him.
A. He used to pay me in the cars when he first began to ride; and then he wanted me to pass him the same as Mr. Lane did. I told him that I could not; that he must see Mr. Gilmore. He saw him, or somebody else did for him, and he got a pass; I think the pass was for two or three months.
Q. Did you ever have any difficulty with Mr. Curtis?
A. Mr. Curtis went down one night when he was intoxicated and noisy, and made trouble in the cars, and the passengers entered complaint. I think that when the train stopped at Hooksett I went to Mr. Curtis and told him that I wanted to see him. He walked out, and I went into the baggage car and he followed me, and I told him that he must ride there the rest of the way. He swore that he would not, and I put the brakeman over him, and kept him there until I got to Manchester, and then set him ashore. I reported him to Mr. Gilmore and my impression is that Mr. Gilmore revoked the pass; or, as it had nearly expired, perhaps he let it run its time. He did not get any other pass after that.
Q. When Mr. Curtis paid fare in the cars, as you have testified, before he got his pass, how was it as to your collecting the extra ten cents.
A. I charged him the extra ten cents.
Q. What did you do with it?
A. I put it into the way bill.
Q. How many fares in all, according to your recollection, did Curtis pay you?
A. I do not remember exactly. I think not more than two or three. He paid for no one but himself.
Q. How was it after the pass expired?
A. He bought tickets after that.
Q. How was it as to substitutes. Did he ever pay any fare for substitutes?
A. He never paid any fare for substitutes to me.
Q. State whether he was in charge of substitutes sometimes.
A. He was a sort of "striker" for some of the rest of them They sent him down to secure men.
Q. How was it with those men—not the men that he was "striking" for, but the men that were *struck*—as to paying fare?
A. They had a man with them that paid the fares, and held the tickets. He had nothing to do with them.
Q. How was it with the man that held the tickets. Did he point the substitutes out to you?
A. Yes; he would show me his tickets, and count them, and point the men out to me. He would go along through the car with me and show me which they were
Q. Was Lane's business the same?
A. Yes; the same.
Q. State your instructions about purchasing tickets to the brakeman below Nashua, and also the practice.

[Objected to by Mr. George.]

*Mr. Rolfe.* If the referees please, I intend to ask Mr. Clough whether it was his instructions to the brakeman, as they came up over the Nashua and Lowell road, to get the passengers who hadn't tickets above, out at Nashua, and let them buy tickets; and what the practice in that respect was.

[Objected to. Objection overruled, and exception taken by plaintiff.

A. I gave the brakemen and baggage master instructions, when passengers came up on the lower roads, to get them out at Nashua to buy their tickets. I gave these instructions to Mr. Thayer and to Mr. Favre. The instructions were either to get the passengers who had no tickets, out to get them, or else to get their tickets for them. Mr. Thayer is running as conductor on the lower road now; I do not know where Mr. Favre is now running. The fare was higher than now, and if I took the extra ten cents it would make it still higher. I think it was some twenty-five or thirty cents higher where a passenger paid twice between Boston and Concord than it was when they paid their fare through; as the conductors below could not take the fare any further than Nashua. The passengers that got in at the way stations without tickets on the lower roads, the conductors only took their pay to Nashua. If they got in between Lowell and Boston, those station agents or ticket masters at those stations had no tickets any further than Nashua. All passengers that got in without tickets at Boston or Lowell, and all those that did buy tickets at the local stations, had to buy again at Nashua, if they went above Nashua.

Q. What was the practice?

A. Mr. Favre and Mr. Thayer both obeyed my instructions and orders in regard to getting passengers out to buy their tickets, or getting tickets for them.

Q. State what your practice was, when you had time, of delaying the trains for the purpose of giving passengers an opportunity to buy tickets.

A. When my trains were on time, passengers would come to me and ask if they had time to get tickets; I would say "yes," and would wait a minute or two until they got tickets. That was my practice. When the trains from below were on time, the passengers had less time to get their tickets, and were sometimes without them; if the trains were late they had more time.

Q. What was the fact as to large and heavy trains about more or less passengers being obliged to pay their fares in the cars?

A. I have taken less money on large trains than I have on small trains.

Q. Give the reason why that was.

A. If I had a heavy train we were generally late starting. For instance, if the northern trains were late we would seldom get away on time. If we had a heavy train, and were late in starting from here, for instance, the passengers would have a better opportunity of getting their tickets; and if we started late we were generally about the same length of time late at the other stations, as we could not make any head time. It would be the same at all the stations between here and Nashua as it was in starting from here.

Q. State whether you made any suggestions to Mr. Gilmore, or any efforts to induce him to make arrangements for preventing persons getting into the cars at this station without buying tickets.

A. I did. I tried to have Mr. Gilmore have the clerks come down and lock one end of the car and stand at the other end and make the passengers show their tickets as they got in. In war times the soldiers would sometimes get in without paying, and when they had once got in it was pretty hard getting anything out of them. I went to Mr. Gilmore several times about it, or at least two or three times—I do not recollect just how many. He used to promise to make such an arrangement, but never did, until the soldiers used to get in by car-loads and refuse to pay, and we could not get anything out of them, and then he established a guard to go on the train for a while—for three or six months, I should think. He used to send six or eight soldiers down on guard. They rode on the trains backward and forward, and stood at the end of the cars, and if anybody offered to get in without a ticket they would send them to the office.

Q. Do you mean anybody but soldiers?

A. No; they had to do only with soldiers. They would not let them in unless they had tickets or transportation. Most of the soldiers had transportation,—two cents per mile.

Q. Did you make this suggestion with reference to other passengers as well as soldiers?—the suggestion to Mr. Gilmore about the clerks coming down?

A. Yes. I said I suggested that he have the clerks come down and see that all the passengers had tickets.

Q. When was this? How long before you left the road?

A. I think it was in 1863.

Q. And from that time forward?

A. Yes. The guard did not go all the time.

Q. What I meant to inquire was, when you first made this suggestion to Mr. Gilmore?

A. That was about 1863.

Q, How was it at Nashua? What were the regulations there as to persons who had baggage, who came in the New York trains? How was it about their getting their baggage on the cars without showing their tickets?

A. They could not get their baggage without showing their tickets. When they did not have tickets they had to go to the office and get them before they got their baggage.

Q. Persons that had no baggage come across and get in the cars?

A. Yes, sir.

Q. State now with reference to the rule requiring you to take ten cents extra of all persons who got into the cars without tickets, how far you enforced that rule; and if you did not enforce it, state at what times, and to what extent you did not enforce it, and why. If there were any difficulties about it, state them?

A. It was always difficult to get the ten cents extra. Passengers would sometimes hand me just the change, and would refuse to pay any more.

*Mr. George.* We might as well raise the question of estoppel here at this point. We are prepared to discuss that question now.

*The Chairman.* I understand that the testimony now going in, is offered with a view of rebutting the suggestion of an arrangement on the part of the defendant to induce passengers to pay in the cars. I understand that Mr. Rolfe is putting this testimony in with a view of showing the efforts that were made to induce people to purchase tickets before getting into the cars. Enforcing the payment of the ten cents extra would of course tend to make passengers purchase tickets. If he did not enforce that payment, of course there should be some reason shown.

*Mr. Rolfe.* That is my object. I do not care now to touch upon Mr. Gilmore's instructions with regard to that rule.

*The Chairman.* I understand that whenever the proposition shall be made to put in the directions of the superintendent, as an explanation or justification of any particular course that was taken,—that then it is desired upon both sides that there should be some discusssion of the point.

*Mr. Rolfe.* We have some few suggestions to make upon that point.

*The Chairman.* I do not know as it is material to the referees when that discussion is had, but it does not appear to be needed now.

*Mr. Clark.* Will the referees allow me to make a single suggestion. I do not understand the matter to refer entirely to instructions that may have been received from the superintendent of the road, but that the ground of our objection goes further, and raises this question: whether or not. Mr. Clough, holding the position of conductor upon the Concord road, can be allowed, for any purpose, to offer this evidence, even for the purpose of rebutting the presumption referred to—that he received money for the purpose of inducing passengers to pay in the cars. Our position goes to that extent.

*The Chairman.* You excepted to that testimony when this class of testimony was introduced.

*Mr. Clark.* It all goes in subject to that exception.

*The Chairman.* The exception was taken particularly to the ruling in of the instructions. That was all that I understood the exception applied to. The testimony having gone in in this way, the point is now made that Mr. Clough cannot now bring in evidence here in any view, either with a view to rebut this point, or in any other view; that he cannot bring in evidence as to his reasons in permitting, at any time, passengers to pass without taking the extra ten cents.

*Mr. Tappan.* Suppose we could show that it was physically impossible for him to collect it; would that evidence be excluded?

*The Chairman.* The referees are all agreed that this evidence is admissible in the same connection. Permitting passengers to pay their fares in the cars without taking the ten cents extra, would certainly remove one of the difficulties in the way of buying in the cars, and there would then be some inducement to buy in the cars. It certainly rebuts any evidence as to Mr. Clough's making arrangements to induce passengers to purchase in the cars. In any instances where he did not take the ten cents extra, it is certainly admissible to show why he did not take it, or that he could not take it.

[Exceptions to the ruling, by the plaintiff.]

*The Witness,*—Sometimes passengers would make just the change to me, and would refuse to pay any more than the ticket office fare, and rather than have trouble with them I used to take it. It was a rule that we could not enforce without a great deal of trouble. If they handed me just enough, and I could not get any more, I used to let them pass rather than have any trouble.

Q. State the difficulty, if you were behind time, or had long trains, in respect to stopping to take the ten cents, or to trouble yourself about the passengers.

A. When I had heavy trains, and any passengers refused to pay the ten cents extra, I did not have any controversy with them. I had not time to do so and attend to the rest of my business.

Q. What would have been the effect if you had attempted to collect it?

A. The effect would be, that I should probably lose a great deal more than I gained, in some other way. If I did not get through the train before I got to the next station, the way passengers would step out, and go off with their tickets. I used to endeavor to hurry through the trains before I got to the first station, so as to get up all the tickets of passengers that got out at that station.

Q. How was it with the heavy trains? If you had to stop and make the change for any considerable number of passengers, what would be the effect in your getting through the train?

A It would be difficult for me to get through the train. I could not get through as many cars before I got to the first station.

Q. In case you did not take the extra ten cents in the cars, in what way did you make your return to the ticket office on your way bills, and in what way did you settle with the road?

*The Chairman.* This question, as to what way Mr. Clough settled with the road—if that is considered as subject to the exception that has been made—perhaps would hardly be admissible testimony, as that raises the question of estoppel.

*Mr. Mugridge.* The proposition is to show how he arranged with respect to fares taken in the cars.

*The Chairman.* (After consultation.) I believe, gentlemen, that the conclusion of the referees is, that they would like to hear from the other side what they may say in answer to the arguments of the counsel for the plaintiff, before they undertake to rule upon this point. We should like to have the matter more fully considered by both sides before we make a final ruling upon the question of estoppel.

*Judge Bellows.* It may be well to state the point before the referees. If I understand the matter correctly, it is proposed to show that certain tickets, which were purchased by Mr. Clough, which were punched and returned to the proper officer of the corporation, did not represent tickets sold in the usual way, and did not in point of fact, represent money which was taken in the cars. That, I understand to be the general view of the matter. Now the point would be, whether when Mr. Clough returned tickets of this kind, representing them to be of that character—the very return of them would represent perhaps that they were of that character—and they were received by the corporation as such, and his way-bills made no mention of them at all—the point would be whether the corporation would be regarded as having changed its position in consequence, so as to bring it within the doctrine of estoppel. Whether the corporation received this return as true, and passed his account in that form, when they might have acted in some other way—when they might have called him to account for the money actually received in the cars in some other form, or required him to account for the ten cents extra, or taken some other steps in relation to it; whether they can be considered to have changed their position in bringing this out; whether they brought this suit upon the assumption that all the money that he took in the cars was represented in the way-bill; and whether that would be acting upon any representations that he made, so as to be a change of position on their part—these are all questions under consideration; but the main question perhaps is, whether this brings it within the definition of estoppel or not.

*Mr. George.* That was the exact point I intended to raise.

*Mr. Mugridge.* That is the point we desire to be heard upon; but we think we could not examine it sufficiently to present our views this afternoon. [Adjourned.]

[SIXTEENTH DAY. Monday, December 21st, 1868.]

The hearing was resumed at 10 o'clock, A. M., and the discussion continued.

*Mr. Mugridge.* Since the adjournment on Saturday we have been able to give some attention to this point. I do not propose to argue, at length, the question involved, but simply to state, briefly, the position which we propose to take in answer to those taken by counsel upon the other side, and the authorities bearing upon those positions. The proposition of the plaintiff, as I understand it, is this: The rules of the road required the conductor to return the money taken in the cars for fare, upon his way-bills; that because Mr. Clough, in instances where it was impossible for him to take the extra ten cents from the passenger who had no ticket, took from him the actual fare, in money, paid it into the hands of a ticket seller of the road, receiving therefor a ticket which he mutilated and returned—that he is by that act estopped from saying that the road received that money in payment of the fares of the passengers. In this action, wherein he is charged *only* for money which he has received and *actually retained* he is to be placed, in a position where he is to be held for money which he has received, *none of which he has retained, but all of which he has faithfully paid over to the road.*

In order to consider our position and theirs, fully, it becomes necessary for me to call your attention to a further proposition which we expect to establish by evidence. In connection with the evidence already offered by Mr. Clough, as to the manner in which he did his business, we propose to show that this course of business, upon the part of Mr. Clough, was directly authorized by Mr. Gilmore, the superintendent of the road; that it had been practiced by Mr. Clough for a long time; was fully known and understood by the ticket sellers on the line; and was never objected to by any person in behalf of the company. There is one further proposition that we have to make in regard to the rules. We propose to show that these rules, which are presented here by the gentlemen upon the other side, and which they claim bind us, as being the rules of the directors of the road, were wilfully violated by three or four of those same directors, by express directions and suggestions made by them to Mr. Clough, which directions he followed, under their authority; and which directions and suggestions were in direct contravention to the rules themselves; and that having themselves violated these rules, they are in no position to undertake now to enforce them as against others.

As I said before, I do not propose to argue these points, but to state them as briefly as I can.

We say that the acts complained of upon the part of the plaintiff, do not constitute an estoppel; that none of the elements that enter into and create an estoppel, are found here. I propose to consider what some of the essentials are which give to an act the force and character of an estoppel.

*First.* It must be made to appear that the act or declaration complained of has in some way directly influenced the conduct of the party by whom the estoppel is to be enforced. (2, Smith's Leading Cases, 644 646). It must be made to appear that the act or declaration complained of, has in some way, directly influenced the estoppel sought to be enforced. That influence must be shown to have been direct and immediate. *Watkins* v. *Peck*, (13, N. H., 360); *Dazell* v. *Odell*, (3 Hill, 219). In *Tufts* v. *Hayes*, (5 N. H , 452,) the doctrine asserted is, that it must appear that the act of the party was one intended to influence others, or to gain some advantage to himself, to constitute an estoppel.

Upon these authorities we submit that the plaintiffs are bound, in the first place to make it appear, by affirmative proof, that their action has in some way been influenced by the acts of Mr. Clough, of which they complain. In other words, it must be shown that they have in some way been called upon by the acts of Mr. Clough, as taken in the premises, to change their position and to do something which they otherwise would not have done. That fact they must show here affirmatively, and the matter cannot be left to inference, or deduction, or suggestion, but must have its foundation somewhere in proof.

*The Chairman.* Do you mean to say that the rule of law is different, in regard to that, from what it is in regard to proof of any other fact?

*Mr. Mugridge.* I mean to say this doctrine of estoppel is one which the law does not favor; and the facts relied upon must be proved and substantiated.

*Second.* It must clearly appear that the injurious influence exercised by the acts or declarations which constitute the estoppel, if not intended, might have been at least forseen by the party, and understood by him. That is, that, at the time he did the act complained of, if he did not intend injurious consequences, he must have been placed in a position where he might clearly have forseen, and understood the consequences. *Morton* v. *Hodgen*, (22, Maine, 137); *Otis* v. *Sill*, (5, Barb., 182); *Cartwright* v. *Gardiner*, (5, Cush., 273); *Cambridge Inst.* v. *Littlefield*, (6, Caven, 216); *Watkins* v. *Peck*, (13, N. H., 360). Without arguing the point that I make here, I want to suggest that Mr. Clough, in this instance, intended no wrong to the Concord Railroad by any act of his; nor could he nor anyone else have understood or forseen any injury to have resulted to the road from his conduct.

*Third.* No estoppel can arise, without proof that a wrong is done intentionally, on one side, and an injury suffered on the other; nor unless the injury be so closely connected with the wrong, that it might and ought to have been forseen by the guilty party. On that point I refer the referees to the following cases: *Copeland* v. *Copeland*, (28, Maine, 525); *Abel* v. *Fitch*, (20, Conn., 90); *Whittaker* v. *Williams*, (20, Conn., 98); *Dyer* v. *Cady*, (20, Conn., 563); *Martin* v. *Angell*, (7, Barb., 407); *Barker* v. *Brown*, (15, N. H., 176); *Wallis* v. *Truesdale*, (6, Pick., 456); *Davis* v. *Sanders*, (11, N. H., 259); *Guernsey* v. *Edwards*, (26, id., 224).

In *Smith's Leading Cases*, (II., 646), the doctrine is stated as follows: "In order to raise an act by a party, from the rank of evidence to the dignity of an estoppel it must not only be shown that the retraction will be injurious to the other, but that the injury results from a course of action induced by the act. There will be no estoppel unless the injury be the direct and immediate result of the wrong. There must unquestionably be an injury coupled with a wrong. Where no injury results, the discussion belongs to the forum of morals and not to judicial tribunals."

In this connection I wish also to refer to the following cases: *Norris* v. *Morrison*, (45, N. H., 499) ; *Austin* v. *Thompson*, (45, N. H., 113) ; *Carpenter* v. *Cummings*, (40, id., 158–169).

The point we make here in this case is: that by the act of Mr. Clough, in returning these tickets in the way he did there has been no wrong shown to have happened to the Concord Railroad Company,— none whatever, not a dollar have they lost; but on the contrary, every single dollar of money that they were entitled to, was received by them, and found its way into the treasury of the company. No actual or theoretical wrong can be shown by the plaintiff to have been suffered by reason of Mr. Clough's making the return of tickets in the way he did.

The Concord Railroad received every single mill that it was entitled to for the fare of passengers. We claim, in the first place, that it is incumbent upon the company to show a wrongful act upon the part of Mr. Clough by which the company has suffered some pecuniary injury, and been in some way, substantially damaged.

An estoppel will never be carried further, in any case, than is necessary to prevent one party from being injured by reliance upon the acts and declarations of the others. I will refer the court to the following cases, where this proposition is very strongly and broadly discussed, and in a way, I think, that entitles it to a very strong practical application to this case: 2 *Smith's Leading Cases*, 644; *Johns* v. *Church*, (12 *Pick*. 307) ; *Kinney* v. *Farnsworth*, (17 *Conn*. 345.)

My suggestion is, that the estoppel will be limited to what is necessary to put the parties in the same relative position which they would have occupied had the act complained of not have taken place.

The doctrine of estoppel is arbitrary and severe. As the effect of an estoppel is to prevent the assertion of rights which are unquestionably valid, or to preclude defence which would otherwise be good, justice requires that it should not be enforced unless substantiated in every material particular. *Carpenter* v. *Hillcock*, (12 *Barb*. 128.)

It cannot arise against a party for an act done, or an omission made in good faith, and under the influence of a mistaken impression as to the nature or extent of his rights either in law or in fact. *Whittaker* v. *Williams*, (20 *Conn*. 98 ; ) *Mackay* v. *Holland*, (4 *Met*. 69 ; ) *Royston* v. *Howie* (15 *Ala*. 309 ; ) *Telghman* v. *West*, (8 *Iredell*, 83.)

The particular application that I wish to make of these authorities, to this case is, that Mr. Clough, as we shall satisfy you by our evidence, acted in perfect good faith towards this corporation, in doing what he did; that he supposed that his every act in the particular complained of was justified by the authority of the superintendent of the road; and if it should appear that he acted under a misapprehension, either in law, or in fact; if we still show that he acted in perfect good faith towards the company, he is not to be punished here for that act. I think we shall have clearly satisfied you of that when the evidence is all presented.

We claim, also, in this case, that if there is an estoppel against Mr. Clough, we have an offset which sets the entire matter at large. The proposition I make is this: that the course of conduct of Mr. Clough in the matter complained of, must have been and was understood and sanctioned by the officers of the road. And that we shall argue to you that this knowledge brought home to the superintendent of the road, and also to the ticket sellers, (but particularly to the superintendent,) binds the road; and on this point we have authorities which to us, seem clear and satisfactory. The course of Mr. Clough in this regard must have been and was understood and sanctioned by the officers of the road; at least Mr Clough so understood it and acted upon that idea.

*The Chairman.* Do you mean to be understood as saying that he was led into that belief by the fault of the officers of the road?

*Mr. Mugridge.* Yes, sir; I take this general position, and claim that for the plaintiff now to deny the propriety of this action, under the circumstances, would inflict a wrong and outrage of a gross and monstrous character, upon Mr. Clough. It seems to me that it is not necessary to enlarge upon that point, that the suggestion is sufficient.

We have another point to which we ask your attention, and which we claim is perfectly conclusive as to this matter, viz: The Concord Railroad Company, so far as Mr. Clough is concerned, are estopped from denying the authority of Gilmore to give the particular directions alluded to, and under which Mr. Clough acted in making the returns in the manner he did. At this point, I wish you to consider for a moment the position that Mr. Gilmore occupied and sustained toward Mr. Clough. He was the superintendent of the Concord Railroad; he was the executive officer of that road; he was the man who in the usage of business, had the management of all the employees of the road, and was the active man in running the road; he was the man from whom Mr. Clough took his directions; he was the man from whom all the employees took their directions; he was the man who hired the conductors, and who turned them off; and was the only medium through which Mr. Clough knew of the board of directors, and he had no knowledge of the action of the board of directors except through the superintendent of the road. There was no other way for him to know them or to become acquainted with their action, save through the superintendent as their mouth-piece. He was only bound by the directions of that board as thus communicated. Mr. Clough therefore had a right, by virtue of the relation which he sustained to him, to assume that the acts of Mr. Gilmore were authorized; and he was in a position where he could act upon no other assumption. For the counsel for the plaintiff to claim that the board of directors of the Concord Railroad Company were the parties who had the active management and control of the affairs of the road, seems to me to be a practical and legal absurdity. We shall not deny but that the general management of the affairs of the railway was in the hands of the directors, but what we do say is that the *practical and every day management*, as far as its general conduct was concerned, was, by the board of directors, put into the *hands of the superintendent* as their executive officer.

Suppose that as Mr. Clough was leaving on the cars, Mr. Gilmore, the superintendent, should come to him and give him a direction, which was known and understood by him to be in conflict with a regulation or rule of the road, must Mr. Clough disregard that instruction? To illustrate, suppose that with the superintendent there were five men who had no ticket, and the superintendent should say to Mr. Clough, "I want you to pass these men over the road." and Mr. Clough should say, "I have instructions, in writing, from the board of directors, not to pass those men, and I cannot do it." Would Mr. Clough be in a position to say that? Is not the superintendent, by virtue of his office, in a position to ask him to pass those men? If the superintendent gives Mr. Clough some material directions with regard to the running of his trains, is he not to obey those directions? Or, must he call together the board of directors, and get their sanction to the instructions before he can heed them. Can the conductor refuse to obey the directions of the superintendent on the theory that he is to be governed only by the president and board of directors of the company? In some particular and isolated cases, that may be true; but suppose the superintendent makes a suggestion as to an every-day occurrence connected with the practical management of the affairs of the road, is Mr. Clough to repudiate that authority, and refuse to listen to the suggestions, under the idea that Mr. Gilmore has no control or direction of his conduct, and that he is only the creature and agent of the board of directors? We say not. We say then, that the Concord Railroad Company, so far as Mr. Clough is concerned, is estopped from denying the authority of Mr. Gilmore to give the particular directions under which we have acted, and under which we justify.

We take the further position, that the open and public acts of Gilmore as superintendent, in managing matters connected with the running of trains, &c., were sufficient, in law, in warranting Mr. Clough to be governed by his authority, to act upon his directions, and to rely upon his assent to a particular course proposed. *Wendell* v. *Abbott*, (45 N. H., 349); *Kent* v. *Tyson*, (20 N. H., 125.)

I wish now to say a single word with regard to the right of Mr. Gilmore to give these directions. In the first place, whether he had any right or not, I say that the plaintiff is estopped, by reason of the relations he sustained to the road, from denying his authority; but we claim that he had authority and right to give these particular instructions. Assuming, if you please, that the general management and direction of the affairs of the Concord Railroad Company, as well as the affairs of all other corporations of a similar character, is in the hands of the board of directors; if Gilmore sustained any particular relation to the road which is known to the law, it was that of agent of the board by which he was appointed; he is the executive officer of that board; he has his existence from them, and he represents them in the management of the affairs of the road; he is put there by the board of directors for a specific purpose, and that specific purpose is to run the railroad. We claim that his position carries with it impliedly an authority to him, from the board of directors, to use such means and measures as are necessary and appropriate to accomplish this object. *Backman* v. *Charlestown*, (42 N. H., 131); *Story's Agency*, sec. 58, and cases cited.

To give my proposition a practical application in this case, I will put it in this way: The superintendent had a right to give special directions to the conductors, to relieve them from the operation of a rule, the enforcement of which in his judgment was found to be impossible, or at least impracticable, and inconsistent with the interests of the road.

*The Chairman.* Do you take the ground that this question of what are the necessary and appropriate means, is a question of fact, or a question of law?

*Mr. Mugridge.* My idea is, that as matter of law he has a right to use the means necessary to carry out the purposes of his appointment; and if it was found necessary by him to give directions to Mr. Clough to relieve him from the difficulties which he was every day encountering, and from the operation of a rule which it was found impossible for him to enforce, he had the authority to do so.

I take the further legal position that where an act complained of was authorized to be done by the superintendent of the road, the plaintiff cannot insist upon that act as an estoppel.

Having shown that the superintendent of the road gave these instructions, the plaintiff cannot now avail itself of the acts done in accordance therewith, as an estoppel. *Pecker* v. *Hoit*, (15 N. H. 143;) *Davis* v. *Sanders*, (11 N. H. 263;) 1 *Greenleaf on Evidence*, 275.

As I have read from the books, the doctrine of estoppel is regarded by the law as a harsh, arbitrary and severe one. It is only recognized by the courts to prevent fraud; and the court will never allow the doctrine to be enforced so as to become an instrument of fraud. It will only allow the doctrine to be set up where, by its assertion and establishment, a great wrong, or a great fraud is to be averted.

As the effect of an estoppel is to prevent the assertion of rights which are unquestionably valid, or to preclude defences which would otherwise be good, justice requires that it should not be enforced unless substantiated in every material particular. *Carpenter* v. *Hilcock* (12 Barb. 128.)

All that I mean to be understood to say upon that point is, that everything that is essential to be made to appear with reference to the enforcement or establishment of the doctrine of estoppel, shall clearly appear. All the material facts shall be made out. There shall be nothing left to surmise, suspicion, or improper inference. The court shall be satisfied, under the ordinary rules of evidence; they shall not go a great ways to *imagine*.

I do not wish to discuss further the practical effects of this doctrine here. My idea is, that the proposition of the plaintiff to enforce it in this case, coming in the way it does, and in connection with the facts that appear here, is perfectly infamous. I do not propose to be heard further upon the question at the present time. I have stated as clearly as I can, without arguing at length, the different positions I take, in connection with the case.

## TESTIMONY OF GEORGE CLOUGH CONTINUED.

Q. (*By Mr. Rolfe.*) I was inquiring, at the time the discussion of these points commenced, as to your practice when you did not take the extra ten cents; and you went on to state what your practice was. Will you now state what your practice was, when it was not practicable or possible for you to take the extra ten cents on fares collected in the cars? State how you managed to square your accounts with the road?

A. When I did not take the ten cents extra, I bought tickets at the ticket office. There were a great many times when it was next to impossible to collect the ten cents extra; and if we had attempted to enforce the rule we would have had a fuss in the cars every day. As a general thing I took the ten cents extra when I made change or got money enough into my hands to keep it back.

Q. (*By Judge Bellows.*) Where did you buy those tickets?

A. I bought them at Concord, Manchester, and at Nashua.

Q, (*By Mr. Rolfe.*) State whether anything was said by you or by the ticket sellers, so as to bring the knowledge home to them as to what you were buying tickets for? [Objected to.]

*The Chairman.* We think the testimony is inadmissible.

*Mr. Tappan.* The referees will please note our exception.

Q. (*By Mr. Rolfe.*) State how long you have been in the habit of accounting for the extra ten cents in this way in cases where you did not take it in the cars?

A. Since sometime in 1859.

Q. Do you recollect when Mr. Corning left the road?

A. Either in 1858 or 1859.

Q. State whether Mr. Gilmore, the superintendent, had knowledge or information in relation to your not taking the extra ten cents, and the way in which you accounted for it?

*Mr. George.* We object to this upon two grounds: first, we object to the question itself; and secondly, we object to its form. We object to the form of the question, even if the matter inquired about were competent.

*The Chairman.* Perhaps the proper way would be to show any communication tending to bring this knowledge to Mr. Gilmore.

Q. (*By Mr. Rolfe.*) State whether you had an interview at any time, (and if so when was the first time) with Mr. Gilmore in relation to this matter of the difficulty of collecting the extra ten cents?

A. At first the rule was that five cents extra should be collected; that was before Mr. Gilmore went on to the road; after Mr. Gilmore went on to the road there was an addition of five cents put on, making ten cents extra. We tried to act according to that rule for a while, and then I had a talk with Mr. Gilmore and told him it was very difficult to carry the rule out. He said that he supposed it was; but he wanted that I should do the best that I could.

Q. What did you do?

A. I did the best that I could.

Q. Will you state what the practice has been since you have been on the road in relation to the directors and their families passing free over the road. [Objected to by Mr. George.]

*The Chairman.* What is the purpose of this testimony?

*Mr. Rolfe.* We propose to show by Mr. Clough that he was never required strictly to carry out this rule of the Concord Railroad Company, by any officer of the road above him; that he was instructed and required only to do his duty as he best could under the circumstances; and we propose to show that he did, as to all the rules and orders of the corporation, do the best that he could. We propose to show that at least five of the board of directors repeatedly requested him to violate these rules in various particulars. The superintendent himself so requested him to violate them. Mr. Clough is here to acknowledge that in repeated instances the clerk of the road has requested him to pass one and another free over the road, and he has done so, in violation of the rules, on the supposition that the director had the right so to request Mr. Clough to pass a person free over the road, and Mr. Clough did not understand that he was violating the rules of the road in any particular by so acting upon the request of the director.

*The Chairman.* We think, now, that the evidence must be confined to showing the obsoleteness of the particular rule or rules which you allege as have been violated. We do not think that evidence tending to show that the directors and their families rode over the road free, is evidence tending to show that the particular rule that is complained of being violated here has become obsolete.

*Mr. Rolfe.* Here is a series of instructions which the conductor is supposed to be required to live up to; and if these instructions say that no person shall pass over the road without a written pass, and one director after another comes and asks to have their families passed, until finally the president and a majority of the board of directors have required the same thing of the conductor over and over again, we claim, that then, such action of the directors tends to nullify the series of rules.

*The Chairman.* We do not so understand it. The disregard of one of those rules is not evidence of the practical disregard or obsoleteness of another one. We may hereafter think the testimony admissible, but we do not think so now.

*Judge Bellows.* The mere fact that somebody else had violated this particular rule would amount to nothing if it stopped there; but if it be shown that the rule is practically abrogated by all the parties concerned, it would then be a different matter.

*Mr. Rolfe.* If it should turn out in relation to the date of the tickets, first in relation to the passes and then in relation to receiving the tickets without date—that the instructions of the directors of the road changed entirely, so that no regard was required to be paid to the date of the tickets—that being on this batch of instructions—how many of those rules would we have to show to be abrogated to affect this one?

*The Chairman.* You will have to abrogate this particular one. That is exactly our ruling.

*Mr. Rolfe.* I suppose then that this testimony may be put in at another time, and for another purpose, we do not propose putting it in now.

Q. (*By Mr. Rolfe.*) State to the referees what Mr. Gilmore said to you in relation to the enforcement of this ten cent rule, when you had your first interview with him with respect to it. State whether the matter was talked over between you and Gilmore, as to how you should make your returns; and if any instructions or suggestions were made by the superintendent, state them also?

*Mr. George.* Our exception covers this same question, I suppose.

*The Chairman.* This testimony may be relevant, at least in one point of view—on the question of estoppel, and as to the misleading of the corporation. The question whether the corporation had notice of this rule being violated, would probably be material; and we are inclined to think, now, that the superintendent must be considered as so far agent of the corporation, that notice to the superintendent would be notice to the corporation; and in that view we think the testimony is admissible. [Exception by counsel for plaintiff.]

A. Our first talk was in 1858 or 1859. I have had several talks with him since. I told him the difficulty of enforcing the ten cent rule, and he replied to me that he supposed that it was difficult; but he said that he wanted that I should do the best I could. That was very soon after the regulations were passed requiring the extra ten cents to be collected. After that, about every time those regulations came out, I had a talk with him about them, but more particularly in 1864.

Q. I asked whether at the time you had your first interview with Mr. Gilmore, soon after the rules came out requiring the collection of the extra ten cents, you received from him any directions or suggestions in relation to how you could make your return; if so, state them?

A. I asked Mr. Gilmore how I should manage with regard to the ten cents extra, if I failed to collect. Our way-bills were printed; the ten cents extra was added to the regular fare as printed on the way-bill. Such being the fact, I could not enter the fare upon my way-bill unless I received the ten cents extra, without losing ten cents myself. He told me to buy tickets at the ticket office, and I did so.

Q. Did he give you any instructions with reference to what you should do with the tickets?

A. He told me to punch them and put them in my collections and return them to the general ticket agents.

Q. I understood you to say that you have had frequent conversations with him since then. Do you recollect the time when Mr. Corning left the road?

A. I do.

Q. If you had an interview with Mr. Gilmore at that time, state what occurred at that interview?

A. About the time that Mr. Corning left the road (I think it was in 1858 or 1859, but I am not quite certain about that,) I told Mr. Gilmore that I was going to resign. He wanted to know my reasons for resigning and I told him that there was some trouble with Mr. Corning, and I did not want to stay if they were not perfectly satisfied with me. He said there was nobody dissatisfied but John H. George, the clerk; that he was bound to get rid of Mr. Corning.

*The Chairman.* It is pretty difficult to see the relevancy of this. I make the remark because the plaintiffs have made a general exception.

*Witness.* He said that the directors were satisfied with me, and that *he* was satisfied with me and did not want me to leave.

Q. State what your supposition was as to the superintendent and the directors knowing your style of doing business?

*Mr. Stanley.* Is his supposition competent evidence?

*The Chairman,* I think that is hardly admissible.

Q. (*By Mr. Rolfe.*) Go on and state when you next had an interview with Mr. Gilmore?

A. The next time I had, was when I talked of leaving in 1861. I bought some shares in a steamboat on Lake Winnipesaukee and I intended to go up there myself. Mr. Gilmore heard of it and came to me and asked me if I was going to leave?

*The Chairman.* It is not obvious how this relates to the matter which you have undertaken to prove—that is, Mr. Gilmore's knowledge of this particular proceeding. Is it coming to that?

*Mr. Rolfe.* It is.

*Mr. George.* Our objections of course covers it all; and I will not interfere with the testimony.

Q. (*By Mr. Rolfe.*) State whether you remained on the road.

A. I consented to stay.

Q. Come now down to the time when these rules were issued in 1864.

A. In 1864, when those rules were sent to me for my signature, I at first refused to sign them. Mr. Gilmore met me one day afterwards and asked if I had signed them. The rules were handed me by the general ticket agent. After they were handed to me I put them in my pocket, and in a few days Mr. Gilmore asked me if I had signed those regulations, I told him that I had not, and was not going to sign them. I told him that no man could live up to them. He said that he knew they could not. I saw him again a day or two afterwards and he said that he wanted that I should sign those regulations. I told him that I could not if I was expected to live up to them. He said that he knew that there was no man that could live up to them; and that he had told the directors so; he said that he wanted me to sign them and do the best that I could, and that he would get my salary raised if I would consent to remain; and it was raised, I think, in 1864.

Q. State whether there was anything said at either of these last interviews about the ten cents extra?

A. There was.

Q. State what?

A. I told him that that was one reason why I would not sign the regulations.

Q. (*By Mr. Rolfe.*) Please state if, after you had signed those rules, Mr. Gilmore said anything to you about seeing the directors?

A. He did, before and afterwards, also.

Q. What did he say afterwards?

*The Chairman.* Having ruled out whatever was said to induce Mr. Clough to sign the regulations, with what view are you now offering testimony as to what took place afterwards?

*Mr. Rolfe.* With a view to show that Mr. Gilmore said that he had seen the board of directors and that the board of directors expressed themselves as satisfied with him and wished him to go on and do as well as he had done, and that nobody desired to have him leave the road.

*Mr. Mugridge.* We propose to show that Mr. Gilmore told Mr. Clough that he would see the directors, and that subsequently he told Mr. Clough that he had seen them, and that they wanted him (Gilmore) to say so and so to him.

*Mr. George.* We object to that testimony.

*Mr. Mugridge.* We suppose that if a man undertakes to be an agent for both parties, what he then does and says is competent testimony.

*The Chairman.* Our view is that Mr. Clough's statement is not legal evidence of the action of the board of directors in modifying their regulations. It takes the form, now, of Mr. Clough's having signed certain regulations and then opened a negotiation with the board of directors, and procured an alteration of the regulations that he had signed. We do not understand that there is any authentic proof of the action of the board of directors in modifying the regulations. [Adjourned.]

[SEVENTEENTH DAY. Tuesday, December 22d, 1868.]

## TESTIMONY OF GEORGE CLOUGH CONTINUED.

Q. (*By Mr. Rolfe.*) After you signed the rules in 1864, please state whether Mr. Gilmore gave you any instructions with regard to taking the extra ten cents in any cases, and if so, state fully what those instructions were. [Objected to by Mr. George.]

*The Chairman.* This I believe is the same kind of evidence which has been already admitted for the purpose of showing Mr. Gilmore's knowledge of Mr. Clough's conduct; it goes in for that purpose.

[Exception by Mr. George.]

A. I had frequent talks with Mr. Gilmore as to enforcing the ten cent rule after the regulations of 1864 came out. I told him that I could not enforce that ten cent rule; that it was impossible to live up to it; that I did not wish to stay any longer, I wanted to get out of the office. Mr. Gilmore said to me that the directors were satisfied with me and that he was himself satisfied with me.

*Mr. George.* This of course goes in with our exceptions.

*The Chairman.* We have not ruled as to this. This evidence is not competent for the purpose of showing that the directors were satisfied with Mr. Clough's conduct; but we held that any conversation that he had with Mr. Gilmore, might go in simply as evidence that Mr. Gilmore knew what was going on.

*Mr. Tappan.* That is exactly what we put it in for. We do not offer it as evidence to prove what the directors did, but simply to show a notice to them.

*Mr. George.* If confined to the purpose stated, we do not except to it.

*The Witness.* Mr. Gilmore said that the directors were satisfied with me, and that he was himself satisfied with me, and he wanted me to go on and make my returns just as I had made them, and be sure that the company got all the money. About that time my pay was raised.

Q. State how you made your returns of the fares you took in the cars when you did not take the ten cents extra. State whether Mr. Gilmore gave you any directions how you should then make your returns?

*The Chairman.* He has just stated that Mr. Gilmore told him to make his returns just as he had been doing before.

Q. (*By Mr. Rolfe.*) What were the instructions that he gave to you, if any, with regard to taking the ten cents extra, and where you could not get it, state what he told you to do?

A. He told me to pay the money into the ticket office, to buy tickets, punch them and put them into my collections and deliver them up to the general ticket agent.

Q. State what Mr. Gilmore, after you signed the rules in 1864, as you have stated, said to you with regard to the action of the directors in respect to the ten cent rule?

*The Chairman.* Has not that been stated fully, already?

*Mr. Rolfe.* No, he has not brought out any fact with reference to that as yet, but I will leave the question.

Q. After you signed the rules, in 1864, what was said between you and Mr. Gilmore with regard to the action of the directors on the ten cent rule. [Objected to.]

*The Chairman.* You understand, of course, that if this fact goes in it is not introduced as showing any action on the part of the board of directors.

*Mr. Tappan.* We do not offer it with that view; we only wish to show notice to Gilmore and to the company; I understand the referees to have ruled already that notice to the agent is notice to the company.

*The Chairman.* It is only competent to show notice to the agent. This temtimony comes in, if at all, simply for the purpose of showing Mr. Gilmore's knowledge of the transactions.

*Witness.* Mr. Gilmore told me that he talked with the directors and that they said that they did not suppose that those rules and regulations could be lived up to.

*The Chairman.* This is not evidence that Mr. Gilmore did ever talk with the directors, or that they said anything of that sort; it is merely a part of Mr. Gilmore's talk tending to show that Mr. Gilmore knew what was going on.

Q (*By Mr. Rolfe,*) In relation to the amount of travel over the road during the hight of travel in 1864 and 1865, state the number of cars run, as you recollect, over the road by the different trains?

A. The 10 o'clock train that we ran out of here in the hight of travel, used to run regularly three passenger cars and a baggage car to Nashua, and a baggage car and two passenger cars to Lawrence, and sometimes three. That was the usual number. They left Concord all in one train and went together as far as Manchester. The train that left here at half-past three ran about four passenger cars and a baggage car to Nashua, and three passenger cars and a baggage car by the way of Lawrence. In the hight of travel, I think that there was a Portsmouth baggage car, and one passenger car.

Q. How was it as to the half-past five o'clock train in the morning?

A. That was a light train. We ran two passenger cars and a baggage car to Nashua, and I think one passenger car and one baggage car went by way of Lawrence; and we put on another at Manchester for Lawrence.

Q. How was it with the up train?

A. The up train had about the same number of cars.

Q. State whether, as a general thing, the same number of cars that went down in the day returned?

A. Yes, sir; we brought the same number of cars back as went down, whether they were full or not. The orders were to bring back all cars to Nashua, whether they were full or empty.

Q. State as to how the trains were filled. Take first, for instance, the morning train?

A. The morning train at half-past five was a very light train. There were very few passengers by that train. The ten o'clock train down, sometimes would be pretty well filled, and then perhaps the cars would not be more than one-half full. As to the half-past three o'clock train down, that would be the same way. Sometimes the cars would be pretty well filled, and then there would be fewer passengers. The half-past ten o'clock train up, when there was a good deal of pleasure travel, would be well filled. In July and August, the cars would usually be pretty well filled; and they would be well filled in going down in September and October, when the mountain travel was going back. During other seasons of the year the cars would not be so full.

Q. How was it with the last train up in the evening—what is called the 8 o'clock train here?

A. The last train up would not be very full, as a general thing. We used to run up on that train from two to three passenger cars, and sent back a car to Nashua, and when they got to Manchester they hitched on the Lawrence train, consisting sometimes of one baggage car and sometimes of two passenger cars. There would be two passenger cars in the hight of travel.

Q. In the hight of travel, when the people were about leaving for the mountains, for pleasure travel, and were returning, state whether the trains were not considerably larger?

A. Yes, sir. I have stated that in September and October, when the mountain travel was going on, the trains were heavier.

Q. State how heavy they were then?

A. They used to have four passenger cars down over the Nashua road, and a baggage car, and sometimes there may have been five passenger cars. The Lawrence train, in the very hight of travel, might have had three passenger cars and a baggage car.

Q. Were there not, at other times, still larger trains than these run? If so, what were the circumstances, and why were they run?

A. Yes; when the soldiers were going we used to put on cars especially for them.

Q. How long a train have you ever known to be run on such an occasion—I mean the whole train, including the one to Lawrence?

A. There might have been sometimes fourteen or fifteen cars.

Q. State whether you recollect ever running a train of sixteen cars?

A. I cannot say. It may have occurred sometimes.

Q. How often was it done?

A. It was not a general thing to run so many cars.

Q. How frequently might or did that happen?

A. At the latter part of the war. I thing it might have happened often. When they would get a company of soldiers together they would send them on.

Q. How was it with the soldiers during the war,—were they furnished with tickets, or transportation?

A. They were furnished with transportation by the government.

Q. Do you know at what rate?

A. At two cents per mile, I think.

Q. Are you able now to form any estimate that will enable you to give the referees an idea of the amount of soldier transportation?

A, There were a great many soldiers passed at that time; I do not know that I could estimate the number. They went off in squads,—sometimes a larger number and sometimes a smaller.

Q. Did they go in cars-full at a time?

A. Yes, sir; we had one, two, and three car-loads of them at a time.

Q. If there were a regiment of soldiers going away, how were they transported?

A. They would go in an extra train.

Q. How would less than a regiment go?

A. They would go on the regular trains.

Q. State whether during the war you have known less than a regiment to go by a special train?

A. No, sir; I think not, I do not recollect of such an instance,

Q. I want to ask you in relation to the travel that paid, and that which did not pay; in other words, who passed free?

A. There was a great deal of free riding at that time.

Q. During what time?

A. During the war.

Q. State whether, during the war, free riding was or was not more frequent than during any other time.

*Mr. George.* We object to this question. We say that the doctrine of estoppel applies to this just as well as to any thing else, because this conductor had signed this regulation, that only people should ride free in a certain way. We say, therefore, that he can be heard to say that he violated this rule.

*The Chairman.* We understand that this comes under the ruling of yesterday morning. It is noted that you claim an estoppel.

*Mr. George.* I do not want to take another exception, but I desire to have my point clearly understood, and then I will not interrupt again. It was the duty, under the rules, of the conductor to charge and collect fare on all passengers except those who were provided with passes in the manner specified in the rules. Our legal position is, that he cannot be heard to say in defense to this action, and that he is estopped from saying, that he did violate these rules, and carry passengers free who had not the requisite passes provided for in the rules.

*Mr. Mugridge.* If the gentlemen will hear the evidence as it comes in, he will see that there is no force to the objection. We propose to show that the cars were filled with passengers who had passes, which the conductor, under the rules, was bound to recognize. That is what we undertake to show.

*The Chairman.* My understanding of the matter, (and if I am wrong I beg the other referees to correct me,) is as that we suggested yesterday, that our course must be, with regard to this matter of estoppel, to receive the evidence and to determine upon the whole evidence that we had, whether that there was a case of estoppel made out, or not, with regard to any particular evidence, and if there was a case of estoppel made out, the party would not have the benefit of that evidence.

Q. (*By Mr. Rolfe.*) State, Mr. Clough, all that you can recollect, during the war and up to the time you left the road, who rode free on passes?

A. All of the directors' families.

Q. Did they have passes?

A. I do not think they had. I was told to pass them.

Q. Did the directors have passes?

A. No; I do not think they had. All the bank officers,—that is, the cashier, the president and most of the directors of the banks I think rode free.

Q. To what extent did the bank directors ride free?

A. My recollection is that pretty much all of them went free.

Q. Did they have passes?

A. I think they had.

Q. Do you refer now only to the bank officers of Concord?

A. No; but to those of Manchester, Nashua, and up north; and about every politician in the state also had a pass.

Q. Politicians of all stripes?

A. No, sir; but there were some of both stripes; they were mostly republicans who rode free. All the army officers, or the most of them, had passes; and a great many soldiers had passes.

Q. From the colonel, how far down in the list of army officers rode free?

A. Down to the captains; and there were a great many common soldiers also who had passes.

Q. How were the soldiers who rode free situated?

A. They would be out of money.

Q. How was it as to soldiers getting in the cars in squads without paying? Were you always able to collect fares under such circumstances.

*The Chairman.* He has already testified to the fact that the soldiers got in, and that he found it difficult to collect the fares and that there was a guard stationed at the car door.

*Mr. Rolfe.* He testified as to that fact existing, but not as to the extent of such travel.

Q. State the extent?

A. Sometimes I have had nearly a whole company get in, and I could not get anything out of them at all. I could not do anything with them.

Q. (*By Mr. George.*) Did they have transportation tickets?

A. No, sir. I recollect several instances where there was perhaps not a full company, but what there was left of them—came home; there might have been thirty or forty; they came home without any money, they could not get paid off, and they wanted to get home, and in such cases they would get in the cars and I could not get anything out of them. When they came back they would be without money also, and so would pay neither way. The excuse would be that the government owed them so much money, and they could not get it, and had nothing to pay with, and I was obliged to let them go.

Q. Didn't you have force enough to enforce the rule of the road?

A. No, sir; I had not.

Q. How have you managed sometimes when a large number did not pay their fares? Have you ever unhitched the cars and left them?

A. I believe I did that once, down at Hooksett.

Q. How many were there aboard then?

A. Very nearly a car full I think.

Q. Let me call your recollection to an instance where Mr. Gilmore was aboard when the soldiers refused to pay. Do you recollect of such an occasion?

A. There was an occasion of that kind, and he told me to let them go, that we could not do anything with them.

Q. How many were there then?

A. There were some fifty or sixty soldiers, I think. I do not know but at that time I set out to unhitch the car, but did not do it; I think Gilmore was aboard, and I asked him what I should do with them, and he told me to let them go; and I rather think that I did not unhitch the car then; but that was done by one of the conductors. Mr. Biddle, I think, unhitched a car on his train. I recollect that I did unhitch a car at Reed's Ferry, once. There was about half a car load of them and they had been drinking; and at Reed's Ferry, eight miles this side of Nashua, we unhitched the car and left them, they soon got sober and we took them up on the last train. I recollect that distinctly. They were in the hind car and I unhitched it and left them.

Q. Let me inquire now in detail how it was about the directors and their families, whether they passed free or held written passes?

A. If the directors were aboard with their wives they would not have any passes; they would just ask me to pass them.

Q. They would not have any passes—for whom?

A. For their wives. They would just say to me that their wives and families were in the cars, and tell me to pass them along.

Q. And you did so?

A. I did.

Q. Was that invariably the case, or were there exceptions?

A. That was invariably the case.

Q. How was it with reference to the clerk of the road?

A. He had a good many with him passed.

Q. Did he and his family pass?

A. Yes, sir.

Q. Did you pass persons at his request?

A. I have.

Q. Can you state the number?

A. I cannot. I do not recollect.

Q. Who were some that you passed; do you recollect any?

A. I think that there was a man by the name of Spofford, that he asked me to pass.

Q. Where did he live?
A. I think he lived in Newburyport.
Q. Do you refer to Richard Spofford?
A. It is the one that used to stump the state.
Q. When was this?
A. If I recollect rightly it was in 1864—in the fall of 1864, I think.
Q. How was it as to the treasurer of the road?
A. The treasurer of the road and his family used to pass free.
Q. The master of transportation, also? Who was the master of transportation most of the time?
A. Mr. Chickering was, and is now.
Q. How was it about the road master and his family?
A. They passed.
Q. And the ticket master?
A. The ticket master and family passed free.
Q. The overseer of repairs in the iron shop?
A. He and his family passed.
Q. The overseer of repairs in the wood shop?
A. He went also.
Q. And the wood agents?
A. Yes; they and their families passed.
Q. The freight conductor and depot master also?
A. Yes; they and their families.
Q. The freight clerk?
A. He went free.
Q. The ticket seller at Concord?
A. He went free, and his family.
Q. The freight agent at Concord?
A, He went free, I believe.
Q. How was it with his family?
A. The families of those connected with the road went free.
Q. The station agent at Manchester?
A. He went free.
Q. Had he any family?
A. No, sir.
Q. The station agent at Nashua?
A, He rode free; and his wife also, when she did go. which was not very often.
Q. The regular mail agents?
A. They all went free.
Q. Did their families go free?
A. I do not recollect about that.
Q. How was it with the expressmen and their families?
A. They usually went free.
Q. Mr. Stearns?
A. Yes, sir.
Q. Did Mr. Sherman?
A. Yes, sir.
Q. Did Mr. Parker?
A. Yes, and his family.
Q. Mr. Todd?
A. Yes, and family.

Q. State how frequently Mr. John E. Lyon, the president of the Montreal road, rode over the Concord Railroad free, during the last six years you were connected with the road?

A. I think that he would average twice per week.

Q. Two trips per week, or only twice over the road?

A. Two trips per week—he rode four times over the road per week. All the station agents on that line of the road also rode free.

Q. How was it with the station agents on the Northern Railroad and the roads above?

A. They all had passes and went free.

Q. How was it with the wholesale dealers in Concord and Manchester—Barron & Co., Casey & Co., and Mr. Hutchins?

A. They had, I think, yearly tickets—free tickets.

Q. Was that on account of their doing a certain amount of freight business over the road?

A. I always supposed so; I never inquired into it.

Q. If Mr. Hutchins, whose firm consisted of himself and two sons, had a season pass, could only one use it, or could all the members of the firm use it?

A. My impression is that the whole firm could use it On some of the tickets I think that whole firms could ride.

Q. How was it with Gust. Walker and David A. Warde?

A. Mr. Warde used to have a free pass; I do not recollect about Mr. Walker.

Q. Was Mr. Warde's a yearly pass?

A. I think it was.

Q. Was his one that either member of the firm could ride upon?

A. I am not quite sure how that was.

Q. How was it with the hotel keepers and their families?

A. They went free; and the barbers went free; they went on written passes,—they were passed by Mr. Gilmore.

Q. What barbers?

A. The one that used to be in the Phenix Hotel,—Wales; he used to have Gilmore's pass.

Q Did Wales' clerks ever pass.

A. I do not recollect about that.

Q. How was it with Mr. Diamond and family?

A. I think they went free.

Q. And Mr. Thompson and family?

A. The same. The keeper of the American House went free.

Q. How was it with Lindsley of the Eagle Hotel?

A. He had a pass.

Q. Did his family?

A. I do not recollect about his family.

Q. How was it with ministers?

A. They went free generally. If they did not go free they only paid half price.

Q. Did they go with passes?

A. Yes, sir; they had passes if they went free. There was a rule for a short time that they should pay half fare, but that was the most we ever got out of them, and a good many of them went free.

Q. Was that by their purchasing a ticket, or by direction of Mr. Gilmore? Was there any order about their going for half fare?

A. Mr. Gilmore gave me orders that the ministers should go for half price, and if there were any of them in the cars that had no tickets, I was to take but half fare from them.

Q. How was it with the members of the legislature?

A. They went free one way.

Q. How large a share of the members of the legislature, if any, had passes?

A. A great many had passes. If they paid at all, they paid but half price, but a great many went for nothing.

Q. State whether the cars, during the first and last days of the week, while the legislature was in session, were more or less filled with members of the legislature?

A. They usually went home on Fridays and Saturdays and returned Mondays.

Q. How was it with the stockholders coming to attend the annual meetings?

A. They were accustomed to come and go from the annual meeting free.

Q. What was your direction from Mr. Gilmore about them, if they did not have passes?

A. They used to send on that day stockholder's tickets to the different stations, but they did not all get them; some got them and some did not; perhaps nearly half of them had not tickets. Gilmore's orders were, if they showed a certificate to pass them; and I did so. It was the same when the Northern meeting was held. The same orders were given then.

Q. State whether on those occasions there was a large amount of free riding or otherwise?

A. There was a great deal of free riding here during the two years that Mr. Gilmore was senator, and during the two years that he was governor.

Q. How was it with the legal profession?

A. They almost all went free.

Q. How many of the legal profession, as far as you remember, in the two counties of Merrimac and Hillsboro paid any fare?

A. I do not recollect but two that did.

Q. Who were they?

A. George W. Morrison and Mr. Foster of Manchester.

Q. Were there any that paid you half fare?

A. Yes, sir.

Q. Which were they?

A. Mr. Stanley and Mr. Clark.

Q. How was it with the doctors?

A. Dr. Peterson used to go free always.

*The Chairman.* Are these cases that you are now giving cases where they had passes?

A. They either had passes or I was instructed by the superintendent to pass them. A majority of them had passes. Dr. Morrill, I think, also rode free. They were both Mr. Gilmore's doctors.

Q. (*By Mr. Rolfe.*) How was it with the proprietors of the hotels at the White Mountains, for instance—Taft, Coffin, and others.

A. Taft and family rode free always. In the winter season they lived at Nashua. He had a pass for himself and family, and I think that all those hotel keepers up around the White Mountains had passes.

Q. How was it with hotel keepers at Nashua?

A. Mr. Wright, who kept the Pearl Street House, had a pass.

*The Chairman.* Don't you think it would save time to go the other way, and give evidence as to those who did *not* have passes?

*Mr. Rolfe.* I should do that, if I thought the referees would not consider such evidence incompetent.

*Witness.* The hotel keepers at Boston, of the American House and their clerks went free. I do not say that *all* the proprietors went free, but Mr. Wright did and his clerks.

(*By Mr. Rolfe.*) How was it about families passing back and forward on the road with passes?

A. There were some that had passes.

*Mr. George.* Do you mean the wives of the gentlemen you were speaking of.

*Mr. Rolfe.* I did not wish to ask the question in that way, for fear it might be ruled out as leading.

Q. How was it with Mr. Gilmore's men, who worked on his farm, and those who worked for him otherwise?

A. They went free generally. He had a lot of Irishmen come up from Manchester—some seven or eight—who worked on his farm digging out the stumps, and improving the land.

Q. Did he bring them up to work once a week, or every day?

A. They would come up the first of the week and go back the latter part.

Q. State in reference to Mr. Wm. E. Chandler. Did he and his family pass free?

A. Yes.

Q. Did Judge Fowler?

A. Yes.

Q. Did Mr. Durkee go free?

A. I think he had a pass.

Q. How was it generally as to the families of the officers of connecting roads—for instance, the clerk of the Northern Railroad?

A. The directors and clerk of the Northern Railroad, and their families, always went free.

Q. And the directors and clerks of the other railroads?

A. I think so.

Q. Did Judge Perkins and his family go free?

A. Yes; and those connected with the printing offices, went free; and the insurance agents used to have a pass.

Q. How was it as to Seth Eastman, and S. C. Eastman?

A. They passed free, and their families likewise.

Q. And Edson Eastman?

A. I think so; I am not quite sure of him.

Q. And Mr. Keyes?

A. He used to have a pass for himself and family.

Q. And J. B. Marston—the one that kept the restaurant at the depot?

A. I don't know about his family, but he went free.

Q. How was it about Mr. Christopher Hart?

A. He used to go free; and his daughters went free. There were some more down at Portsmouth that went free.

Q. How was it with the governor and his honorable council? Did Governor Gilmore's council go free?

A. I think they did; likewise Mr. Noyes, and Mr. Geary, and Mr. Hackett and his son, of Portsmouth, and there were several men there whose names I don't recollect, but I should know them by sight.

Q. How was it with General Hunt?

A. He used to have a pass to Nashua.

Q. Did you ever know General Hunt to pay his fare?

A. I never did.

Q. Was he frequently on the train?

A. Yes; he used to be there quite frequently. He had passes on all the connecting roads.

Q. How was it with the staff of Governor Gilmore?

A. They all went free, as well as I can recollect.

Q. And how was it with Governor Smyth and his staff?

A. He had a free pass. I don't recollect about his staff.

Q. Did Mr. Scripture go free?

A. Yes; he used to have a pass.

Q. They kept the two hotels in Nashua?

A. Yes.

Q. Who went free at Manchester?

A. I don't recollect whether the hotel keepers did or not; I know that Mr. Shepherd, who keeps the Manchester House, always paid his fare. Captain Harrington had a pass, and I think Mr. Currier had a pass. Mr. Kidder used to ask me to pass his clerks and his wife.

Q. Did you?

A. Yes.

Q. How frequently?

A. One of his clerks went quite often. I supposed he was going through the country to sell his flour and grain. B. F. Martin and family likewise had a pass.

Q. How was it with E. A. Straw?

A. He was a director on the Lawrence road and went free.

Q. Did his family?

A. He had no family, I think.

Q. Of these persons I have inquired about, how was it with Mr. Straw, was he not frequently on the train?

A. He rode frequently.

Q. How was it with B. F. Martin?

A. He and his family rode. He was frequently on the train.

Q. How was it with Justin Head?

A. He went free, and his brother likewise.

Q. How was it as to their being frequently on the train?

A. They rode very often. Mrs. Head was on the road a good deal.

Q. How was it with Joseph T. Goss?

A. He used to have a pass. but I am not sure about his family.

Q. Was he frequently on the train?

A. He rode quite often. Waters, of Hooksett, rode free.

Q. How was it with Geo. Riddle?

A. He had a pass, but he did not ride so frequently as some others.

Q. How was it with the legal profession at Manchester, Nashua and Concord? did they ride frequently, or otherwise?

A. They rode very often. I used to see them frequently in the cars.

Q. State, if it is within your recollection, how many of what are usually termed dead-heads have you seen on a single train?

A. I have seen as many as fifty or sixty.

Q. How frequently would that happen?

A. That would happen quite often in certain seasons of the year.

Q. In what seasons?

A. In the summer time, at the hight of travel.

Q. How would it be in the winter season, before the election?

A. There would not be so many then.

Q. How was it as to the train at all times, ahout there being more or less dead-heads?

A. There were more or less on every train. There was scarcely a train on the road but what had more or less dead-heads.

Q. As near as your recollection serves you now, how would it average? What would be the more and what would be the less?

A. I should think that it might average from fifteen to twenty on a train.

Q. When you say fifteen or twenty, do you speak in regard to the employees of the road?

A. No, I do not. I mean besides them.

Q. In addition to this, state what was the amount of free passes used by the employees of the road?

A. There were not a great many of them who passed free. The section hands that worked on the road. It was necessary sometimes for them to go free from one part of the road to another, and sometimes, when a freight train broke down, they had to send a force out to get it off the track. They would all pass free on the passenger train.

Q. How was it as to persons giving passes below, and persons giving passes above on the roads?

A. I don't understand you.

Q. For instance: the superintendent or other officers of the Lowell road, or the agents of the Vermont Central—were there any passes from either of these gentlemen? If so, how frequently?

A. A great many passes were given by the agent of the Vermont Central road, and a great many to persons that went up through here to the north.

Q. Were these persons, or any of them, known to you, or were they strangers?

A. A great many of them were strangers; the majority of them were strangers.

Q. In addition to these, you state that the superintendent, president and the board of directors individually requested you to pass people, do you, and that you did pass them? State to what extent, and who of the board of directors, if any, thus requested you to pass people.

A. Mr. Kidder requested me to pass persons without a written pass. This was in 1864. Mr. Beesley asked me to pass others. Mr: Crocker has been up with his family; I recollect that at one time he had six or eight with him going to the White Mountains.

Q. How was it with the president, Judge Upham.

A. He has asked me to pass persons, but not a great many.

Q. Who of the directors did not ask you to pass persons.

A. I believe Mr. Spalding never asked me to pass anybody, neither did Mr. Stickney of Boston. I don't think either of these gentlemen ever asked me to pass anybody. I never saw Mr. Stickney's pass on the road. I have seen Mr. Spalding's, but very seldom.

Q. State in relation to the tickets which you took up signed by Mr. Gilmore and countersigned by Mr. Spalding. Do you recollect taking up any?

A. Yes. When I spoke of passes from Mr. Spalding, I meant his own individual pass. [Counsel produces package of 28 passes.]

Q. Do you recollect taking up these?

A. Yes, I took them up.

Q. Were there any others that passed over the road?

A. There were other passes that did not come into my hands.

Q. Where did they go to, and what should you estimate as the number that went on your train?

A. I think that some went up above Nashua. It strikes me that some went up on the Plymouth road, and that one or two went to Boston.

Q. How was it about going up over the Concord, Claremont, and Connecticut Valley road?

A. I think quite a number went up there.

Q. There is a great variety of punches on these—are they all yours?

A. They are different punches—they are not all mine.

Q. Where was the counsellor convention held?

A. It was held at Manchester.

Q. Did any part of those passes come from the Wilton road, round to Manchester and over the Connecticut Valley and to Manchester, and from Claremont to Manchester?

A. Yes.

Q. And those going up on the train to the convention would be punched on the other roads, and with a punch different from yours?

A. Yes.

Q. Do you recollect whose punch this is? [Showing punched ticket.]

A. I do not.

Q. Did the men that came to the convention come on your train, and if so, was it not necessary when they passed one way to punch the ticket?

A. Yes.

Q. Where were the tickets taken up?

A. These were taken up between Manchester and Nashua and between Concord and Manchester.

Q. These were taken up on the train going down from Manchester to Nashua, and on the return train from Manchester to Concord?

A. Yes, sir.

Q. State at what age you left your father's house?

*Mr. George.* That is beginning pretty early; I suppose the point is to show what he was worth when he was on the Concord Railroad.

*Mr. Tappan.* We desire to show not only what he was worth when he went on to the Concord Railroad, but we want to show how he obtained it, and we want also to show his careful and industrious habits from his youth up—which we claim will be a material element in this case.

*Mr. George.* We say, that the only material point is, what he was worth when he was on the Concord Railroad.

*The Chairman.* Do you mean to be understood as objecting to the testimony?

*Mr. George.* We do.

*The Chairman.* We think this testimony is admissible, as it tends to show what Mr. Clough was worth at the time of his connection with the road. [Exception to the ruling by Mr. George.]

Question repeated.

A. I left my father's home in 1832.

Q. Where did you then live?
A. In Epping, N. H.
Q. What was the value of your worldly goods then—what property did you have?
A. I had nothing but what I had on and a small bundle, probably a couple of dollars would have covered it all.
Q. What was your age then?
A. I was fifteen years old.
Q. Where did you then go to—into what employment?
A. I went to Newmarket first, and stayed there over night with a brother of mine who was stopping there at that time.
Q. What did you next do?
A. A man came in with an ox team. He lived at Atkins, and had been to Great Falls and was on his return home. His name was Noyes. I don't know how I got introduced to him. At any rate, I jumped into his team and went with him, and stayed with him until the next spring in 1833.
Q. How long a time was that?
A. Six or eight months, as near as I can recollect.
Q. Did you accumulate anything during that six or eight months—and if so, what?
A. I don't think I had anything more than my clothes and board—If I had, I don't recollect how much. I left there in the spring of 1833 and went to Newburyport. I let myself there to work in the hotel. I was in the hotel about a year, and then I went into the stable and worked, and tended to the horses.
Q. What were you doing in the hotel?
A. I was chore boy there, and did all that I was called upon to do. I used to bring in the wood, black the boots, scour the knives, wait upon the table, carry baggage and do all that I was called upon to do. I worked in the hotel one year, and then went into the stable, and worked for the same man. His name was Drake. I stayed in the stable a year.
Q. What pay did you receive there the time you were there?
A. In the hotel I commenced with $10 a month, and the perquisites that I used to get would amount to more than my wages.
Q. After you went into the stable what pay did you receive?
A. I believe I got about the same pay in the stable. I think I got $16 per month.
Q. Did you make anything besides your monthly pay?
A. I got perquisites there.
Q. In what way?
A. The people that put up their horses there used to give me something. I used to wash the wagons and clean the harness, &c., for which I received pay in addition to my monthly pay.
Q. How long a time in all were you engaged with Mr. Drake in the house and in the stable?
A. Two years.
Q. What did you expend for clothes and boots and shoes?
A. My expenses for them were very small. The boarders used to give me second-hand clothes.
Q. How much, during those two years, were you able to lay by and save?
A. I laid by $300.

Q. Did you invest it, and how?

A. I put $200 into one of the Newburyport savings banks.

*Mr. Rolfe.* I desire to ask the counsel upon the other side if they will take a copy of the record of the bank, signed by Mr. Hall, the secretary of the institution, in proof of the deposit, or whether we shall have to summon him.

*Mr. George.* I shall ask you to prove your case, unless you will agree that any evidence will be received here which the referees believe to be true and as bearing on the case. I cannot consent to waive legal objections on the one side while being held to the strict letter of the law on the other.

*Mr. Tappan.* We have frequently during the trial admitted facts to save you the trouble of proving them.

*Mr. George.* I am not aware of it. I cannot agree to waive any of our legal rights except in cases where the referees will consent to hear anything that they may believe to be true.

*Mr. Mugridge.* We offer the record of the deposit made by Mr. Clough in the savings bank, at Newburyport, at that time.

*Mr. George.* We object. It is not legal evidence of any fact.

*Mr. Mugridge.* We have the certificate here and desire to put it in.

*The Chairman.* The certificate does not appear to be legal evidence of any fact.

Q. (*By Mr. Rolfe.*) You deposited $200 in the savings bank?

A. I did.

Q. What did you do with the other $100?

A. I left Newburyport and went home.

Q. When was that?

A. In the spring of 1835, I think. I then went home and gave $50 to my mother and kept the rest in my pocket.

Q. How long did you stay at home, and where did you next go?

A. I stayed at home two or three days. In 1835 I went to Raymond and let myself to Stephen Osgood. I let myself there to Mr. Osgood for $10 per month to attend to his stage and horses, with the understanding that I should have the first chance to drive his stage. I attended to his horses for three or four months, and he then put me to driving on one of his lines. I commenced on the line from Chester to Pittsfield. He had at the same time another stage from Lowell to Dover through Chester. The Pittsfield stage connected with the Lowell and Dover stage. The travel increased, and he was obliged to run two stages to Lowell from Chester, where we connected with the Dover line, and I drove through to Lowell from Pittsfield.

Q. Was there any variation then in your wages?

A. He gave me more pay when I went as driver. When I drove from Chester to Pittsfield he gave me $15 per month, and when I drove through to Lowell he gave me $20 and my board. I drove for Mr. Osgood, and was in his employ four years.

Q. Did you receive any other pay, or have any other source of income, except your monthly pay?

A. I did. I used to carry produce down on my coach from Pittsfield, and on the line of the road down through Deerfield and Chester.

Q. What kind, and what amount?

A. I used to carry poultry, butter, eggs, mutton, veal, pears, dried apples and cranberries. I used also to carry up stuff on the return trip.

Q. What did you carry up?

A. I carried up fresh fish.

Q. Who did the express business on the line?

A. I did. I used to carry money packages and bundles.

Q, To whom did that pay belong?

A. It belonged to me.

Q. State whether all the perquisites belonged to the driver?

A. They did.

Q. Were there any other articles except fresh fish that you carried up?

A. I don't think of any now.

Q. What did your pay amount to besides your monthly salary?

A. What I got in that way amounted to considerably more than my wages.

Q. During the time that you were at work for Mr. Osgood, how many days were you absent from your duties?

A. Ten or twelve days, I think.

Q. On what occasions?

A. I was sick.

Q. Were you absent on any other days?

A. No, sir; I never lost a day for four years except the ten or twelve days when I was sick.

Q. Did Osgood deduct that from your pay?

A. I don't remember.

Q. During the time you were at Drake's, how much time did you lose?

A. I don't remember losing a day; I don't think I did.

Q. At the time you left Osgood's employ, how much property had you, and what was its value, and how situated?

A. I had over $2000 I think, and very near $2500.

Q. That included all that you made prior to going with Mr. Osgood?

A. Yes.

Q. What did you do when you left Mr. Osgood's employ?

A. I bought half a stage line from Lowell here.

Q. When was that?

A. In 1839.

Q. Do you recollect the season of the year?

A. I think it was in the spring of 1839.

Q. How was your property invested?

A. I had it in notes at that time.

Q. Had you withdrawn this from the bank?

A. I had.

Q. What did you do then?

A. I bought Peter Dudley's interest in the stage line from Lowell here. He owned the teams on this end of the route, and a man by the name of Rodney Parker owned the teams on the lower end of the route.

Q. What did Dudley's half consist of?

A. It consisted of twelve horses, a stage coach, a sleigh and blankets and harnesses. For these and the twelve horses I paid him $1200.

Q. Did you pay him all that money down?

A. I paid him $1000 down, and the rest a very short time afterwards. I had money enough to pay him, and more too, but it was out at interest. Mr. Dudley was not particular about all the money, but I called it in, and settled with him very soon after that.

Q. What next did you do?

A. I went on and drove my stage myself.

Q. Did you drive through?

A. I did. Mr. Parker paid me for driving his team, and very soon I bought him out.

Q. What was your pay from Mr. Parker?

A. He paid me $1 per day, and kept me at his end of the route. He paid me besides $1 per day for the use of the coach, making my wages two dollars per day. The line ran up one day and back the next. It left Lowell on Mondays and came down on Tuesdays, up Wednesdays, down Thursdays, up Fridays and down Saturdays.

Q. You say that you bought Parker out. How soon after that?

A. I think about six months after I bought into the line. It might have been longer.

Q. What did you pay him, and what did you have of him?

A. I paid him $1500 for his share. He had but eight horses and harnesses and blankets.

Q. When did you pay Parker?

A. I paid him pretty soon after that. I don't think I paid him all down at the time, but I paid him very soon after I bought him out.

Q. How were you employed from that time up to the opening of the Concord Railroad?

A. I drove my stage until I went on the cars, in 1842.

Q. Are you able to state the amount of money that you made, net, during the first year that you drove over the Mammouth Road, or about the amount?

A. I think the first year I made about $1500.

Q. How much the second year?

A. About $3000.

Q. How much the third year?

A. I made the last year I was staging about $4500.

Q. Did you have anything during the third year, by mail contract or otherwise, that you did not have before?

A. Yes; I got $500 per annum on a mail contract. That was in 1841—the last year that I drove the stage.

Q. I want now to ask you how the fact was during the time that you were in the employ of Osgood, whether you spent any money in going to places of amusement, balls, dances, &c.?

A. No, sir; I never did.

Q. State whether at the time you left Osgood's employ, you were indebted to anybody?

A. I was not.

Q. State whether up to the time you left his employ you lost any money, or had made any bad debts?

A. No, sir; never a cent.

Q. As to the amount of the produce and poultry you carried over Mr. Osgood's line of stages, did the fact of your carrying produce and poultry over that line occasion any remark in relation to the line.

[Objected to.]

*Mr. Mugridge.* If his line had any particular name, may he not be asked what that name was and what gave rise to it?

*The Chairman.* This particular question is leading; but of what consequence is the question? Mr. Clough has shown us just how he managed business and how much he was worth. Is it not a waste of time to go any further in that direction?

*Mr. Mugridge.* It may be, but we want to show that in consequence of the amount of poultry carried over that line, the line was known as the "chicken line." If the referees think it of no consequence, we do not want to urge it.

*The Chairman.* We do not want to take up any more time than is absolutely necessary.

Q. (*By Mr. Rolfe.*) During the time that you were in Mr. Osgood's employ, did you use any ardent spirits, or have you up to this time?

A. I never drank a pint of liquor in my life.

Q. How is it about purchasing it for others to drink?

A. I never bought a cent's worth for anybody else.

Q. You say that the first year you made $1500, the second year $3000, and the third year $4500. Please describe the travel over the route during those years?

A. I used to have all that I could carry. The travel increased right along every year. This Concord Railroad was building at the time, and the travel increased so that I was obliged to put on another stage from Lowell to Manchester.

Q. During what year was this?

A. 1841.

Q. What was the fare from Lowell to Concord?

A. Two dollars.

Q. If a passenger got in at Lowell to go to Manchester, and stepped out, and another took his place, what was the fare?

A. I used to get $2.50. From Lowell to Manchester, the fare was $1.50, from Manchester here, $1.

Q. What was the number that you usually carried on the stage the first year—what the second, and what the third?

A. I think I could safely say that I averaged about ten or twelve passengers through. Perhaps not quite so many the first year, but I think it would average that during the three years. I mean in the stages that I drove—the through line, not the line that I ran from Lowell to Manchester. That is separate.

Q. I want you to state the relative amount of travel during the three years. You say that it increased. How much did it increase?

A. It increased so that I had to put on another stage the last year.

Q. During the last year, how many passengers were you accustomed to carry on your coach?

A. I think twenty-four was the most I ever carried.

Q. How frequently have you carried so large a number, or near that number?

A. It used to happen quite often during the last year.

Q. Who drove your extra stage?

A. His name was Edward Sargent. He is the man who now runs an express from Lowell to Boston.

Q. What was the extra—what was the number of horses?

A. I think he drove what we call a spike team, three horses on a team. He had what we call a six passenger coach. There was what we reckon eleven seats—six inside and three behind the driver, and one on each side of him. That is what we call a six passenger coach—a spike team.

Q. During that time what did it cost per day for keeping and shoeing the horses?

A. We used to calculate that sixty cents per day would keep the horses and shoe them. I got my horses shod by the year. I paid from five to six dollars each year. I used to get my horses kept in some parts of the road for two and three pence per day. I believe I did not pay fifty cents for the horses with the exception of each end of the road. If I had six horses at a time, I paid three dollars for keeping them and myself. If I had but four horses they would charge me two dollars for the team and driver, and three dollars for six horses and the driver.

Q. During the time you were engaged in staging for yourself, did you lose any horses?

A. No, sir; I never lost a horse when I was the owner.

Q. During the last year or two that you drove the stage, where did you keep your horses and how did you manage them?

A. The last year that I drove stage I owned a place in South street. I used to put up my horses there and take care of them myself.

Q. Were you sick at any time during the three years that you drove stage for yourself?

A. No, sir; I was not away from my team more then two days—calculating one trip.

Q. On what occasion was that?

A. It was on the occasion of my marriage. I was married on the 15th of October, 1840, and I was then away two days. I got a man to drive for me one trip. That is the only time I was away from my teams.

Q. During the time you were driving your own stage, what income did you receive besides the fares of passengers and the mail money?

A. I had more or less packages every day—money packages and express matter.

Q. Can you state the amount you received from all sources daily or monthly?

A. I think it would average $3 per day—from $2 to $3.

Q. What did you do with your money as often as you accumulated any? In what amounts and for how long?

A. When I got $50 or $100 ahead I used to put it out at interest.

Q. Up to the time when you went into the employ of the Concord Railroad, what bad debts did you make, if any?

A. I do not think of any.

Q. During the time you were staying, what sums of money did you spend for amusements—balls, dances, &c.

A. I paid out $5 for the old annual stage ball in 1839. That was only one year. That was the year before I was married, and I have not been to a ball since I was married.

Q. Did you save your way-bills?

A. I kept my way-bills right along regularly, but they were burnt when my house was burnt. They were packed in a trunk in the attic.

Q. State now what was the value of what property you had when you went on to the Concord Railroad, what was the amount?

A. I had somewhere between ten and twelve thousand dollars, I think about $11,500

Q. State what that property was?

A. It was my stage line from here to Lowell, my house on South street and notes at interest.

Q. What did you pay for your house on South street, and when did you buy it?

A. I paid $600 for it,

Q. When did you buy it?

A. In 1841, I think.

Q. How did you dispose of your stage property, and what did you realize out of it?

A. I sold two teams and a coach to a man by the name of Stone, between Manchester and Lowell. He gave me $2000—paying me $500 down, and carried the mail contract, and he made the payments as fast as they became due.

Q. What did you do with the rest of it?

A. The rest of it I sold after the cars got out. The cars in the first place ran to Manchester, and stopped there for a while, and I let my stage to a man by the name of Clifford, to run from Manchester to Concord until the cars came up here. He paid me $4 per day, and he paid the expenses.

Q. For how long a time was that?

A. The cars came up to Manchester in July, 1842, and they ran in here, I think, in either October or November following.

Q. State what you did with the balance of your stage and what you got for it?

A I sold one six horse team and harnesses for $600 to a man by the name of Pratt. I had one pair of horses that I wintered over, and sold them next spring with the coach.

Q. Who had them.

A. I sold them in Boston to a man by the name of Long.

Q. What did you get for them?

A. I sold them for $600, including the coach and harnesses.

Q. At the time you went on to the Concord Railroad, had you made any bad debts or lost any money in any way?

A. I don't rememember that I had.

*The Chairman.* I wish to make a remark to the counsel before we adjourn. This case has occupied so much time, and there have been so many rulings and exceptions, that we think it is very important that before the testimony closes, each party should furnish us with a list of all the exceptions they propose to insist upon; and we wish it understood that we shall consider all those exceptions waived which are not furnished to us before the testimony closes. This is with a view to consider all the ruling and all the exceptions for the purpose of determining whether there are any exceptions that can be waived.

Q. (*By Mr. Rolfe.*) Of what did your property consist at the time that you went on the railroad in 1842? I speak in relation to the time when you had sold to Mr. Stone the line below.

A. My property consisted of my house here in South street.

[Objected to and allowed.]

Q. Will you proceed and state, Mr. Clough?

A. It consisted of a note against Daniel S. Kimball, of about $6000.

Q. (*By Mr. Mugridge.*) Of where?

A. Lowell. And notes against Samuel Stone and others for $3000, and the upper part of my staging from Manchester up here.

Q. (*By Mr. Rolfe.*) For which you realized how much?

A. Fifteen hundred dollars, I think. Sold six horses for $600, and a pair of horses and coach for $600, and sold the other horses, I think, for about $300, if I recollect right.

Q. Now, Mr. Clough, will you state when you sold the property on South street and what you received for it?

A. I sold the property for $1500. I don't recollect the dates exactly. I will tell you in one moment. [Examining memoranda.] $1500

Q. You paid for it how much?

Q. (*By Mr. Tappan.*) When did you sell it?

A. In 1848.

Q. For how much?

A. Fifteen hundred dollars.

Q. (*By Mr. Rolfe.*) Were there any improvements on it?

A. No, sir; I think not of any account; set out some trees. I paid $600 for it, and sold it for $1500,

Q. What was the next piece of property that you purchased, and when did you purchase it, and what did you give for it, and when did you sell it, and the price?

A. I bought the place called the Sanger place in 1846.

Q. Well, go on.

A. I gave $300 for it. I sold it in 1847 for $600, bought it back again for $600, kept it awhile and sold it for $800 in 1848.

*Mr. George.* I should like to know what the witness is testifying from.

*Mr. Rolfe.* As I understand it, it is a memorandum that he has made.

*Mr. Mugridge.* It is a memorandum made under his direction and certified by him to be correct; the amounts are put there in that shape as a convenient way of getting at them, that is all.

*Mr. George.* I object.

*The Chairman.* Mr. Rolfe, you must show by your witness what kind of a paper you have got there.

Q. (*By Mr. Rolfe.*) Will you state how that paper happened to be made out?

A. That paper was made from the dates of the deeds when I sold land,—the deeds when I bought them—those places.

Q. State whether it was made by your direction?

A. It was made by my direction.

Q. State whether, at the time it was made, you did or not know it to be true?

A. I did know it to be true. It is impossible for me to recollect all my transactions.

*Mr. George.* I object. It seems that this is a collection from some deeds which should be produced:—apart from another objection which I have.

*The Chairman.* If I understand, the objection you make is one which is not yet in evidence. Mr. Clough says that the paper was made under his direction, and that when he made it, he knew it was true.

*Mr. George.* He says it was made from deeds.

*Witness.* It was made from the dates of deeds.

*The Chairman,* But he says he knew it to be correct at the time.

*Mr. George.* If the referees will allow me, perhaps I can obviate the difficulty by asking one or two questions.

Q. (*By Mr. George.*) In whose hand-writing, is that paper?

A. My nephew's.

Q. Were you present when he made it?

A. I was.

Q. Where was it made?
A. At my house.
Q. From what was it copied?
A. From the deeds—from the dates of deeds, and from my recollection of what I had bought and what I had sold it for.
Q. That is, you didn't take anything from the deeds except the dates?
A. Well, they took what I sold them for—the price.
Q, You took the consideration from the deeds, and the dates from the deeds?
A. And from my recollection.
Q. From the deeds and from your recollection?
A. Yes, sir.
Q. From both?
A. Yes, sir.

*Mr. George.* Very well; now I object.

*Mr. Mugridge.* Because the deeds may be evidence, that does not exclude the memorandum.

*Mr. George.* He says he took the memorandum from the deeds *and* from his recollection.

*Mr. Mugridge.* That is simply a memorandum made by him of the time when he bought this property, and when he sold it; made from records and from his recollection; a memorandum which he knew to be made right at the time, it being made under his direction. He simply put it in that shape that he might testify to it readily, without referring to the deeds which may be produced here, if the gentlemen desires. It is merely to refresh his recollection as to the numerous dates.

*The Chairman.* We think that the objection certainly would prevail, if it is insisted upon.

*Mr. George.* If they will agree to hand the papers in, I am willing they should go on. It is not merely a technical objection.

*Mr. Mugridge.* The deeds are entirely at your service.

*The Chairman.* You can go on with the examination, it being understood that the deeds are to be produced before the cross examination.

Q. (*By Mr. Rolfe.*) Go on, Mr. Clough. What outlays did you make upon that?
A. Not any that I recollect of.
Q. Do you recollect what rents you received, if any?
A. No, sir; I do not. I think I got $50 a year on the rent.

*Mr. George.* I respectfully submit to the referees that Mr. Clough's answer to that inquiry shows how improper it is, testifying from a memorandum of that kind. Mr. Clough's first answer is that he don't recollect, that he had made any outlays or received any rents.

*Mr. Mugridge.* The gentleman assumes that he don't read from his memorandum, or at least he represents that he don't. There is the memorandum, subject to their inspection. [Objection overruled.]

Q. (*By Mr. Rolfe.*) Now, Mr. Clough, will you proceed and state with reference to the "Morrison house," when you bought it, and all the circumstances connected with it—where it is situated?
A. The Morrison house—I bought a farm and moved the house down to Thompson street—I think it was on Thompson street. I bought it for $50, and I paid $50 for moving it, and I paid $75 for the land, and paid out $250 for fitting it up.
Q. Well, go on as fast as you can, Mr. Clough?

A I rented it seven years, I think, for $50 a year, I sold it for either $500 or $600.

Q. (*By Mr. Clark.*) Rented for how many years?

A. Seven, I think.

Q. (*By Judge Bellows.*) Sold it for how much?

A. Either $500 or $600, I don't recollect which.

Q. (*By Mr. Rolfe.*) That was on Thompson street, in this city?

A. Yes, sir. At the south end.

Q. Now as to the Shanks house, when did you buy that?

A. I bought the Shanks house in 1855. I swapped a pair of horses and wagon for that house, which cost $300. I sold it in 1856 for $900. [Subsequently corrected.]

Q. And what rent did you get out of it?

Q. (*By Mr. George.*) How much did you say you sold it for, sir?

A. Seven hundred dollars. I rented for either $50 or $60 a year; I don't recollect which.

Q. (*By Mr. Rolfe.*) You may state with reference to the house on Thompson street.

A. The house on Thompson street I bought in 1849, just as it is, a double house, for $1225. That is what it cost me.

Q. (*By Judge Bellows.*) When did you say it was?

A. 1849. I rented it from 1849 to 1862 for $100 a year, a tenement. That makes $200 for the house.

Q. (*By Mr. George.*) Mr. Clough, I simply desire to ask you if the rent of the house is written down on the memorandum that you have?

A. It is.

Q. From what source? By whom was that written?

A. The same man that wrote the rest of it.

Q. From what source was that information derived?

A. From recollection.

Q. From whose recollection?

A. My own.

*Mr. George.* Then I object to his using the memorandum if he swears from recollection. There is no use in using a memorandum for that purpose.

*The Chairman.* He says that he made the memorandum from his recollection, when he knew it was true, and I suppose if he is using a memorandum, and he now wishes for it to refresh his recollection, it is perfectly admissible, I think.

Q. (*By Mr. Rolfe.*) Mr. Clough, aside from your memorandum, as to the house on Thompson street, could you now remember what you rented that house for?

A. I reduced the rents in war time, and put them up again soon after—the latter part of the war.

Q. Well, now, do you recollect how that is, aside from this memorandum?

Q. (*By the Chairman.*) Can you state now from your recollection, as to this matter of rent, without looking at your paper, or do you want that paper?

A. I could tell a great deal better.

You can use it then.

A. I rented it from 1849 to 1862, for $200, from 1862 to 1864, at $85 a tenement; from 1864 up to date, for $200 a year.

Q. (*By Mr. Rolfe.*) Well, you own that now, do you?

A. Yes, sir.

Q. Well, you may proceed in relation to the other property in that vicinity?

A. The house on the corner of Thompson and Jefferson streets.

Q. (*By Mr. George.*) That is another one, or the same one?

A. Another one. I built that house in 1853. Cost me $1050. I reduced the rents the same time and put them up the same time that I did the other house.

Q. What were they rented for?

A. Rented from 1853 to 1862 for $100 a tenement.

Q. (*By the Chairman.*) Were there two tenements in this house?

A. Yes, sir. From 1862 to 1864, $85 a tenement. Rent them now for $125, each tenement.

Q. (*By Mr. Rolfe.*) Any other buildings in that vicinity?

A. Next, house on Jefferson street.

Q. Near by, is it?

A. Right there, close by; cost $600.

Q. Did you build it?

A. Yes, sir; I bought it and moved it on to the place; bought it for $300 of Thomas Pillsbury and moved it. It cost me $200 more to move it, and fit it up.

Q. Do you own that now?

A. I do. I rented it in 1862 for $75 a tenement. I think that was in 1853 or 1854. Rented it from 1862 to 1864 for $65 a tenement; since 1864 for $85.

Q. The next is the Brown house?

A. I don't think that is down.

Q. Well, you may go on and state about the Jamaica Plain property?

A. The Jamaica Plain property I bought in 1853.

Q. In Massachusetts, near Boston?

A. Yes, sir; three miles out. I bought it in 1853, for $1800. There was about three acres, or 140,000 feet.

Q. What have you done with it since?

A. I have sold up to two years ago, $5381.

Q. (*By Mr. Mugridge.*) How much have you got left?

Q. (*By the Chairman.*) Where was that property?

A. It was down near Jamaica Plains. It was upon the Providence and Worcester Railroad, down by the gas works—right by the gas works.

Q. Who did you buy it of?

A. In the first place, I loaned some money; I loaned the money, $800, to a man by the name of Joseph Palmer.

That is all; mere curiosity; I happened to know the locality, and felt a little curiosity?

A. I think I have half of it left. I have sold since February 1866.

Q. Have you half, or less than half?

A. About one-half left, I think.

Q. Give a description of the value?

A. It shows here I have the value of about 70,000 feet left.

Q. Now what is the comparative value of what you have left?

[Objected to, and objection waived.]

A. It is worth more.

Q. (*By Mr. Rolfe.*) I simply desire you to state whether the land that is left is worth more than that which has been sold?

A. Yes, it is.
Q. Mr. Clough, what has been the outlay on that land?
A. Nothing but taxes, and my agent, a man that sold, may have had something.
Q. What were the taxes for the first five or six years?
A. There was only some $4, or $5. That is, it has increased from year to year, until it is $20. The last tax that I have was $30 to $34, $34 seems to me it was, if I recollect right.
Q. Have I inquired of you about the house on Downing street?
A. No.
Q. Let me inquire of that?
A. That is the Shanks house, I think.
Q. You mentioned that was on Thompson street. Have you mentioned that?
*Mr. Clark.* You didn't say what street it was.
A. I think it was on Downing street. It was on the south end of the city.
Q. Now you may go to the Brown building, if you please?
A. The Brown building was built in 1859.
Q. (*By Judge Bellows.*) By you?
A. Yes, sir.
Q. (*By Mr. Mugridge.*) Cost how much?
A. It cost $520. I stated in my deposition that it cost $490; and I find a bill for extra work since then, and I have added it.
Q. (*By Judge Bellows.*) Does that include the land?
A. The land I hired. I received $200 a year rent for the same since 1859.
Q. (*By Mr. Rolfe.*) What have been your outlays, and what rent have you paid for the land?
A. I think the first two or three years I paid $25 a year. For the last three years I paid $35 for the land, and $6 a year for the water.
Q. (*By Judge Bellows.*) You say the last year $35 dollars for the land?
A. Yes, sir; $35 a year.
Q. Mr. Clough, you may go on now to the Masonic Temple. State what that cost?
A. The Masonic Temple, one-half of it cost $15,500.
Q. When was it erected?
A. In 1858.
Q. Does that include the Allison property?
A. No, sir, it does not;
Q. What income have you received from the Masonic Temple since it was erected?
A. That includes the store where Parker was, and the Allison Block.
Q. Then you may include them in the cost?
A. I bought the Allison building about the same time, about 1858.
Q. (*By Judge Bellows.*) Did you buy one-half of that, or the whole?
A. Bought one-half of that; Corning owned the other half.
Q. (*By Mr. Rolfe.*) What did that cost?
A. I paid, I think, $3800.
Q. For the whole?
A. For the whole; yes.
Q. So that your half amounted to $1900?

A. Yes, sir.
Q. How was it about the Bartlett store? Was that reckoned in with the $15,500?
A. Yes, sir.
Q. Now, sir, what rent have you received from that?
A. I have received between $9000 and $10,000.
Q. (*By Mr. Mugridge.*) In all?
A. My half.
Q. (*By the Chairman.*) In what year did you say you bought it?
A. In 1858.
Q. (*By Mr. Rolfe.*) Do you mean to say that you bought it in 1858, or only erected it in 1858.
A. Became the owner of it.
Q. The land you had purchased before?
A. We owned the land—yes, sir.
Q Does that include the Allison building, or rents that you received on the Allison building?
A. Yes, sir; on the Allison building, and the Bartlett store too, I think.
Q. Now, Mr. Clough, have you any more real property there?
A. I have the Page place.
Q. State about the Page place.
A. The Page place cost $2000 in 1854.
Q. (*By Judge Bellows.*) Cost you?
A. Cost $2000 in 1854.
Q. (*By Mr. Rolfe.*) What has been the income, and what the outlays?
A. In the first place it cost me $1800; and I laid out about $200 on it; and I call it that it cost me about $2000. There is about an acre and a half of land. There is a man lives on it, a poor man, and I don't get any rent from it. I get some fruit off from the land, and get some hay. I sold that property once for $3000; that is, a man had the refusal of it for that; and if he did not take it he was to give me $100. He didn't take it. And I don't consider that I get a great deal of income.
Q. Whether the income of that would pay the interest on what it cost you?
A. I should think it might.
Q. Any other rent?
A. I bought some land in Manchester of Capt. Walker, in 1849 or 1850, and made some over $500 on my transactions.
Q. How did you obtain the land where your house now stands on the railroad ground?
A. That land I got in the way of trade. It didn't cost me anything. I swapped, or at least I bargained with a man by the name of Cleaver; I swapped houses with him, and his wife opposed the trade, and didn't want to move down where I used to live. He said he would give me the back land where Dr. Morrill now lives, on River street.
Q. Fronting the railroad square?
A. Yes, sir. I built a house there in 1849. The house cost me $2600. I lived there some twelve or thirteen years; lived there until 1861.
Q. (*By Judge Bellows.*) Cost you how much?
A. Twenty-six hundred dollars

Q. And you lived there how many years?
A. Twelve or thirteen years; until 1861.
Q. Until you moved to your present residence?
A. Yes, sir.
Q. (*By Mr. Rolfe.*) Were there two tenements,—two houses?
A. Yes, sir.
Q. Did you occupy both?
A. One. Mr. Elliot occupied the other part.
Q. How much rent did you receive during that time?
A. Not more than $200 or $300.
Q. During the whole time?
A. During the whole time.
Q. After you moved up to the place where you now live, for what did you rent them.
A. I rent that house now for $200 a tenement.
Q. (*By Mr. Clark.*) That is on Railroad square?
A. Yes, sir.
Q. What was the rent, when you first moved out?
A. I believe it was $175 for a year or two after I moved out; and then I put the rent up to $200.
Q. (*By Mr. Rolfe.*) State as to your Dunbarton farm?
A. Bought in 1858; sold in 1866. I have got $275 more than I gave for it.
Q. (*By Judge Bellows.*) When did you sell it?
A. I sold it in 1866.
Q. (*By Mr. Rolfe.*) What did you give for it, and what did you sell it for?
A. Well, I have not got it down here; I have got a deed for it. I sold it for $2000; bought it for $1700; $1725 I think it is.
Q. Now, what outlays did you make on that, and what income did you receive?
A. I rented it for six per cent. and taxes; except one or two years, when I carried it on myself.
Q. While you carried it on yourself?
A. I carried it on, but I got at that rate, or more.
Q. Did you make any outlays on it?
A. No, sir; but I bought the land and built the house. It was a small house.
Q. (*By Mr. Mugridge.*) What year was that?
A. The year that I bought it?
Q. Yes, sir.
A. It was in 1858.
Q. (*By Mr. Rolfe.*) And you built a house on it, and the cost to you was $1725?
A. The land and house, I think, cost $1725.
Q. Sold it for $2600?
A. Sold it for $2600.
Q. You may proceed to the next item that you have on your memorandum?
A. The Orange farm next.
Q. Now go on and state in relation to that?
A. I bought that for $1000.
Q. (*By Mr. Clark.*) When did you buy that?
A. Bought in 1859.

Q. (*By Mr. Rolfe.*) What did you do with it? How long did you own it, and what did you do with it?

A. I owned it until February, 1867, I think. I swapped—I think it was in February 1867—I swapped it for some real estate here in Concord.

Q. What did you receive out of it for the time you had it?

A. I let it for six per cent.; and then I had the pasture land.

Q. Without paying the taxes?

A. Yes, sir.

Q. Then you had your pasture besides?

A. Yes, sir.

Q. And what else? Did you get any income besides from your pasture?

A. There was a maple orchard on it. I got a barrel of syrup every year.

Q. Did you say how much you rented that pasture for?

A. I used to take for pasturing—used to get about $75 or $100 a year, perhaps.

Q. Proceed to the next, and state what it is?

A. The Manchester Print Works.

Q. What did you pay for that, and how long did you keep it?

A. Bought that in 1852, three shares, $750 a share. The par value was $1000 a share.

Q. What dividends did you receive on it, and when did you sell it?

A. I received $1260, dividends. That is the statement from Charles Emory, treasurer—$1260 dividend.

Q. When did you sell it?

A. Sold it in 1862, and bought it in 1852. I sold it for $1200 or $1500 a share. I don't exactly remember any further than that.

Q. (*By the Chairman.*) In 1862, you say you sold it.

A. Yes, sir.

Q. (*By Mr. Rolfe.*) What was your next transaction, Mr. Clough?

A. There is the Concord Railroad stock—if I have that here.

Q. Well, proceed to the Concord Railroad stock. How many shares of Concord Railroad stock had you at the time this suit was brought?

A. I had four hundred shares.

Q. Are you able to state when you bought in and what amounts you paid?

A. I have the statement of my railroad stock from 1855; but I have owned stock since 1846. But previous to 1855, I cannot find the transfer book to get it off. I have got all there was at the office. May 1st, 1855, $22.50.

Q. (*By Mr. Mugridge.*) What is that?

A. Dividend.

Q. (*By Mr. Haile.*) How many shares did you own at that time?

A. Well, let me see,—twelve shares that would be; I didn't give the number of shares, but merely—I just take these from the treasurer's books. The interest on that amount, up to 1866, to $14.85.

[Objected to interest and omitted.]

May 1st, 1855, $22.50; November 8, 1855, $37.50; May 14, 1856, 37.50; November, 1856, $37.50.

Q. Now, Mr. Clough, what was the par value of that stock?

A. Fifty dollars a share.

Q. Now will you state as near as you can the average cost of these four hundred shares of stock?

A. I bought my stock as low as $33 a share, and from that up to par; and the four hundred shares, I think it averaged about $40 a share.

Q. What was the per cent. that you received on your stock, from the time you bought it up to the time this suit was brought?

A. I received eight per cent. with the exception of one six months, I think I didn't get but three for these months, at the rate of six per cent.

Q. How was that paid?

A. Four per cent. semi-annually?

Q. Soon after you went upon the Concord road, did you have any dealings with James M. Elkins? And, if so, what; and state the result?

A. In 1843, I loaned James N. Elkins $3000.

Q. (*By Mr. Mugridge.*) He was ——

A. He was conductor on the Lowell and Nashua road, and afterwards appointed to come up here. I think he was appointed to come on this road in 1843, and ran until he was appointed superintendent of the Montreal road—opposite train to me.

Q. Well, proceed and state about that?

A. Mr. Elkins wanted some money to go into a stock speculation—stock and other speculations. He said to me that if he had some money he could make some, and he made a trade with me; he should pay me six per cent. for the money, and give me a quarter of what he made. I think it ran along until 1847 or 1848; he made for me about $5000, including interest and all on the money that he hired.

Q. Did you state whether he settled with you from time to time, or how that was?

A. Yes, sir; he settled with me once in two or three months, and gave me his note, for I didn't want the money at that time. If I wanted any money to use I used to call on him.

Q. State as to your transactions in Boston, Concord and Montreal stock?

A. I think it was after he left this road, he and I bought a couple of hundred shares, a hundred shares apiece of the Montreal stock. I think it was the preferred stock.

Q. When was this?

A. In 1848—1847 or 1848.

Q. State the result of that transaction?

A. We got either fifteen or twenty per cent. in advance on that stock.

Q. (*By Judge Bellows.*) You sold it, did you?

A. Yes, sir; he sold it for me.

Q. (*By Mr. Rolfe.*) State as to your transactions in Old Colony stock?

A. I bought some Old Colony stock in 1843 or 1844. It was either twenty or forty shares, and I don't recollect which. Got an advance of ten per cent. on that.

Q. In relation to your steamboat transaction?

A. Bought in 1861, one hundred and eighty-three shares of boat stock—"Lady of the Lake."

Q. What did that cost you?

A. That cost me $30 a share, and the par value was $50.

Q. How long did you hold it, what did you receive out of it, and for what did you sell it?

A. I bought in 1861, and sold it in 1863. I sold it for $40 a share. Got $400 a year out of it: making $800 in all.

Q. Well, the western tickets?

*Mr. George.* Not with a view of objecting to the testimony proceeding, I desire to object to this evidence on the ground of estoppel. A rule was adopted that the conductors, after a given time, should not have any business outside of the legitimate business of the road. My idea is that the doctrine of estoppel applies also to that.

Q. (*By Mr. Rolfe.*) Will you proceed to state in reference to the western tickets?

A. We sold the western tickets three seasons.

Q. (*By Mr. Mugridge.*) What do you mean by the western tickets?

A. For instance, from here to Chicago, which was put into my hands, and I got a commission for selling them.

Q. (*By Mr. Stanley.*) What seasons were those?

A. I have not got the dates here; I have got them at home, I think.

Q. (*By Mr. Mugridge.*) From where to Boston?

A. From here.

Q. Tickets that went over your road?

A. Yes, sir.

Q. Part way?

A. Yes, sir.

Q. (*By Mr. Rolfe.*) What commission did you receive on these tickets?

A. I received two dollars. They were tickets that were issued from here, and had coupons for the different roads that they passed on—coupons for the Worcester road, and coupons for the Western road.

Q. Who issued these tickets?

A. Well, I had some on the Great Western, through Canada, and some Lake Shore. A man by the name of Allen, I think, furnished them to me.

Q. (*By the Chairman*) Do you mean to say that the tickets that passed over the Concord road from here were not issued by the Concord road?

A. Yes, sir. They were not issued by the Concord road. They were issued by other roads with the permission of the Concord road, I suppose.

*Mr. George.* Of course all this evidence goes in under my objection, on the estoppel ground. I suppose it is understood all around that this is evidence which is liable to objection on the ground of estoppel; that it is not necessary for me to take the exception as it goes along?

*Mr. Mugridge.* I don't suppose it is. There is no harm in it, however.

*Mr. George.* In regard to this sale of these western tickets. I claim that Mr. Clough is estopped on two grounds; first, on the ground that he should not use or sell any tickets not received from the general ticket agent; second, because there was an express rule passed that the conductor should not have business outside of the road. It is only with a view of estoppel.

*The Chairman.* They are offering this evidence to rebut your testimony. That is to show, not that he has not got the money of the Concord Railroad. That is not to show that he has accounted for what they have got, but simply to show how he got certain money.

*Mr. George.* Now, I say that he is estopped from that.

Q. (*By Mr. Rolfe.*) Now, Mr. Clough will you state what amounts you received during these three years from the sale of western tickets?

A. I got $2 on each ticket, commission. I think I sold about two or three hundred a season.

Q. Now, what years did you sell these, Mr. Clough?

A. I have not got the dates here.

Q. Have you a memorandum?

A. I have got one at home. I think one or two of the years is on my memorandum.

*Mr. George.* Mr. Rolfe, for the purpose of this inquiry, I am willing that he should refer to the deposition for this purpose. You may read them to him.

*Mr. Stanley.* It is interrogatory 367.

*Witness.* I had some of these books, but I have lost them. Must have been burned at my house, I think. I don't remember.

Q. (*By Mr. Rolfe.*) Do you recollect, Mr. Clough, what years they were besides 1857?

Q. (*By Mr. George.*) How late did you sell them?

A. I sold them, I think, in 1858; whether it was 1859, I cannot say; or 1856; I think I sold them three years.

Q. (*By Mr. Mugridge.*) You think you realized about $1500 in commissions?

A. Yes, sir.

Q. (*By Mr. Rolfe.*) Now, will you proceed to the matter of cows?

A. I kept—when I lived on Railroad square—I kept sometimes from three to five cows and sold milk.

Q. How many years did you keep cows?

A. I think it was about twelve or thirteen years.

Q. What kind of cows did you keep, and how much milk did they give, and for how much per quart did you sell the milk?

*The Chairman.* Is it worth the while to go into the details of that?

*Mr. Rolfe.* I should like to have him state so that the referees may infer.

Q. (*By Mr. Rolfe.*) You may go on and state what you estimate the profits?

A. I estimate about forty dollars a year on a cow, net profit.

Q. Taking the number of cows that you kept, how much do you estimate that you made net? You may refer to your deposition, if there is no objection? [Objected to.]

A. I have not got any other figures.

Q. You may state in relation to deserters?

A. I secured seven or eight deserters during the war.

Q. Well, sir; what did you receive apiece for securing them?

A. Thirty dollars. I got $30 apiece.

Q. During what years did you receive them?

A. I think it was in 1864 and 1865; 1864, I think, most of them.

Q. Well now, sir, will you state as to your trade in produce on the cars, and the amount you realized. State first in relation to merchandize, etc., that you sold in the cars—what is termed the "peanut trade," I think?

A. When the road first opened, I hired a boy and put him on some seven or eight years. I paid the boy from 50 to 75 cents per day. He took from $1 to $6 per day. I think the profit would be about $1.50 per day.

Q. Are you able to state the amount that you estimate that you had?

A. A dollar and a half a day, I said.

Q. Now you may state in reference to the chicken and produce trade.

A. I used to carry chickens, turkeys, geese, butter, eggs, apples, cranberries and chesnuts. Sent potatoes down on the cars occasionally.

Q. By freight?

A. By freight trains. Brought back fish—fresh fish. The estimate I made, about $500 or $800 per year,

Q. For how many years—how long did that continue, Mr. Clough?

A. I should think about five or six years.

Q. Did you make anything in tobacco speculation? If so, what amount, and when?

A. I loaned George Eaton some money, and he bought some tobacco. My profits were from $800 to $900. He paid the interest on the money, and allowed me for the profits.

Q. What year was this? Do you recollect?

A. I don't recollect it. I had the whole statement here from Mr. Eaton himself. I will find it so as to bring it in to-morrow.

Q. Let me ask you about the gain you made on what is called the buckskin horses, when you bought them, and when you sold them?

A. The buckskin horses, I paid $400 for them in 1865, and sold them for $450.

Q. (*By Mr. George.*) When did you sell them?

A. I sold them to the hospital folks; sold them in 1866, I think.

Q. (*By Mr. Rolfe.*) Did you work them?

A. I kept them to work all the time.

Q. Won't you state as to the gain on the Hutchinson horse, what you bought him for, and what you sold him for?

A. I paid $80 in 1858; sold him in 1859; sold him for $150.

Q. You may state as to the Silver horse, when you bought him, and when you sold him?

A. The Silver horse I paid $125 for in 1850; sold him for $200; sold him in 1850.

Q. How long did you keep the Silver horse, Mr. Clough?

A. I kept him, I think, about six or eight months.

Q. Did you work him during the time?

A. Worked him—yes, sir.

Q. You may state that in reference to all the horses you ever had?

A. I kept no horses but what I worked.

Q. You may state in relation to the John H. Pearsons horse, when you bought him, and when you sold him, and for how much you sold him?

A. The Pearson horse, $150; bought him in 1853, in the spring, and sold him in two or three months; sold him for $275.

Q. What time did you buy the Milford? For how much did you sell him, and how much did you realize on him?

A. A hundred and seventy-five dollars in 1854; sold him for $275. I bought him in the summer of 1854. I think I sold him that fall.

Q. Now the gain on the D. C. Watson horses, if anything? When did you buy them, and when did you sell them?

A. The Watson horses I gave $400 for in 1856.

Q. What season of the year? How long did you keep them?

A. I bought them, I think, in the spring; sold them along in September, I think.

Q. What did you get for them?
A. Got $600.
Q. Now the William B. French, when did you buy him? When did you sell him? And what was the result pecuniarily of the transaction?
A. I bought him in 1852 for $130, and sold him for $300.
Q. (*By Mr. Stanley.*) Sold him when?
A. I think I kept him through the winter.
Q. (*By Mr. Mugridge.*) How much did you say you sold him for?
A. Three hundred dollars.
Q. (*By Mr. Rolfe.*) The Clifford horse?
A. The Clifford horse cost $200.
Q. When?
A. In 1850.
Q. When did you sell him, and what did you get for him?
A. I kept him through the winter, I think; sold him for $400 next spring.
Q. Allison horse, No. 2?
A. I had two horses of Allison.
Q. The lowest priced one you may state?
A. Paid $187 for him in 1853; sold him for $350,
Q. (*By Mr. Mugridge.*) Sold him the same year?
A. Yes, sir; bought him in the summer and sold him the same year.
Q. Allison horse, No. 1?
A. Allison horse, No. 1—I supposed I paid $250 for him, but Mr. Allison says I didn't pay him but $210. The bill was made out $250, but he said that I didn't give him but $210 for him. I did not keep him but a short time; I sold him for $600.
Q. Who did you sell the Allison horse to?
A. I sold him to a Boston man. His name was Pratt. He was a man who used to take down a great many horses, and traded in horses altogether.
Q. To whom did you sell the Pearson horse? Do you recollect?
A. I sold him to a Boston man; I believe he was a West India goods trader there; I forget his name.
Q. Was he the one that the law-suit was about?
A. No, sir; he was taken sick and died; got a cold down there.
Q. To whom did you sell the D. C. Watson?
A. To a couple of persons by the name of Pierce.
Q. To whom did you sell the Clifford horse?
A. The Clifford horse went to New York. I don't recollect his name. He was a large pacing horse.
Q. (*By Mr. Mugridge.*) Did you say he was fast?
A. Yes, sir.
Q. To whom did you sell the Hutchinson horse, if you recollect?
A. The Hutchinson horse went over to Dunbarton, I think.
Q. Who had him,
A. I forget the man's name.
Q. Who had the Silver horse?
A. I don't remember who had.
Q. Did Mr. Silver have it?
A. Yes, a man by the name of Silver, here in town.
Q. Do you recollect who had the Milford horse?
A. He went to Nashua the last I heard of him.
Q. You may state in relation to the Rebel horse?

A. The Rebel horse I bought of Plum Whipple, here in Concord. He was a rebel horse and captured, and went out to war again; he was taken out there and never got back.

Q. (*By Mr. Stanley.*) For how much?

A. One hundred and twenty-five dollars, I believe.

Q. (*By Mr. Rolfe.*) What did you sell him for, and how long did you keep him?

A. I sold him for $200. I kept him for six months, I think, or thereabouts.

Q Will you now state the amount you estimate you made on selling coupon tickets?

A. I believe I estimated somewhere about $2000 in my deposition.

Q. You may state what tickets these were?

A. These were tickets from Montreal and from the western part of New York.

Q. From Detroit?

A. Might have been some from Detroit,—Detroit and west of that.

Q. Over what roads?

A. Ogdensburg road, I think, and Vermont Central and Northern.

Q. Now you may state about those tickets—the circumstances in relation to your selling them?

A. When I first began to buy them, they wasn't dated at all After a while they dated them good for thirty days.

Q. State whether you purchased any of these tickets previous to your having an interview with the superintendent in relation to them?

A. I found that there was other people on the street here buying them, and I got one and carried to Mr. Gilmore.

*Mr. George.* Of course our exception here applies.

*The Chairman.* This is one of your cases of estoppel?

*Mr. George.* Yes, sir.

*The Chairman.* And this conversation with Mr. Gilmore was for the purpose of showing notice to him.

*Mr. Mugridge.* Yes, sir; showing notice to the road.

Q. (*By Mr. Rolfe.*) Well, sir; proceed.

A. I carried one of these tickets to Mr. Gilmore, and showed them to him, and told him that they were sold here on the street and bought, and asked him if he had any objections to my buying them when they came along. He said no, not any. but not to take any west of Detroit.

Q. To what points did these tickets that you bought, most of them, go?

A. They went to Boston, most of them. Some of them went over the Manchester and Lawrence road; some by the way of Nashua.

Q. What share of them, so far as you recollect, went to Boston?

A. The biggest half, I should judge.

Q. From where?

Q. (*By Mr. Mugridge.*) From what stations?

A. Some from here; some from Manchester,

Q. (*By Mr. Rolfe.*) Any from Nashua?

A. No, I think not. They used to have tickets come off from these upper roads, the Northern and Montreal; sometimes when passengers got off at Nashua. Passengers wanted to go to Worcester and had a Boston ticket, and I would take it of them.

Q Did you make anything out of that arrangement?

A. No.

Q. Did it merely to accommodate the travelling public?

A. Yes, sir.

Q. For what length of time—from what time to what time—did you deal in these coupon tickets? Can you state?

A. As near as I can recollect. I sold them until I left the road. I should think it was perhaps five or six years.

Q. How were the tickets marked that you dealt in during the last part of your being on the road?

A. They were marked good for thirty days.

Q. Have you any such tickets now?

A. I believe I have one.

[Ticket shown; read by Mr. Rolfe. "Issued by Grand Trunk Railway to Boston. Good for one first class passage, for only 30 days from date of issue"]

*Mr. Rolfe.* The date of this ticket seems to have been scratched out.

Q. (*By Mr. Rolfe.*) Have you any more, Mr. Clough?

A. No.

*The Chairman.* I do not know any particular object there can be in presenting these and examining them.

*Mr. Rolfe.* Simply to show the kind of tickets that he dealt in.

*The Chairman.* Do not spend much time then.

*Mr. Tappan.* We think it is important to show the amount of these sold for thirty days.

Q. (*By Mr. Rolfe.*) Previous to those issued good for thirty days, was there any limitation on the face of the ticket?

A. No, sir; some were not dated at all.

Q. Now, Mr. Clough, did you ever buy any tickets that are termed spent tickets?

A. No, sir; I never did; never bought one nor sold one, after the date was run out.

Q. Now, Mr. Clough, you may please commence and state your losses if you can, now. In the first place, if you are able, state the amount of taxes you have paid, up to the time this suit was brought.

A. [Stated from the deposition.] $3841.17.

Q. Have you a memorandum so that you can state the expense of living since 1860, up to the time you were sued?

A. From what time? From the time I went on to the road?

Q. No; from 1862. What amount over and above what your salary was?

A. My salary was $70 a month. I think, from 1861 up to the time this suit commenced, it cost me over $2000—$2500 perhaps.

Q. (*By Judge Bellows.*) A year?

A. Yes, sir. Since 1861.

Q. (*By Mr Rolfe.*) You mean since December, 1861, or since January? What date did you place it?

A. I think it was in August that I moved up on to the hill.

Q. (*By the Chairman.*) From that time you think it would cost you $2500 a year?

A. Yes, sir.

Q. (*By Mr. Stanley.*) In excess of your salary?

A. No.

Q. (*By Mr. Rolfe.*) Do you know how much you paid for internal revenue? Or, state the amount first that you lost by Cooper, and when you lost it?

A. I lost about $2000 by Cooper.

Q. When do you date the loss?

A. I do not recollect exactly, but I think it was perhaps 1864; somewheres thereabouts; I could not tell exactly,

Q. When was it? About the time you went into your new house upon the hill?

A. It was after that.

Q. Let me ask you if the loss had not accrued at that time—at the time you purchased the Cooper house?

A. Well, I was thinking it was since that; I am not sure.

Q. When were the notes signed on which you lost the money? were they not signed as late as 1857 or 1858?

A. I think they were.

Q. When do you date the last? When did you have to pay the notes?

A. I think it was 1860.

Q. Are you able to state the amount you lost in the flour speculation?

A. I believe I lost about $1600 or $1700 there.

Q. When was that?

A. That was 1863, I think.

Q. Can you state the amount you lost by John Gass?

A. Two hundred dollars.

Q. When was that that you let him have that money?

A. I should think it was as long ago as 1856 or 1857.

Q. The amount was just $200 was it?

A. Just $200.

Q. State the amount you lost by George Watson, and what time?

A. I believe that was about $400 or $500. I think it must be 1857 or 1858. I don't know but it was as late as 1859.

Q. Did you lose anything in the California speculation? And if so, how much, and at what time?

A. It was either $500 or $1000; I forget which. I guess it was $1000,

Q. You lost just what you paid him? When was that?

A. During the early stages of the California excitement. That was as long ago as 1845, I should think, or 1846.

Q. Mr. Clough, will you state as to whether it was about the time of the first California excitement?

A. It was. I think it was longer ago than that, I think it was when I ran through to Boston.

Q. Did you run through to Boston as late as 1848 or 1849?

A. Yes, sir. I think it was 1848 or 1849.

Q. Let me inquire in relation to the amount you gave away to O. C. Rogers?

A. I gave him $1200.

Q. When was that?

A. I should think that was in 1863.

Q. Did you lose anything by J. M. Duffendorf, and if so, how much, and when?

A. Lost a couple of hundred dollars.

Q. When was that?

A. That was 1863 or 1864.

Q. When did you let him have the money?

A. That was longer ago than that, I think, that I let him have the money. It might have been in 1862.

Q. When you gave your deposition, did you estimate the loss on the account of C. H. Ham?

A. I did. I have reckoned that since. That has recently been paid.

Q. How recently was it paid?

A. Paid a month or six weeks ago.

Q. Are you able to state, Mr. Clough, the amount you have lost by loss of income on the Milford Springs property, while you owned it? In the first place will you state what you paid for the property?

A. It cost me $2500 or $3000. I ain't certain which.

Q. When did you buy it?

A. I don't recollect; I have got the deed at home.

Q. What income did you receive from that? Or did you receive income enough to pay the interest.

A. I never received any income from that place.

Q. That is the way you lost the interest on what you paid? When did you say you bought it?

A. I paid for it—I should judge I had owned it seven or eight years.

Q. At the time you was sued?

A. No, from now, I mean.

Q. (*By the Chairman.*) Have you got it now?

A. No, sir; I have put it away.

Q. You had when you were sued?

A. I swapped for some other property, some real estate in town.

Q. (*By Mr. Rolfe.*) Were you owning that as long ago as 1854?

A. I think I owned a quarter of it at that time, and bought another quarter afterwards.

Q. When did you buy the other quarter?

A. I bought the other quarter, I think, perhaps in 1857.

Q. Mr. Clough, during the first six years you were on the Concord Railroad, of what did your family consist?

A. Myself and wife, until 1846. My oldest boy was born in 1846; and my next boy was born in 1850—the second one.

Q. During the first several years you were on the Concord Railroad, did your family consist of any one except yourself and wife until your children were born?

A. No, sir.

Q. Did you have any domestic in the family?

A. I think I didn't until 1850.

Q. State whether, in the house, a memorandum for the first few years, after you went on the Concord Railroad, of the amount of your family expenses was kept by any one?

A. I think my wife kept it for a year or two.

Q. State, sir, what was your entire family expenses during several years after you went on to the Concord Railroad, and for how many years?

A. I think my family expenses the first years I was on the road wasn't over $300 a year.

Q. That is up to 1850? For the next ten years, from 1850 to 1860?

A. I think it cost me more.

Q. Well, what amount, with reference to your salary?

A. I think that I lived within my salary up to 1861; since that it has cost me more, considerably.

Q. (*By the Chairman.*) You spoke of your family expenses. Does that include your expenses away from home when you were on the road?

A. Yes, sir.

Q. (*By Mr. Rolfe.*) When you ran through to Boston, or up to 1849, as you have testified, did it cost you anything for board at the other end of the road?

A. No, sir; they didn't charge me anything at hotels.

Q. How was it when you ran through to Nashua?

A. Well, when I stopped at the Pearl Street House, I never paid anything; but I paid at the Indian Head, to Mr. Gillis. I insisted upon paying it; he didn't want to take anything. I insisted upon his taking the money; and I paid him. When I stopped at the Pearl Street House—Mr. Wright's,—he charged me nothing.

Q. You may state, if you have any means by which you can state it, the amount you paid for insurance?

*Mr. George.* Between what times?

*Mr. Rolfe.* All that he ever had. [Insurance waived.]

Q. Mr. Clough, what are those tickets? [Showing tickets.]

A. Those tickets seem to be on the Boston and Maine road, most of them—Lewiston, Boston and Maine; South Newmarket to Haverhill; Exeter.

*The Chairman.* For what purpose do you want these?

*Mr. Rolfe.* We want to ask Mr. Clough if he ever had these tickets.

A. Exeter to Durham, on the Boston and Maine; Great Falls to Dover, and to Lewiston; Danville Junction to Lewiston; Exeter to Newmarket; Rockport to Boston; Great Falls to Dover; South Newmarket Junction to Newmarket.

Q. (*By Mr. Tappan.*) How many are there?

A. There are ten.

Q. (*By Mr. Rolfe.*) Mr. Clough, I want to ask you if you ever had any such tickets in your possession?

A. Never did; no, sir.

Q. Look at these. [Showing other tickets.]

A. [Reads.] Newmarket to Dover; Lowell to Nashua; Boston to Lowell; Boston to Lowell; Boston to Lowell; (package tickets); Boston to Lowell; (package); Boston to Lowell; Lowell to Nashua; Lowell to Nashua; Lowell to Nashua; Boston to Lowell; (package); Nashua to Boston; (package); Nashua to Boston; (package); Lowell to Nashua; Nashua to Lowell; (package tickets).

Q. Mr. Clough, did you ever have any of these.

A. No, sir.

Q. Take these tickets. [Showing other tickets.]

A. [Reads.] Boston to Salem; Boston to Salem; Boston to Salem; R——— to Gloucester; Boston to Salem; Boston to Salem; Eastern road,

Q. Have any such tickets as these?

A. No, sir. There are six, [continues reading.] Boston to Lawrence; Boston to Lawrence; (these are package tickets;) Boston to Lawrence; Boston to Lawrence; Boston to Lawrence; Boston to Lawrence; Boston to Lawrence; and all packages.

Q. Did you ever have any of these?

A. No, sir; none.

Q. Look at these, [showing tickets.]

A. [Reads.] Manchester to Lawrence; Manchester to Lawrence; Lawrence to Manchester [seven times repeated;] Lawrence to Derry; Lawrence to Manchester—nineteen.

Q. Let me ask you, Mr. Clough, if these tickets could come into your hands?

A. No, sir; they were taken up on the Lawrence road before they got to Manchester.

Q. Let me ask you if you ever had anything to do with these except when you run over that branch of the road?

A. There was some seven or eight years ago—no, it is longer than that. [Reads again.] Great Falls to Salmon Falls; Great Falls to Berwick; P. S. & P. to Junction; Great Falls to South Berwick; Great Falls to Salmon Falls. There are eight.

Q. Did you ever have any of these?

A. No, sir; never. Boston to Salem, two.

Q. Ever have those?

A. No, sir. [Reads again.] Lowell to Lawrence; Lake Village (package tickets;) Lowell to Lawrence; Lowell to Lawrence. There are four.

Q. Ever have any of those?

A. Never.

Q. You mean to say that you never had any such tickets in your hands before you had them?

A. Never. [Reads.] South Newmarket Junction to Dover—two.

Q. Ever have any such as these?

A. Never.

Q. Look at these.

A. [Reads.] Portsmouth to Suncook; Portsmouth to Suncook; Concord to Portsmouth; Concord to Portsmouth; Concord to Portsmouth; Newmarket to Junction; Concord to Portsmouth.

Q. How many of these could come into your hands in the regular course of business?

A. Newmarket Junction—these tickets were counted out to us and charged. Concord to Portsmouth—I could not have them. Concord to Portsmouth and Portsmouth to Suncook—that one, we used to have them counted out to us and charged to us.

Q. (*By Mr. Mugridge.*) Where did these go to?

A. Newmarket Junction—we took fare from here to Newmarket Junction, and we gave them a ticket. We gave them a ticket and we had to account for them. There are six Portsmouth tickets that I could not have.

Q. Couldn't you have the Suncook ones?

A. No, sir; they go from Concord to Portsmouth; they are collected on the Portsmouth road; oh, no, there are two Portsmouth to Concord which might come into our hands in the regular course of business.

Q. How are these two?

A. Portsmouth to Concord. That came down off from the Contoocook river. There couldn't either of these. Bradford to Concord—that couldn't come into our hands. [Reads.] Boston to Salem, (package;) Boston to Salem, (package.) [Repeated three times again.] Boston to Lynn—eleven from Boston to Salem and one from Boston to Lynn.

Q. Ever have any of these tickets before?

A. Never saw one of them before. These are tickets that the conductors used to carry on the Concord road and sell in the cars. These [illegible]e counted out to us and charged.

Q. (*By Mr. Tappan.*) Was there what was called a conductor's ticket?

A. Yes, sir.

Q. Good where?

A. From here to Boston.

Q. On any of the stations?

A. Yes, sir.

Q. So that you had these and disposed of them to anybody? You had to pay for them at the ticket office?

A. Yes, sir. They were collected on the other end,—collected below Nashua.

Q. How many of these were there?

A. Three of them. [Reads.] Boston to Gloucester; Boston to Salem, (Eastern R. R.;) Salem to Newburyport; Gloucester to Boston; Boston to Newburyport; Boston to Newburyport. (Eastern R. R.;) Eastern R. R.; Boston to Newburyport, Eastern R. R.; Boston to Gloucester. Eastern R. R.; Boston to Newburyport; Boston to Gloucester; Salem to Newburyport; Boston to Gloucester; Salem to Newburyport; Boston to Newburyport; Rockport to Boston.

Q. How many?

A. Sixteen.

Q. Did you ever see any of these before?

A. Never saw them before in my life. [Reads.] Boston to Lowell, (package ticket;) Nashua to Lowell; Manchester to Lawrence. There is one that would come into our hands; three that could not.

Q. That is ——

A. That is on the Montreal road going to Manchester.

Q. Marked on the back "White, G."

A. That is the station that I bought of Mr. White.

Q. What was that passage for?

A. From Boston to Lawrence over the Maine road,

Q. Won't you describe the tickets as you go along, so that the referees may know what kind of tickets they are?

A. That ticket [indicating] goes down over the Lawrence—on the Montreal road, Concord, N. H. and Lawrence.

Q. That is on the Montreal road?

A. Over the Montreal road to Lawrence. That [indicating] is over the Montreal road to Manchester; that [indicating] is over the Montreal road to Manchester; that [indicating] is over the Montreal road to Lawrence; that is over the Montreal Road to Lawrence; that is over the Montreal road to Manchester; that is over the Montreal road to Lawrence; that is over the Montreal road to Lawrence; that is over the Montreal road to Lawrence; that is over the Montreal road to Lawrence; that is over the Montreal road to Nashua; that is over the Montreal road to Lawrence; that is over the Montreal road to Nashua. Tickets over the Montreal road to Lawrence: eleven.

Q. (*By Mr. Tappan.*) Tickets that you could have?

A. That I could not have.

Q. (*By Judge Bellows.*) Those to Manchester and Nashua you could have?

A. These to Manchester and Nashua I could have. There are two to Nashua and three to Manchester.

Q. (*By Mr. Rolfe.*) If the referees will pardon me, in my opening I spoke of coupons that came over the Montreal Railroad. There was a package that I went out of the room and tried to find. This is the package of which I spoke. I said that they must have been dated in 1866, because there were some dated on the 4th of the month, and if they were in 1865, they would have been dated on Sunday.

*The Chairman.* These checks bear the date of 1865 on the face of them.

*Witness.* These tickets bear no date anywhere.

*Mr. Rolfe.* There were two of these dated by a man by the name L——, dated in 1866.

Q. (*By Mr. Rolfe.*) Mr. Clough, I don't know whether I have asked you the date on which you left the road and ceased running as conductor.

A. The day that I left the the road?

Q. Yes, sir.

A. Left the road January 14th or 15, I believe—January, 1866.

Q. Did you have anything to do with buying tickets after that?

A. No, sir.

Q. There is a ticket, Mr. Clough, won't you state what tickets those are?

A. Dover to Boston—no, sir, I never saw it; [Reads] Salem to Lawrence, 4.

Q. Did you ever see any such tickets as these?

A. Never saw them; this is the first time.

Q. These go over the Essex road. There is one, [handing ticket.]

A. Boston to Salem, Eastern R. R., J. Prescott.

Q. That one, Mr. Clough? [designating.]

A. Coupons worthless and detached from this ticket, I never saw that before. To Nashua, P. M. and P. S. I never saw a ticket like that before.

Q. You say you never saw such a ticket as that before?

A. No.

Q. There is a ticket? [designating.]

A. Portland District, G. T. R. Please keep in sight. I never saw that before. [Adjourned.]

EIGHTEENTH DAY. Wednesday, December 23d, 1868.]

The hearing was resumed at 9 o'clock P. M., and the testimony of Mr. Clough the defendant, was continued.

## TESTIMONY OF GEORGE CLOUGH CONTINUED.

Q. (*By Mr. Rolfe.*) Mr. Clough, will you state the amount of taxes that you have paid in Concord, in all?

*Mr. George.* We should prefer that the amount of taxes in Concord should be put in year by year.

A. I paid $3,841.17, ever since I owned real estate in Concord.

Q. (*By Mr. George.*) To what time?

A. From 1841 to 1866.

Q. (*By Mr. Rolfe.*) State the amount of taxes that you paid?

*Mr. George.* I should like to know if Mr. Clough has any knowledge of his own.

*Mr. Rolfe.* Allow me to say that Mr. Clough has stated this in his deposition. He took many months to prepare these matters.

*The Chairman.* Mr. Clough testified yesterday from certain papers, which, after some talk, he was permitted to testify from, after objection.

Now he is going on in the same way. If there is any objection, it will probably be shown that it is a paper that he has a right to look at.

Q. (*By Mr. George.*) Mr. Clough, who prepared for you the statement of your taxes? Who made the paper? In whose handwriting is it?

A. I had my nephew go up to the city clerk's office with me and take the amounts off.

Q. You had your nephew do it?

A. Yes, sir. And I presume it is correct.

*The Chairman.* Of course that, if it is objected to, is not evidence. It is mere hearsay evidence.

Q. (*By Mr. Rolfe.*) Mr. Clough, have you paid your taxes all along, since you have been in town?

A. I have.

Q. Won't you state the amount of taxes and expenses that you paid on the Jamaica Plain property. In the first place state whether you are able to state from your recollection aside from any memorandum that you have?

A. I paid $225 Jamaica Plain taxes.

Q. Are you able to state, aside from any memorandum, the amount that you paid?

A. I could not recollect to the dollar, but my impression is that it was somewhere between $200 and $300. I think I stated in my deposition when they took that, that it was $200. My recollection is, I believe, that it was somewheres between $200 and $300.

Q. Mr. Clough, will you state your expense of living over and above your salary from January, 1862 to January, 1866.

*The Chairman.* He has stated that. He has stated his expense of living; and he has stated that his salary was $70 a month.

Q. Will you state now, Mr. Clough, the cost of what property you owned at the time this suit was brought, as near as you are able to ascertain?

A. It cost me about $91,000, I think.

Q. State the amount of your indebtedness at the time this suit was brought?

A. It was over $19,000.

Q. [*To the Referees.*] I asked the question in relation to the amount of interest that he lost on the Milford Springs property, and my recollection is that he was not able to state. [*To the Witness.*] Have you calculated the amount of interest?

A. Yes, sir. It was between $1400 and $1500.

Q. (*By Judge Bellows.*) I should like to have Mr. Clough, for my own satisfaction, state how much of the money that he took in the cars —give some idea how much money he took in the cars which was represented by the tickets which he purchased?

A. I made, when my deposition was taken—I made an estimate somewheres from, I think, $50 to $75 a month.

Q. Is that your idea now?

A. I think that ain't far out of the way.

Q. You are speaking now of the money you took in the cars, and that you accounted for by the tickets?

A. Yes, sir.

Q. (*By Mr. Stanley.*) Tickets that you purchased of the ticket sellers?

A. Yes, sir.

Q. (*By Judge Bellows.*) Does that apply to the whole time that you were on the road?

A. Yes, sir; this continued from some time in 1859, as long as I remained on the road.

Q. (*By Mr. Rolfe.*) I would like, Mr. Clough, to have you state how you managed it, that is, when you purchased the tickets, and how the business was done?

A. It was done, sir, when I did not take the ten cents extra.

Q. When and where did you buy them, and how did you keep your accounts?

A. I bought them at the ticket offices in Concord, Manchester, and Nashua.

Q. When did you buy these, Mr. Clough, in relation to the time you took the money?

A. I used to keep a memorandum in my pocket, (one of these,) and when I took a fare that I didn't take the ten cents, I put the amount down, and also a fare that I did take the ten cents extra, and when I got through I transferred on to my way-bill, and if there was any there that I didn't take the ten cents extra, I bought tickets at the office.

CROSS EXAMINATION.

Q. (*By Mr. George.*) I understood you to state that you were married in 1840?

A. Yes, sir.

Q. Had your wife any property?

A. Not of any account. She had a small sum left her—$300 or $400.

Q. Has she had any property from that time to this, except what came from yourself?

A. No, sir.

Q. You have spoken of the expenses of your family between given years. Will you be kind enough to state how much, as part and parcel of this expense, is included in the expense of clothing during these years. Confine yourself from 1861 to 1865?

A. I made an estimate when my deposition was taken.

Q. Let me suggest, I am not inquiring about your deposition or any estimate there made. My question is, what do you estimate? You have testified here that your expenses were $2000. My inquiry is, what do you estimate your clothing during these years when your family expenses were from $2000 to $2500?

A. It would be a matter of guess-work. I could prepare an estimate.

Q. No, sir, I don't want your estimate.

A. I don't think I could make any estimate unless you give me time. When my deposition was taken, I made an estimate.

Q. You have stated that your expenses were from $2000 to $2500 during these four years?

A. Yes, sir.

Q. Your clothing and your wife's clothing was a pretty large item in the two thousand dollars? At any rate it was an item?

A. It might not be very large.

Q. What was it as part and parcel of the $2500?

A. That is what I couldn't tell.

Q. Can't you tell that it was part and parcel of the $2000?

A. Yes, sir; I think it was.

Q. Have you any doubt about it?

A. I don't know as I have.

Q. Now, I ask you, what part you estimate that it would cost for yourself, your wife's, and your children's clothing?

A. It is impossible for me to tell.

Q. Make the best estimate you can?

A. I don't consider my wife a very extravagant woman; I always got her what she wanted.

Q. Wasn't your estimate of $2000 to $2500 made up of the various expenses of your family: clothing, food, help, and the various items which go to make up the family expenses?

A. I suppose it was.

Q. Now, sir, how could you arrive at this $2000 to $2500 without a basis, or a source from which those expenses are derived?

A. I suppose I did at that time, but I don't remember what it was. It was made up two years ago, when my deposition was taken, and I don't remember.

Q. You swear here now that your family expenses cost you from $2000 to $2500 a year, in your judgment. Now, sir, that was your estimate then, when you swear to this?

A. Yes, sir.

Q. Now, how much did you estimate your clothing in making that up?

A. I put it all together; I did not separate it; I didn't know how much it would be.

Q. State, as near as you can, the best estimate of your clothing. How much do you estimate that item?

A. Give me time and I will

[Objected to as having been already answered.]

*The Chairman.* I don't know as there is any objection to asking the question.

Q. Now, sir, have you any papers whatever from which you can make an estimate, or have you made any estimate from those papers. Have you got a memorandum from which you made a single estimate you have given.

A. It is mostly from recollection I made up my real estate operations; and I have the deeds to show. And on my stock, I have got a record of the dividends on my stock taken from the treasurer's books.

Q. Have you the memorandum anywhere of the cost of a single building that you have testified to, or of the repairs on a single building that you have testified to—the original memorandum, I mean?

A. I have bills at home for the repairs, I think.

Q. What repairs?

A. Why, for repairing houses, whatever was done—painting and the natural wear.

Q. Have you testified at all with regard to a single repair?

A. I don't recollect whether I have or not.

Q. How many dresses had your wife, sir, at the time of the fire in 1865, silk dresses that cost sixty dollars and upwards?

A. I don't think she had many.

Q. Do you know?

A. I don't recollect. The best dress she had was a present to her from Eben Sutton, of Danvers. That was the best dress she had, I think.

Q. Did you estimate in your deposition (and now I will show you the deposition as far as that inquiry is concerned), that it cost you $40 a year during a certain portion of the time, and $60 during the balance of the time for your personal clothing? I will hand you the deposition. Did you swear in your deposition,—"I estimate my clothing for fifteen years at $40 per year, and for the last five years, I estimate it at $60 per year; the clothing of my wife I call $50 a year for twenty years. The clothing of my oldest son, George, I estimate for fifteen years, $20 per year, and for the last five years, $40 per year; clothing for Charles, $25 per year, and one year $40." Did you testify to that?

A. I think I did. I think that is correct.

Q. [Reading.] "The clothing of my niece, Eliza Clough, something over two years, something over $75. Spending money I estimate at $25 a year for self and family." Mr. Clough, in whose name was your house and more or less of your personal property insured at the time it was burned in 1865?

A. When I bought that place——

Q. Mr. Clough, be kind enough to answer my question which I asked you?

A. In my wife's name.

Q. What proportion of the furniture and of the articles of personal property were burned up?

A. All the furniture in the second story, all above the first floor was burned; in all the chambers, in the parlor, and a portion of it in the parlor chamber, that was taken out and damaged.

Q. Was there any actual burning on the lower story? I am not talking of actual damage either by smoke or water. Was there any actual burning in the lower story?

A. I think there was not. All burned above, except ——

Q. What room did you and Mrs. Clough occupy?

A. Occupied one of the chambers.

Q. Up stairs?

A. Yes, sir.

Q. Was that chamber burned?

A. Yes, sir.

Q. Was the chamber itself burned up? I don't mean damaged by smoke and water?

A. It was burned clear down to the floor—the main body of the house. The L, I think, was burned. The main body of the house, or the back part of the main body of the house, was burned clear down to the chamber floor, and the front was brought down to the first floor I think.

Q. What time in the day did the fire occur? About what time?

A. Three o'clock.

Q. Three o'clock in the afternoon?

A. Yes, sir.

Q. Was a large portion of the personal property carried out safe? Was the fire a very slow one? Was it in the ceilings?

A. No, I believe there wasn't anything—very little if anything except on the first floor. We got the furniture out of the parlors.

Q. Your clothing saved?

A. No, sir.

Q. None of it?

A. Nothing only what I had on my back, I think. There might have been an overcoat in the entry; I think there was.

Q. Your wife's clothing saved, any of it?

A. I think there was some saved.

Q. How much?

A. I don't remember now.

Q. Can you give any idea?

A. No, sir; I cannot. It was all damaged.

Q. Did you put in a claim for your clothing under your wife's policy and receive pay for it?

A. I got the insurance on it.

Q. Did you put in your claim for your clothing and receive pay for it?

A. I don't remember. [Objected to.]

Q. Will you state whether you put in a claim for your clothing under your wife's policy, and received pay for it from the insurance company?

A. I don't remember, sir, how that was.

Q. Can you state whether you did or not?

A. I cannot. The policy, I suppose, embraces all the furniture and wearing apparel; I don't recollect how it was. I gave the property to my wife, the real estate and furniture, and the policy was taken out in her name.

Q. Won't you tell me whose property this was? one pair black broadcloth pants, $16, &c., [reading list of clothing.] Won't you tell me whose property that was?

A. I presume it was mine, a portion of it, and my children's.

Q. Then, sir, did you, or did you not put in a claim under your wife's policy, and receive money therefor from the insurance company?

A. I had nothing to do about making up the claim. There was a man that was appointed to appraise the property. I didn't put in any claim myself, that I recollect of, at all.

Q. Was there a claim put in, to your knowledge, and did you receive the money for it, under your wife's policy?

A. I received the money for the insurance, but I have no recollection of putting in the claim.

Q. You recollect that a claim was put in for your claim?

A. I don't recollect anything. I suppose that the insurance policy covered all the wearing apparel. There were people appointed to appraise damages, and my wife went up to Mr. Rolfe's and stated as near as she could as to the damages. She was before him two days, I think. I had nothing at all to do with the appraising of the damage; I wasn't around with the man at all, when the damage was appraised.

*Mr. George* then offered in evidence the policy of insurance, dated August 22d, 1862, issued to Mrs. Eliza Clough. Also the schedule of loss, return and receipt ($5000.)

Q. Mr. Clough, was any of your wife's clothing saved outside of this schedule?

A. I think not; I don't think there was.

Q. Did you buy your wife a fur cloak, sir?

A. I did.

Q. Did you buy your wife a fur cloak about 1860, or the year before?

A. I bought her a fur cloak in 1860.

Q. What did you give for it?

A. Three hundred dollars.

Q. That included in this schedule?

A. No, sir; that was saved.

Q. Of course Mrs. Clough saved the clothing she had on the day of the fire?

A. Yes, sir.

Q. You mean to say that none of her clothing was saved?

A. I don't recollect that she had anything saved but what was damaged; I don't recollect how that was; very little, I think, if any.

Q. Were the clothing and the articles saved carried down to the South Church?

A. They were.

Q. I notice there is nothing put in here for bonnets for Mrs. Clough. There is nothing in the schedule, as I recollect?

A. I don't know whether there was any saved or not.

Q. She had bonnets at that time?

A. I presume she had.

Q. Is your experience that Mrs. Clough's bonnets cost money?

A. Yes, sir.

Q. Did Mrs. Clough have a cloak, a winter cloak, one or more?

A. I don't remember; I don't think—she had a fur cloak? whether she had any other or not, I don't remember.

Q. Mrs. Clough had boots and shoes, as people ordinarily do?

A. Yes, sir.

Q. Any claim in here for boots and shoes?

A. I don't recollect of any; I don't recollect of hearing any read.

Q. Your statement in this application that it is understood that the insurance was to Eliza R. and the personal to George Clough, was without your knowledge and consent?

A. I don't remember how that was. I supposed that the insurance policy made out to either of us embraced the whole of the wearing apparel for the family.

Q. My question was whether that clause in the statement of loss was without your knowledge and consent?

A. I suppose that I must have known it at the time. I seldom make a policy. I always trust to my insurance agent.

Q. This was notice to the company under oath of the application of your wife?

A. I have no recollection of presenting the schedule.

Q. My question is whether or not the statement of your wife, that it was understood that the insurance on the real estate was to your wife, and the personal property to you?

A. I don't seem to remember about that now.

Q. Now, sir, will you state at the time of the fire how many carriages there were. Will you state how many, at the time of the fire—how many horses and how many carriages, and how many harnesses you had?

A. I don't recollect whether I had more than one horse or not. I don't remember whether I had more or not. I had a top buggy I think. I don't know but I had a two-seated covered carriage. I don't recollect; I had a Concord wagon, I think, at that time.

Q. I think your return was on May 1st, 1866. There is one 1864. This is merely horses and carriages owned during the year 1864, without designating the time. I will take them both. Did you, sir, in 1864 have one carriage valued at not exceeding $100?

A. I think so.

Q. What carriage was it sir? What kind of a carriage was it?

A. I don't recollect whether it was a buggy or a Concord wagon. I think it was a Concord wagon that I bought at auction about that time.

Q. Did you have but one carriage valued at $100 and not above $200.

A. I don't recollect now.

Q. Will you say whether you have in your internal revenue return, and in which of them? I believe they are running through the same. The same three carriages are, I believe, Mr. Clough, if I am not in error, are returned up to the time of this suit, and the year following. Now, won't you state what the carriage was that you call $100 and not over $200?

A. I think it was a buggy without a top.

Q. Was it a riding, livery stable buggy?

A. I don't know. I was a buggy without a top, I think.

Q. Was it for other than riding purposes?

A. It was a pleasure buggy.

Q. Now, sir, we will take the third carriage which is valued here at—here are two carriages valued at $200 and not above $300. Won't you describe these?

A. I don't recollect about the carriages. I usually kept a top buggy, and one without a top.

Q. What were the two carriages? [Showing schedule.]

A. I cannot tell.

Q. What were these? [Indicating.]

A. I don't recollect.

Q. Had you a two-horse carriage with two seats?

A. I don't think I had one at that time. I don't remember. I wouldn't want to state.

Q. How many horses did you have?

A. I had one or more. I don't know whether I had more than one at that time or not.

Q. Has there been a time within ten years, a month at a time ——

A. I had a pair of buckskin horses at the time I built my house.

Q. Has there been a time within ten years prior to the time that your property was attached, that you didn't have three carriages, and, on the average, three horses? If so, when was it?

A. I think that I haven't had more than one horse. Perhaps, part of that time I have bought and sold horses to work—team horses.

Q. Did you own a pair of little $750 roan horses that Mr. Dumas afterwards owned?

A. I had them a short time—yes, sir.

Q. What did you do with them except to ride around?

A. I didn't keep them but a short time.

Q. What did you do with them, except to drive them around as fancy horses?

A. I didn't use them much. I only had them about a month.

Q. For how many years did you own the roan horse formerly owned by Plummer Whipple?

A. I owned her a few days, I guess.

Q. Can you tell for how many years you owned a double, two-horse carriage, valued at between $200 and $300?

A. I bought a double carriage in 1857 or 1858 for $250. I sold that carriage in 1861 or 1862, for $400; and I believe I haven't owned a double carriage since.

Q. Can you tell whether you didn't own these in 1864 and 1865 and 1866?

A. I think they were cheap carriages; that is my idea now.

Q. And you put them in between $200 and $300. You call them cheap?

A. They had been used. They were not new carriages.

Q. Now, sir, what was the number of horses, carriages and harnesses that you had at the time your property was attached in 1866—at the time this suit was commenced?

A. I had a pair of buckskin horses and one driving horse.

Q. Any others? Any colts?

A. I don't recollect whether I had two or three carriages at that time; I hadn't any double carriage, since I sold the one I mentioned. I think it was in 1862 I sold it. I haven't had a double carriage I think, since.

Q. Did you have any colts,—any young horses at that time?

A. One.

Q. When your property was attached?

A. I think not.

Q. Quite positive?

A. I don't remember now whether I had.

Q. Did you give Mr. Gilmore a carriage? If so, when?

A. Mr. Gilmore sent up and got a carriage, and I never got it back again. It wan't more than $40.

Q. A carriage which you had in your hands, Mr. Gilmore sent up and got, and you never claimed it afterwards?

A. I said I never got it afterwards.

Q. Did ever you claim it afterwards?

A. Yes, sir; told him to bring it back, but he never did.

Q. Why didn't you go and get it?

A. Didn't happen to.

Q. I want to know why you didn't?

A. I let it go until he died, and then I thought I wouldn't claim it.

Q. When was it?

A. When he died?

Q. No, no, no; when you let him have the carriage?

A. I think it was in 1865; I am not quite sure.

Q. I will ask you how many shares of Concord Railroad stock you have to-day, either in your own name or in the name of other people?

A. I have between 900 and 1000 shares; I believe the number is 974.

Q. What is the present market value of that stock?

A. I believe that the last time I saw a sale it was 72.

Q. How many of those shares stand in your own name?

A. I believe that 400 do.

Q. Then, if I reckon correctly, you have $71,028 worth of stock of the Concord Railroad,—400 shares of which is in your own name. In whose name is the rest?

A. I have 200 shares in the Merrimack River Savings Bank,

Q. (*By Judge Bellows.*) That is, they are standing in the name of that Bank?

A. Yes; pledged there as collateral security for money. I have 200 shares at the Amoskeag Savings Bank, also pledged for money. I have 94 shares pledged to Charles H. Carpenter.

Q. (*By Mr. George.*) Can you state how much they are pledged for in each case, as you go along?

A. In the Amoskeag Savings Bank, $6000; in the Merrimack River Savings Bank, $10,000; to Mr. Carpenter, $4600; making in all $20,600. I believe the remainder is in my own hands. 150 shares has been pledged to the Union Bank, that has never been transferred back to me, although the debt has been paid.

Q. You have accounted for 940 shares You stated that you had 974 shares. Where are the other shares?

A. I am not quite sure whether it was 974 shares in all, or not. At any rate, the balance stands in my name, with the exception of what I have mentioned.

Q. Then there are 580 shares standing in your own name, and in the name of the Union Bank, subject to no liability by pledge?

A. I do not think that I have so many as that. I can ascertain exactly the amount.

Q. Have you 100 shares of that stock standing in the name of Henry P. Rolfe?

A. No, sir; none at all.

Q. Did you buy 100 shares of stock, and was it paid for by you, and the certificate made out to Henry P. Rolfe? Please answer the question as directly as you can, and then explain?

A. Mr. Rolfe had one hundred shares of stock, and I had his note.

Q. Will you answer my question? Did you buy 100 shares of stock, and pay for it, and was the stock made out in Mr. Rolfe's name?

A. Yes, sir.

Q. Now explain if you wish?

A. I had some United States bonds; I pledged them to the Amoskeag Savings Bank, of Manchester, and bought that stock and let Mr. Rolfe have it, and he gave me his note and pledged the stock as security.

Q. When was this?

A. This was some two years ago, I think.

Q. Was it immediately after this suit was commenced, or within a month or two after?

A. It was before the next annual meeting.

Q. This suit was commenced in March, and the annual meeting was on the 29th of May; was it between those dates?

A. The suit was commenced in February, and it was not long after that time.

Q. It was between that time and May?

A. I think so.

Q. You have not said anything about any United States bonds in accounting for your property, have you?

A. I have none now.

Q Did you mention anything about any United States bonds when you were asked what your property was at the time you left the road?

*Mr. Mugridge.* The question implies that he had some at that time, while the fact is that he had none.

*Mr. George.* He has sworn that he had.

*Mr. Mugridge.* He has not sworn any such thing.

Q. (*By Mr. George.*) Did you have any bonds of the United States at the time you left the employ of the Concord road?

A. I had.

*Mr. Mugridge.* He has not sworn so before.

*Mr. George.* He has in his deposition.

*Mr. Mugridge.* My objection was that you assumed what had not been proven in this case.

*The Chairman.* It seems to be now in the case.

*Mr. Mugridge.* It is now; and there is no objection.

Q. (*By Mr. George.*) What was the amount of the United States bonds that you had when you left the road?

A. Ten thousand dollars. When I got my insurance money I put it into United States bonds; and when I bought the stock that I let Mr. Rolfe have, I pledged $6000 of it to the Amoskeak Savings Bank, and with the other $4000 I paid the bills on my house, that I built on the spot where the old house was burned.

Q. You say that you hired some money at the bank and bought 100 shares of stock, and that the stock was taken out in Mr. Rolfe's name. Did Mr. Rolfe then give you his note for the exact purchase money of the stock?

A. He gave me his note for that stock.

Q. For the exact purchase money of it?

A. For what was paid for it, I think.

Q. Was it for the exact amount that you paid for that 100 shares of stock?

A. I think it was at somewhere about 61.

Q. Have you any means of telling?

A. I do not think that I have.

Q. Where is Mr. Rolfe's note?

A. I do not know. I have taken the stock back, and given him his note, I suppose.

Q. You have taken the stock back, and given up his note, and that squared it, I suppose?

A. Yes.

Q. And that stock is included in the 974 shares of stock to which you have referred?

A. Yes.

Q. When was that transaction of giving up the note?

A. I think it was about a year ago.

Q. Had Mr. Rolfe paid you any interest on that note, or had you collected the dividends on the stock?

A. I do not recollect, but I think he paid me the interest on the note.

Q. Who collected the dividends on the stock while it stood in Rolfe's name.

A. I do not know whether he collected them, or I did. If I collected them he gave me an order.

Q. , Can you tell which way it was?

A. I think he paid the interest. The bargain that Rolfe and I made was that he should pay me the same interest that I had to pay at the bank.

Q. What interest did you pay at the bank for the money?

A. I think it was seven and three-tenths per cent. but I am not quite sure of that.

Q. Did you take Mr. Rolfe's note at six per cent. for exactly the amount?

A. I do not know whether I took his note at six per cent., but the agreement was that he should pay me the same interest that I paid at the bank.

Q. What other stock had you? Did you have stock in the name of Col. Tappan?

A. Yes; five shares, I think.

Q. When was that purchased?

A. At the same time.

Q. Did you take Col. Tappan's note?

A. No, sir; I charged it to him.

Q. Where and how did you charge it to him?

A. [Producing account book.] I charged it on this book:—May 27, paid Tappan $25, retaining fee; March 17, paid him $166; May 27, 5 shares Concord Railroad stock, $353.75. It was at 70¾, I think, per share.

Q. What was your object in buying five shares of stock and putting it in Col. Tappan's name?

A. I wanted him to attend the meeting. That was one object.

Q. The annual meeting of the Concord Railroad?

A. Yes, sir.

Q. Did you wish him to attend the meeting as your counsel?

A. Yes.

Q. What other stock, if any, did you buy, that has not been referred to here?

A. I had 83 shares of Judge Minot,—or I had the money of him to buy it with.

Q. Did you pay for that?

A. I had the money of Judge Minot, and the stock was left in the bank as collateral security for the money.

Q. Do you refer now to what is denominated in your deposition as the Shaw stock?

A. No; I think that was separate.

Q. What became of that 83 shares?

A. Judge Minot bought it, subsequent to the annual meeting.

Q. Did you buy 100 shares of stock which was taken in Mr. Shaw's name?

A. Shaw and I bought it together, or we signed the note together, and I left the stock as collateral.

Q. Did you pay for that stock?

A. We got the money at Minot's bank. Mr. Shaw was the principal on the note, I think, and I signed it with him. After that, Mr. Shaw paid the note and took the stock.

Q. Did you not lose on that stock?

A. I think I did some.

Q. How much did you lose?

A. I do not remember now. It was somewhere from $200 to $600 I think, that is my impression now.

Q. (*By Judge Bellows.*) You made that loss yourself, on the Shaw stock.

A. Yes.

Q. (*By Mr. George.*) Can you not tell nearer than that how much the loss was?

A. No, sir.

Q. Did you have anything to do with the purchase of any other stock?

A. I believe that I once bought 12 or 13 shares here at auction. I forget now what auction it was.

Q. When was that?

A. It was about the same time, I think,—before the annual meeting.

Q. How much did you pay for that stock?

A. I do not remember.

Q. State as nearly as you can?

A. I think the stock was selling at that time at from 60 to 65.

Q. Didn't you pay exactly $70.25 for that stock, and was it not stock that was sold by Mr. L. D. Stephens' Administrator?

A. Yes; I recollect that it was, I think I paid 70, or thereabouts.

Q. In whose name was that stock pledged?

A. Mr. Lincoln bid it off. I paid for it, and took Mr. Lincoln's note at that time.

Q. For how much?

A. I do not remember what the note was for.

Q. How much did you lose? and what was the agreement on which it was put into Mr. Lincoln's name, if any?

A. I think I did not sell that stock, but kept it.

Q. What has become of Mr. Lincoln's note?

A. He transferred that stock to me, and I gave up his note.

Q. That stock is included in the 974 shares?

A. I think so.

Q. Is there any question about it?

*Mr. Mugridge.* We should like to know the object of this testimony. We do not see its materiality. If it is material we do not object to it.

*Mr. George.* The whole object of this testimony is to show the object of Mr. Clough's property at that time; at least that is a sufficient object. When I say the whole object, I do not mean to preclude myself from any legitimate and proper use of the testimony; but that is a sufficient object of the showing.

*The Chairman.* I do not see why these enquiries are not relevant. They tend to show the amount of property that Mr. Clough had in 1866, when he left the road. If the evidence is competent in any point of view, that probably is enough to determine the present ruling. Whether it can logically be made to apply to any other matter, will be for consideration when it shall be so offered.

Q. (*By Mr. George.*) What has been the lowest price of the stock of the Concord Railroad from 1860 down to the present time?

A. I do not know as I could state that.

Q. Has the market value of the Concord Railroad stock, since 1860 been less than $60 per share?

A. I do not know whether it has or not.

Q. State according to your best recollection. You have bought it repeatedly from 1860 to the present time, have you not?

A. I bought it before 1860, also.

Q. I am talking of from 1860 down to the present time. During that time has the market value of the Concord Railroad stock varied between $60 per share and $75 per share?

A. It has been working up. I could not tell the prices from year to year.

Q. My question was whether it had varied between those two prices per share. Has it been below $60 per share since 1860?

A. I do not remember.

Q. Have you any recollection that it has?

A. I cannot tell. My impression is that it was lower than that as far back as 1860.

Q. Is there any other stock that you had anything to do with?

A. No, sir; I do not think of any other.

Q. Had you anything to do with giving indemnity to people to buy stock with, or did you make any agreements of that kind?

A. No, sir.

Q. Had you any correspondence or negotiations with any parties with regard to giving indemnity of any sort against loss or purchases of stock, within three months of the time of the annual meeting?

A. I did not indemnify anybody.

Q. Did you have any conversation with any person to procure indemnity to be given?

A. There was a bond executed I believe, and the other conductors signed it.

*Mr. Mugridge.* I think this relates more to the "ring" than to Mr. Clough's property.

*Mr. George.* Suppose it relates distinctly to the "ring;" is it inadmissible upon that account? Suppose it is offered here, as was suggested when the case was opened, with a view to show that after this suit was commenced, there was an attempt to operate upon the board of directors; is it not admissible?

*Mr. Tappan.* We do not object to anything that they may desire to show with regard to this "ring."

*The Chairman.* The question that occurs to us is, whether an investigation as to that matter would not be taking more time than is necessary.

*Mr. Tappan.* We leave that to the referees, entirely. We rather court investigation as to that matter.

*Mr. George.* You shall have it; we will put in Mr. Clough's deposition when the proper time comes.

Q. (*By Mr. George.*) State whether or not, at the time the other conductors were sued by the Concord Railroad, and their property attached, you executed bonds of indemnity against loss, to induce people to buy stock; and if so, at whose suggestion you did it.

A. There was a bond of that kind executed. It was suggested by Mr Gilmore.

Q. Who gave the bond?

A. George Noyes, Joe Langley, Jim Jones and Jim Eaton signed the bond.

Q. They were the other conductors?

A. Yes, sir.

Q. Were they all sued?

A. I believe so.

*Mr. Mugridge.* If this is to be regarded by the referees as material, we want the bond itself produced.

*The Chairman.* If you make that as an objection, undoubtedly the bond must be produced.

Q. (*By Mr. George.*) I will now ask you if you had an interview with Mr. Gilmore, with regard to procuring a bond of indemnity. If so, state when it was, and what was said, and where it was?

A. Mr. Gilmore spoke to me, at his house I think, in the spring of 1866, after the suits were commenced.

Q. State what that interview was, and who were present?

A. I think he was alone, if I recollect rightly. I told him that there were parties who would buy stock, if they could be assured of not losing anything. He wanted to know who they were, and I told him that James W. Johnson would buy some, and that James R. Hill would also buy some.

Q. Do you mean Mr. Hill, the father-in-law of the son of the conductor, and the father-in-law of the young man who actually run as conductor on the Lawrence road?

A. Yes. There were some 400 or 500 shares bought, I believe.

Q. Tell what Gilmore said to you, and what you said to him?

A. I do not recollect any further conversation at that time with Mr. Gilmore. He said that he would send for those men, or that he wanted that I should see them, and see if they would stand in for any loss.

Q. What men?

A. The conductors.

Q. Did you see them?

A. I did.

Q. Did they agree to stand in for the loss?

A. They did.

Q. How was that loss adjusted? How was it paid, and how much was it?

A. I do not know anything about it. I had nothing to do with the signing of the bond, and do not know how it was settled up.

Q. Have you once sworn that that loss was adjusted by the conductors giving their notes, immediately after the annual meeting?

*Mr. Mugridge.* If he has sworn to it in his deposition,—read that.

Mr. George read from Mr. Clough's deposition, the following:

"About a week before the annual meeting of the road, Gilmore wanted me to buy 100 shares of stock with him. I refused. He said that Parker of Manchester would let us have money. I said I would not buy it with him, but would take 50 shares separate from him. It was bought and fifty were set off to me, but never were transferred, and I never heard anything more from it. I agreed to stand in for the loss on it, and that was all."

Q. By whom were said shares bought, of whom taken, and in whose name?

A. That I do not know. They were bought, I understood, of Parker, who was going to furnish the money. All that they wanted was to have somebody stand in for the loss if the stock should drag after the meeting.

Q. Who do you mean by "they"? State the names of all referred to?

A. I referred to Parker.

Q. Didn't Gilmore also want some one to stand in for the loss? and didn't you have frequent consultations with him about that time as to the purchase of stock?

A. I did.

Q. Did you so swear in your deposition?

A. I think I did. I recollect now more distinctly about it.

Q. How many interviews did you have with Governor Gilmore, between the time of the commencement of the suits, and the annual meeting; and what was the subject matter of said interviews?

A. I used to go down to his house occasionally. I believe that I never met him anywhere else. He said that he wanted to turn out enough of the old directors to get rid of John H. George, the clerk; that he was the man that made all the trouble on the road.

Q. You wanted to do the same thing pretty bad, didn't you?

A. Yes, I did. It would not have hurt my feelings at all to have seen them turn out a man that had abused me for the last seven or eight years, as you have.

Q. Will you state whether you were in consultation with Gilmore from that time up to the time that there was an application made to the legislature to investigate the affairs of the Concord road?

A. No, sir; I do not think that I had anything to do with Mr. Gilmore about the legislature. I have no recolection of it now.

Q. How often were you in consultation with Mr. Gilmore between the time these suits were commenced, and the time of the annual meeting; and how much stock did you arrange with him to buy to control that meeting? State how many shares of stock were purchased with a view of controlling the annual meeting, by yourself, in connection with Mr. Gilmore, and the other conductors?

A. I think I voted on from six to eight hundred shares; I do not recollect exactly.

Q. That was what *you* voted on. How much was there bought by the combination?

A. Mr. Johnson had, I think, 250 shares; James R. Hill, I think, had 150; Sam Bell had 100; Messer had 200, I think, or at least I heard that he had; I do not know the fact; Mr. Parker, the present treasurer had 200 or 300 shares, I understood. I do not know the fact however, but merely state what I heard that he had at the time the ring was formed.

Q. Was an application made to the legislature for the appointment of a committee of investigation with regard to the affairs of the Concord Railroad, in the June following?

A. I think there was.

Q. Did that application pass the House of Representatives, three hundred to three or four?

A. I believe it passed the House and Senate both.

*Mr. Tappan.* If this matter is gone into, we should be permitted to to explain, on our side, just what was done.

*The Chairman.* It is not obvious to us how it can be gone into unless you permit it.

*Judge Bellows.* We have come to the conclusion that as neither party objects to the introduction of evidence of this character, we will not, on our own motion, undertake to exclude it; but at the same time, so far as regards any action before the legislature, we are unable to see how that has any direct bearing upon this question. So far as regards any combination between Mr. Clough and Mr. Gilmore, to stifle this suit against Mr. Clough, or anything that goes to show any collusion between them in respect to this transaction, we think it would be admissible, legally; but we are unable to see anything in the suggestion that evidence about the proceedings before the legislature, would bear upon that question, at present. We think that an enquiry into the general affairs of the Concord Railroad, before the legislature, would not have such a bearing upon this suit, as would make it a matter of any special importance or materiality. Although we should not undertake to say that the counsel might not go into that, and offer evidence in relation to it, we should expect the inquiry to be confined to matters which distinctly show the connection between the defendant here, and Mr. Gilmore, with a view to putting an end to these proceedings, and bearing directly upon these proceedings; and we should listen pretty impatiently, as I think, to any testimony that was confined to the general subject of the condition of the Concord Railroad, or the way in which it has been managed, and that did not bear especially upon this case. I think that is the general view we have arrived at. It is pretty evident to us, from what has been said, that the admission of testimony as to what has been done in the legislature, opens a pretty wide field, and would be likely to be entered into by both parties with a good deal of zeal, while it would afford but little, and perhaps no light upon the subject before us, as we now understand it. With these suggestions we are disposed to allow the counsel to proceed, and offer such evidence as they may suppose will bear upon the questions more immediately before us.

Mr. George desired leave to put in at this time, the testimony of certain witnesses in behalf of the plaintiff, who were present. Which was objected to as being an interruption of the proceedings.

Leave was granted by the referees.

Col. J. S. Kidder was called in behalf of the plaintiff, and duly sworn, and testified as follows:—

## TESTIMONY OF J. S. KIDDER.

Q. (*By Mr. George.*) Were you one of the directors of the Concord road? If so, what years?

A. I think I was elected in 1853 and was a director until 1866—thirteen years.

Q. You may tell about how far the directors were knowing to Mr. Clough's practice of taking money from the passenger in the car and purchasing a ticket with it, which he returned to the office?

A. I have no knowledge myself, or individually, nor of the board of directors.

Q. When was the first time you ever heard of such a thing?

A. It was about the time that the prosecution was commenced against Mr. Clough, I learned that there were tickets.

Q. When was the first time that you ever heard that this was the way money taken in the cars and returned on the way-bills was accounted for?

A Not until to-day.

Q. How was it with regard to prior to this prosecution, prior to the discharge of Mr. Clough, about yourself individually or the board of directors, so far as you have any knowledge, having any notice that Mr. Clough or any of the other conductors were selling tickets that had been once used?

A. I never had any knowledge of it.

Q. So far as you knew, how was it with the board of directors?

A. So far as I knew the board of directors never heard of its existence.

Q. When was the first time that that fact of tickets being sold that had been once used, was called to your attention?

A. It was about the time that the prosecution was commenced against Mr. Clough.

Q. You may state, if you please, the circumstances briefly?

A. There was a gentleman informed me— [Objected to.]

Q. From information given you what did you do?

A. I employed a man to go to an individual in Manchester, and buy some tickets,—a gentleman that was in business, that had nothing to do with the railroad in any manner that I know of. He bought me a ticket over the Concord Railroad from Manchester to Lawrence.

Q. (*By Mr. George.*) Will you now state what you did with the ticket—of whom it was purchased, and what you did with it?

A. Of whom it was purchased?

Q. Yes.

A. It was obtained from Mr. Weston.

Q. Were you there when the ticket was purchased?

A. I was not.

Q. Did you put your initial there?

A. Yes sir.

Q. What did you do with the ticket?

A. I passed it over to Mr. Gilmore.

Q. And what time was that with reference to our finding the tickets on Mr. Whitcher?

A. It was a short time previous? I forget how long. It might have been a week before; I think it was some days before. It is my impression. I do not know as it was more than two or three days.

Q. Now, sir, I will ask you how it was about you, or the board of directors, so far as you had knowledge, having any knowledge whatever, or information, with regard to the sale of tickets of the Concord Railroad which had been once used.

*Mr. Mugridge.* I object, so far as any information through other directors is concerned.

*The Chairman.* I think that it is possible that there might have been a meeting of the board of directors, and this communication might have been made to them in the presence of all of them. When Mr. Kidder is asked of his knowledge, he can say whether he did or not. It is possible that there might have been information or communication to them all together.

Q. (*By Mr. George.*) Whether prior to this you had any knowledge, or the board of directors, as far as you are aware, had any knowledge that tickets of the Concord Railroad, that had been once used, were sold?

A. I had no knowledge of it.

Q. Immediately after this was a meeting called of the board of directors?

A. I think there was.

Q. Were you present?

A. I was.

Q. Mr. Kidder, I want to ask you if you had any knowledge, or the board of directors, so far as you are aware, had any knowledge that the conductors were selling coupon tickets over their road?

A. I had no knowledge of that fact; don't know that the directors had any.

Q. I was speaking of the board, so far as you are aware.

*Mr. Mugridge.* We object on the same ground.

Q. Was there a suit brought against the Concord Railroad while Clough was on the road for Mr. Horace Johnson being put off the cars, for using one of these tickets. [Objected to.]

*The Chairman.* I don't think any more testimony is wanted to prove that.

Q. Mr. Kidder, were you present at the meeting of the board of directors, held at Boston, October 26th, 1865, when this vote was passed? I will ask whether when that vote was passed I was present? What representations were made, and who made them?

A. My impression is that you were not present.

Q. What representation was made?

A. The representation was made by Mr. Gilmore.

Q. What was it?

A. The conversation ran upon the conductors and their taking money that they did not return as they should have done. And Mr. Gilmore made the statement that there was no doubt that the conductors stole fifty thousand dollars a year—or less than that; more than fifty thousand dollars, I think, was his statement.

Q. State whether or not that was the occasion for the passage of that vote?

A. That was.

Q. Do you know whether he said anything about Mr. Clough particularly? [Objected to.]

Q. Now, sir, you say I was not present. When did I reach that meeting?

*Mr. Tappan.* We don't object to that question.

A. The train gets into Boston about half-past one, or quarter-past.

Q. Do you remember how that was about the directors repassing the votes so that I could record them?

A. After you got there—my impression is that Mr. Spalding was there first, and that the votes were repassed after you got there, so you could record them.

Q. What did he say about Mr. Clough? [Objected to and objection withdrawn.]

A. He suggested that Mr. Clough was the ring-leader; and if he was out of the way the other conductors would probably be more honest.

Q. I will now ask you the question whether you ever heard him make any representation to the board of directors of the Concord Railroad prior to the commencement of these proceedings?

A. I don't know that I ever did.

Q. You spoke about the rules, Mr. Kidder. The rules have been already put in evidence, 1865 and 1864, and so on. What was the occasion of the repassing of these rules, and making them more stringent?

A. It was in order to make the conductors more honest, and in order to check them as far as possible from taking money, and in order to have people buy their tickets at the office, rather than pay in the cars.

Q. Do you remember the rules which were passed, and that the conductors were required to sign them on the back of the rules?

A. I recollect there was something passed of that kind.

Q. I want to ask you who presented these rules to the board for adoption—at whose suggestion they were adopted, if you recollect?

A. My impression is that they were presented by Mr. Gilmore.

Q. Mr. Kidder, I want to ask you if you had any knowledge, or, so far as you knew, the board of directors were aware that Mr. Clough had ever carried chickens and poultry, etc., over the road?

A. No, sir.

Q. When was the first time you heard of that?

A. To-day.

Q. Whether you knew, or the board of directors knew, so far as you know, that he had a boy on the cars peddling peanuts?

*Mr. Mugridge.* We raise the same objection to this, as to what has been Col. George's connection with the matter.

Q. I want to know, sir, whether you ever knew, or whether the board of directors so far as you know, knew of Mr. Clough's employing a boy to go on the cars and sell peanuts.

A. I did not.

Q. When was the first time you ever heard of that?

A. To-day.

Q. So far as you know, were the rules—these rules which I have shown to you, these last rules here, and the other rules preceeding—ever annulled, waived or modified, except by the substitution of one rule for another?

A. No, sir.

CROSS EXAMINATION.

Q. (*By Mr. Mugridge.*) Did you understand that these were not infringed upon or broken over by anybody that they were intended to apply to?

A. No, sir; I believe I did not make that statement.

Q. Haven't you broken these rules?

A. Not often.

Q. Do these rules forbid you as one of the directors from passing persons upon the cars?

A. Yes, I think they do?

Q. For the last five years that Mr. Clough was upon the road as conductor, did not you in numberless instances direct him to pass men over the road?

A. I think I did not after these rules were passed here.

Q. Do you mean that Mr. Clough did not after 1864, in numberless instances pass men ever the Concord road at your suggestion?

A. My impression is that he did not, sir.

Q. Did he not in any instance?

A. Yes, sir; he has.

Q. Didn't you know, when you gave that direction to Mr. Clough that you were acting in defiance of a rule that you had helped to make?

A. I think that I never passed anybody or asked him to pass anybody after these rules were passed.

Q. Do you mean to state that you did not, after 1864, and before Mr. Clough left the road, order him to pass people in the cars, in defiance of the rules?

A. Yes, sir, that is my impression.

Q. And you state that in no case after 1864, did you order him to pass any one over the road.

A. My impression is that I did not.

Q. Are you quite confident?

A. Yes, sir.

Q. And you mean to have the referees take that as positive recollection on your part?

A. Yes, sir.

Q. Did you, sir; before these rules were passed direct Mr. Clough to pass people over the road?

A. I have done it.

Q. Did you have a right to do it, under any existing rules at that time?

A. I supposed I had. I took that liberty at any rate.

Q. Didn't you know as director of the Concord Railroad, whether you had or not?

A. I supposed I had the right, and I took that right occasionally.

Q. Do you mean that it is uncertain in your mind whether you had a right to give these passes or not?

A. I supposed, when I gave the passes I had a right to do it.

Q. Didn't you understand at that time whether or not you had a right?

A. Well, I supposed I had a right.

Q. Will you show a rule of the road that gave you that right?

A. I don't know that there is.

Q. Don't you know that ever since you have been a director, there has been a rule expressly to prohibit that thing, and havn't you yourself voted for these rules?

A. I presume that I voted for most of the rules that were passed.

Q. Since you have been a director of the Concord Railroad, hasn't there always been a rule that prevented directors from passing persons in the cars? How is that, sir?

*The Chairman.* If I understand it, the rules have been put in the case. The question therefore is not to prove the rules, but whether he remembers or not.

Q. Let me ask you, sir, if you did not understand when director of the Concord Railroad, that there was a rule in force on the road which prohibited a director from passing a person on the road?

A. Well, I could not say previous to this; I could tell by examining the rules that were passed.

Q. You mean to say as illustrative of the knowledge that you have of the rules, that you don't know whether such a rule had been passed or not?

A. I don't remember particularly.

Q. If there was such a rule as that passed did you help make it?

A. Yes, sir.

Q. Didn't you do it in open violation of the rules?

A. I guess I did.

Q. Have you taken your family with you over the Concord Railroad, and have you been in the habit of doing it ever since you have been a director?

A. I haven't always.

Q. Have you sometimes?

A. Yes, sir.

Q. Had you a right?

A. Yes, sir.

Q. Where did you get your passes?

A. I had a pass over that road—I had one of Gilmore and one from Stark, and one over the Boston and Maine road and several others.

Q. Have you ever taken friends to Boston, and passed them by word of mouth?

A. I have.

Q. Had you a right to do this?

A. I supposed I had at that time.

Q. Then your knowledge of the rules of the Concord Railroad led you to believe that you had a right to pass persons in that way?

A. I supposed I had.

Q. You are a flour dealer I believe?

A. Yes, sir.

Q. Have you been accustomed to send your runners, who sell flour, over the Concord Railroad on your pass?

A. We don't have any runners.

Q. Have you ever sent any man in your employ over the Concord Railroad on your pass?

A. I think I have.

Q Did you think you had any right to do that?

A. I supposed I had.

Q. Now, I want you to read from that book of 1865, and say who passed that vote [designating]?

A. Well, I don't know that I could tell positively. My impression is that Judge Upham did.

Q. Will you state that Judge Upham did do that?

A. No, sir.

Q. Will you state that John H. George didn't do it?

A. No, I won't. It might have been before the meeting of the directors. My impression is that it was Judge Upham.

Q. But still you won't say that George didn't do it?

A. No, sir.

Q. Have you any positive recollection in regard to it, any way.

A. With regard to the writing?

Q. Yes, with regard to the writing. Who wrote that note?

A. No, sir, I have not.

Q. Now, I want to ask you if there is anything in that record to indicate that Mr. George was not present from the time the meeting was called, up to the time it was closed?

A. I don't know that there is.

Q. I suppose that in his absence, the clerk *pro tem.* would have made it out?

A. Sometimes he would and sometimes he would not.

Q. Wouldn't that appear on the record?

A. Yes, sir.

Q. Wouldn't an accurate, truthful representation of the fact appear there?

A. Not necessarily. If there was a clerk *pro tem.* chosen, and before the meeting dissolved, Col. George came, and the votes repassed, I suppose it would be proper for him in this case.

Q. Now, sir, you say that Mr. Gilmore made a statement in regard to the conductors stealing $50,000. Did ever Mr. George make that statement to you?

A. I don't know about the $50,000.

Q. Hasn't Mr. George, time and again, represented to you that the conductors were stealing large sums from the Concord Railroad? And didn't he, about the time that vote was passed?

A. I have frequently heard him make the suggestion that he had no doubt the conductors were stealing.

Q. Have you frequently?

A. Occasionally; I don't know as frequently, or not. It was talked over at our meetings very frequently.

Q. Let me ask you if he has not made these representations to the directors of the Concord road prior to the passage of this vote frequently?

A. I think very likely he may have made that statement, or that the conductors were taking money that did not belong to them, and did not return it.

Q. And has he represented to the board of directors that these amounts were large amounts, and that Mr. Clough was the chief thief?

A. I don't know that he has.

Q. Hasn't he discriminated between Mr. Clough and the other conductors?

A. I don't know as he has.

Q. Let me ask you if Mr. George ever used to take pieces of paper out of his pocket to elucidate the fact that Mr. Clough had taken $200,000 of the funds of the Concord Railroad?

A. I don't know.

Q. Didn't he ever take pieces of paper, and argue that they showed a certain result, which was that Mr. Clough had stolen $200,000 from the funds of the Concord Railroad?

A. I don't think that he ever did, until after the suits were commenced.

Q. I am talking about that time. Didn't Mr. George, after the suit was commenced, pull from his pocket slips of paper, and show you certain figures upon them, which he said demonstrated conclusively the fact that Mr. Clough had stolen $200,000?

A. I don't know that he did.

Q. Did he ever show you anything at all about that?

A. I have no recollection of it.

Q. Will you say that he never showed you anything in regard to that?

A. I say I have no recollection, unless he made it to the board of directors after this thing commenced.

Q. Now, didn't Mr. George, at the meetings of the board of directors, produce certain figures, and certain statistics which he said to the board of directors demonstrated conclusively the fact that Mr. Clough had been stealing $200,000?

A. I don't know as he did.

Q. Did you ever hear him give that statement?

A. Not in my hearing.

Q. What did you mean by saying that he never did, "unless he did it after these suits were brought?"

A. I don't know but he might have made some representation. I don't recollect about it.

Q. What statement might he have made in that direction?

A. Well, I don't recollect myself.

Q. Will you state, sir, that he did not make such statements?

A. I would not say that he did not.

Q. Now, I want to ask you where the board of directors held their meetings in Boston?

A. They sometimes held them at a room on Tremont Row, when I first became a member of the board; and afterwards they held them at the Revere House sometimes, and American House sometimes.

Q. How much did it cost the old board of directors for liquors drank at the Revere House and other places where they held their meetings during the last year?

A. I have no idea, sir, at all.
Q. Did it cost them more than $1500?
A. I could not tell.
Q. Will you state that it didn't cost them more than $1500?
A. They must have done a good business.
Q. How much was thus paid?
A. I have no knowledge; I never saw any of the bills.
Q. Mr. Kidder, I want your best information as a director of the road, if it didn't cost more than $1500 for *rum?* I am not talking about the dinners, but for the rum alone, during the last year?
A. I am not able to swear anything about it. I will give you my opinion. My opinion is that it did not.
Q. How much is your opinion worth?
A. I don't know.
Q. Let me ask you if the occasion didn't happen many times, when they held a session, that they were four-sevenths of them so drunk that they didn't know what they were about?
A. No, sir.
Q. Now, I understand you to say that the object in passing these rules was to keep the conductors honest? Was that so?
A. I don't know as I made that statement exactly. It was similar to that; might have been exactly that. To prevent them from taking the funds that belonged to the road, and appropriating it to their own use; to induce people to buy their tickets at the office rather than buy them in the cars?
Q. Let me ask you, sir, if Mr. Gilmore ever gave you any tickets on the Concord Railroad?
A. He never did to my knowledge. I don't think he ever gave me a ticket.
Q. You are quite positive of that?
A. I am quite positive.
Q. Did you ever ask Mr. Clough to pass persons?
A. I never have, sir, excepting when I have been in the cars.
Q. Have you in those cases?
A. I have, sir.
Q. That is when you have been aboard a train and had a friend, and Mr. Clough was conductor, and that friend has had no ticket?
A. Yes, sir.
Q. (*By Mr. George.*) Prior to 1864?
A. My impression is that I haven't since that vote was passed.
Q. (*By Mr. Mugridge.*) Is your impression so strong that you will swear to it?
A. That is my impression.
Q. (*By Mr. George.*) In how many instances did ever you do it in your life? A few or a good many? Just give the best idea you can to the referees?
A. Well, during my directorship I might perhaps have passed a hundred; perhaps not more than fifty; perhaps not so many as that.

## TESTIMONY OF GEORGE CLOUGH.

### CROSS EXAMINATION RESUMED.

Q. (*By Mr. George.*) I will ask you if your counsel, Mr. Rolfe, Mr. Mugridge, Col. Tappan, and Mr. Bell,—I will ask you if your three counsel were up to the legislature to prevent an order for investigation, Mr. Mugridge, and Tappan, and Rolfe, all three present, and yourself?

A. To prevent the passage of what?

Q. The resolution appointing a committee of investigation?

A. I believe they were employed for that purpose by the new directors.

Q. Were they all present?

A. They were not my counsel for going before the legislature.

Q. They were your counsel in this suit at the time?

A. They were.

Q. Now, sir, did you have interviews with Mr. Gilmore, or did your counsel have interviews with Mr. Gilmore, with regard to that matter?

A. Not to my knowledge.

Q. Have you stated to any one that Mr. Rolfe, your counsel, was employed by Mr. Gilmore, and that he agreed to give him $400?

A. No, sir.

Q. Made any statement of that kind to any one?

A. No, sir.

Q. At that investigation, at the state house, will you please state whether I read before the committee of the senate, where the hearing was, in your hearing and presence, the returns of conductors on the Portsmouth road, and the Manchester and Lawrence road, and the Concord road.

A. I don't think I was present at the hearing before the committee.

Q. Were you not up in the third story?

A. No, sir; I was not in the senate room at all.

Q. In the room over the senate room, up where the hearing was?

A. No, sir. I was not at any of the hearings.

Q. Were you present that night at the state house?

A. I was, but I didn't go into the room where the committee was.

Q. Mr. Clough, I want to ask you a question with regard to 50 shares of your stock. I want to know if 50 shares of stock is in the name of Charles H. Bartlett?

A. Yes, sir; I was going to correct that.

Q. Charles H. Bartlett of Manchester?

A. Yes, sir.

Q. How happened it to be put in his name?

A. I borrowed some money of him.

Q. That is 50 shares in addition?

A. Yes, sir.

Q. What did you pay for that stock?

A. That stock I have owned for some time.

Q. How much did you pay for it?

A. I don't recollect.

Q. State as nearly as you can?

A. I can't tell anything about it.

Q. When did you put it in Bartlett's hands?

A. I put it in Bartlett's hands—I have got the date here in my pocket. [Refers to memorandum.] This don't make but 974 in all. November 2d, 1868, I borrowed $3000, and pledged 50 shares of Concord stock.

Q. What did you do with that stock?

A. I paid my debts with it.

Q. (*By Mr. Tappan.*) When was that?

A. November 2d, 1868. I owed my brother some money, and I sent him up $2000 of that that I borrowed of Mr. Bartlett.

Q. That is all the Concord Railroad stock you have got in any body's name?

A. I think so.

Q. Don't you know?

A. That is all.

Q. What became of the shares that you put in Col. Tappan's name.

A. I have got that charged to him.

Q. That is not reckoned in the amount?

A. No, sir.

Q. You have got that charged to him on your book?

A. Yes, sir.

Q. Now, sir; on that 974 or 990 shares, as the case may be, I want you to put down how much you owe on it—how much incumbrance there is on it?

*The Chairman.* He has stated that, has he not, already.

Q. Then so far as that stock is concerned, you own, subject to a pledge of $23,000. Now, sir, what notes have you got of any persons or bonds or securities?

A. I will furnish you with all my notes, if you will give me time. I will bring them down this evening or to-morrow morning.

Q. Haven't you any means of stating the notes, bonds, and securities that you have?

A. I have no bonds.

Q. Now, sir, your house on Warren street, that house is how far from Judge Minot's house?

A. Just a little ways above.

Q. Mr. Fletcher lives a little this side; any house between?

A. One house.

Q. What did your house cost you to build it—this new house—just the building itself?

A. I believe it was between $12,000 and $13,000.

Q. Made any expenditures since?

A. No, sir.

Q. Haven't you enlarged the house?

A. No, sir; I put on some wings.

Q. Didn't you state in your deposition that it cost you $15,500?

A. I mean that the furniture and all cost me $15,500.

Q. Did you keep any account of that?

A. I had the bills.

Q. Mr. Kilby built it?

A. Yes, sir.

Q. Is it the finest house in Concord?

A. I don't know; it suits me very well.

Q. What did your furniture cost you in your house?

A. Well, sir, my furniture, aside from the furniture that I bought in the first purchase, didn't cost but little.

Q. You mean the furniture aside from the furniture you saved at the fire?

A. No, sir, I mean the furniture that I bought when I made the first purchase of the house that was burned. It was very well furnished then.

Q. You saved that from the fire didn't you?

A. Some of it; not all.

Q. Take your parlor carpet, what did you give for that?

A. That carpet was in the first purchase of the house. The first house, every room in it was carpeted. The parlor was furnished; the parlor chamber was furnished. That was in the first house. I paid $8000 for that.

Q. (*By the Chairman.*) That is the house that was burned?

A. Yes, sir.

Q. (*By Mr. George.*) Did you build on the same place?

A. I did.

Q. Did you save that carpet?

A. I did.

Q. That is on your present room?

A. Yes, sir.

Q. What is the carpet in your library, or reading-room?

A. I don't recollect. In the library, I believe that was saved; I think it was. It is a small one.

Q. Will you state, sir, whether or not at the time your personal property was attached you procured receipters for it to the amount of $9000?

A. I procured receipters; I don't know to what amount.

Q. Wasn't the amount $9000?

A. I don't know.

Q. You don't know whether it was or not, or what it was? Well, you have got a piano; what did you give for the piano?

A. Two hundred and fifty or three hundred dollars.

Q. You don't know how much, whether $250 or $300?

A. I don't know which.

Q. Mr. Clough, how much did you expend for clothing in the year 1865. Take that identical year, how much did you spend for yourself and your family?

A. I don't know.

Q. You can tell, when you sold a horse twenty-five years ago, whether you sold it for $125 or $150; but you can't tell how much you paid for clothing in 1865. [Objected to as argumentative in form, and improper.]

*The Chairman.* I suppose the argument cannot very well be taken back.

Q. I want to know if you can give the referees any idea of what it cost you for clothing in 1865?

A. Well, I should think it might be a pretty good bill, that being the year when the house was burned.

Q. How much do you think it would cost in 1865, as your estimate?

A. For myself alone.

Q. No, sir, for yourself and your family?

A. I can make an estimate. It might be $700 or $800, and perhaps—

Q. Perhaps how much more?

A. I cannot tell.

Q. Can't you form any idea? Will you state whether or not you presented in your deposition $1000 worth as assets not paid?

A. It may be so. I am inclined to think that is so.

Q. Now, sir, do you own the Sherman Hotel?

A. Yes, sir, I have a deed of it.

Q. When did you buy that, and how much did you give for it?
A. I swapped a farm up in Orange, and my interest in the Milford springs.
Q. What did you call the value of the Sherman Hotel at the time you made the bid? Bid it off at auction, did you not?
A. Yes, sir.
Q. How much?
A. Eighty-five hundred dollars, I think it was. If I recollect right that is what it was. And on that I owe $3000.
Q. Then you own one-half the Masonic Temple on Pleasant street?
A. Yes, sir.
Q. And one-half of the Allison property on Main street, adjoining the Masonic Temple?
A. Yes, sir.
Q. And then you own the Masonic Temple, between there and Judge Bellows?
A. I own half of it.
Q. Now, sir, I want to know if ever you kept, or have got anywhere, bills of the items of the expense of all these buildings?
A. Yes, sir; Mr. Corning has, I suppose, at his house. He kept the bills.
Q. He kept the bills of the cost of those buildings, did he?
A. Yes, sir.
Q. When did you see them?
A. I haven't seen them since the time it was built; I believe not.
Q. How much did it cost to build the addition to the Masonic Temple, on Pleasant street, which was built subsequent to the building of the Temple itself—those two stores on Pleasant street?
A. It may have cost $6000, I think.
Q. Do you know anything about it?
A. I took off a statement from my book at the time my deposition was taken.
Q. What book?
A. That Mr. Corning kept—kept the rents on.
Q. Can you produce that book here, sir? Do you mean to say that that book contains a statement of the cost of these two buildings?
A. I think so.
Q. Are you certain that it does?
A. [Refers to paper—no answer.]
Q. What did you give for the Allison property?
A. Thirty-eight hundred dollars.
Q. What did you do with it? Did you lay out any expense on it, build a building on it, or alter the building or make any expense on it?
A. Made some repairs on it.
Q. Made some alteration on it?
A. Yes, sir, I think I did.
Q. How much did it cost you to alter the building?
A. I don't know. Corning had the bills.
Q. Did you rebuild that building?
A. No, sir; didn't lay out but very little on it.
Q. How many stores are there on Pleasant street, towards Judge Bellows'.
A. Both buildings?
Q. Yes, sir. How many stores are there?

A. Three now.
Q. How much did you say the three stores cost you ?
A. That building cost, I believe, $6000.
Q. Have you any personal knowledge of what that cost ?
A. I had it taken off of Mr. Corning's books.
Q. How much did you pay out for the Masonic Temple ?
A. I think we let it out by the job. It cost $12,000 or $13,000. It was built when materials were very cheap.
Q. Do you know anything about the cost of that ? Have you got any items of the cost of it ?
A. I think I have got the specifications of it at home.
Q. Have you got any items of the cost of that building ?
A. That would be the cost, the amount to be paid.
Q. Have you got any items ?
A. I have the specifications. We made our proposals and the lowest bidder took it, and I think the sum was between $12.000 and $13,000, and on the brick store beyond.
Q. (*By the Chairman.*) Is that the cost of building the Temple. or the cost of the building and the land ?
A. The cost of building the Temple.
Q. Don't you own the land ?
A. Yes, sir.
Q. (*By Mr. George.*) Can you bring your contract showing the cost of the Masonic Temple building ?
A. I will, sir ? I believe I have got it.
Q. You think it was $12,000 or $13,000 ?
A. That is my impression.
Q. If the building beyond cost $6000, that would be $18,000 or $19,000 ?
A. Yes, sir.
Q. Then if the building this side cost you $3800, that would be some $23,000 ?
A. That is for the whole.
Q. Then there would be your repairs ; and then there is the land. Now, sir, we will go to the next piece of property. Then you own the "Page place," do you not ?
A. Yes, sir.
Q. That is up by the asylum ?
A. Yes, sir.
Q. How much land did you say there was there ?
A. About an acre and a half.
Q. A small house, was it ?
A. Yes, sir.
Q. What did you say you paid for that, and when ?
A. I bought that in—bought a portion of it of Col. Grover, who was executor of the will of Col. Page, and a portion of it of the son of the widow. I think the land and the house cost me $1800 ; and my impression is, I laid out some $200.
Q. And you say that a man lives on that without rent ?
A. Yes, sir ; I haven't got a cent of rent out of him. He is a poor man.
Q. Then you own a house on Railroad square ?
A. Yes, sir ; a double house there.
Q. That double house which faces about half way between the bridge ?

A. Between the bridge and the freight depot. I have lived there twelve years myself.

Q. What other estate have you? You have the Jamaica Plain estate near Boston, to which you have testified?

A. I have a house on High street, near the Tom Tandy house, which cost $300. Then I have another small house that cost me $200—a small house with two rooms in it, that I bought a year ago.

Q. Well, sir?

A. Then I have got two lots that I bought about two months ago, down on South street. I moved a building on to one of them, the building that we call the Brown house, that I moved from opposite the brick stores there,—opposite the Pettingill house.

Q. What did you pay for that?

A. I paid $200 for one lot, and $250 for the adjoining lot.

Q. You have moved a building down?

A. Yes, sir.

Q. And that cost you?

A. Three hundred and seventy-five dollars.

Q. Have you repaired it, and fitted it up?

A. I got some men to work on it.

Q. How much have you laid out on that?

A. I let the job out for $500. I havn't paid it yet.

Q. Have you paid any portion of it?

A. No, sir.

Q. Any other property sir, or real estate?

A. No, sir.

Q. You have 37,000 feet of land at Jamaica Plain?

A. Yes, sir.

Q. You have no other real estate, or any other interest in real estate now?

A. No, sir, I think not.

Q. The other bonds and mortgages you will bring?

A. The notes I will bring down. There are 75,000 feet of land at Jamaica Plain.

Q. Have you any stock in any corporation except the Concord Railroad?

A. No, sir, I think not.

Q. Have you any bank stock?

A. Not a dollar.

Q. Have you any other personal property except the notes, which you are to bring, and the Concord Railroad stock, and your furniture? If so, specify what. Either in your hands or anybody's else?

A. I have got two horses; I have got a covered buggy, and a buggy that ain't covered, and a Concord wagon. That is all that I think of now.

Q. Go now, if you please, to the Dunbarton farm for a few moments. How long did you own that farm, and how large a farm was it, and what did you do with it?

A. In the first place, when I bought it, there was no buildings on it. I bought it, I think, for $1200, the land. There were forty acres of land. And I let out the job of building a small house and barn on there. And I took the timber off the land. I think I paid $1000 or $1200 for the land; I let the building out, and built a small house and barn, and for the house I paid $400, and I took the timber off the land, and I paid $150 for the barn.

Q. Have you got any memorandum from which you can state what that barn cost?

*The Chairman.* He has just stated that.

Q. My question now is with regard to the cost of the house and barn. Have you any memorandum showing the cost of it, sir?

A. No, sir; I have not. I only state it from recollection.

Q. Did you state, upon your direct examination, that you could not tell?

A. I think I did not state so.

Q. Did you state, upon your direct examination, that you could not tell how much that house and barn cost you?

A. I don't remember it, if I did. I recollect I let the house out for $400; and then I took the lumber off the land. And I think the barn cost $100 and I let them take the lumber.

Q. Did you testify upon your direct examination that you paid $1725 for the farm?

A. That is what it cost me altogether.

Q. That would be $525 for the house, if you paid $1200 for it?

A. I think I didn't pay but ten. I think I paid $200 down, as there was a mortgage.

Q. Now, of whom did you buy the Dunbarton farm? Was it a cash trade or property exchange?

A. It was a cash trade.

Q. Of whom did you buy the Orange farm?

A. Of S. K. Ford.

Q. Was that a cash or a dicker trade?

A. That was a cash trade. I don't recollect whether I paid down at the time.

Q. What I meant was whether you exchanged other property for it?

A. No, sir.

Q. Now, sir, when did you sell the Orange farm and the Dunbarton farm, and to whom did you sell, and how was the trade? Was it for a cash or a dicker trade?

A. That was a cash trade. The Dunbarton farm was a cash trade. The Orange farm I swapped for the Sherman House, and some of my interest in the Milford springs, and I gave $2000 or $3000 to boot.

Q. Have you been to any expense for the Sherman House since you bought it?

A. I let it to a man who does his own repairing inside.

Q. Did you build a brick stable?

A. Oh, yes, I built a brick stable; I thought you meant on the house.

Q. How much did it cost?

A. That shed and stable cost about $3000.

Q. That makes the Sherman House property cost you $11,000 instead of $8000—$11,500?

A. Yes, sir; that is so.

Q. Any other expenditures?

A. At that place?

Q. Yes, sir; or about that place?

A. I don't think of anything. When the house got on fire one time, the insurance paid that and made it good.

Q. So that was no expense to you?

A. No, sir.

Q. Have you laid out anything except that?

A. I think that is all. Mr. Chesley pays the expenses.

Q. Did you make alterations in the house?

A. No, sir; I think I did not; my impression is that Mr. Chesley does his own alterations.

Q. Can you state whether that is so or not?

A. I don't recollect.

Q. Now, sir, is there any other—or we will come to the Orange farm. You sold that farm when? Won't you fix the date?

A. The Orange farm, I swapped it for the Sherman House; I traded for that—it will be, I think, two years—I think I took a deed for the Sherman House two years ago this present month, I think.

Q. That would be in the last part of the year 1866. Now, did you return, in your income returns for 1864–5–6, one dollar of income from either the Orange farm or the Page place?

A. There is no income from the Page place. A man lives in the house—Mr. Elliot—who is not able to pay the rent; and I let him live there.

Q. I want to know whether you did return a dollar of income from either of these three places?

A. I don't know, sir.

Q. Did you ever?

A. I didn't have an income.

Q. Did you swear in your income return, that you did not have any income from these two places?

A. I could not tell. Those are made out by Mr. Rolfe. I have left the whole thing to him, and told him to make it out.

Q. Did Mr. Rolfe, when you made that return, certify that he was the nearest justice to you, and that he had no connection with the business? There is a little mistake.

A. In what?

Q. When Mr. Rolfe certified?

A. I don't know.

Q. Doesn't Judge Minot live nearer to you?

A. Yes, sir.

Q. Doesn't Mr. Fletcher?

A. Yes, sir. I don't know anything about that; I had no hand in it.

Q. This property that you returned was yours,—not your wife's?

A. These insurance policies, I seldom look at after I get them out. I let one man do my insurance. And I supposed all the time that the household furniture and wearing apparel was included, and mentioned it was a matter that I didn't pay particular attention to on the policy.

Q. Then, sir, if you swear in your income returns that you had no profits from farming operations during either of the three years—if you swear to that, it was because somebody else made out your income returns?

A. That is one thing, and probably it was an offsett to the Milford Springs, where I never got anything.

Q. Well, was there a loss at the Milford Springs?

A. Yes, sir; $1400.

Q. What do you mean by a loss of $1400?

A. I lost the interest on what it cost me during the time I had it.

Q. Did you operate the Milford Spring house?

A. No, sir.
Q. Did you make any loss outside of the question of interest?
A. I let that to a man, and I never got anything but the taxes.
Q. Did you furnish him with any means?
A. No, sir.
Q. Did you ever make any interest out of a lease?
A. I don't recollect whether I did or not; I don't think I did.
Q. Whether you made an absolute loss besides?
A. I can't tell; I don't remember.
Q. Make any additions or repairs to the building, or anything of that kind?
A. There was an addition put on, I believe, at that time. I owned a quarter of it for a short time, and then I bought another quarter.
Q. Were there additions made to the building?
A. There was an L put on to it.
Q. How large was the L?
A. I had only a quarter interest.
Q. What was the expense of putting on the L? It was a tavern house, wasn't it?
A. It was a tavern house; it was bought at a very low rate.
Q. How large was the L?
A. The L might be thirty feet long, perhaps, and thirty wide.
Q. How many stories high?
A. Only two, I think.
Q. You paid quarter or half the expense, didn't you?
A. I paid, at that time, only quarter; I owned quarter at that time.
Q. Who did you buy that of?
A. I bought of Enos Blake, I think.
Q. That the original purchase?
A. The original purchase was of a man by the name of Wallingsford. I believe the original purchase was $2000. I don't know but it is $2500.
Q. Is the deed here?
A. I don't know whether it is or not.
Q. The original purchase was $2500 or $2000?
A. It strikes me so.
Q. Have you the means of ascertaining?
A. I have got the deeds.
Q. (*By Judge Bellows.*) Was that for your quarter of $2500?
A. Yes, sir; I think so, I only owned a quarter at first, but I can't state what we did give for it at first.
Q. (*By the Chairman.*) Can you state whether this $2000 or $2500 was the price of the whole or the price of quarter?
A. That is my impression—price of the whole.
Q. (*By Mr. George.*) Then you put on the L while you owned only a quarter?
A. Yes, sir.
Q. Who owned the other three quarters?
A. Mr. Dunklee, Mr. Hoyt, and Mr. Blake.
Q. Mr. Dunklee, Hoyt, and Blake owned the other three quarters?
A. Yes, sir.
Q. What did you pay Mr. Blake for his quarter?
A. I think there was $1000 for each quarter, and for the L, and digging the well and cementing it.

Q. How much did you expend for fixing the spring and fixing the ground?

A. That is all; we merely dug it and set it in cement stone.

Q. How much did it cost?

A. I can't tell.

Q. Did you not attempt to run that house for the parties' interest who owned it?

A. Never.

Q. And you say you sold that to the party of whom you bought the Sherman House, and you put that in, and you put in your Orange farm?

A. And paid $3000.

Q. And you call the whole $8500; that is, you call the Orange farm and Milford springs $5500; and you call the whole $8500?

A. Yes, sir.

Q. Now, I want you to go to the Winnipesaukee steamboat company. Of whom did you buy that stock?

A. Captain Walker.

Q. What did you pay him, cash or property?

A. I swapped real estate for it.

Q. What real estate?

A. A piece of land in what is called the "Eleven lots."

Q. What did you mean by saying that you paid 30 and sold it for 40?

A. I paid 30 and sold it for 40.

Q. Didn't you make a dicker?

A. Yes, we made the price of the stock on the real estate.

Q. What did you give for that real estate? And of whom did you buy it?

A. In the first place, Joe Wyatt bought it, and I bought Wyatt out.

Q. What did that real estate cost you?

A. I think that real estate cost me $2000; somewhere from $2000 to $2500, I think.

Q. Well, do you know?

A. My impression is that it is somewhere in that neighborhood.

Q. Build any buildings on that real estate?

A. Yes, sir.

Q. What building did you build?

A. I built, in the first place, a barn; and I finished that off into a house, and built another barn.

Q. Do you know how much it cost you to build that and finish it off into a house.

A. I think that barn and house cost me about $1200.

Q. I want to know if you have any means of knowing—any memoranda by which you can tell what that cost?

A. I have nothing but recollection.

Q. Did you contract it out, or did you contract it by the day?

A. I think that I found the lumber; I dont know whether by the day, or contracted it out at that time.

Q. How many years ago is that?

A. I should think it was 10 or 12 years ago, and I don't know but more.

Q. Was it the same barn as Mr. Thomas Stewart owns?

A. I think so.

Q. Nice barn. Isn't it one of the best barns there is?

A. Oh, no; it is a very good barn.

Q. If I understand you, you cannot state from any means that you have, what these buildings cost, or what you laid out upon them, except from memory?

A. That is all. I state from recollection, as near as I can recollect.

Q. Have you any memorandum or anything else by which you can state how much that building cost, and how much you paid for work and lumber?

A. I bought the lumber; It was when lumber was very low; I could buy it for $6 or $7 a thousand.

Q. Have you any means by which you can show what that building cost?

A. I don't think I have.

Q. How many acres of land have you there?

A. Eleven or twelve.

Q. How much did you pay for it?

A. I think it was $100 an acre.

Q. You bought the land and then put the buildings up?

A. Yes, sir.

Q. The land you called what?

A. Twenty-five hundred dollars; that is what I think I put it in for.

Q. Did you lay out anything in bringing up that land?

A. I put manure on it somewhat.

Q. Did you improve it—lay out money on the land?

A. I used to, I believe.

Q. What else did you pay towards the steamboat stock?

A. I paid the rest in money.

Q. How much money?

A. I paid the difference; I think I put the farm in for $2500; I paid the difference in money, what the stock would come to at $30 a share.

Q. (*By the Chairman.*) How many shares were there?

A. One hundred and eighty-three.

Q. And you paid the rest in money?

A. I paid the rest in money,

Q. To whom did you sell?

A. I sold to Capt. Sanborn, the man that is there now.

Q. How did you get your pay from him?

A. I got money.

Q. After you were burned out, where did you board with your family for the year following?

A. The Eagle hotel.

Q. How much per week did you pay for your board?

A. Thirty-five dollars a week.

Q. How much for your washing?

A. I don't know; my wife settled that.

Q. You can't tell how much you paid for washing any more than you can tell how much you paid for your clothes.

A. No, sir; it is mere matter of estimate.

Q. Isn't pretty much all this testimony upon matters of estimate? Can't you estimate the value of your washing about as easily as what your house and barn cost you?

A. Part of it is record, I think. My wife usually paid for the washing.

Q. You gave her the money?

A. Yes, sir.

Q. How much did you pay for—what was your subscription for the minister, when Mr. Parker was here?

A. I think one or two years I paid $100 a year. My tax commenced on $6, and it run up.

Q. How much did you contribute to build the South church?

A. I have got one pew that cost me $150.

A. I want to know now how much you paid for building the new South church.

A. I subscribed for pews. I subscribed for two pews; the pews were sold at auction, and bought for choice—I got the second choice in the church. I think I paid $10 for the choice; and I think my pews cost me somewhere about $300, all I put in.

Q. How much did you pay? Didn't you subscribe $500?

A. I think I subscribed for pews. My impression is that we bought pews.

Q. Did you subscribe and then pay for the pews?

A. That might be; I don't recollect; that might be true.

Q. Didn't you subscribe $500? And you have got that property now?

A. Yes, sir.

Q. Now, sir, how much did you use to pay out for contributions.

A. I didn't pay a great deal for contributions.

Q. Did you state in your deposition that you paid out $1600?

A. I don't know. I suppose I have in my life-time.

Q. Now, sir, how much did you contribute for political purposes?

A. Well, there was two or three years that I paid something.

Q. How much? How much is the most that you paid in any one year for political purposes?

A. I think I have paid perhaps $50.

Q. Do you mean to say that you never paid but $50 for political purposes in any one year?

A. I wouldn't say that I didn't pay more.

Q. You mean to say that is all you can say about it?

A. That is my impression.

Q. Will you say that you havn't paid $300 a year for political purposes?

A. Yes, sir.

Q. What is the most that you will be willing to say you have paid a year?

A. I may have paid $75. I don't know as I did that.

Q. What year was it that you paid the most?

A. I don't know.

Q. As nearly as you can state—that is, when your friend was running for Governor?

A. I think it was Gilmore at that time.

Q. Did you subscribe for any purpose for Mr. Gilmore? Did ever he get you to subscribe to pay money for any of his purposes?

A. I don't recollect as I have. What do you mean? For what?

Q. I mean any way that swallowed up money?

A. I don't think I did.

Q. Did you contribute to the diamond pin that was given to Gilmore?

A. I know there was one. I don't know whether I subscribed or not. I know they came to me for it.

Q. Can you remember whether you paid $100 exactly?
A. There was a talk about it, but I can't tell.
Q. What is your best impression in regard to your giving just $100?
A. I couldn't tell.
Q. Wasn't there eleven of you who contributed $100?
A. Stephen Kenrick came to me; whether I paid anything on it I can't tell.
Q. What is your prevailing impression with regard to your giving?
A. Well, it may be so.
Q. Is that your prevailing impression?
A. Well, I don't hardly think it is? but still it may be so. I won't swear that I did.
Q. That was at the time that these awful resolutions were got at?
A. I can't say whether it was or not. I didn't state the year, I believe, positive.
Q. Did you make any contributions for volunteers at the time the war first started—at the time it was fashionable to make contributions in that direction?
A. I don't know whether I did or not.
Q. Did you?
A. I don't think I did.
Q. Are you able to state with any more positiveness than that?
A. No, sir.
Q. Now, sir, did you use to make contributions towards the support of Mrs. Elliot and her family?
A. Mr. Elliot lived in my house ever since 1849. And I received from him, I think, about $200 rent. That is all the rent I ever received from him. He was a man that was sick a great deal, and he was very poor.
Q. My question was whether you made any contributions towards the support of Mrs. Elliot and her family?
A. I have since she died, towards the support of the children. He lives in my house now. The little girl, I think, I found in clothes some.
Q. Did you buy $1000 worth of coal stock of Mr. Gilmore?
A. I had $1000 worth of coal stock, but it was Mr. Gilmore's.
Q. It was of his brother-in-law?
A. Yes, sir.
Q. What did you do with that coal stock? Did you put that in among your losses?
A. Yes, sir; I believe so.
Q. Have you made an allusion to coal in the statements of losses that have been put in here? When did you buy the coal stock?
A. I bought some other fancy stock of Whipple.
Q. How much?
A. I believe I had $400. I believe it cost $400, and I swapped with Mr. Whipple, and gave him $600 to boot. He said that he had bought of a man by the name of Boynton, in Boston. I knew nothing about Gilmore or anybody else at that time.
Q. It was a Manchester Coal Company in which Mr. Gilmore was connected in selling stock?
A. Yes, sir, I guess so.
Q. So that you made, in your two operations, a total loss of $1000.
A. Yes, sir.
Q. Now, sir, did you have any copper stocks?

A. Yes, sir.
Q. Who did you have your copper stock of, and how much?
A. I had them of Seth Greenleaf.
Q. How much did you buy?
A. Bought $1000.
Q. Where was the copper stock?
A. That was up in Littleton somewhere.
Q. That copper stock valueless?
A. It is said to be good.
Q. Have you represented, over and over again, that it is entirely valueless.
A. I don't know much about it, only what I hear.
Q. Have you alleged, over and over again, that it was entirely valueless?
A. I have been oftener told that it was good stock than that it was bad.
Q. Then if you have $1000 worth of valuable copper stock, it should be added to your assets?
Q. (*By Judge Bellows.*) In what mine is it?
A. I forget the name of it.
Q. (*By the Chairman.*) When did you buy that stock?
A. I bought that about two years ago, I think.
Q. Before you went off the road or after? You owned it at the time you left the road?
A. Yes, sir.
Q. Now, sir, did you have an interest in the sand bank out in Connecticut?
A. Yes, sir.
Q. How much?
A. About $1500.
Q. Who did you buy that of?
A. A man down in Milford, [Conn].
Q. How much did you pay?
A. About $1500, and sold it for the same.
Q. When did you buy it?
A. I think it was nine or ten years ago.
Q. When did you sell it?
A. I sold it, I think it must be five or six years ago. [Adjourned.]

NINETEENTH DAY. Thursday, December 24th, 1868.]

The hearing was resumed at 9 o'clock A. M., and the cross-examination of Mr. Clough, the defendant, was continued.

## TESTIMONY OF GEORGE CLOUGH CONTINUED.

Q. (*By Mr. George.*) What did you state you paid for the Sanger place?
A. Three hundred dollars, I think.
Q. And you state that your testimony is founded upon your deeds?
A. And from recollection.

*Mr. George.* I will put in the consideration. Among the deeds furnished is a deed acknowledged before Franklin Pierce, dated in 1846, to Abner and Mary Clough. The consideration is $300. Then there is a deed dated ——— 9th, 1847, of the same premises, a quit-claim deed;

consideration $600.

*Witness.* I bought it for $300 and sold it for $600; and bought it back for $600 and sold it for $800. I sold it to Nat Baker, I think.

Q. What did you swear you gave for the Jamaica Plain property?

A. Eighteen hundred dollars; I think it was between $1700 and $1800 that it cost me.

Q. What is the condition of the mortgage? [Dated Feb. 6th, 1852.] The annexed assignment of a judgment against Mr. Palmer on which a levy was made for a consideration of $1200. That is dated July 7th, 1853.

*The Chairman.* Who is that deed from?

*Mr. George.* This second deed of assignment is from Amos Binney, to recover judgment from Amos Palmer.

Q. Have you any explanation to offer with regard to that?

A. Yes, sir.

Q. What is it?

A. I loaned Mr. Palmer $800. He reckoned, I think, one year's interest in at the time I took the mortgage. He gave me a mortgage of that land in Jamaica Plain for security of that mortgage.

Q. That all the explanation you have to offer?

A. I found, after I took that mortgage, an attachment on it. I sent Mr. Rolfe to see about it and investigate it. He went to see counsel there and also to examine the matter. He said that the attachment was not good and could not hold; but it did. He got out an execution. I was going to contest it, and Mr. Rolfe went down and bought the execution at a discount. So it made the land cost somewhere between $1700 and $1800. I think I called it $1800 in my assets.

Q. Then the statement in the assignment that you paid $1800 for the assignment isn't true?

A. It isn't; I didn't pay that for it.

Q. What did you pay for that assignment?

A. I think it was somewhere over $1000. I don't recollect exactly the sum.

Q. Now, sir, did you pay Mr. Palmer $100 afterwards to release his interest?

A. I did, to get a quit-claim deed from him and his wife.

Q. In addition to the other?

A. In addition to the $800, and in addition to what I paid Mr. Binney. I paid Mr. Palmer $100 to sign a quit-claim deed.

Q. Did you know how it happened to be put into the assignment that in consideration of the sum of $1200 paid to me in hand?

A. No, sir, I don't know how that came in. I know I didn't pay that. You mean Mr. Binney?

Q. Yes, sir.

A. I bought a discount on the execution that was somewhere between $800 and $1000.

[Deed of Mr. Binney offered in evidence by Mr. George.]

A. All I paid to Mr. Binney was what I said before.

Q. Can you tell how the consideration to the amount of $1200 happened to be put into the deed as the consideration of the deed to you?

A. No, sir, I cannot.

[Description of the note from Mr. Palmer to Mr. Clough put in evidence.]

Q. Have you that note, now?

A. I don't know whether I have got the note or not.
Q. If you have, will you produce it?
A. I will.
Q. Now, what did you swear you paid for the Warren street place.
A. About, $8000. [Deed read, having a consideration of $1200.]
Q. Any dispute about that being the same premises?
A. That is the same premises.
Q. The consideration of that deed is $1200. Have you any explanation to offer in regard to that?
A. All I paid for it was $8000.
Q. You cannot explain with regard to the consideration?
A. I believe that Mr. Cooper suggested to have the deed made in that way. I don't quite recollect about that. I know I didn't pay but $8000 for it.
Q. Now, have you ever stated that you gave $9000 for that house on Warren street?
A. Never.
Q. Never stated that?
A. No, sir.
Q. You swore in your direct examination, if I understood you correctly, that you lost $2000 by John D. Cooper?
A. Yes, sir.
Q. Have you ever stated, in any other way, that you lost a different sum by him?
A. I don't know as I have.
Q. Did you swear in your deposition that you lost $2500.
A. That and the interest, I think, came to $2500. I don't quite recollect now.
Q. Won't you state, while I am looking at this deposition, how long you were making up your statement of property, and with whom you made it? How many months or weeks were you in making up your statement of property, and who assisted you in making it up?
A. I think my nephew, Marland B. Clough, and Mr. Rolfe.
Q. Did you state there [showing deposition]—"Lost by John D. Cooper $2500 and $750 interest"?
A. That seems to be so now.
Q. What was your business with John D. Cooper, and what was the occasion of the loss?
A. No business with him except to sign his note that was not paid.
Q. You and Mr. Corning signed his notes together, did you? If so, to what extent?
A. I think Mr. Corning signed the note with me, and Mr. Gilmore too. Mr. Gilmore was on with us.
Q. For how much were you liable for John D. Cooper?
A. I think that I was for some $15,000 or $20,000, or more.
Q. How much of that $15,000 or $20,000 did John D. Cooper pay?
A. I think it was all paid except one note of $6000, and Mr. Corning made up $2000, and I $2000, and Mr. Gilmore $2000. That is my impression now.
Q. How many were there in when he failed?
A. He didn't fail; he put his property into our hands to get our pay,
Q. This house which was conveyed to you, was that put into Mr. Corning's hands by Mr. Cooper?

A. Yes, sir; he and Gilmore took the property and sold it and paid his debts as far as it went. There wasn't enough. There was a $6000 note, and the interest on it.

Q. Wasn't there a large amount outside?

A. There were some debts; I don't remember how many.

Q. You took this deed from Mr. Corning as growing out of the transaction?

A. Yes, sir; I bought the house of Mr. Corning.

Q. And you received the title from Mr. Corning? Or your wife did?

A. Yes, sir.

Q. Now, have you stated to this tribunal all the real estate you own now, or that you have an interest in?

A. I don't think of any more now.

Q. I find a deed here, from Moses Humphrey to Eliza R. Clough, of a tract of land situated in Concord, on the southerly side of Warren street, so called. The deed is dated in 1861. You own that land now?

A. No, sir.

Q. Nor your wife?

A. No, sir.

Q. When did you dispose of that?

A. I disposed of it very soon after I bought it. I bought it sometime in 1862, or 1863 perhaps. I don't recollect the time. I bought the place above for the sake of getting more back land. I took off what land I wanted and sold the place.

Q. I will read the deed, and see if this refreshes your memory [reads the deed]. You mean to say that you don't own that?

A. Yes, sir.

Q. (*By Judge Bellows.*) Did I understand you that you reserved a strip from this?

A. I bought the house and land adjoining me, for the sake of getting more back land—of William H. Bailey. Then I took off what I wanted for back land, and sold the remainder to Emery Humphrey.

Q. (*By Mr. George.*) This is a deed of land from Humphrey to you. The question is whether you own that back strip?

A. Yes, sir.

Q. And you paid $300 for it?

A. Yes, sir.

Q. You say you bought the Bailey house, and you wanted to get some back land. How much back land did you add to your homestead?

A. A small strip, that he allowed me $300 for.

Q. Then you sold the Bailey house?

A. Yes, sir; I gave $3000 for the Bailey lot, and reserved a piece of land that he allowed $300 for. I sold him the place for $2700.

Q. And you paid $300 for the strip?

A. Yes, sir.

Q. Now, is there a piece of land called the Rolfe land?

A. That is on Spring street.

Q. You own that now?

A. No, sir.

Q. When did you sell that?

A. I sold that about a year ago, I think.

Q. How much did you sell it for?

A. Fifteen hundred dollars.

Q. (*By Judge Bellows.*) Where did you say that was?

A. On Spring street.

Q. (*By Mr. George.*) Do you own any land in Dorchester, N. H.

A. No, sir.

Q. When did you sell that?

A. Six or seven years ago, I should think, as near as I can recollect.

Q. Do you own that land that you received from William A. Nichols.

A. That is the Tom Tandy house that I gave in the other day. I sold it to him and was obliged to take it back. I think it cost me $300. I bought it in the first place for $280, house and land; and I laid out something on it; not a great deal. I sold it to Mr. Nichols—I think it was $600. He paid me $50 down, and secured the remainder by mortgage. I bought it back in just a year for the same. He paid me $50 down, and lived there a year, and I bought it back for the same.

Q. (*By Judge Bellows.*) For six hundred?

A. Yes, sir.

Q. (*By the Chairman.*) What did you say was the original cost?

A. I paid $280 for it in the first place, and then I laid out something on it.

Mr. George offered in evidence the deed of Mr. Nichols, the consideration of the deed being $600.

Q. (*By Mr. George.*) Do you own any land in Holderness?

A. I used to, but I have sold that.

Q. When did you sell that?

A. I haven't owned any property there for twelve or fifteen years. I don't know as I am correct about the time.

Mr. George offered in evidence the deed of James P. How and wife, dated May 2d 1861, for a consideration of $400. [Objected to.]

A. My first impression was that it was a dozen or fifteen years ago, but he says the deed is dated since. The deed is dated in 1861. Well, that is the time I sold it, probably.

Q. That is a deed to you?

A. I certainly don't own any land there now.

Q. Have you sold any land since 1861, there.

A. I don't remember as I have. I was thinking it was a man by the name of Howe that I sold to. I don't recollect the least thing about that deed—not the least thing.

Q. What was the consideration paid for the land on which the Masonic Temple stands?

A. I don't recollect; it was somewheres I think, $4000 or $5000 for the land.

Q. Who did you buy it of?

A. Bought of Mr. Pettingill, I think.

Q. How much did you buy in the first instance?

A. I think I bought a quarter of it the first time. That is my recollection about it.

Q. Won't you tell me how much you paid for the land on which the Masonic Temple stands, and how much you reckoned that amount when you made an estimate of the cost of the property?

A. I think it was somewheres in the neighborhood of $4000 or $5000 that the land cost then. I can't recollect now.

Q. What did you call it in your estimate of the cost of one half of the property?

A. When I made up my estimate I think I called it $2600, or $2500; I don't recollect.

Q. How much land did that embrace? Or, from whom did you receive the title?

A. I think I received the title of a portion of it of Mr. Pettengill; and then I think there was one lot in the rear of the Masonic Temple that came from Wyman.

Mr. George offered in evidence the deed of H. P. Rolfe, dated 4th December, 1856, for a consideration of $1725, the same being for one quarter part of the land occupied by the Masonic Temple.

Q. Was that the amount paid to Mr. Rolfe for a quarter part of the land?

A. I don't recollect whether it was or not.

Q. What did you pay Mr. Pettengill for the quarter part that you bought of him?

A. I don't recollect.

Q. If you paid $1725 for one quarter, and $1725 for the other quarter, that would make $3450 for your half of the land?

A. Yes, that would.

Q. Will you state whether that was all of the land of the Masonic Temple that you bought?

A. I think I bought some of Mr. Wyman, in the rear part there.

Q. What did you pay him?

A. My impression is that it was somewhere between $1200 and $1600, but I am not sure. That is my recollection now.

Q. And you own one half of it?

A. Yes, sir.

Q. What Wyman was that?

A. Gard Wyman.

Q. In making up your estimate of the Masonic Temple property, what did you call that?

A. I think it was $15000.

Q. What did you call this piece of land that you bought of Wyman?

A. I told you what I recollect I paid.

Q. What did you call it in your estimate?

A. I don't recollect now.

Q. If you gave $1725 for each of two quarters, that would be $3450? That would be rising $4000 instead of $2500?

A. Yes, sir.

Q. That makes about $4000 for the whole land?

A. I meant the land that the Masonic Temple stands on; I didn't mean the back land.

Q. Have you said one single word about these stores as distinguished from the Temple.

A. I have given in the stores, but not the land; the land that the stores stand on in the rear I entirely forgot.

Q. Then you had said nothing about it before?

A. I didn't recollect anything about it; I didn't put that in at the time.

Q. In your estimate—both in your estimate and in your deposition—of one half of the Masonic Temple, did you include the three stores, and did you include the land on which they stand, and did you call it $15,500? In your estimate of $15,500 of the cost of the Masonic Temple, did you include the three stores situated on the land bought of Wyman and Pettingell, and the cost of the land itself?

A. I included the stores that are now on there.

Q. Exactly; you lumped it altogether, and included the three stores and the building, the Masonic Temple itself?
A. Yes, sir; I think so.
Q. And had you said one word with regard to the land on which these three stores stood, or made any distinction between these stores and the Temple itself?
A. I think the land that went to the Masonic Temple and the land that I bought of Wyman includes all the land there. That is my impression.
Q. But didn't that land—both pieces of land—cost somewhere in the neighborhood of $6000?
A. I included everything, I think, with the exception of that which I bought of Mr. Wyman.
Q. Aren't these three stores on that?
A. There are some wooden buildings; these brick stores were not on there at that time.
Q. And these brick stores stand on the land that you bought of Wyman?
A. Yes, sir.
Q. Didn't the land on which this Masonic Temple stands, and on which these stores stand, cost in the neighborhood of $10,000?
A. No, sir; I think not.
Q. Did they buy their land any cheaper than you, or sell it?
A. I don't know. I bought of Mr. Pettengill and Mr. Wyman.
Q. Mr. Corning bought of Mr. Pettengill?
A. I think he did.
Q. Did Mr. Corning buy his title originally from the same quarter that you bought of?
A. I think he did.
Q. And if he did, and you gave $1725, that would make $3450?
A. Yes, sir.
Q. Then if you paid from $1200 to $1600 for the other, that would make a little rising $8000?
A. I think so.
Q. Was there any other land?
A. I don't recollect now; it appears to me there was a small strip I bought of Mr. Fellows; a very small strip.
Q. Do you know how much it was?
A. I don't recollect now.
Q. Any other?
A. I don't remember.
Q. And yet, with that statement, you make an estimate of $15,500? That is the basis?
Q. *(By Judge Bellows.)* What do you say is the cost of the Fellows strip?
A. I don't know whether it was the Fellows strip or merely some foot of land; there was a small strip that we wanted to get. It seems to me it was somewhere from $100 to $200; but I am not quite sure about it.
Q. *(By Mr. George.)* Did you have some real estate of H. M. Robinson?
A. Yes, sir.
Q. What did you pay for that?
A. It cost some $1225.

Q. What became of it?

A. I own it now.

Q. Hadn't mentioned that before?

A. Yes, sir; that has been put in; put in for $1225.

Mr. George put in the deed of the same, showing a consideration of $1907.

A. I had a piece of land over on the Plains here, that cost $25; and I swapped that land and gave Mr. Robinson $1200 to boot; and he made the deed out in that way. He didn't want to make out a deed at that price, because it would lessen the value of real estate there. He owned real estate along there. That is the explanation on that; that is clear in my mind; I have got the deed here now somewhere. [Deed of the land mentioned, read by Mr. Clough, showing a consideration of $25.] At that time Mr. Robinson was in want of some money; and I drove a pretty good trade with him. I gave him the land on Pine Plains and $1200 in money.

Q. Now, I will ask you again if you had any further interest or property in land that you have not stated to this board of referees?

A. I don't think there is; I don't recollect that there is.

Q. Do you say there is not?

A. I don't think there is.

Q. Have you any memorandum of papers showing the cost of the house on Jefferson street or Thompson street, about which you testified in your direct examination?

A. The house on the corner of Thompson and Jefferson streets?

Q. Have you any memorandum or original papers showing the cost of those houses?

A. I don't think I have. I let it out by the job, the one that is on the corner of Jefferson and Thompson streets.

Q. To whom?

A. Mr. Whipple, I think. I think he was the man.

Q. And you have no papers?

A. I had some, but they were burned at my house.

Q. You own the furniture in the Sherman House?

A. I have got a mortgage of it?

Q. Who bought it and paid for it?

A. I own that furniture.

Q. Who bought it and paid for it?

A. I bought that furniture of Mr. Dunklee.

Q. What did you pay for it?

A. I don't know whether I bought there, or whether I got Chesley's note and got a mortgage of it. I think Chesley bought it, and I let him have the money. It was mortgaged to Mr. Dunklee; and I think that I took Chesley's note, and took a mortgage of the furniture.

Q. Who paid for that furniture?

A. I paid Mr. Dunklee.

Q. How much did you pay him?

A. I can bring the statement.

Q. I want to know how much you paid for it. I believe there has been no allusion to that?

A. I think I have got a claim on the furniture for $1300 or $1400; but I cannot say certain. I can bring in the documents this afternoon.

Q. I asked you to bring in all your deeds and memorandum of sales.

A. I did not bring that for the reason that I bought it when my property was attached.

Q. Didn't I ask you to bring them of all the property you had at the present time. and haven't we been talking about it ever since?

A. I have got a list of the notes.

Q. Did you leave any other notes at home?

A. None that I know of; I have had a schedule in my pocket.

Q. Did you pay Mr. Dunklee $1300 or $1400 in cash?

A. I think I did; I think I borrowed part of the money of Mr. Nutter, $600. My impression is that I settled with Mr. Dunklee in some way; I don't recollect now how.

Q. Have you made additions to that furniture since you bought that property?

A. I have not.

Q. Have you furnished money for the making of additions to that furniture?

A. No, sir.

Q. Neither directly nor indirectly?

A. No, sir.

Q. Now, have you stated, so far as you know, all your property? I am talking about the present time?

A. I don't think of any more now.

Q. Will you say that you haven't an interest to the extent of thousands of dollars in addition to what you have testified?

A. Yes, sir; I think so.

Q. Have you any interest in Mr. Barter's firm here, Barter, Clough, & Pillsbury?

A. Yes, sir.

Q. Have you an interest in that concern?

A. Yes, I put in $5000.

Q. Have you mentioned that before?

A. No, sir. That was paid in less than a year ago.

Q. I expressly stated in my question that I was inquiring about your present property?

A. I didn't so understand it at the time. I understood that I was to give an account of property up to the time the attachment was made.

*The Chairman.* Mr. Clough explains that when he made his statement on the direct examination that he undertook to give an account of property at the time the suits were commenced; and he understands now that you are examining him as to the property he has got now.

*Witness.* I supposed that the property that I had invested, or had since my property was attached was not going in; I don't understand it so.

*The Chairman.* Mr. George informs you that he is inquiring about what you have now. And I believe it is proper.

Q. (*By Judge Bellows.*) What time did you say you invested this $5000.

A. Last April.

Q. (*By Mr. George.*) You invested $5000 in cash?

A. Yes, sir.

Q. Have you any other interest in that concern? You understand now that I am inquiring as to your other present property, and others that you have stated to the board of referees. Now, understand me as including your present interest. I am inquiring as to your present property?

A. Yes. I had an interest in a shoe store; that is, I aided my nephew in buying out half of a shoe store here.

Q. How large an interest have you there?

A. I think it was about $4000. Mr. Piper and I bought out for my nephew. They owed Mrs. Corning about $4000, and I took that note. I was to stand in for that note. I didn't advance any money at that time.

Q. Have you advanced any since.

A. I have let my nephew have some money.

Q. How much?

A. I think it is about $1300.

Q. Paid Mrs. Corning the note, or any part of it?

A. No, sir.

Q. Thirteen hundred dollars cover your advance to your nephew?

A. I think so, yes. I have signed notes with him to raise money.

Q. You haven't furnished any cash beyond $1300?

A. I think not.

Q. What other interests have you, sir?

A. I have a note from Mr. Noyes, I sold out that concern. I think his note is $4000.

Q. (*By Judge Bellows.*) You sold out this shoe business?

A. Yes, sir.

Q. To Mr. Noyes?

A. Yes, sir.

Q. (*By Mr. George.*) What Noyes?

A. George.

Q. (*By Judge Bellows.*) You have his notes for $4000?

A. Yes, sir.

Q. Is that in addition to the Corning note that you have assumed?

A. Assumed to pay?

Q. Yes, sir. Is that in addition to the Corning note that you have to pay? Are you to pay Mrs. Corning yourself?

A. Yes, sir; I am to pay the Corning note.

Q. (*By Mr. George.*) The interest and all, I think, is about $4000. Have you any interest in any other property?

A. We have finished a couple of stores down here, that cost us some $400 or $500. That belongs to me to pay. I don't think of anything else now.

Q. Have you any steamboat stock?

A. I guess I have one share; cost me $30.

Q. Have you any acqueduct stock?

A. No, sir.

Q. When did you sell that?

A. I sold that about two or three months ago, to Mr. Natt White.

Q. You spoke of a lot in the cemetery. What did it cost you to fit that up?

A. My impression is that it cost me about $300 altogether; putting the stone around.

Q. How many watches have you?

A. Got one.

Q. What did you give for that?

A. Cost me $200, I believe.

Q. How many watches have your boys that you got for them?

A. One apiece.

Q. What did you pay for them?

A. Those are cheaper watches; I don't recollect; I think $60 or $70; somewhere along there.

Q. Are those watches referred to in your return of your inventory—your internal revenue return?

A. Yes, sir; I don't know but they were more than that.

Q. Has Mrs. Clough a watch?

A. Yes, sir.

Q. How much did you give for that?

A. I don't remember; perhaps $100 or that matter; I don't recollect.

Q. I see you have sixty-two ounces of silver plate returned in addition to the forty ounces exempt by law?

A. Yes, sir.

Q. What did you pay for that?

A. I can give you an explanation of that, I think. In 1848—I was running into Boston at that time, and there was a prize offered by the Olive Branch, of a silver tea set, valued at $350, to the person that would get the most subscribers in a given time. I went at it, getting subscribers, with a good many others; and I happened to be fortunate enough to get it. I got some two or three hundred subscribers, and I received the prize. I can give you a list of the articles, with my name engraved on them: "Presented by T. F. Norris, of the Boston Olive Branch, Feb. 1, 1848." That is on the articles; one silver teapot; one sugar bowl; one cream pitcher; one dozen tea spoons; one dozen forks; two table spoons; two desert spoons; two sugar spoons; two salt spoons. That is what the set consisted of, my wife had—this was solid silver—my wife had, that her mother left her, ten silver tea spoons; two table spoons; two desert spoons; one sugar spoon and mustard spoon. That was a present from her father, and was her mother's. We have four goblets that she bought of her own money, and presented one of them to me and one to each of the children, and kept one herself. That is the story about the silver.

Q. There is a hundred ounces of silver you returned. That would be eight pounds I suppose. The balance you bought?

A. That is all the solid silver I have in my house. We have some plated, and some German.

Q. You returned a hundred ounces, did you not,—a hundred and two ounces?

A. Mr. Odlin came up and looked it all over.

Q. You took an oath that it was true, did you?

A. The silver was all put out for him to see.

Q. Well, I want to know if you took the oath?

A. I suppose I did.

Q. Have you any other interests in property, direct or indirect, at the present time?

A. I don't think of anything now. If I had understood how this matter was coming in, I could have been prepared.

Q. Can you give me your notes that you have at the present time?

A. I supposed it was the notes that were back of 1866, and I had a schedule of them; and that is what I supposed you wanted.

Q. Then we will go on to another branch. You have made a statement of losses. Do you mean to say that those are the only losses you have made.

A. Yes, sir.

Q. You made a statement of losses to Mr. Rolfe. You mean to say that those are the only losses you made.

A. I calculated to make it all right up to that time.

Q. To what time?

A. Up previous to Feb. 1st, 1866.

Q. I want you to state all the losses you have made up to this time. If you have got any additions to make, make it, if you please.

A. Yes, sir; I will make them up if you will give me time.

Q. You have stated your losses upon your direct examination?

A I have previous to 1866.

Q. Can you tell what losses you have made since?

A. I could not tell; I can tell some of them.

Q. Tell them as far as you know them?

A. I lost, I think, about $1300 by my nephew here in trade.

Q. Well, sir, go on?

A. I lost $100 by Mr. Remick.

Q. George L. Remick, you mean?

A. Yes, sir; he is a trader; I don't recollect his given name.

Q. (*By Judge Bellows.*) You are speaking now of the time since February, 1866.

A. Yes, sir. He is the man that had a portion of the Rolfe lot, and gave me his note; and he failed since that time.

Q. Any other, either before or since 1866, that you have not stated in your direct examination?

A. I don't remember now whether there is anything more or not; I don't recollect.

Q. Did you ever testify before Judge Bellows, one of the referees in this case, in regard to losses from John F. Neally.

A. There was a matter left with Judge Bellows.

Q. Did you state to Judge Bellows, in that connection, the loss he had made, and how much it cost?

A. I don't remember.

Q. Did you lose anything by John F. Neally?

A. It couldn't be a very large sum; I don't remember how much now; it was left with Judge Bellows, and I think I have got his documents; I will bring them in if you want to see them.

Q. Did you not testify to what you had furnished John F. Neally, and on other occasions that you have said nothing about?

A. I recollect testifying before Judge Bellows, but I don't now remember what it was.

Q. Did you testify to what you had lost by John F, Neally.

A. I think I might have lost a small amount, but I could not tell how much; I think it was a small amount.

Q. Did you buy any gold stock of anybody? If so, of whom, and when did you buy it, and how much did you give for it?

A. I bought $750 worth of gold stock of Mr. Gilmore.

Q. Was it what was called the Messigemet?

A. Yes, sir.

Q. That was dead loss wasn't it?

A. Yes, sir; that has gone up.

Q. Was it 1864,—about the time you subscribed for the diamond pin?

A. I don't remember subscribing for a diamond pin; I didn't say that I did.

Q. You remember there was such a thing?

A. I recollect there was something said about getting up a diamond pin, but whether I paid anything towards it, I can't remember, I might or I might not; I cant say.

Q. Did you lose anything by the Pawtuckaway Bank stock?

A. I believe there was about $250 there.

Q. When did you lose that?

A. I think I have got it down here; I am not certain. I will give you the dates, if I have them.

Q. I want to know about the date of the Pawtuckaway Bank stock?

A. That was sometime ago. That was,—I should think it might be along in 1858 or 1859—I should think, but I can't recollect. I have got that down somewhere, the whole of it.

Q. I will ask you of losses, if any, by Moses Fellows?

A. My impression is that I didn't lose much; I don't recollect how that was.

Q. How much did you lose, do you think?

A. I don't recollect how that was; I think it was small, if anything.

Q. Moses Fellows owed you at the time of his failure?

A. No, sir; I had an interest in a stable there one time; I think that is the way I happened to get his note. I sold out a stable, and took his note in part pay, if I recollect right.

Q. Has that note ever been paid.

A. I think I got my pay; I think a man by the name of Haskell paid it. I think he did.

Q. Did you not lose from $300 to $500 in that transaction.

A. I don't think I did.

Q. Will you say you did not?

A. I can't say that I did not, but my impression is that the loss there was small. I think the note was some $800; I ain't quite sure about it, but I was thinking that a man by the name of Haskell paid it.

Q. Did you sue it and get out an execution?

A. I don't know how; it strikes me that I didn't sue it; I can't tell.

Q. I was asking you yesterday with regard to the sand bank speculation. Was that a bank out in Massachusetts or Connecticut that was alleged would make glass?

A. No, sir.

Q. Did you buy it of Robert Moore?

A. No, sir.

Q. Was he the one who negotiated the trade?

A. He was the one that told me about it; I went out there and bought it.

Q. Did Mr. Moore and Mr. Collis, and others have a similar bank, and was it adjoining?

A. It was in the same town. It was probably less than an acre, right in the town of New Milford.

Q. You paid $500 for it?

A. I did.

Q. How many times did you go out there,

A. Twice. I went out and bought it. This Robert Moore sold it. and forged the deed. I found it out and I arrested Moore. I went out there and the man that bought it of him. Moore took Brad. Cilley out with him to settle the matter; and it was settled. Cilley, I think, paid; that is, paid me for the land.

Q. Paid you how much?
A I think it was $500.
Q. You state that was the amount?
A. That is my impression.
Q. Is your impression very distinct?
A. I think so.
Q. Won't you say that Mr. Cilley paid you just the amount of your expense?
A. No, sir; no such thing; I got pay for the land.
Q. How many years did you own it?
A. It might have been perhaps three or four; something like that; I cannot tell exactly.
Q. Wasn't it eight or ten?
A. I don't think it was.
Q. You spoke of the California speculation. Did you send some people out to California in the early gold fever, and find them with money?
A. No, sir; not that I recollect of.
Q. Say you didn't?
A. I don't think I did.
Q. Can you tell whether you did or did not?
A. I don't believe I did.
Q. Can't you tell absolutely?
A. There was a party of us got out a cargo of stuff and sent it to California, that I had an interest in.
Q. How much did you put in?
A. I think I put in $2000 and got back $1000. Robert Pecker was the man.
Q. Didn't you furnish people money to go out to California?
A. No, sir.
Q. Who else went into that California speculation that Pecker had to do with, besides you, here in town, that put money in?
A. I think a fellow by the name of Jones. I think they sent him out. Pecker was the man.
Q. Can you state who were your associates in putting money into this?
A. I don't remember.
Q. Do you mean to swear that you got fifty per cent. back?
A. I believe I do.
Q. Are you willing to state that as an absolute fact?
A. Yes, I think I will. I know I lost $1000; and if I put in $1000, I lost the whole of it; if I put in $2000 I got back half; I know I lost $1000.
Q. If you put in $4000 how much did you lose?
A. I didn't put in as much; I put in $1000 or $2000, and I lost $1000.
Q. Did you make any memorandum of your loss there?
A. I believe not.
Q. Have you any memorandum of any sort?
A. I don't know as I have.
Q. Have you seen any in the last fifteen years in relation to it?
A. I don't know as I have.
Q. You have two sons, you have already testified?
A. Yes, sir.

Q. Your sons are now 21 and 19 years old?
A. One 22, and the other 18.
Q. You have had these boys away to school, haven't you?
A. One of them has been away; and the youngest has been to Hanover one term; that is all.
Q. One is in college?
A. No, sir; he is in the Chandler school.
Q. How long was one of your boys at St. Johnsbury to school?
A. I don't recollect how many terms he was there.
Q. About how long? About what period of time?
A. I should think perhaps one or two terms; I don't recollect; perhaps it may be more.
Q. One of your boys at Burlington to school?
A. Yes, sir.
Q. How long?
A. I think he was there a year.
Q. Either of your boys at the Military Institute at Merrimack to school?
A. Yes, sir.
Q. How long?
A. I think one or two terms.
Q. Can you tell how long?
A. I don't remember.
Q. Have either of your boys been in the way of earning anything up to this time?
A. Not a great deal.
Q. You may state it if they have?
A. The oldest boy is in a bookstore here; been there a year and over.
Q. And with that exception, is that all the earnings?
A. Pretty much all, I think; small, if anything.
Q. I will ask how much you paid for physicians' bills?
A. I don't know; haven't had a great deal of sickness.
Q. You were sick here, were you not, down on South street? When you were down on South street, did you pay Dr. Gage a bill?
A. I don't recollect.
Q. Were you not sick down on South street, and attended by Dr. Gage?
A. I don't recollect that I ever had him.
Q. You don't recollect Dr. Gage ever attending upon you?
A. No, sir.
Q. Can you state how much you paid for physicians' bills?
A. I don't know as I can.
Q. Have you had a large amount of cemented pavement laid about the walks and grounds connected with your house?
A. Yes, sir.
Q. How much was the bill?
A. I think the bill was about $200.
Q. Will you say it was not $500?
A. Yes, sir, I think so.
Q. How much have you paid for legal services? Have you paid Mr. Rolfe for example?
A. I haven't settled with Mr. Rolfe for—I don't know as I have for five or six years.

Q. Have you paid him any money, and if so how much?

A. Yes, I have let him have some money, I think.

Q. How much?

A. I don't remember how much.

Q. State as nearly as you can?

A. I think I have got his note for what he owes me; but I don't recollect how much it is.

Q. I ask you how much you have paid him for services. He has stated that you are very intimate, and he has done your business. And he has been your counsel in lawsuits, has he not?

A. He has.

Q. I want to know how much you have paid him?

A. I could not tell.

Q. Give the best statement you can make, how much money you have let Mr. Rolfe have for which you have not got his note, for the last twenty years? Will you answer the question as near as you can?

A. I cannot tell.

Q. I want you to state as near as you can, the amount that you haven't got his notes for to-day?

A. I believe I have got a few charges on my book.

Q. Mr. Clough, did you understand my question? If you didn't I will repeat it. How much money have you paid Mr. Rolfe for the last twenty years, that is not represented by notes that you hold against him to-day?

A. I cannot tell.

Q. Won't you state as near as you can?

A. It would be guess work. It might be $1000; call it that; it will be all guess work.

Q. Isn't it a much larger sum than that?

A. I don't think it is.

Q. Can't you state?

A. I can't unless I have the bills.

Q. Have you never paid him any money apart from bills?

A. No, sir; I have no recollection.

Q. Ever let him have any money—made him any presents?

A. No, sir.

Q How much have you paid other lawyers?

A. Paid Mr. Tappan $25. That is all I think of now.

Q. Pay Judge Minot a retainer?

A. Haven't paid him anything yet; I expect there is a bill there.

Q. I was asking about what you have paid in cash?

A. I don't recollect anything else except what I paid Col. Tappan for retainer.

Q. Haven't you paid Ayer and Foster money?

A. I don't know about Mr. Ayer. Foster sold some land down there; I don't know but he had a commission on that.

Q. Didn't give Foster anything for a settlement by an execution on the account of Moses Fellows when Moses Fellows failed?

A. I think that Mr. Foster has done some business for me; but I don't know how much it was.

Q. Have you paid Mr. Morrison anything?

A. I don't recollect as Mr. Morrison done any business for me at all.

Q. Didn't Mr. Morrison bring a suit for you on which an execution was recovered in the Tasker and Fellows matter? And didn't Foster subsequently make an adjustment?

A. Mr. Morrison has done, I think, all my business that I had to do in Manchester; but I don't think his bills were very large.

Q. Did you tell Mr. George W. Morrison when you adjusted that, that you would make it right when he had occasion to pass over the road?

A. No, sir; I never told him any such thing; and he or any other man that states so states false. I understand that Mr. Morrison made that statement; but it is false.

Q. I understood you to state that your insurance cost you, for all the time up to the present time, $341 and some cents?

A. I think when I was making up my property——

Q. I haven't asked you for any explanation. I am merely asking your attention to this fact. I see in your estimate for 1866, you have stated your insurance that year to be $80.99.

A. It seems to be so there.

Q. I want to know how you make out your insurance to be $300, if the insurance on your property was for one year, $80. This was in 1865, and the return was made in 1866?

*Mr. Tappan.* How long does that policy run?

*Mr. George.* This is a policy paid by the owner.

Q. How do you make your insurance $341 and some cents and your return for one year, $80?

A. Those statements were made out, some of them by Mr. Rolfe, and some of them by my nephew; and they both know about my matters as well as myself; and I always told them to make it up correctly; and at the time I was making up my business assets, and taking out the interest and so on, I went to Mr. Keyes, and got him to take off the statement as near as he could. He done all my insurance. And he made it. I have given it, I think, in making it out, 300 and some odd dollars

Q. You mean to say, and take the position, that it was not true, and that you swore it was true; and that you swore it to be true because you supposed somebody else made it up true? Which position do you mean to take?

A. I cannot tell.

Q. Well, take another item, The next is repairs,—repairs of buildings, average paid out, for the preceding five years, including amounts paid for permanent improvements thereon, $577.66. Was that true, sir, or wasn't it true?

A. I don't know, I am sure. As I said before, these were made out by other men, and I supposed it was right.

Q. Which position do you take now; whether it was true or was not true?

A. I cannot tell, but my intention was to have them right.

Q. You swore to that return as being true, did you not?

A. I suppose I swore to them.

Q. According to this statement, you paid on the average, the five preceding years, in round numbers, $3000 for ordinary repairs? If this statement is true, which you swore to, you paid for the five preceding years $3000 in round numbers? Now, if this statement is true, you paid out for the five preceding years $28,088.30 for ordinary repairs on your buildings?

A. I don't think there is that amount paid out; I don't know how it is.

Q. You swore that that was so, didn't you?

A. I guess so, by that.

Q. And adding the preceding year, it makes out $34,065.96 for the ordinary building, without improvements. Now, you have put in what you allowed the aggregate amount of taxes you have paid. You have stated here that the national, state and local taxes paid in a single year were $10,016.25. That is considerably more than a quarter part of what yeu have stated—I take it you gave the same statement as in the deposition—you stated that you paid in all the taxes you had ever paid?

*Mr. Mugridge.* I don't think that Mr. Clough has made any such statement. He stated that he paid so much to the city of Concord; and the certificate shows it.

*Mr. George.* I stand corrected.

Q. What taxes did you pay, apart from the Jamaica Plains land. and poll taxes, and the Orange farm, and apart from the taxes on the Page place, except what you paid in Concord.

A. I don't know that I paid any?

Q. Now, sir, the taxes on the Jamaica Plains land (according to your statement,) and the taxes on the Orange farm and the Page place, amounted to how much in a single year? Take the year 1864.

A. I think on the Jamaica Plains property in 1864 there was paid $34 or $35, if I recollect right. I have got bills of these taxes, I think, at home.

Q. About how much? No matter about a dollar or five dollars?

A. I think it was about $34 or $35 a year on the Jamaica Plains land.

Q. And how much on the Orange farm, and how much on the Page place?

A. The Orange farm, I rented that for five years. The man paid the taxes and gave me so much.

Q. Then there are included in the amount of taxes paid by you but $1025 that you swear you paid, are there?

A. No, sir.

*The Chairman.* Is there any doubt that the taxes paid by the tenants are just as much a part of the rent as anything else.

*Mr. George.* As there is no rent put in here at all—as I shall come to in a moment—it is perhaps not very material.

I wish to state the four income returns that we have here of 1864, 1865, 1866 and 1867. The amounts are, that he has sworn to as paid taxes: for 1864, $882.08; for 1865, $1016.25; for 1866 $769.14; for 1867, $815.11; making $3482.58.

Q. Your taxes were pretty light for the balance of your life, if the aggregate paid in for these years was $3400?

A. The taxes were taken off from the city clerks' book; and I suppose they were all in.

Q. I am assuming that your statement is correct. Now, sir, we will come to another point. In 1864 you have sworn that the amount actually paid for insurance of homestead is $100. Have you so sworn there?

A. I should think so.

Q. Have you any doubt about it?

A. I made oath to these papers, I think.

Q. It is a paper stating that the actual amount paid for homestead was $100?

A, Yes, sir.

Q Who paid the insurance on your Masonic Temple? That was insured all the time? It never has been uninsured? Has it ever been uninsured?

A. Not to my knowledge.

Q. For what sum?

A. I believe I got $7500 on my lot.

Q. What is the rate of insurance that you are to-day paying for that insurance?

A. I think it is one per cent.

Q. You mean one per cent. a year?

A. Yes, sir.

Q. That would be $75. How much is the Allison property insured for?

A. I think my part is insured for $2000. That is my impression.

Q. What do you pay on that?

A. I pay the same.

Q. Do you pay the same on a wooden building that you do on a brick building?

A. It is my impression that it is one per cent., but I may be mistaken.

Q. Are you certain that it is not one and a half per cent. that you pay for that? There would be $95 a year if you are correct. How much are you paying on your house, and for how much is it insured? I mean the house that is in Mrs. Clough's name?

A. My house and furniture are insured, I think, for $15,000.

Q. How much on the house, and how much on the furniture?

A. I don't know how it is divided. I don't recollect.

Q. Will you be kind enough to bring your insurance policy on your house and furniture when you come down this afternoon?

A. I will.

Q. How much are you paying for that?

A. My impression is that it is one per cent. for five years.

Q. That would be then, if you are correct, $30 a year. Now, will you state how it was that in 1864 you paid $100 on your homestead—for what that payment was? What was the amount of insurance?

A. I don't know; I cannot tell anything about it.

Q. Are your buildings on Thompson street insured?

A. I believe they are.

If you will put in all your policies as they exist to-day, it will perhaps save time.

I can bring them all in. I have got some in the bank. The Masonic Temple is in the bank.

Q. These are in stock companies—your policies?

A. I think they are, all of them.

Q. I will pass the insurance matter, because if Mr. Clough brings his policies, these will show. Mr. Clough, you said in your examination, that Mr. James R. Hill bought a lot of stock in the Concord Railroad—150 shares, I understood you—upon a bond given to him and signed by the other directors to indemnify him against loss. Will you state whether that stock, after the annual meeting, was transferred to you?

A. Yes, sir; I bought it.

Q. What did you give for it?

A. I think it was 61.

Q. Do you know? Have you any memorandum—any minutes?

A. I have not.

Q. How long after the annual meeting before that stock was transferred to you?

A. I can tell by the date of the note I have. That stock was bought in November, 1867. I gave my note to the Merrimack River Savings Bank for $10,000. That is the money that bought that stock.

Q. Mr. Clough, it is agreed on all hands that the transfer books of the corporation may be used without authentication. I have here a schedule that, May 1st, 1867, James R. Hill conveyed, by certificate No. ——, 150 shares of stock. How was that paid for?

A. The May notes—that $10,000 that I borrowed at the savings bank, and then six months' interest from that time.

Q. (*By Judge Bellows.*) You bought first in May 1st.

A. I bought it May 1st. I paid six months' interest in advance.

Q. (*By Mr. George.*) And the loss between what Mr. Hill took it for and the amount you bought it for, how much was that?

A. I don't know; no trade of mine, that wasn't.

Q. Did you state in your deposition that they agreed together to pay the loss?

A. I should not know anything about it. I never had a cent's worth of interest in it. I never so stated in my deposition, or anywhere else. I paid the loss on the Shaw stock.

Q. I understood you to state that Mr. Spalding knew that you were carrying this express matter backward and forward on the train; and that you had conversation with him in regard to it?

*Mr. Mugridge.* If the gentleman means to say that Mr. Clough testified to that, he is mistaken.

*The Chairman.* The recollection of the referees is this: that Mr. Clough thought Mr. Spalding knew of it when he was doing it, and that he afterwards had a talk about it with Mr. Spalding, and he told him he had better stop.

Q. (*By Mr. George.*) Do you mean to be understood that Mr. Spalding, or either of the board of directors, or the superintendent, knew with regard to your carrying express matter on the train?

A. I suppose that Judge Upham knew it at the time, but I don't know that fact.

Q. Did you pay anything to the road for the privi.ege of doing express business over it?

A. Never.

Q. Mr. Spalding was, at that time, president of the road, was he not?

A. He was.

Q. Won't you state the conversation which you allege you had with Mr. Spalding on that subject?

A. It was one day that it was my train down at ten o'clock from here, and Judge Upham says to me that Mr. Spalding was coming up and wanted to see me—wished me to stop over a train. In the morning, Mr. Spalding came up; came up in the morning train. I was in Judge Upham's office at the time, when he came in; and he said that there had been a complaint entered to him that I was carrying express matter and produce, and such stuff. He wished I wouldn't carry any more; said if I carried the others might want to. He did'nt care about what I had carried, but wished I wouldn't carry any more. That is what the conversation was, as near as I can recollect. I never did carry any after that.

Q. And prior to that time you had carried express matter to the extent that you had made $5200, according to your estimate? Is that so?

A. I should have to look at my deposition.

Q. (*By the Chairman.*) `When you speak of express matter, does that mean stuff that you carried on your own account there?

A. Yes, sir.

Q. Or packages you took for other people?

A. There were some packages that I took for others, but mostly for myself.

Q. Stuff that you bought and carried to market?

A. Yes, sir; stuff that I carried to market.

Q. (*By Mr. George.*) Including both kinds—that that you carried for other people and that that you carried for yourself, but mostly for yourself? That is correct, is it?

A. Yes, sir.

Q. There it is. sir [showing statement]; chicken and produce trade $5200?

A. Yes, sir; that is so.

Q. Then you mean to be understood—I repeat the question—that before you allege Mr. Spalding spoke to you, as you have stated, you had carried these various articles to such an extent that the profit had amounted to $5200, according to your estimation?

A. Yes, sir; according to what I had estimated at the time.

Q. I will take the peanut business. As I understand your testimony, you say that you paid a boy or boys upon the cars, and kept that boy upon the cars selling peanuts, and the other articles which you enumerate. And you estimate that your profits from that was $3276?

A. Yes, sir; that is the case.

Q. Won't you explain how that is made up?

A. I think I estimated it a $1.50 a day profit. How I came to suggest that, I cannot tell.

Q. Now, sir, was there any person connected with the railroad, so far as you know, that knew you had a boy on the cars?

A. Yes, sir.

Q. Who of the directors, or what employee of the road, knew that you had a boy on the cars?

A. They always have had a boy on the cars.

Q. Who knew that you had a boy on the cars?

A. I don't know whether they knew it or not.

Q. Did ever you tell any of the directors or the superintendent of the Concord Railroad that you were hiring a boy.

*Mr. Mugridge* inquired what was the object of this testimony?

*Mr. George.* I distinctly desire to put it on one ground; that he is liable to the road for the fare of the boy and the profits. Because if he did this thiug without the knowledge of the directors of the road he is liable in one way or another for the profits that he received and the fare of the boy. And I put it on the second ground, if it should be held that we cannot recover under this form of action, that this attempt to account for his property in this way is perfectly absurd and ridiculous.

*Mr. Mugridge.* If the counsel puts in on the last ground, I withdraw the objection. But that wasn't the way he stated it before.

Q. (*By Mr. George.*) Now, Mr. Clough, I want to ask you if there was a single director or officer on the road who had any knowledge, so far as you are aware, that you were employing boys to run on the cars to sell peanuts?

A. I don't know whether they knew it or not.

Q. Did you ever tell anybody so, prior to the commencement of this suit?

A. I don't know whether I did or not.

Q. Did you pay anything, for having that boy ride backward and forward in the cars, to the corporation?

A. No, sir.

Q. Mr. Clough, you have referred to a number of people with whom you had negotiations; Mr. Raymond Kimball, for one. Is he dead?

A. Yes, sir.

Q. Mr. True Garland. Is he dead?

A. Yes, sir.

Q. Stephen Osgood. Is he dead?

A. I don't know; I think he is.

Q. Robert E. Pecker, is he dead?

A. Yes, sir.

Q. I will ask you, who were those boys that made this $3200 selling peanuts?

A. There was one boy, from Manchester, by the name of Clough. Then I believe there was a boy by the name of Osgood.

Q. Where are they?

A. I don't know where they are.

Q. How many years since you have seen either of them?

A. I don't know as I have seen them since they left.

Q. How many years since they ran?

A. They ran on the road six or seven years for me.

Q. How long since you have seen either of them?

A. I don't kuow as I have for ten years.

Q. Don't know where they are?

A. No, sir.

Q. Know whether they are living or dead?

A. I don't.

Q. Mr. Clough, was there a suit brought agaist the Concord Railroad of the fact of which you were aware, by Mr. Noyes putting Mr. Horace Johnson off the train for attempting to ride on a ticket that he bought in Chicago? [Objected to.]

*Mr. Mugridge.* The objection is that they cannot prove a suit in that way, and what it is brought for.

*Mr. George.* That is not the point at all.

*Mr. Mugridge.* So far as anything is implied from the question, that is implied, and nothing else.

*The Chairman.* We do not think, in this case, it is necessary that the existence of that case should be proved in the first instance, before inquiries are made about it. The case has been cited here, perhaps, on both sides. We do not think it is necessary to prove, by the clerk of the court, the existence of the suit before Mr. Clough can have knowledge of it.

Q. (*By Mr. George.*) Mr. Clough, were you aware of the suit that was pending between Mr. Noyes and Horace Johnson?

A. I heard that there was a suit of that kind.

Q. You knew of the existence of that suit at the time?

A. I heard about it.

Q. Were you present in court at the time that was being tried?
A. I have no recollection of it.
Q. Were you not stopping off the train, and were you not up in the morning to see Judge Upham?
A. I don't think I was.
Q. Will you say you was not?
A. I don't think I was
Q. Was the matter of that suit a matter of talk between you and the officers and directors of the road?
A. Yes, sir; I suppose we talked about it at the time.
Q. You knew the fact that the road were taking the ground that Mr. Noyes did right in putting Mr. Johnson from the cars?
A. I suppose they were, or else they would not have stood a lawsuit.
Q. You had no doubt about that fact at that time?
A. I supposed they were.
Q. You knew that the road were justifying the action of Mr. Noyes in putting Mr. Johnson off the cars?
A. I supposed they thought so.
Q. That was your supposition at that time?
A. I think very likely. I don't know what it was at that time.
Q. Have you any doubt about it, sir?
A. I suppose that the company thought they were right or else they wouldn't defend the suit.
Q. Do you mean to swear that at the same time you with Mr. Gilmore's assent were buying and selling tiekets to people?
*Mr. Mugridge.* There is no such position taken in the case.
A. I understood that it was an old ticket, a spent ticket, that Mr. Johnson had.
Q. What do you mean by spent ticket?
A. That the date had run out.
Q. Did you understand at that time that the road took the position that under their rules tickets were only good on the day of their date, and for such further time as would enable the passenger to reach his destination by the continuous trains of the road? Did you understand that was the ground on which the suit was brought, and Mr. Johnson opposed? Will you look at that. There is the very ticket on which he rode. Won't you look at that and see what you meant by spent ticket?
A. I should suppose that if Mr. Johnson or any other man rode on that, on the 18th day of October, it would not be good.
Q. Why not?
A. Because it has gone by the date.
Q. What date?
A. The 17th. These are long tickets—western tickets.
Q. Go on and explain what you mean?
A. I have nothing more to say about it as I know of.
Q. You say that you should suppose if he attempted to ride on that on the 18th day of October, he would have no right to ride on it. Now why not?
A. Because the date has run out, and it was a western ticket.
Q. What do you mean by the date's running out?
A. I mean that if a person rode on that after the 17th—that is what I mean.

Q. I do not understand what you mean by that. Won't you explain yourself?

A. For instance, if a man should have this ticket in his possession after the 17th of October, 1860, it is not supposed it would be good.

Q. Why not?

A. Because the date had run out.

Q. Why because the date had run out? Do you mean by the date's running out that it was used one day after. Why because of that? Why because the date had run out, sir?

A. Because that was the object of dating these long tickets, and to have them ride on them while they were good.

Q. Can a man come from Chicago to Boston in one day?

A. That probably was dated——

*The Chairman.* You understand by that that the date that is put on there is the date of the ticket, or the date on which it is used?

A. These tickets generally said "good for thirty days," but that ticket don't seem to state so.

Q. (*By Mr. George.*) Was it because there was a rule of the road that you say it would not be good?

A. I never saw—I don't recollect seeing a ticket dated in that way.

Q. I am asking now in regard to these tickets. I ask you if it was because there was a rule of the road regulating the tickets that you say it would not be good?

A. We never paid attention to dates on our local tickets. Probably these tickets were given with the understanding that they should have time to go through on. All the western tickets that ever I saw or had, if there was any date at all, it was dated,—"good for thirty days,"

*Mr. George.* This is the rule that was actually passed, the paper that was served upon Mr, Clough, under the rules of 1860. All tickets over the Concord, Manchester and Lawrence roads, said tickets shall only entitle the passenger to a passage on the day of their date, provided that joint tickets shall be good for such further time as shall enable the passenger to reach his destination by the continuous trains of the road.

Q. (*By Mr. Mugridge.*) Was that rule in force at the time of this suit?

A. Yes, certainly.

Q. (*By Mr. Tappan.*) Was there any rule at this time—in the instructions of 1860—prohibiting the using or selling of these coupon tickets.

*Mr. Mugridge.* The case don't show that it was a rule of 1860.

Q. (*By Mr. George.*) Won't you look and see if that was the notice served on you? [Objected to.]

*Mr. Tappan.* In one word I will just state my position. In the absence of any rule to the contrary, these tickets, I hold were good in the hands of anybody until they had spent their force.

*Mr. George.* They had spent their force.

*Mr. Tappan.* That is a matter of argument.

*The Chairman.* Precisely what is the question put?

Q. I want to ask you if you received the original, of which this is a copy, from Mr. Gilmore?

A. I think it is.

Q. Is there any doubt that it is?

A. I think there isn't any.

Mr. George read that the notice was dated Dec. 18th, 1860.

Q. And yet, sir, notwithstanding that paper was served upon you, you mean to testify here, do you, that you continued to buy and sell coupon tickets up to the time that you left the road? [Objected to.]

A. I did, by permission of Mr. Gilmore.

Q. How long after Mr. Gilmore wrote you that letter, or had that letter served upon you before you received his permission to buy and sell coupon tickets over upper roads?

A. I had a talk with him, after 1860; I don't recollect precisely at what time; and told him that I was selling these tickets. He said that if they were not out of date it was all right; but not to sell them after the date had run out.

Q. But the rule required, did it not, sir, that they were to be only good on the day of their date, and on such continuous trains as would enable the passengers to reach his destination?

*Mr. Mugridge.* Is the witness to be called upon to construe this rule?

*The Chairman.* Perhaps you had better state the rule.

Q. [After reading the rule.] That is what the rule provided, is it not?

A. Well, if the tickets were not dated, or if they were dated good for thirty days, and were used during that time, they were supposed to be good. And I had permission from Mr. Gilmore to buy and sell them.

Q. Did you understand that the object of that rule, at the time, was to prevent their use for local tickets? [Objected to on the ground that the question had been directly and properly answered.]

Q. Did you understand that the object of the rule was to prevent the substitution of through for local tickets?

*The Chairman.* This rule here is in these words:—"As soon as practicable all tickets over the Concord, Manchester and Lawrence roads shall be dated on the day of the sale, and only entitle the holders, &c.," As soon as practicable after this rule. Now that seems to recognize that tickets might continue for sometime to be used, and to be good for some particular time; and it would be the duty of the conductor to recognize them. And it seems in that view of the case that Mr. Clough has answered the question. It seems to me that the question is answered, Col. George.

*Mr. George.* These rules, if it please the referees, went into operation on the 1st of January.

*The Chairman.* If you want to get at some other matter, why don't you put another question?

*Mr. George.* I want to ask Mr. Clough if he did not understand the object of the rule to be to prevent a person going and buying a through ticket and substituting it for a local fare.

Q. (*By Mr George.*) Did you understand the object of the rule was to prevent the substitution of a through ticket for a local fare?

A. I did not understand by the rules that we were prohibited taking tickets within the date.

Q. I don't ask what you understood about other matters. I ask you if you understood one of the objects of the rule was to prevent the substitution of a through ticket for a local fare,

A. I don't think I did understand it so.

Q. Have you with you now any notes of indebtedness to you? If so, please produce them.

Witness hands to counsel a package of notes of which the following is a list:—

Note of J. P. Eaton; dated March 29, 1866, for $136, with annual interest. Endorsed, "April 9, 1866, received $23.25 on this note."

Note of Edward Shanks; dated July 1, 1856, for $100. Several endorsements of payments. Due, July 29, 1865, $59.12.

Note of Henry F. Chickering; dated January 2, 1864, for $200, payable on demand with interest annually. No endorsements.

Note of J. D. Cooper; dated July 27, 1861, for $30, payable on demand with interest. No endorsements.

Note of Eliza A. Morse, Martha Morse, and others; dated December 30, 1863, for $100, payable on demand, with interest annually.

Note of D. T. Whipple; dated June 28, 1864, for $125, payable on demand; interest annually. Endorsed March 16, 1866—received interest up to January 28, 1866.

Note of D. T. Whipple and Harriet S. Whipple; dated March 17, 1866, for $900, payable two years from date; interest semi-annually. Endorsed Sept. 11, 1866—received on the within note $273; Feb. 11, 1867, received $132, on the within note.

Note of I. W. Wilmeth and Harriet W. Wilmeth, Bloomington, Ills., for $500; dated Oct. 13, 1862; due two years after date; interest at the rate of ten per cent. from date. Endorsed—Oct. 17, received the interest on the within note to Oct. 13, 1863, being $50; also two other endorsements, receipting interest to Oct. 13, 1865.

Note of J. H. Wilmeth, Bloomington, Ills.; dated Feb. 26, 1863, for $500; interest at ten per cent. endorsements of interest paid to Feb. 20, 1866.

Note of Patrick Glennen; dated April 1, 1867, for $189.96; interest payable semi-annually, at six per cent. Endorsed—June 27, received on the within as principal, $40.

Note of John McDonald; dated April 1, 1867, for $366; payable one year from date with interest semi-annually, at six per cent.

Note of Minerva B. Fuller and Charles A. Fuller; dated January 17, 1867, for $837.50, payable $100 annually, with interest. Endorsed—Jan. 28, 1868; received $50, on within note; Sept. 5, 1868; received $100 on within note.

Note of W. B. Hoyt; dated January 17, 1867, for $137.50; payable $25 annually with interest. Endorsed—January 18, 1868; received $35 on within note.

Note of Michael McDonald; dated April 1, 1867, for $467.25; due one year from date; interest semi-annually. Endorsed—received July 23, 1867, on the within note, $100.

Note of George F. Sewell; dated Dec. 4, 1868; $307.25, with interest.

Note of Mary H. Smart; dated July 17, 1867, for $134; interest semi-annually.

Note of David Hurley; dated July 1, 1867, for $352.86; interest semi-annually at six per cent.

Note of George Noyes for $4000.

Q. Have you any other notes or evidence of debt, or of money owing to you from any person, directly or indirectly besides these?

A. No, sir; I don't think I have.

Q. I understood you to say that you had Mr. Rolfe's note. Perhaps I misunderstood you?

A. You did misunderstand me. That note has been paid.

Q. Did you not state that you had loaned him money and had his note? or did you mean that you had had it, but that it was now paid?

A. I don't think I stated so. I had a note against him for $6000; but that has been paid.

Q. At the time your deposition was taken you stated that you had notes for money loaned, amounting in all, to $15,205.39. Are you able to state what those notes were at that time?

A. I have a schedule here of all my notes; but a portion of those notes have been paid since that time.

Q. Here is a note of John McDonald's, for $915.68, which has not been produced here, nor is it marked paid on this memorandum. Has that been paid?

A. I think it has been paid.

Q. Here is "P. Sweeney—$872.24," not marked paid. Why has not that note been produced?

A. That may be at home. I supposed I had all my notes here.

Q. "Thomas McGuire—$357.06."

A. That is one of the Jamaica Plains notes, and I think it must be in another package.

Q. Here is M. E. Donahoe's note for $571.54.

A. That must be in the other package.

Q. Patrick Glennen's note—$225.36. Where is that?

A. That must be in the same package.

Q. Don't he live in Concord?

A. No, sir; all the notes you have read to me are Jamaica Plains notes.

Q. John McGlennan—$261.61. Have you that note?

A. I think I have all those notes at my house.

Q. James Munson—$916; where is that?

A. I think it must be at my house.

Q. Can you tell whether he has paid you that amount of money since your deposition was taken?

A. I do not think it has been paid.

Q. D. M. Dearborn—$444. Has that n ote been paid or protested?

A. That note I paid away for a pair of horses.

Q. At the time your deposition was taken, hadn't you had those horses for a long time? and did you not testify about them?

A. I gave a note for a pair of buckskin horses, and I think that was the note.

Q. My question is, did you not have the horses on hand at the time the deposition was taken and did you not then testify about them?

A. I don't remember?

Q. Do you say now you gave that note for those horses?

A. I let Mr. Rolfe have a note; it was my impression when I spoke about it before, that that was the note.

Q. You have not produced it here, and it has not been paid, has it?

A. I can't tell.

Q. Has the note of Patrick Baker, for $1280, been paid?

A. No, sir; it is in Mr. Rolfe's hands to collect.

Q. Does anybody owe you anything besides these notes?

A. I don't think of any one else who does.

Q. Does anybody owe you now anything that is not represented by the notes?

A. No, sir.

Q. Have you any other property besides?

A. I don't think that I have any other property besides what I have mentioned.

Q. Neither directly nor indirectly?

A. No, sir.

Q. Please give your present indebtedness, excluding the notes signed on account of Barter & Co., where you have an interest in the store?

A. I hired that money and paid it in there. I mortgaged some of my real estate to do it. I mortgaged the Masonic Temple for $7500, and $5000 of it went into the firm of Barter & Co. I have here a list. [Produces paper.] Amoskeag Bank, $6000; Reuben Lake, $3000; Perley Clough, $2100,

Q. Did you testify that the money that you hired from Mr. Bartlett went to pay your brother, Perley Clough?

A. In part. I did not pay him all.

Q. What was the original note?

A. I think the note, in the first place, was for $6000. I bought one hundred shares of stock with him.

Q. Did you send some money to pay that note with that you handed to Mr. Bartlett?

A. I did.

Q. How much?

A. Eighteen hundred dollars.

Q. How much was the note before you sent the $1800?

A. My impression is that it was $6000.

Q. Then that would leave $4200 still due, and the interest?

A. I think that is the amount, I paid some before that.

Q. In whose hand-writing is that memorandum?

A. In my nephew's. [Witness reads from memorandum]. $6900 was the original amount of the note. I paid, Feb. 15, 1868, $3000; Oct. 31, $1800, which would leave due, $4100.

Q. The next item is the Merrimac River Savings Bank, $10,000. Has any of that been paid?

A. Nothing but the interest.

Q. C. H. Carpenter, $3100; C. H. Capenter, $1500. Have either of these notes been paid, or any portion of them?

A. No, sir; nothing but the interest.

Q. When was this statement made?

A. It was made within a week or two.

Q. Since this hearing commenced?

A. Yes, sir.

Q. First National Savings Bank, $3200; First National Savings Bank, $7500. Where did that $3200 go to?

A. I don't recollect. I think, perhaps I bought some stock with it.

Q. Did any part of the $7500 go to Barter & Co.?

A. Five thousand dollars did.

Q. C. H. Bartlett, cash $3000. Was that the indebtedness that you pledged 50 shares of stock for?

A. Yes, sir.

Q. W. W. Elkins, $200. What was that for?

A. For money.

Q. H. S. Shattuck, $3360. What was that for?

A. He had a mortgage on the Sherman House for $1500, at the time I traded for it, and I gave my note to Mr. Dudley for $1500 more, and he bought that note of Mr, Dudley and sold both notes together, and holds both notes against me.

Q. Mr. Bickford, $150. What is that for?

A. For money.

Q. H. S. Shattuck, $350, What was that for?

A. A note that Shattuck has had on me for some time. I don't recollect the transaction now. He has had the note, I think, four or five years. I don't recollect exactly what the transaction was.

Q. E. S. Nutter, $672. What was that for?

A. Borrowed money.

Q. What did you borrow that for?

A. I paid that money on account of the furniture that is in the Sherman House.

Q. At the time you gave your deposition you had $1700 of money in hand, you stated. How much money on hand have you now?

A. I guess that I have got about $50.00. No, I don't think I have as much as that.

Q. You mean to say that you have substantially no money on hand, or on deposit?

A. I have none in my pocket.

Q. And none on deposit?

A. I have none.

Q. If you have any more liabilities, you may state them and produce them?

A. That is all, I believe.

Q. What you have stated then constitutes yonr liabilities?

A. I think so,

Q. I understood you to say that Mr. Elkins, a conductor on the Concord Railroad, is dead?

A. Yes, sir.

Q. Have you any memorandum or papers of any kind showing your transactions with Mr, Elkins?

A. I have not.

Q. Did you swear by anything then besides your mere memory of the matter?

A. I have no record.

Q. Your suggestion was very general. Can you state any facts any more specifically than you have already stated them, in regard to your alleged transactions with Mr. Elkins?

A. I can state what they were.

Q. Can you state in detail any more specifically than you have?

A. I think I can tell all the operations between us.

Q. Any more specifically than you have?

A. I think so.

Q. Then please do so?

A. I loaned Mr. Elkins $3000, in 1843, for the purpose of trading in stocks, or anything else that he wished.

Q. I do not want you to repeat what you stated before, but I want to know if you can give the details any fuller than you did before?

A. I can tell you what the transactions were between us. What more can I do?

Q. My question is a plain one. You have made some general statements with regard to your transactions with Mr. Elkins. I now ask if you can give the details any more specifically than you have already done?

A. I will go on and tell you, if you wish me to. I loaned him $3000 in 1843. I think he came on to the Concord Railroad at that time. He said that if he had some money he could make something. He knew that I had some, and he came to me for it, and made a proposition that if I would loan him $1300 he would give me one quarter of the profits and pay interest on the money. I let him have $3000, and he was to divide the profits. Sometimes he would do so once a month, and sometimes once in three months. If I did not want the money at the time, he would give me his note. The matter ran along, I think, for five or six years, and then we settled. The amount of money he made for me in that length of time, including interest on the notes, was about $5000. Whether a little more, or a little less than that, I can't say. He had a broker in Boston, by the name of Plummer, that did his business. He did not belong to the brokers' board, but was a street broker, who made in that time over $20,000.

Q. My question was whether you could state in detail any more than you do?

A. I cannot give any further details. I told it just as it was.

Q. (*By the Chairman.*) Do you know now what stock was bought and who it was sold to?

A. He bought a good deal of Amoskeag stock and sold it in the country, and sold a good deal in this town. He bought a good deal of Long Island Railroad stock also. I don't know whether he paid me much on the Old Colony stock that he bought or not; I cannot remember. Mr. Plummer, his broker, used to pick up stock for him, and he would sell it. Mr. Elkins ran on the road from Boston here, and would sell it for him. He bought some land in Woburn, I think, where he made something.

Q. (*By Mr. George.*) At the time you gave your deposition you stated that you had carriages, carts, &c., costing $600. State whether you have them now?

A. No, sir; I have two horses now. I believe I stated my carriages yesterday.

Q. What became of the sheep you then had?

A. I sold them when I sold my farm.

Q. Were they included in the sale of the farm?

A. No, sir; I think not.

Q. Whom did you sell the sheep to?

A. I let Mr. Dunklee have the sheep.

Q. What for?

A. I think I sold him all the stock and the farm for $2500.

Q. Then you sold the sheep in connection with the farm?

A. I think that I did.

Q. What other property did you sell on the farm in connection with it?

A. I think that was all. I think there were 100 sheep.

Q. Do you know whether you sold any other personal property or stock there?

A. I believe there was no other stock.

Q. Did you have stock upon your Orange farm?

A. That is the one I am speaking of.

Q. Did you have any other stock beside sheep?

A. No, I think not,

Q. Do you know Martin V. B. Edgerly?

A. Yes, sir.

Q. Did you ever sell him tickets on the Concord road by $3 or $5 worth at a time?

A. Yes, I sold him tickets at half price. I don't remember the amount.

Q. Have you sold him from $3 to $5 worth at a time?

A. I don't think I ever sold him more than $3 worth at a time.

Q. What tickets did you sell him?

A. I sold him the tickets that I had collected.

Q. Tickets that had been once used?

A. Yes, sir.

Q. Did you ever sell Lewis W. Clark tickets by the $3 worth at a time?

A. No, sir.

Q. Did you ever sell him any?

A. I have taken half fare of him.

Q. Did you also sell Lewis W, Clark tickets in the Manchester depot which had been once used?

A. No, sir; never in my life.

Q. Did you ever sell him any tickets?

A. I never sold him a ticket outside of the cars.

Q. Did you ever sell him tickets that had been once used inside of the cars?

A. I have taken his fare and given him a return ticket.

Q. My question is, did you ever sell Lewis W. Clark tickets that had been once used?

A. I have.

Q. Did you ever sell Mr. C. W. Stanley tickets that had been once used?

A. Yes, sir.

Q. Did you ever sell Mr. William Whittle tickets that had been once used, or give him a return ticket?

A. I have no recollection of doing so.

Q. Will you say that you did not?

A. I do. I have no recollection of it.

Q. Will you say that you have not?

A. I don't think I ever did. I say that I never did unless I had orders to.

Q. Did you sell William Whittle tickets that were once used, or give him a return ticket when he paid you fare in the cars?

A. I don't recollect anything of the kind.

Q. My question is whether you ever did or not?

*Mr. Mugridge.* Can he say anything more than he has?

*The Chairman.* He can say whether he will or will not swear positively?

*Witness.* I will not swear positively, but I have no recollection of doing so.

Q. (*By Mr. George.*) Did you ever sell L. B. Clough, of Manchester, tickets at half fare, or tickets once used?

A. I have no recollection of it.

Q. Will you say that you did not?

A. I will not state whether I ever did or not. There were people that I had orders to sell to; I don't recollect whether they were among them or not.

Q. Who were the other people?

A. I can't tell now.

Q. Can you mention any of them?

A. I can't say who they were.
Q. Did you sell tickets to other people, that had been once used?
A. If I had orders to do so.
Q. I do not ask about your orders, but about the fact. State whether you did or not?
A I could not state.
Q. What do you mean by saying that there were others that you had orders to sell to?
A. I think there were. I have no doubt that there were.
Q. Can you tell who they were?
A. No, I don't remember.
Q. Do you know Mr. Lafayette Robinson, of Manchester?
A. Yes, sir.
Q. Have you sold him tickets that had been once used?
A. I don't recollect that I ever did.
Q. Do you say that you did not?
A. I don't think I ever did.
Q. I ask you if you will say that you did not?
A. I will not say that I did not.
Q. What do you mean by saying that you might have sold him some coupons?
A. I mean coupons to Boston. The tickets—the tickets that I used to buy.
Q. Have you sold Mr. Robinson tickets that had been once used, and had been rode upon, and you had collected but did not punch?
A. I don't recollect whether I did or not.
Q, Did you ever sell any to Mr. Orison Hardy?
A. I don't know as I know him?
Q. Here is a note given him for $522, in the schedule?
A. I never took any fare of him. That is to say, he always had tickets that he bought at the ticket office, I suppose.
Q. My question is, did you ever sell him any tickets that had been once used?
A. Never.
Q. You can say that, can you?
A. Yes, sir.
Q. Can you say that with regard to Robinson?
A. No, for I don't know. I think I have let Robinson have tickets at half price.
Q. How many tickets have you sold him at a time?
A. Never more than one at a time, and I gave him a return ticket.
Q. Perhaps your recollection is refreshed with regard to Mr. William Whittle?
A. No, sir.
Q. Did you sell to W. P. Newell, of Manchester, tickets that had been once used, or give him a return ticket?
A. I don't think I ever did. I have no recollection of doing so.
Q. Will you say that you never did?
A. I don't think I ever did.
Q. Will you swear that you never did?
A. I have no recollection of it.
Q. My question is, will you swear that you never did?
A. No, sir, I would not.

Q. For how many years has Mr. Lewis W. Clough, or C. W. Stanley, or Mr. Robinson, or Mr. Edgerly, paid you fare in the cars? Did you give them a return ticket?

A. I should think four or five years.

Q. Did you ever do the same thing for Mr. True Eaton, of Pittsfield?

A. No, sir.

Q. Do you know him?

A. I do.

Q. Did you ever sell Mr. True Eaton a ticket or tickets that had been once nsed?

A. I never did.

Q. Did you ever sell to David S. Carr tickets that had been once used?

A. I don't think I ever did.

Q. When Mr. Carr was in the substitute business here, did you sell him tickets that had been used?

A. I don't think I ever did.

Q. Or give him return tickets if he paid you in the cars?

A. I don't think I did.

Q. Are you willing to swear that you never did?

A. I don't recollect anything of the kind.

Q. Are you willing to swear that you never sold Mr. David S Carr a ticket that had been once used, or give a return ticket when he paid his fare in the cars?

A. I have no recollection of having done so.

Q. That is not my question. I ask whether you are willing to state absolutely that you never did?

A. I have no recollection of ever having done so.

Q. Do you know Mr. J. S. Carr?

A. I know there are two brothers by the name of Carr.

Q. Are you willing to state that you never sold him tickets at half price, or that you never gave him return tickets when he paid you in the cars?

A. I state the same as regards others. I can't remember of ever having done so.

Q. I should like to know about how many people there were that you had orders to sell tickets to at half price, or to give return tickets to when they paid you in the cars?

A. I recollect three, distinctly.

Q. Can you recollect any more. I want to know how many there were?

A. I don't think of any more now.

Q. Then if you sold those other tickets for half price in the cars, you did it contrary to instructions or without directions?

A. I don't know whether I did or not.

Q. Don't you know if you had instructions to sell to three people and you sold as many more, those that you sold to other than these three as to whom you had instructions, were sold without instructions?

*The Chairman.* He has not yet said that he had orders to sell to only three. He has said that he remembers three distinctly, but he does not sa y that he had no instructions to sell to any others.

*Witness.* A good many of those men were riding during the war times and Mr. Gilmore used to make it a practice to come along in the cars to me and tell me to pass those men at half price. I don't remember all of them, now.

Q. Assuming that there were but three that you were instructed to pass and give return tickets to, tickets that had been once used, then all that you gave to the others besides the three, you gave without instructions and in violation of instructions, did you not?

*Mr. Mugridge.* I submit that a man is never bound to testify upon an assumption of that kind.

*The Chairman.* I should think he was not obliged to.

Q. (*By Mr. George.*) I will now ask you about E. M. Topliff, of Manchester, and a lawyer. Do you know him?

A. Yes, sir.

Q. Did you ever sell him any tickets, or give him return tickets when he paid you in the cars?

A. I believe I did, once or twice.

Q. Did you have any instructions from Gilmore to sell him tickets?

A. I don't recollect whether I did or not.

Q. Did Mr. Topliff ever pay anything more than half fare to you at any time whenever he rode over the road during the time when you were passing Stanley, Clark and Edgerly? Was it not your uniform practice to give him return tickets?

A. No, sir; I think not.

Q. What were the occasions when you sold Mr. Topliff tickets or gave him return tickets?

A. He was one of Gilmore's friends—one of his runners.

Q. Do you mean to say that you gave to Mr. Gilmore's friends and runners tickets at half fare, or tickets that had been once used, or gave them return tickets?

A. That was Gilmore's orders at the time he was governor—to favor all those politicians of that kind and all bank men.

Q. Did you give the bank men return tickets when they paid their fare in the cars?

*Mr. Mugridge.* We want the referees to understand that we make the same objection to this testimony that we have already made—that the road is estopped of availing itself of this evidence.

*The Chairman.* We understand the objection that has been made.

Q. (*By Mr. George.*) What bank men did you give tickets that had been once used or give return tickets when they paid you fare in the cars?

A. There were not many that paid fare anyway.

Q. If any I want to know who they were?

A. I don't remember now.

Q. Did you ever sell David Cross any tickets or give him a return ticket when he paid you fare in the cars?

A. I don't think I ever did. I have passed him.

Q. Will you state that you did not sell David Cross or give him return tickets, or tickets that had been once used.

A. I will say the same that I do of the others. I have no recollection of it.

Q. What officers of the banks in Manchester did you ever pass or give return tickets to?

A. I don't know as I have passed them on my own account. I think they had passes from Gilmore.

Q. Many of them?

A. Mr. Harrington was one.

Q. How many people in Concord were there that you sold tickets to at half fare, or sold tickets that had been once used, or gave return tickets to?

A. I don't recollect now.

Q. How was it with Mr. George Eaton?

A. I never sold him any tickets.

Q. George Eaton is the man that you were concerned with in the tobacco business?

A. Yes, sir.

Q. Did you never give him any tickets?

A. No, sir.

Q. Did you ever pass him?

A. No; I don't think I ever did.

Q. Did Mr. George Eaton, during the time you were connected with him, pay one cent of the fare to the railroad company?

A. I have passed him through the direction of Mr, Gilmore.

Q. Was that the time when he was engaged with you in the tobacco speculation?

A. He had a pass from Mr. Gilmore at that time.

Q. Did Geo. Eaton ever, during the time you were connected with him in the tobacco speculation, pay you one cent of fare when he rode with you?

A. He had a pass from Gilmore and he did not pay fare to me. He had a pass from Gilmore during all the time that he and I were together.

Q. Did Geo. Eaton ever pay one cent of fare when he rode on your train on the Concord Railroad.

A. He did not pay any fare, because he had a pass from Mr. Gilmore.

Q. Did he ever at any time when he rode on your train pay one cent of fare?

A. Yes, sir; he has had, sometimes, regular tickets, that I suppose he got from the regular office.

Q. Do you mean to say that you never passed him unless he had a pass from Gilmore or tickets? Do you say that you never supplied him with tickets?

A. I never supplied him with any tickets.

Q. And you swear that you never passed him unless he had a pass from Gilmore or a ticket?

A. I will not say but that I have passed him a few times.

Q. Did you ever give a return ticket or pass to your insurance agent, Mr. Keys?

A. I never did.

Q. I was asking you with regard to the bank directors to whom you gave return tickets, as you stated you were ordered to do. I want to know who of the bank directors had passes, or received return tickets from you?

A. I didn't say that I gave return tickets to any of the bank directors. I said that they had passes.

Q. What ones?

A. Mr. Currier and Mr. Smith.

Q. Mr. Currier is the treasurer of the Portsmouth road, isn't he?

A. I don't know.

Q. Governor Smith is another?

A. I don't think of any more in Manchester that I passed.

Q. Then all but Governor Smith (assuming that Currier was treasurer of the Portsmouth road) were entitled to a pass according to the usages of the road, on account of their positions?

A. A good many of those men that had passes to and from Manchester, I don't know whether they were bank directors or not.

Q. Is there anybody at Nashua that you used to give return tickets to?

A. I don't think of any now.

Q. Did you ever give any return tickets to Mr. Barrett, of Nashua, a lawyer there?

A. I don't think I ever did. He had an old ticket once, but I had a quarrel with him in the cars about it.

Q. Did you ever sell any tickets to Mr. William Barrett that had been once used, or give him return tickets on payment of money to you in the cars?

A. I have no recollection of having done so.

Q. Will you say you never did?

A. There were people that used to come into the cars that Gilmore has told me to pass down and give them a ticket back, but I do not remember who they all were. Gilmore would come along to me with a man and say pass this man to Nashua or Manchester, or give him a ticket back, if he was a man that he wanted to pass both ways, and did not want to give him his pass. There were a good many of that kind of men that went over on Gilmore's account.

Q. (*By the Chairman.*) Why didn't Gilmore want to give them his pass?

A. He didn't want to give so many passes out. He said that the directors were finding fault with him because there were so many passes given, and he did not want to give them. There were times without number that he would tell me to pass persons in that way—to pass them down. and if they did not come back on my train, to give them a return ticket, and I don't know but that some of the men that have been mentioned here may have been passed by me in that way.

Q. (*By Mr. George.*) Then Gilmore wanted to cover up the fact that he was passing so many people over the road?

A. It looked like it.

Q. Do you mean to state that he told you that the directors were finding fault with him and therefore he wished you to give to them tickets to supply the places of passes?

A. He said that the directors thought that there was a great deal of free riding to Boston, and he requested me to bring the passes to him when I collected them, and not to allow them to go to the ticket agent, where they should have gone.

Q. If you carried passes to him instead of carrying them to the ticket agent, how could anybody know with regard to the passes except Gilmore and yourself?

A. That was only during the last three or four years that he wanted me to do that.

Q. Do you mean to be understood here that the object was to keep the fact of these passes being used from the notice and knowledge of the directors?

A. Yes, sir.

Q. Did you ever sell tickets that had been used to William Moore, of Merrimac?

A. No, sir.
Q. Did you ever give him a return ticket?
A. I never did.
Q. Are you certain of that?
A. I don't know that I ever did. I don't know that I know the man. I may know him by sight and not by name.
Q. Have you ever given a Mr. Moore, of Merrimack, tickets that had been once used?
A. No, sir, I never did.
Q. At the time Robert Moore was interested in a speculation there, did you ever let him have tickets that had been once used?
A. No, sir.
Q. Did you ever pass him free?
A. No, sir.
Q. He always paid his fare?
A. I don't know but that I passed him over the road. I think I did pass him when he went out to Milford over the Concord Railroad.
Q. How was it with Gill. Howard, of Nashua. Do you know him?
A. I do.
Q. He kept a billiard saloon in Manchester, did he not?
A. I don't know as to that.
Q. Didn't you sell him tickets that had been once used?
A. I don't think I did.
Q. Are you willing to say that you did not repeatedly sell him tickets that had been once used?
A. I have no recollection of having done so.
Q. Will you say that you have not?
A. No, sir.
Q. Will you say that you have not given him return tickets when he paid you fare in the cars?
A. I have no recollection of it.
Q. Will you say that you have not done so?
A. No, sir.

Mr. Stanley here read a schedule produced by the witness showing the amounts for which the houses owned by Mr. Clough were severally insured; the rates of insurance, and for how long insured.

Q. Do you pay for insurance on the Sherman House?
A. I do not; the man that hires the place pays the insurance and taxes.

*The Chairman.* What is the object of this testimony?

*Mr. George.* It has a double purpose: First, Mr. Clough has stated that all the insurance that he ever paid amounted to $341, and we desire to contradict that statement. Another purpose is to show the value of the property which he has himself insured for these amounts, with a view of contradicting him as to the value of his property.

## TESTIMONY OF JAMES T. MOREY.

Q. (*By Mr. Rolfe.*) Where do you live now?
A. In Montreal, Canada.
Q. For how long have you lived there?
A. About two and a half years.
Q. Did you formerly live at Franklin?
A. Yes, sir.

Q. In whose employ were you there?
A. In the employ of Mr. James Colburn.
Q. What was James Colburn's business?
A. He was a dealer in produce, and had a store for country goods.
Q. Were you a clerk in that store?
A. Yes, sir.
Q. State whether you were a clerk in his store at the time George Clough was running as conductor?
A. I was.
Q. For about what length of time did Mr. Clough run as conductor to Franklin?
A. As near as I can remember, I think it was near the space of one year.
Q. Did Mr. Clough have any trading with Mr. Colburn? and, if so, what kind of trading, and to what extent?
A. He had a good deal of trading at that time in produce; poultry, potatoes, butter and cheese, and farm produce generally.
Q. Was Mr. Colburn a very lame man?
A. He was.
Q. Will you state who delivered the articles that were sold to Mr. Clough at the depot?
A. I delivered them myself, unless it was a cart load of produce.
Q. State, as near as you can recollect, the extent of Mr. Clough's dealings in those articles with Mr. Colburn?
A. If I remember rightly, he used to come there about three times every week, and would buy some $300 or $400 worth at a trip.
Q. What did the produce consist of?
A. Potatoes, butter, cheese and poultry.
Q. For how large a portion of the time that Mr. Clough ran to Franklin did this continue?
A. In very warm weather he would buy butter and cheese; when it moderated he would buy the other articles as they would be coming in from the farms.
Q. With reference to the time that Mr. Clough was in the trade: how long did it all continue?
A. Perhaps a year. I should think about a year.
Q. Where did the most of this go?
A. If he bought a large lot, it would go to the freight depot. If it was a tub of butter or cheese, it would be carried to the passenger station.
Q. How would it be with poultry?
A. It went as freight, if I remember rightly. There were boxes of 400 or 500 pounds each.
Q. (*By Mr. George.*) When did the cars run to Franklin?
A. I could not tell you when they began to run there.
Q. Did they not run to Franklin in the winter of 1847?
A. I don't remember.
Q. Do you mean to state that Mr. Clough run from Franklin through to Boston and back for a year?
A. If I am not mistaken, he did.
Q. Do you mean to state that?
A. I mean to state that he came to Franklin three times per week.
Q. Did he run to Franklin when the cars run beyond Franklin?
A. No, he did not.

Q. When did the cars go beyond Franklin?
A. I could not tell.
Q. Do you remember whether they stopped at Franklin more than three months?
A. I don't remember, but I think it was about a year.
Q. These transactions happened twenty years ago, or more, did they not?
A. They happened at the time the cars first came to Franklin.
Q. That is full twenty years ago?
A. If I am not mistaken, it is.
Q. What is your impression?
A. I have no impression about it, only that I was there when the cars first came to Franklin.
Q. What train did he come up in?
A. I don't know.
Q. What time of the day did the train arrive in which he came?
A. One train used to come into Franklin about 2 o'clock.
Q. Which train did Mr. Clough run?
A. I think he used to come in the 4 o'clock one day.
Q. What do you mean by coming at 4 o'clock one day? Did he come at different times on different days?
A. I think at that time there were three conductors going through, —Mr. Perkins, Mr. Clough, and another conductor who is now dead—Mr. Dole.
Q. What trains did Mr. Clough run?
A. As I told you before, he used to come in at 4 o'clock in the afternoon some part of the time, and go out again in the morning.
Q. He used to stay at Franklin over night, did he?
A. I think he did.
Q. Are you certain of that?
A. I think he stayed there more or less at nights.
Q. Have you seen, or have you any memorandum or minutes with reference to these dealings?
A. I have not.
Q. Who else used to buy stuff of Mr. Colburn?
A. We had a good many customers.
Q. Please name some of them?
A. He used to ship a good deal to Lowell, I think.
Q. Tell some of the customers who bought goods there?
A. I don't know now that I could tell who they were.
Q. Can you name any of them?
A. No, sir; I cannot.
Q. Can you mention anybody, and the particulars of his dealing?
A. I don't think that I can.
Q. (*By Mr. Rolfe.*) Is there any particular reason why you remember about Mr. Clough?

*Mr. George.* I have not asked any question which entitles to a re-examination of this witness.

*The Chairman.* I don't know that there is any objection to the witness being re-examined. Of course the counsel will not go over the same ground that he has already been over.

Q. (*By Mr. Rolfe.*) Is there any other reason why you recollect about Mr. Clough?

A. We considered him a great customer, and sold him a great deal. We calculated that what we took in one day we would sell to him when he came up again.

Q. (*By Mr. George.*) You can't remember a single other customer beside Mr. Clough?

A. He was the largest customer that we had.

Q. Can you tell one single other person that you dealt with?

A. I could not recall their names.

Q. For how long were you in that store?

A. I was in that store from three to four years.

Q. What is your business in Montreal?

A. I keep a livery stable.

## TESTIMONY OF HORACE JOHNSON.

Mr. Stanley offered to read in evidence the report of the case of Horace Johnson *v.* the Concord Railroad Corporation, (46 N. H. R. 213.)

*Mr. Mugridge.* What is the object of this offer?

*Mr. George.* The object is to show that there was a case, the fact of which Mr. Clough knew, against Mr Noyes, for putting Mr. Johnson out of the cars for attempting to ride upon a coupon ticket, which will be produced here.

*Mr. Mugridge.* We object to that. Proving the existence of the suit, is no proper or competent way of showing knowledge of the suit home to us. Our witness has been interrogated and has sworn to all the information he has on the point; and the offering of the record referred to has no tendency to contradict him in any way.

*The Chairman.* We understand that this suit of Johnson *v.* The Concord Railroad, was an action which was brought against a conductor for putting a man out of the cars, who was trying to ride on a spent ticket. The question, if I remember rightly, which was determined in that case, or which was one of the important questions in that case, was whether the railroad had the right, in the way in which they attempted to do it, to limit the time which that ticket had to run, and under the circumstances to enforce that limitation. It is, therefore, perhaps correct to say that it was a case where the railroad was contesting the right of a passenger to pass on a ticket, which by their regulations, was a spent ticket; and we understand Mr. Clough to say that that was his understanding of the case; that the railroad at that time were contesting the right of Johnson to pass on a ticket, which by their regulations, was a spent ticket. This testimony, if it went in, would have a tendency to prove Mr. Clough's knowledge of the resistence of the road to any such use of the coupon tickets, but perhaps would have no such tendency to contradict any idea that the superintendent had authorized any such use. But according to our recollection, there is not a particle of evidence in this case tending to show that Mr. Clough ever sold a spent ticket. Mr. Clough states that Mr. Gilmore authorized him to sell tickets which had time to run, which were not spent tickets; but that he told him to sell no others; and he states that he did buy and sell tickets which were not spent, and had time to run. In another point of view, the testimony as to the using, the buying and the selling of tickets which were not spent. is important; and on that point we have an opinion which we are willing to give at any time it may be required. We think that this testimony, about the existence of this case, and the contest that was made about it, has no tendency to prove that there was any resistance to Mr. Clough, or to disprove what Mr. Clough says about buying and selling tickets which were not spent tickets; and therefore we think the testimony is inadmissible. If I remember rightly, the referees have already communicated their view with regard to the matter of buying and selling tickets which were not spent. I think that they have intimated to the parties here, that Mr. Clough, being an agent of the road, and an agent to sell the tickets of the road, and the evidence tending to show that the buying of these coupons at a low price, and selling them for full price, was injurious to the road, and would deprive the company of their profits, the road would have a right to adopt his acts, and make him its agent for that purpose, and to claim from him the profits; and very likely it would be able to hold them in that matter, but we do not think the testimony in this case has any tendency to prove that.

*Mr. George.* I think it would be well to read this deposition of Mr. Clough's.

*Judge Bellows.* As I understand the evidence, it was offered for the purpose of enabling the referees to make an inference that Mr. Gilmore gave no such authority.

*Mr. George.* That is one point.

*Judge Bellows.* And that Mr. Clough had knowledge of the fact of this action. Now, so far as regards the selling of tickets which were spent, Mr. Clough does not state that he had any authority to do so from the corporation, in any way. He had authority only to sell such as were in force. Therefore if he sold any that were wholly spent, it was without authority, and this evidence in that point of view, would be a matter of no importance, because he does not pretend that he had any authority to sell a spent ticket.

*Mr. George.* I beg leave to suggest in reply to that, that these tickets were all, every single one of them, spent tickets; that every ticket that Mr. Clough took, as matter of fact, was a spent ticket. We say further that the object of the rule was clearly and simply to prevent the substitution of through tickets for local fares. Nobody can conceive of any other object of the rules, because, otherwise, the local business of the road would be very largely diminished. Mr. Clough makes the statement here that Mr. Gilmore told him he might sell some of those tickets in contravention, as we say, of the rule, and of the reason of the rule if you please; and as bearing upon that question, we say that both the notice served upon Mr. Clough by Mr. Gilmore, and the fact of the existence of that suit, is competent. The only way that the road was imposed upon, was not by the passenger laying over on one of those tickets, but it was by the substitution of that ticket for a local fare; and that is what we say was the fraud upon the road.

*Mr. Mugridge.* That is the inference that you make from the evidence that is in; we take an opposite view from that. Do I understand that the referees have considered the question, and have intimated an opinion as to whether or not there was any fraud practiced upon the Concord Railroad by the purchase and sale of tickets that were not spent, or within thirty days of their purchase?

*The Chairman.* The present ruling has nothing to do with that.

*Mr. Mugridge.* I thought that the referees intended to make a suggestion as applying to that point.

*The Chairman.* Not at all. I should have said that the rule, notice of which was served upon Mr. Clough, orders that these things shall be done "as soon as possible;" that is, the rule distinctly contemplates that the old order of things may go on some time longer.

*Mr. George.* The order was that as soon as practicable, the tickets should be dated. In order to date the tickets, it was necessary that the ticket offices should be supplied with stamps.

*The Chairman.* But the material matter is not with respect to the tickets sold over the Concord road, but those sold over the western road.

*Mr. George.* But all the offices had to procure stamps in the same way.

*The Chairman.* But this rule contemplates that some time would elapse before it went into operation.

*Mr. George.* That time was fixed in Mr. Gilmore's notice to Mr. Clough, that the rule would go into absolute effect on the first day of January following.

*The Chairman.* The idea now, as I understand the matter, is this: that in neither point of view is the testimony in regard to this case admissible, because it does not relate at all to the selling of tickets which were not spent, and Mr. Clough avows that he had no authority to sell tickets that were spent.

*Mr. George.* In his deposition, in reply to interrogatory 178, Mr. Clough says: "My way-bills would have been larger if I had not bought tickets at the office with the money that I took in the cars. Another thing, I used to buy Boston coupons that were brought from the north, and sold them."

Interrogatory 185 is: "In answer to interrogatory 178 you say you bought 'Boston coupons.' Please state what you mean by 'Boston coupons,' how many you bought, what you paid for them, and for how much you sold them, and how you disposed of them. Ans. People used to buy tickets up north, and could buy tickets through to Boston just as cheap as they could to here. I used to buy their tickets of them. Some used to stop here, and some at Hooksett, and some at Manchester. I used to buy them for from 50 cents to $1 apiece. I have had them given to me. There was a family of eight once got out at Manchester, and gave me their tickets to Boston, saying they didn't want them any longer, as they stopped there; and I used to buy them. My impression is that I bought from 150 to 200 per year. I sold them for regular fare. I disposed of them in the cars. If passengers wanted tickets to Boston, I used to sell them if I had them."

"Int. 186. Was not this buying of tickets, and selling them for local use, a direct fraud upon the road, and were not said tickets, both upon the face, and by the rules of the road, not transferable, and good only to the original purchaser, and on the day of their date, and for such additional time as would enable the passenger to complete the passage? Ans. I did not understand it was any fraud to buy those tickets and sell them again. Gilmore knew I was buying and selling them. I considered them good for one seat to Boston. I think they were not 'not transferable.' I don't think they were on their face good only to the original purchaser and for such time as would enable persons to complete the passage; but I am not certain about that."

"Int. 187. Did you not put a person off the cars who claimed to pass on one of these tickets, and were you not sued, and did not the corporation defend your suit, and sustain your action? If yes, state who it was, and when, and what kind of a ticket was attempted to be used. ANS. I have no recollection of being sued for putting off a man. I recollect having a fuss about it. I recollect putting a man out who had one of those western tickets, but I think it was a very old ticket, which looked as if it had been rode on over the road once before. I think it had been punched."

"Int. 188. Did you ever refuse, or were you ever instructed to refuse to take these western coupons when they had passed into second hands? ANS. I did so refuse when they were of old date. I was instructed to so refuse such coupons of old or long standing date."

"Int. 189. If said coupons were good in second hands, and they had never been rode upon, what difference did it make whether they were of old or of recent date? ANS. On these western tickets we didn't get but little any way. Our proportion was very small. I think I have heard Gilmore say that the road didn't get more than 25 or 30 cents on them. I never had anything to do with the western tickets. The tickets I bought were from western New York and Canada way. They sometimes used to get by us on these old ones. We might skip them in some way. When they were very old I did not use to take them. I did take them unless they were very old, rather than have a fuss with them. I do not know why they are not just as good, if old, as if of recent date, provided they are good in second hands and never have been rode upon." [Adjourned.

[TWENTIETH DAY. Friday, December 25th, 1868.]

The hearing was resumed at 9 o'clock A. M.

## TESTIMONY OF GEORGE CLOUGH CONTINUED.

Q. (*By Mr. George.*) Since you were discharged from your position as conductor, have you made any addition to your property, with the exception of the ordinary increase of the property from rents and income,—that is, since February or March, 1866? If so, what?

A. I do not think of anything now.

Q. Then your property now, so far as the amount is concerned, is the same that it was in 1866, adding the difference between your income and your expenses, and deducting the losses that you have testified to, between these two periods?

A. I think so.

Q. Have you a niece in your family? and have you supported her? and, if so, for how many years?

A. I have. I think she has been there about five years.

Q. You have found her her clothing and supported her entirely?

A. I have.

Q. (*By the Chairman.*) How old was she when she came to your house?

A. I think she was seventeen years old—sixteen or seventeen.

Q. (*By Mr. George.*) What kind of a fence have you around the house where you live, and what was the expense of the fence around your lot and connected with it?

A. It is the same fence that I bought when I bought the place.

Q. You have not made any addition to the fences?

A. I have painted them; that is all.

Q. Do you know what the expense of painting the fences was?

A. No, sir; I do not.
Q. How large a lot is it?
A. I think it is not far from a quarter of an acre; it may be a little more.
Q. You have been to considerable expense in setting out trees and flowers, and arranging the grounds, have you not?
A. No, sir; the trees were all set out when I bought the place.
Q. How about the stables connected with the place? Have you built new stables, or built additions to the old ones?
A. I have put some wings on to the stable.
Q. How large is the stable connected with your residence? State about its size?
A. I think it is about twenty-five feet by thirty or thirty-five.
Q. Including the wings?
A. No, sir.
Q. I want the size of the wings and all. How large are the wings?
A. I should judge they were about from eight to ten feet each way.
Q. Is the stable clapboarded and painted?
A. Yes.
Q. Is it painted the same as your house?
A. It is.
Q. Is the stable lathed and plastered, or finished in any particular way?
A. There is one room that is plastered; and that is the same as it was when I bought it.
Q. In what wood is the stable finished?
A. It is finished mostly with spruce, I think.
Q. Is it painted, or stained inside?
A. I think it has had one coat of lead-colored paint.
Q. How are the stalls arranged?
A. They are common stalls.
Q. With racks, or patent feeding apparatus?
A. With patent feeding apparatus.
Q. How many stalls are there with patent feeding apparatus?
A. Five, I think.
Q. Who put those in?
A. I did.
Q. Have you a furnace connected with your house?
A. I have steam.
Q. What was the expense of that?
A. I think it was about $800.
Q. When was that put in?
A. It was put in when I built the house.
Q. The stables were not burned when the house was burned?
A. No, sir.
Q. What was the expense you laid out on the stables?
A. I should estimate it at somewheres near $1500.

*Mr. George.* I will now read a copy of the purchases and sales of stock, by Mr. Clough, of the Concord Railroad Company, furnished by the treasurer, which it was agreed should be read without objection.

[Statement read.]

In this case—October 31, 1868, when you bought 100 shares of stock, you simply altered the certificate, or had the certificate divided up.

A. I will explain how that was. The stock was transferred to the Merrimack River Savings Bank, where I expected to get money; but Governor Smyth said that Mr. Bartlett had $3000 that he wanted to let, and I got of him; and as the stock had already all been transferred to the bank, I had it transferred back to myself. There was a certificate of 150 shares, and I had 100 shares transferred back to myself.

Q. There was then no purchase and sale of stock on the 150 shares?

A. No, there was not.

Q. When was the stock of the Concord road the lowest? Are you able to state about what it was worth in 1846?

A. My impression is that it was then at par.

Q. Was it not worth $75 per share then?

A. No.

Q. From 1842 to 1850, did not that stock pay ten per cent.? And did not the directors then make an extra dividend to cover ten per cent. on the investment on the road, from the time that the investments were called in?

A. I do not think the stock was far from par at that time.

Q. Did the stock pay ten per cent. interest, from the time that the road was built, up to the time of the opening of the Lawrence road, right along?

A. I think it did. People were more afraid of railroad stock at that time than they are now.

Q. Was not the second track built about that time? and were not the rights to subscribe sold for $10 or $15?

A. I do not know; the rights were sold, but I do not know at what price.

Q. Then if the rights were sold the stock must have been above par?

A. Yes, if the original stock was not below par.

Q. Have you ever heard of the original stock being below par?

A. Yes; it has been as low as $33.

Q. When did you buy it as low as $33?

A. In 1856 or 1857.

Q. I am now talking about 1846, when the Lawrence road was opened. Do you not know that up to that time the road divided ten per cent. and then made an extra dividend to cover ten per cent. upon the amount invested from the time the investments were made?

A. I do not recollect about it.

Q. You say that in 1856 or 1857, you bought stock as low as $33. Will you state how many shares you bought as low as $33?

A. I do not remember.

Q. In 1857 was not the market price of the stock at the brokers' board, $38.75?

A. I do not think so. I know I bought some stock at $33 per share, but how much I cannot tell. I bought it, I think, all the way along, from that up to par, about that time. My documents at home will show that.

Q. Will you bring in everything that you have showing the price of the stock?

A. I will.

Q. It seems that on February 2, 1846, you transferred to the Merrimack River Savings Bank, twelve shares of stock, which was the first stock you had. Was that a sale to the Merrimack Bank, or was it a collateral transfer.

A. I think it was to borrow money.

Q. February 5—there was a transfer of twelve shares to the bank. Was that a sale or a collateral transfer?

A. I cannot say whether it was a sale or not. I bought stock, and sold it, and I also pledged stock.

Q. "C. A. Evans and J. A. Elkins, 16 shares;" was that a sale or a transfer?

A. I think it was a sale.

Q. Are you positive about it?

A. That is my impression.

Q. "1859, May 11, Joseph A. Gilmore, 7 shares." Was that a sale or a transfer?

A. I cannot tell, I should think it was a sale.

Q. If you will be kind enough to bring in any papers that you have, showing what you paid for this stock, it will save further increase upon that branch of the subject. I omitted to ask you upon one point. What was Daniel Raymond Kimball's business?

A. He kept a livery stable.

Q. In the rear of the Merrimack House, at Lowell?

A. Yes.

Q. Had he any other business?

A. He used to let and hire money.

Q. But as far as you know, he had no other business except keeping a livery stable, unless he hired or let money?

A. No, sir.

Q. Did you undertake at one time to carry on the livery stable business in Manchester?

A. I had an interest in one there.

Q. How much did you make or lose in that operation?

A. I think I came out about square, but I do not recollect now certainly.

Q. When was that?

A. I think that was in 1846 or 1847.

Q. Was John Floyd your partner?

A. He and Captain Walker were my partners.

Q. Did you not lose money there?

A. I do not recollect. We lost but little, if any. There were three owners.

Q. How large a stable was it? How many horses were there, and how much had you invested in it?

A. My impression is that there were about twenty horses.

Q. Do you remember how much you had invested there?

A. I should think not far from $4000.

Q. Was there $16,000 in all, or only $4000 in all?

A. I think there were $4000 invested in all.

Q. Did not John Floyd burst up, and did you not lose all of your investment in that transaction? And was he not poorer than a church mouse when he came out of it?

A. He hadn't much when he went in.

Q. Did he have anything when he came out?

A. I do not think that he had much money.

Q. He was as poor as a church mouse when he came out, was he not?

A. I do not think that he had a great deal.

Q. He was the managing man, was he not?

A. Yes; he stayed there.

Q. What was James Whitcher's business?

A. In wooden ware, brooms, &c., I think.

Q. Did he make brooms, clothes wringers, sap buckets, mop handles and wooden ware of that general description? Was not that his occupation?

A. I do not know that he made mop handles, or clothes wringers, nor do I know but that he did.

Q. Did he make clothes dryers?

A. I think so.

Q. He made wooden ware of that general description?

A. I think so.

Q. How many sap-buckets did you have of him during the time that you were letting him have tickets,—be they more or less?

A. I had three hundred sap-buckets of him at one time.

Q. Did those go up on your Orange farm?

A. Yes.

Q. Did you have brooms, pails, and a clothes dryer of him?

A. I did; and when I got into my new house I had some brooms and pails of him.

Q. Didn't he furnish you the wooden ware for your new house, in the line of his business?

A. I say that I had some brooms and pails of him at the time I went to keeping house the last time.

Q. Did you have a clothes dryer of bim?

A. Yes; and I owe him for it now. I told him to send his bill when he sent the things, but he did not, and I have not seen the man since. When I see him I intend to pay him. I have the list of the stuff in my pocket, now, and I intend to pay him as soon as I see him. I have seen him but once since then, and he was standing on the platform at Hooksett, and I was in the cars, which were in motion.

Q. Have you sent nobody else to see him?

A. No.

Q. Do you mean to say that you did not send Mr. Rolfe down to see him?

A. He went on his own hook.

Q. Didn't you know that he was going?

A. I do not think that he told me anything about it until he got back. That is my impression.

Q. Your deeds are somewhat mixed. In my examination of them I am unable to understand them precisely. Will you be kind enough to select all the deeds that you have of all of your real estate property—of all that you have purchased from 1842 down to the present time, and let me have them?

A. They are all here in this bundle.

*Mr. George.* We offer these deeds, so that there may be no question about the matter. If there is no objection, the referees may consider as in, such deeds as the witness shall produce.

### RE-DIRECT EXAMINATION.

Q. (*By Mr. Rolfe.*) You were asked in relation to the deed of Howe and wife to you; and I think that you stated that you did not own that land at this time, and was not able to explain about it. State whether you previously owned a farm in Holderness?

A. I did.
Q. State whether you deeded all of that property to Mr. Howe?
A. I believe I did, a part of it.
Q. State whether you sold the whole farm to Mr. Howe?
A. I think I did.
Q. And took a mortgage back?
A. Yes.
Q. Did, or did not Mr. Howe want to sell a certain part of it to two men by the name of Webster?
A. Yes.
Q. How was that transaction managed?
A. Mr. Howe deeded it back to me, and I deeded a portion of it to Mr. Webster.
Q. And this is the land that you afterwards sold to Mr. Webster, that was deeded to you by Mr. Howe?
A. Yes.
Q. At the time you gave your deposition two years ago, state whether you made a detailed statement of all that you then owed?
A. I calculated to at that time.
Q. Did you do it, according to your best recollection?
A. I did. I did not know that I left anything out.
Q. When you were enquired of the other day in relation to your losses, and your outs, did you give in the statement you made, what you had paid as internal revenue tax the year before?
A. I believe not.
Q. Please look at the statement and see how that is. Look on the statement and see if you do not find the internal revenue tax paid?
A. I now find that I did estimate that. It is here—$441.05.
Q. Then if you neglected to state that, the other day, you did it by mistake?
A. Yes; I supposed that was in my deposition.
Q. Have you anywhere a detailed statement of your indebtedness on the first day of February, about the time this suit was bought?
A. I have. [Witness produces it.]
*Mr. Rolfe,* I think, if I am not mistaken, that the witness has stated the amount that he owed, but not in detail. I propose to have him state now, in detail, the amount of his indebtedness at the time this suit was brought.
*Mr Stanley.* Is it the same as in his deposition?
*Mr. Rolfe.* I do not think it is in his deposition.
*Mr. Stanley.* There is a detailed statement of what he owed, in his deposition, which is to be put in.
Q. (*By Mr. Rolfe.*) I want to ask you if you put in that estimate any physiciansj bills?
A. I did not.
Q. Why not?
A. I forgot it; it did not occur to me at the time.
Q. Who has been your family physician?
A. Dr. Carter has been our regular family physician.
Q. Have you employed Dr. Morrill some?
A. Yes; and Dr. Moore.
Q. You were enquired of yesterday as to the amount of your physicians' bills. Will you state now, as near as you are able, the average amount that you have paid for physicians' bills?

A. I should estimate it at somewhere from $15 to $20 per year, during the time I have lived in Concord.

Q. You were enquired of yesterday as to whether you had ever employed Dr. Gage, and you said that you did not think that you had. Upon recollection what have you to say?

A. I think that I did once, when I first came to Concord.

Q. Can you state the amount of his bill?

A. I do not recollect precisely; I think perhaps, it was $15 or $20.

Q. Since you have been connected with the railroad, what, to your knowledge, has been the custom of the Concord road in relation to having a boy, or other person, ride on the cars and peddle sweet-meats, apples, oranges, &c.?

A. They always had one. I never knew the time when they had not.

Q. (*By Mr. Rolfe.*) State the fact, during the time that you were on the road, as to a boy running in the cars?

A. There was a boy running on the road and I employed him.

Q. For what length of time did you employ him?

A. My impression is that he run six or seven years.

Q. During what years,—from the first opening of the road?

A. Yes; from 1842 to 1848 or 1849.

Q. In what trains?

A. He used to run with me a part of the time, and I think he run on the other trains a part of the time.

Q. How frequently?

A. Every day.

Q. What has been the fact since then, and up to the time you left the road, about boys running on the trains?

*Mr. George.* I wish to renew my objection to that. I suppose that may be objectionable upon another ground. The objection that we now make is, that the fact that a boy ran subsequent to his employing one, would have no legal tendency to show any fact with regard to their running previously. Subsequent custom does not tend to show preceding custom.

*The Chairman.* We think that this evidence, in connection with the evidence that is already in, is admissible.

*Mr. Rolfe.* Proceed, and state what has been the fact since you left the road?

A. The fact is that they have always had a boy on the trains. I never knew a time when there was not one. There were two,—a newspaper boy and a peddler boy.

Q. State the fact as to their paying fare?

A. I never knew of their paying fare. I never had any order to make their pay, and I do not think they ever did pay anything.

*Mr. Tappan.* I would like the referees to make a note of the fact that the whole deposition of Mr. Clough is in the case.

*Mr. Clark.* We did not say that it was in, but that we proposed to put in in.

*Mr. Tappan.* I understood you to put in the whole depostion.

*Mr. George.* We shall not put in the whole deposition; but I want to examine the deposition with a view of seeing whether there is anything in it that would make it objectionable as a legal matter. We shall probably put in the whole deposition; if we do not, it will be because we suppose it will be objectionable, because of some legal technicality. But everything that pertains to this, we will put in. We do not propose to be bound to put in the entire deposition. We propose that the detailed statement of property shall go in; and I will say in addition to what I have said, that unless there is a legal technicality against putting in the entire deposition, we shall put the whole of it in.

*The Chairman.* On the suggestion that the deposition, if already in, was put in by mistake, I think the parties can, if they desire, modify their action. It being understood now that this detailed statement of property will be put in with the deposition, do you, Mr. Rolfe, care about going on now with the details?

*Mr. Rolfe.* I suppose that if the deposition is put in for that purpose, there is no object in going on.

*Mr. George.* I wish it understood that only that part which relates to the detailed statement is now put in.

Q. (*By Mr. Rolfe.*) You were enquired of yesterday in relation to some notes that you had on the list, and whether or not they were paid. Will you now turn to the list that you showed yesterday, and state whether or not you have been able to find the notes of which Col. George enquired.

A. I have not been able to find them. I think that my agent at Jamaica Plains must have those notes. Some of them have been paid.

Q. Are they all Jamaica Plains notes?

A. Yes; all of them.

Q. How is it with John McDonald's note?

A. That has been paid. I found a document last night that satisfied me that that has been paid.

Q. How is it as to the other notes?

A. I supposed that I had them at the time, but I think now that Mr. Dudley, my agent there, must have them. They pay every year, more or less, and Mr. Dudley indorses the payments.

Q. You have not the other notes that are mentioned on the statement?

A. I have not.

Q. What amount has been paid on them, as near as you recollect?

A. I think there are none of them but have more or less indorsements on them, but how much has been paid, I cannot tell,

Q. Indorsements of money paid to you?

A. The money is sent to me by Mr. Dudley, the man who tends to my business there.

Q. Give a list of the Jamaica Plains notes?

A. John McDonald, that has been paid; P. Sweeney; Thomas McGuire; M. E. Donohue; Patrick Glennen; John Glennen and James Morrison.

Q. Now come down on this list, to the note of David Dearborn?

A. That has been paid.

Q. When was it paid?

A. I think it was paid a year and a half ago.

Q. State how, and in what manner the note was paid?

A. Mr. Rolfe collected it for me. It was paid in money.

Q. State whether there was a suit against the surviving partner of that establishment?

A. There was.

Q. Do you recollect who was the man that paid the money?

A. I think it was his partner?

Q. You have spoken about a note that was given in consideration of a pair of buckskin horses. What about that note?

A. That note was Ira Morse's note for something over $400. My first impression was that it was this Dearborn note that was given, but I find that it was not.

Q. State whether you gave me the note and I paid you the balance between that and the horses.

A. You did.

*Mr. Rolfe.* We may wish to call Mr. Clough again, and would like to save that privilege.

*The Chairman.* Of course; and perhaps the other side may wish to examine him further.

## TESTIMONY OF ISAAC SPALDING,

Q. (*By Mr. George.*) Will you state for how many years you were one of the directors of the Concord Railroad?

A. Twenty-five years, I think.

Q. From what time to what time?

A. From the commencement of 1866.

Q. State how it was about your being also treasurer of the road?

A. I was treasurer of the road.

Q. From what time to what time?

A. The same time.

Q. Were you the first treasurer?

A. No; Mr. Peter Carter received one dividend, and died, and I succeeded, and remained treasurer for several years.

Q. You were subsequently president of the road?

A. Yes, sir.

Q. Do you remember the years when you were president?

A. I think I was president 20 years or more. I was president over 20 years.

Q. You were president from 1846 to the time Judge Upham was elected?

A, Oh, yes. I was director all the time.

Q. Judge Upham was elected in 1857. Mr. Spalding, I want to ask you how it was with regard to your having—or, were you present at the time the rules were adopted in 1864 and 1859? Were you present?

A. At all the meetings.

Q. I will ask you how it was so far as you have any personal knowledge, and so far as you are aware of any knowledge on the part of the directors as to those rules being waived, modified, or annulled. [Same exception by deft. as previously.]

A. No, never.

Q. Mr, Spalding, I will state, Mr. Clough has stated here, that in the early years of the road, prior to 1849, he carried express matter, butter, eggs, etc., down over the road, and brought up articles over the road, from which he estimates a profit of $5200. From 1846 you were president of the road and director; now I want to know if you ever heard of any such business?

Q. (*By Mr. George.* Will you state whether you had any knowledge of Mr. Clough's doing any express business during the time you were president?

A. No, sir.

Q. Had you any knowledge of his carrying stuff over the road; and if so state what?

A. Except when Judge Upham was superintendent, I was coming up here, and the Judge said that Mr. Clough had brought some fish, I think he said ; as I understood, it was but one or two that he brought. And he wished me to talk with him about it; and I went and talked with him about it. It wasn't anything of any consequence, and I told him that he could not be allowed to carry stuff.

Q. With that exception, have you knowledge of his carrying anything over the road? Had you any knowledge?

A. I had not.

Q. And had the board of directors, so far as you are aware?

A. No, sir.

Mr. Mugridge called the attention of the referees to the fact that exception was taken to the introduction of all this testimony, the same as in the previous objections.

Q. How was it about that being the only time that your attention was ever called to it?

A. That is the only time.

Q. Mr. Clough has testified, sir, that he had a boy upon the cars, paying him a certain sum. [Objected to.]

*The Chairman.* It will be enough if you simply ask the question. The counsel on the other side object to your reciting the testimony of Mr. Clough or anybody.

Q. Had you any knowledge, or had the board of directors, so far as you are aware, of Mr. Clough's employing a boy to run upon the cars to sell peanuts?

*Mr. Mugridge.* We take an exception to this, on the same ground as previously; and the further exception that it is entirely immaterial in this connection. [Admitted provisionally.]

Q. Had you any knowledge on that subject?

A. I had not.

Q. When was the first time you ever heard of it?

A. Yesterday, I believe,

Q. So far as you are aware, had the board of directors any knowledge?

A. No, sir; never heard any mentioned.

Q. Mr. Spalding, had you any knowledge, or had the board of directors, so far as you are aware, that the conductors, Mr. Clough, or any of the other conductors, did not return the money taken in the cars upon their way-bills; but did take the money, and buy fresh tickets at the ticket office, and punch them, and return them in their collections as representing the money. Had you any knowledge of any such transaction as that? [Same exception as before.]

A. No, sir.

*The Chairman.* If this matter is excepted to, there is one limitation perhaps. I suppose this should be confined entirely to Mr. Clough. Before, we have ruled out everything that did not relate to Mr. Clough.

Q. When was the first time you ever heard of any such thing as that?

A. Day before yesterday.

Q. Had you any knowledge, sir, or had the board of directors, so far as you are aware, that Mr. Clough was buying up coupons of upper roads where the fare is the same to Manchester, Boston and Concord?

A. No, sir.

Q. Mr. Spalding, I want to call your attention to this. Do you remember, sir, a series of rules that were adopted in 1864, which were required to be signed by the conductors?

A. Yes, sir; I am not certain about the year; but I presume it was 1864.

Q. You remember the fact?

A. Yes, sir.

Q. Now I want to ask you who presented those rules to the board?

*Mr. Mugridge.* How is that material.

*The Chairman.* Well, it was permitted to go in to gratify Col. George.

A. You or Judge Upham, I believe it was you, I am not certain. I don't know whether we did or did not modify them. There might be some circumstances that would recall it. It is gone from my recollection now.

Q. Mr. Spalding, I want to know whether you had any knowledge, or the directors, so far as you are aware, of Mr. Clough selling tickets that had been used once. I want to ask you whether you had any knowledge, or the directors, so far as you are aware, of Mr. Clough's selling tickets that had been once used, and that he took the money that he received from these tickets, and bought fresh tickets and punched them and returned them with his collections?

A. No, sir.

Q. When was the first time you ever heard of that?

A. Day before yesterday. These have an agreement on the back.

Q. Yes.

A. I now recollect something about it. I believe it was offered before; I am not sure. I believe this with an endorsement on the back, they were to sign it, weren't they?

Q. Yes, sir.

A. I believe that was the fact.

Q. Now, Mr. Spalding, I want to ask you—on the 24th of October, 1865, there was a meeting of the board of directors in Boston, when an order for investigation was made—I want to ask you upon what representation, and who made it, that representation was made, [Mr. George reads from the records of the corporation,—"present the whole board, except Mr. Peaslee." Vote also read.] Do you remember the circumstances of that vote being passed?

A. I know it was passed.

Q. Mr. Spalding, was I present at that meeting?

A. I think not; I was clerk *pro tem.* and Col. George came in before the meeting adjourned, and the votes were passed over again, that he might record them instead of myself.

Q. Will you state upon what representation, and who made the representation, that vote was passed? What was the representation and who made it?

A. What?

Q. Who was the foundation and reason of that vote being passed?

A. I don't recollect.

Q. Was any statement or representation made to the board about the irregularities of the conductors?

A. No, sir. I don't know as I understand your question.

Q. Did anybody make any statement to the board as to what the conductors were doing?

A. They had several meetings, and there were some intimations that there was fraud. I cannot say who made the representations.

Q. Do you remember Mr. Gilmore's making the representation that the conductors were stealing to a large amount?

A. Oh, yes.

Q. What did he say.

A. He said they had stolen $50,000 worth. He had considerable to say about it; and the directors thought there was something wrong about it.

CROSS EXAMINATION.

Q. (*By Mr. Mugridge.*) What is your age?

A. I am sixty-eight or nine.

Q. Let me ask you if representations were also made to the board by Col. George that the conductors were stealing large amounts of money at different times?

A. I should think there might have been some; I don't recollect.

Q. Let me ask you, Mr. Spalding, if Col. George at different times before and since that time represented to you that the conductors were stealing large sums of money from the Concord Railroad?

A. He has since.

Q. Has he told you that he was satisfied that Mr. Clough had been stealing more than $50,000 a year. If so, will you state it?

A. That was Gilmore, I think.

Q. Hasn't Mr. George?

A. I don't remember; he thought that the conductors were stealing money from the Concord road.

Q. Did he ever undertake to show you from papers that he had, at the time he talked with you, how much they were stealing?

A. I should think that he had something about the different amounts; I don't know what; I couldn't say what.

Q. Where did he carry the papers that he showed you? Can you tell?

A. No, sir.

Q. Do you remember?

A. No, sir.

Q. Where were you when you had this meeting, when these things were said?

A. At the Revere House, Boston.

Q. Did the Colonel talk pretty freely about this?

A. He explained them.

Q. He is apt to talk pretty freely when he gets a going?

A. Yes, sir.

Q. Do you know who wrote that vote passed October 26th, 1865?

A. No, I do not; I don't remember who wrote the vote.

Q. Who ordinarily wrote the votes?

A. Mr. George wrote some, and Mr. Upham wrote some.

Q. You are familiar with his style; hear this, and see if this is his style. [Reads the vote,] Now, sir, from hearing that read, who should you say wrote that vote?

A. I don't know but Col. George made the first draft. I don't know but it was amended afterwards.

Q. You say he made the first draft?

A. I don't know, but I think he did; whether he made it or Judge Upham, I don't know.

Q. Now, Mr. Spalding, about the peanut boy. Have you ever known a time since you knew anything about the Concord Railroad, but that there has been a boy on the train selling candy and sweetmeats, etc.?

A. There is generally some such thing.

Q. Let me ask you if there sometimes have been two? I don't mean that they both sold peanuts; but hasn't there been one that sold peanuts and candies, and one that sold papers?

A. Yes, sir; there have.

Q. Now, did you ever know of one of those boys paying anything to the railroad for the privilege of running? As president or director did you ever know of a single fare being paid by a boy who ran on the train to sell peanuts.

A. Never did. They might or might not; I don't know.

Q. You didn't know of anything of that kind?

A. No, sir.

Q. Have you acted, Mr. Spalding, sometimes as superintendent?

A. No, sir.

Q. Never in that capacity?

A. No, sir.

Q. But you have acted as president?

A. Yes, sir.

Q. Now, let us go to another matter. You say that you do remember that you had an interview with Mr. Clough once; at the request of Judge Upham, was it not?

A. Yes, sir.

Q. Do you remember when that was?

A. I do not know.

Q. What would be your impression, as near as you can state?

A. It was before Judge Upham went to Europe.

Q. He went to Europe during Gen. Pierce's administration?

A. Yes, sir.

Q. Did George Clough stop over that day for the purpose of seeing you, and having an interview with you?

A. I suppose he did. I think it was his day to go down.

Q That is, he should have gone to Nashua that day, but remained for the purpose of having this interview with you?

A. Yes, sir; the conversation was very short.

Q. Now, didn't you understand,—hadn't it been stated to you by Judge Upham before you had this interview with Mr. Clough, that Mr. Clough had been carrying produce over the road to the market below?

A. He stated to me that he had violated the rules.

Q. In what particulars?

A. In carrying something, some fish, or something that he took with him. I talked with Mr. Clough, but it did not seem to amount to anything.

Q. Did you understand that Mr. Clough stayed over, and that Mr. Upham requested that interview simply for carrying fish over the road?

A. I thought there was more to it than there was. I thought he had been violating some of the rules. I think he said he had brought two fish.

Q. I suppose you are not certain as to the exact words of the conversation?

A. I am not.

Q. And you would not undertake to come in here and repeat the exact words that were used, or the exact extent of the transactions you were called upon to consider with Mr. Clough?
A. No, sir.
Q. You would not attempt to testify with assurance with regard to it?
A. I am sure I had this conversation.
Q. I mean as to the details of the conversation?
A. No, sir; I didn't think it was of much consequence.
Q. Now, I want to ask you some other questions. There has been some evidence here regard to collusion. I want to ask you if you ever saw those tickets—"not good unless countersigned by Mr. Spalding." [Showing a package of tickets or passes.]
A. Yes, sir.
Q. These were issued upon occasion of your being a candidate for councilman?
A. Yes, sir; Mr. Gilmore gave me fifty, and I used about half.
Q. Was it at your suggestion?
A. No, sir; they were sent to me.
Q. It was an original movement on his part, which you knew nothing about?
A. I knew nothing about it.
Q. Why did you understand that he put on,—"not good unless countersigned by Mr. Spalding?"
A. I suppose for that purpose; because I was a candidate for councilman.
Q. What position were you occupying?
A. I was director of the road.
Q. Did you understand that Mr. Gilmore, in the first place, had any right to issue these tickets for this purpose?
A. I don't know as I ever understood anything about it.
Q. Did it occur to you, when you signed these, that it was his right to do anything of this kind?
A. I don't know as there was any prohibition.
Q. Didn't you understand that the special object of issuing these was to further your interests as a political candidate?
A. I did.
Q. Did you think that Mr. Gilmore had any right to do that?
A. I didn't think it was prohibited.
Q. Did it occur to you that he violated the rules of the road?
A. It didn't occur to me.
Q. Did you know of any special reason why that was put on?
A. No, sir.
Q. Did Mr. Gilmore undertake to shoulder off on to you part of the responsibility of these passes?
A. I don't know what he did.
Q. Didn't you understand that Mr. Gilmore was unwilling to do it entirely as his own act, and was unwilling to issue them unless you would take your responsibility as director?
A. I don't know as I did.
Q. Didn't you have a talk with him about that?
A. No, sir.
Q. You say that was put in without your knowledge, and without any talk between you and him?

A. Yes, sir.

Q. And it was not put in there to shoulder part of the responsibility on you, and you had no conversation to that effect? Do you say that?

A. Yes, sir; I never had any conversation about it.

Q. Did it occur to you, when you signed these tickets, that Gilmore had transcended his authority at all?

A. I don't know that I did; I don't recollect that I did.

Q. How many of these did you countersign?

A. I should think about half.

Q. Didn't you sign all the fifty?

A. No; I think I burned up ten or fifteen of them.

Q. You signed all except what were burned up?

A. Yes, sir.

Q. It didn't seem to you that you were doing anything in violation of the rules of the road?

A. I thought he was not prohibited from issuing the tickets himself.

Q. Did it occur to you, when you countersigned these passes, that you and Gilmore were doing something that the rules of the road did not authorize?

A. I suppose it might have done; I think it did.

Q. You did think that and understand it. Now let me ask you if, after you signed them, they were used to a certain extent?

A. Yes, sir; it seems so, by that.

Q. Now, wasn't it understood by you and Gilmore that Gilmore should not issue them alone, but that you and Gilmore should issue them together?

A. They were sent to me without my knowledge.

Q. And you never asked him to do it?

A. No way, shape, or manner.

Q. There are 28 of these tickets. Do you mean to say there were not but two or three more?

A. I know there were 50 sent.

Q. How large was your councillor convention?

A. I do not know.

Q. You don't remember how many there were present?

A. No, sir.

## RE-DIRECT EXAMINATION.

Q. (*By Mr. George.*) I want to ask you if any tickets were sent to you before or afterwards?

A. No, sir.

Q. I neglected to ask you one question—which is involved in the other perhaps. Had you any knowledge, so far as you are aware, or had the board of directors, so far as you are aware, that the conductors, when any man paid in the cars, gave him a return ticket, either that had been once used or otherwise?

A. No, sir.

Q When was the first time that you ever heard of that? [Not answered, the witness having stated previously.]

Q. (*By Mr. Mugridge.*) You was not a candidate for councillor for but two years, were you?

A. No, sir.

The counsel for the defendant then offered in evidence the deposition of Mr. Charles H. Clifford, taken Sept. 7th, 1868; and the same was read by Mr. Rolfe.

It was agreed by the counsel of either side that the certificate of the market value of the stock of the road at the brokers' board in Boston, might be put in without requiring the calling of the secretary to certify the same, provided such statement be competent evidence of itself to be admitted. [Adjourned to Tuesday, Dec. 29, 1868.]

[TWENTY-FIRST DAY. Tuesday, December 29th, 1868.]

## TESTIMONY OF N. G. UPHAM.

Q. (*By Mr. George.*) Judge Upham, you were formerly superintendent of the Concord Railroad, and subsequently president for many years? Won't you be kind enough to state your first connection with the road, and what your relations were?

A. I have been connected with the road as director, superintendent, or president, since the road was opened, down to within about two years—two or three years.

Q. Well, sir, won't you state from what time to what time you were superintendent?

A. I have been superintendent since three or four years after the road was opened.

Q. Wasn't it earlier than that?

A. It was from the early commencement of the road that I was superintendent.

Q. This road was opened in the fall of 1842; now could you tell?

A. I think I was superintendent in 1843.

Q. And continued to be superintendent until 1856 or 1857?

A. Yes, sir.

Q. And when you ceased to be superintendent you became and remained president up to the time you left the road?

A. Yes, sir.

Q. Judge Upham, I will ask you whether you were aware, or, so far as you know, any of the board of directors were aware of Mr. Clough, or any of the other conductors, doing express business, carrying articles on the cars?

A. I was not aware myself that the conductors, any of them, were concerned in doing what might be regarded as an express business.

*Mr. Mugridge.* We want to have a distinction between express business and the business which Mr. Clough has described himself as having done here.

Q Were you ever aware of the conductors carrying articles such as butter, cheese, and provisions?

*The Chairman.* I suppose this testimony is confined particularly to Mr. Clough.

*Mr. Mugridge.* We except to any testimony in regard to other conductors.

Q. I will put you this question—whether you were aware, or whether the board of directors, so far as you know, were aware of Mr. Clough's carrying produce,—butter and cheese, and articles of that character—backward and forward?

A. I never understood that he carried butter and cheese and matters of that kind. I think that the express agents made complaint at some time that he brought fish up, or something of that sort, by special trains; and there was some little complaint made about it, and the matter was discontinued. That was in the early history of the road; and I supposed that anything of the kind was limited to something of that character. I was not aware that Mr. Clough, or any one else, was concerned in doing what might be regarded as general express business. I think when he first went upon the road, he might have done some business in small articles; I don't recollect of hearing anything said except about bringing fish, or something of that sort. But for a long time—I think that the complaints from that cause were continued a very short time—this whole matter ceased. It was in the early history of the road. That is my recollection of it.

Q. I will ask you whether any permission was ever given, directly or indirectly, for conductors to carry articles of that kind? [Objected to.]

*Mr. Mugridge.* Mr. Clough has not claimed that there was any permission. He does not say that the superintendent, or anybody else, ever authorized him to do it.

*The Chairman.* Do you understand, Mr. Mugridge, that there is evidence in this case tending to show that the conductors were permitted to do this?

*Mr. Mugridge.* Yes, sir.

*The Chairman.* Then the question will be, I suppose, whether that evidence can be rebutted here or not. The ruling now stands on this ground, that the defence claim here that there is evidence tending to show. The plaintiff is putting in evidence to rebut that evidence, whatever it is, which tends to show permission. On that ground we think it is admissible.

*Mr. Mugridge.* Is that evidence competent as bearing upon the general question?

*The Chairman.* I understand that the position has been taken here by the defence very distinctly, that the superintendent has very great power; that he is a controlling man; that he has very great power; and that his position would authorize many acts. It seems to me, therefore, that if the defence claim that there is evidence tending to show that the superintendent permitted these things to be done, it must be proper to rebut it by the evidence of the superintendent. I cannot see how it is not.

*Mr. Tappan.* By the evidence of the superintendent?

*The Chairman.* Why, certainly—if the superintendent authorized it.

*Mr. Tappan.* Is this offered to rebut evidence of permission?

*The Chairman.* He was superintendent at the time; and I understand that this is offered to rebut this evidence of permission. I don't exactly see why it is not admissible, in that point of view now.

Q. (*By Mr. George.*) I will ask whether any permission was ever given to Mr. Clough?

A. I don't know that Mr. Clough ever asked any permission to take any articles of any kind over the road; I have no recollection of having any conversation with Mr. Clough in reference to that subject. I do not now recollect of having spoken to him at all about it. I think that Mr. White, the express agent, was a little sensitive about some things that were done by the conductors; and he made some complaint. There was some talk in reference to practices of carrying articles that he thought might interfere with him, or come in collision with him in some way. And I suppose it came to the knowledge of the conductors, and I heard nothing more about it after that. I have no recollection of conversing with Mr. Clough in relation to his carrying any articles, and personally I have no knowledge.

Q. Whether any leave—of course it follows—any leave was given, so far as you know?

A. No; I was not aware that any leave was asked or given.

Q. How it was with regard to Mr. White's having an exclusive contract for the express business over the Concord Railroad during this time? [Objected to.]

Q. Now will you state whether you had any such knowledge, or the board of directors, so far as you know?

A. Knowledge of what? Won't you repeat?

Q. Knowledge that Mr. Clough was purchasing tickets of upper roads and selling them for the local fares over the Concord road?

A. I never heard any intimation of that kind while I was superintendent or president; and I had no suspicion of any transaction of that sort; of course could give no permission to anything.

Q. It is, of course, unnecessary to ask if, so far as you knew, any of the other directors knew?

A. I never heard from any of the directors that a practice of that kind existed while I held office on the Concord Railroad, either superintendent or president,—on the part of one of the directors.

Q. Judge Upham, I will ask you whether you had any knowledge, while you were superintendent or president, or whether any of the other directors had, so far as you are aware, of the conductors selling tickets which had been once used—selling them a second time?

*Mr. Mugridge.* We take the same exception, of course.

A. I had no knowledge of any practice of that kind until some testimony was taken, after suits had commenced; not aware of any practice of that sort; didn't suppose such a practice existed, if it did exist.

Q. Whether you had any knowledge, or any of the other directors, so far as you are aware, had any knowledge, of conductors taking money received in the cars, and going to the ticket office and purchasing fresh tickets and punching them and returning them for the collection? Whether you heard of that?

A. I never heard any intimation of that kind from any quarter, until since this hearing has been going on.

Q. Whether you had any knowledge, or, so far as you are aware, the directors had any knowledge, any of them, that the conductors took the money received in the cars for fares, and, instead of returning it to the general ticket agent on the way-bills, took that money and bought tickets at the ticket office and returned them in their collections?

A. I was not aware of a practice of that kind; never have heard any of the directors speak of anything of the sort.

Q. And, so far as you are aware, they had no knowledge of that kind?

A. They never communicated the knowledge to me, at any rate. I had no knowledge myself, and I had no suspicion myself, because I think there would have been some conversation on the subject if they had had any knowledge of the kind. I supposed the money taken in the cars was returned as usual; I wasn't aware that ——

Q. What do you mean by usual?

A. Returned to the cashier.

Q. That is, on the way-bills?

A. Yes, sir. I supposed that the money taken in the cars was returned there. I was not aware that any substitute was adopted in place of buying tickets of any kind.

Q. I want to ask you now the same question that has been asked Mr. Spalding and Mr. Kidder, whether you were present at the meeting— or, I want to ask you how it was about your having any knowledge, or the board of directors, so far as you are aware, that Mr. Clough employed boys on the cars, to ride upon the cars, and sell peanuts for his private benefit?

A. I was not aware that Mr. Clough had any pecuniary interest in anything of that kind. I think I have heard him speak of some lads that had the privilege of going on the cars, as being very well behaved boys, that they were poor, etc.—something of that kind; but I did not know of any pecuniary interest, if there was any. There have been boys upon the cars distributing papers; occasionally other articles; there was a man some time used to go with parched corn. I am not certain but that he may have paid something to the road for that privilege. It never occurred to me that this traffic of the boys was of much account; and I did not know that there was any pecuniary interest on the part of the conductors in reference to it.

Q. I want to ask you if you were present at the meeting held October the 24th, 1865, at the time this investigation was ordered into the affairs of the road, into the manner in which the conductors had done their duty? Were you present at that meeting?

A. Yes. sir; I was present at the time when the vote was passed.

Q. Will you state upon whose representation, and who made it, that vote was passed?

A. Well, Mr. Gilmore at that time—at that meeting—and at some other times along—some other meetings before and subsequent to that time,—made some round statements about losses that he thought the road had incurred; and seemed to speak with a good of confidence in regard to them; and Mr. Crocker introduced a resolution that inquiry should be made by the counsel of the road.

Q. What was Mr. Gilmore's statement? [Objected to.]

*Mr. Mugridge.* We understand the referees to rule this to be competent so far as to what his actions were. I understand that ruling to go no farther than to the fact that a communication was made by some person to the board. But I did not understand the referees to rule that the particular communication was to be introduced here.

*The Chairman.* Judge Bellows understands as I do, that what was said about Mr. Gilmore went in by consent. [Objection waived.]

Q. Now will you state what Mr. Gilmore said?

A. He stated that he had not any doubt that the road had lost more than $50,000 a year by money that was taken in the cars by conductors and not returned to the road. He made some broad statements of that kind, and seemed to persist in it, and it occasioned a good deal of surprise to many of the directors.

Q. You say Mr. Crocker introduced a vote, or made a motion for a committee of investigation?

A. Mr. Crocker made a motion that the counsel of the road be directed to inquire into the subject in some way; and the vote was passed; and at that time you were not present; you were delayed for some cause or other. At a subsequent time you came in, and the vote was reaffirmed.

Q. You recollect who was clerk *pro tempore* in my absence?

A. Mr. Spalding was usually appointed clerk *pro tem.*, and I think he was at that time. I believe there was an expectation that you would be in before the directors separated. And in this instance, and in one or two other instances where there had been some delay, votes had been passed and were re-affirmed after the regular clerk came in, in order that he might make the record. That was the case at this time.

Q. I will ask you, sir, if you remember, if you will please state who informed me first of the passage of this vote when I got to Boston, and of the proceedings of the directors? Who was the first man that informed me? Do you remember?

A. I don't recollect with regard to that. I think you came in, and shook hands all around with the board, and had general conversation about what had been said and done. I think all the directors talked more or less about it.

CROSS EXAMINATION.

Q. (*By Mr. Mugridge.*) Judge Upham, you say that Mr. Gilmore made certain representations to the road with regard to the amount of money that the conductors were taking. Let me ask you if you ever heard Col. George express any opinion to the board on that subject?

A. Well, I think I have.

Q. What representations did he make in that direction—before these suits were brought, I mean?

A. At that time, I think that Mr. Gilmore's representations on that subject exceeded those of any other person.

Q. Let me ask you whether Col. George did make representations to the road on that subject, and if so, what they were,—before these suits were brought, I mean?

A. After the passage of the resolution and before the suits?

Q. Yes, sir. I will ask you in the first place, if you will allow me to withdraw that question, whether Col. George made any representations at the time of the meeting on the 24th of October, on that subject, at the time the resolution was passed, if you remember?

A. Well, I don't recollect further than this. I cannot recollect what words he said, but I think he concurred in the idea at that time, that money had been taken that had not been accounted for; but when he understood that the matter was to be submitted to the counsel of the road for investigation, he inquired what course he should take; and his conversation was rather in that direction than otherwise. There wasn't a great deal.

Q. Let me ask you whether, before this time, the matter has been alluded to in general by Col. George, as to the conductors appropriating funds?

A. Well, sir, I am not aware that he was in the habit of speaking on that subject, or speaking in reference to it, in any peculiar or marked manner.

Q. Well, without reference to any particular manner, let me ask you if he has frequently made that suggestion to the board of directors, antecedent to the meeting of the board in October, 1865—suggestions to the effect that I have indicated?

A. Well, I don't know; I have no recollection of his suggesting it to the board that ——

Q. Well, to you, sir, as president of the board, did he ever make the suggestion?

A. I am not aware that he made any suggestions or further inquiries. The only measures taken in that direction by the board were, from time to time we passed some resolutions to require an extra amount to be taken in the cars, or something of that kind. And at those times it was thought to be very desirable, for every reason, to check the payments in the cars.

Q. Now, let me see if I cannot refresh your recollection. Mr. Corning left the road some time before Mr. Clough did ?

A. Yes, sir, he did.

Q. Did Col. George make a suggestion to you at the time Mr. Corning left, that the conductors were appropriating the funds improperly ?

A. Well, there were some complaints that the rules were not enforced and some suspicions that the road did not receive all that belonged to them at that time.

Q. Let me ask you this question. Were not these suggestions thrown out by Col. George to you as against the conductors ?

A. Well, I don't think he ever undertook to come formally.

Q. I don't mean that he came in a formal and ceremonious way to you as president of the road. But do you mean to swear that he did not make these complaints and throw out these intimations to you, against the conductors antecedent to October 24th, 1865 ?

A. I have no doubt that he made a statement that he thought there was not a full return of the amount collected on the cars, but whether he undertook to state any specific sum. I don't think he did.

Q. Now, I don't mean to inquire for that. Let me ask you whether Col. George, after the vote of October 24th, and prior to the commencement of the suits, undertook to demonstrate the fact to you, by the figures that he carried about, that George Clough had stolen $250,000.

A. I don't think he ever made any statement to me.

Q. Did he say that he could demonstrate by certain figures that he carried that George Clough had stolen $250,000 ?

A. I think he said that he could show, and he had reason to believe from the evidence he had got; but as the $250,000 ——

Q. Do you mean to say that he didn't say that he stole $250,000, and that he could show it by a comparison of the way-bills ?

A. I don't recollect his stating the figures as high as that.

Q. How high did he set it ?

A. I think he stated large sums, but didn't put any definite figures.

Q. I understood you to say, Judge Upham, that since you have known anything of the Concord Railroad, there has been a boy upon it who has peddled peanuts and sweetmeats and pop-corn and the like of that ?

A. Not exactly.

Q. Well, what did you say ?

A. I said there had been boys upon the road.

Q. I don't mean the same boys, of course. Has there not been upon the Concord Railroad, since you have been connected with it, a boy or boys peddling sweetmeats, pop-corn, &c., sir ?

A. I think the pop-corn administration commenced very long after the administration of the Concord Railroad.

Q. Leave out the pop-corn—peddled sweetmeats and candy, oranges and apples ?

A. The first I recollect of, there was quite a large sized man came on to the road.

Q. Hasn't there been a boy who has run upon the train, who has peddled articles of some kind ?

A. It hasn't been a permanent institution. I should think likely that a boy, carrying papers and selling some small articles of different kinds, had been running on the road a considerable portion of the time.

Q. Did you ever know a boy to pay a cent for this privilege?

A. I don't know; it was considerably a charitable matter.

Q. Did you ever know a charge to be made by the road against such a boy?

Q. No, I never did. I am not certain of it.

Q. Wasn't he understood to be there, acting in that capacity for the accommodation of the passengers riding upon the road, as well as for himself?

A. Sometimes there was complaint made of their being an annoyance; but if the boys were well behaved, they were ordinarily permitted to run. I rather think not more than one to a train.

Q. I don't care how many run. Did you ever know of one paying for the privilege of riding in the cars?

A. I never knew of their paying anything. It was regarded as a matter of charity in part; perhaps as a convenience to passengers in part, where they were not annoying, and where they were well behaved. We had to look after them a little, but when the conductors said that they behaved well, and were good clever boys, we ordinarily let them run without anything being said.

Q. Now, I want to ask you just one question more. You have said that Mr. Gilmore came to the board of directors and made a statement to the effect that the conductors were stealing $50,000 a year. Did you ever tell George Clough what you thought the reason was why he made that statement? If so, what did you say that reason was? [Objected to.]

Q. (*By Mr. Mugridge.*) Judge Upham, do you know why Gilmore made the statement that you repeated here to the referees?

A. He appeared to make it in great sincerity; and he insisted upon it as his belief.

Q. That ain't the question.

A. That is all I know about it. I don't know as I know.

Q. You answered in regard to the manner in which he made the statement, I made an inquiry as to that. Do you know why Gilmore made that statement?

A. Of course I don't know.

Q. Have you stated to anybody that he made it for the purpose of covering his own rascality and rottenness?

A. I can state the conversation in relation to the subject.

Q. The question I put you is this: Have you stated to any person that Mr. Gilmore made that statement to the board of directors for the purpose of covering his own rascality and rottenness,—using these two terms?

A. No, I did not make a statement covering that in the way you state it. If you want to know, I will state that at the time this resolution was passed Mr. Gilmore apparently seemed to be acting from his own knowledge and impression, and made statements to the board there; and there was no apparent excitement or malice, it seemed to me. But after the investigation went on, and at subsequent meetings, there were new declarations. There were not only declarations that came out before the board implicating the conductors to a certain extent, but also implicating Mr. Gilmore more or less. The effect of the investigation was to attract more or less attention to Mr. Gilmore, and certain acts of his that had been brought to the knowledge of the board; and at these subsequent meetings Mr. Gilmore seemed to enter with a great deal of zeal into this matter of fraud on the part of the conductors. But after this investigation went on, now combinations and alliances were formed, and they were surprised at the different talk that Mr. Gilmore had before the board apparently, and what he had outside. I have no doubt that I stated that he persisted in the accusation, and continued after the investigation went on, and that his object in part may have been to screen himself from attention. I recollect having a conversation of that kind with Mr. Clough, or with Mr. C——, something of that character, that after circumstances came out that implicated Mr. Gilmore to a considerable extent, and his persisting so strongly as he did, and enlarging so much on the action of the conductors, it occurred to me that he wanted to show his zeal to the directors in the course he was taking, so as to screen himself in part. I have no doubt that that in part was his object at the later stage of this proceeding. That is my opinion at the present time. At the later stage of the proceeding., the fact is, a great many things were developed that occasioned suspicion on the part of the directors.

Q. What were some of them?

A. In the first place we looked into his accounts.

Q. What did you discover there?

A. We discovered that there was undoubtedly a large defalcation on the part of Gilmore.

Q. To what extent?

A. To a considerable extent; I could not tell.

Q. Give us your best judgment.

A. I supposed that if we got into a fight with Mr. Gilmore at that time, it was in his power to withhold funds from the road to the amount of at least $50,000. The election was coming off, and he seemed to think that his position was rather dangerous anyway. I have no doubt he had that impression. I did all I could for the last six months to induce him to bring up his accounts to the fullest extent. I was more alarmed as the facts came home to my knowledge as to other particulars. I devoted my time to bringing up his tickets, and succeeded in bringing up most of some $5000 or $6000.

Q. Let me ask you if he didn't practically turn out to be a defaulter in the sum of more than $50,000?

A. No sir; I don't believe any such thing.

Q. How much do you think it was?

A. I never undertook to form an opinion.

Q. What was your best opinion at that time? Wasn't he practically a defaulter to the amount of $50,000?

A. I know that his defalcation was misrepresented to a considerable extent.

Q. Did you direct this suit to be brought at the time it was brought? Now that is a plain question?

A. Of course I did not undertake to direct in regard to this matter at all; for this reason, that the resolution required that the counsel of the road should take such measures as he deemed proper.

Q. Are you quite sure that that is the resolution?

A. That is my impression, that the counsel of the road was directed to make certain inquiries in reference to money; whether he was instructed to institute suits or not, I don't know, but I understood, so far as I recollect, that the matter should be submitted to the counsel, leaving Mr. Gilmore and myself out of the question. For I thought possibly some collision might arise between us—perhaps between Mr. Gilmore and myself, some way or other; and I understood the matter to be left with the counsel.

Q. And you understood that the counsel commenced a suit on his own arbitrary motion?

A. I don't recollect that I had any conversation with the counsel except after the disclosure of Whitcher.

Q. Did you direct this suit at the time it was brought? Do you remember giving any such direction?

A. I don't think I directed the suit.

## TESTIMONY OF SYDNEY H. CARNEY.

Q. (*By Mr. George.*) State whether you are a brother of Geo. H. Carney?

A. I am.

Q. Practice in Boston?

A. Not practicing in Boston; have been for six or seven years.

Q. I want to ask you whether you were desired by your brother, Major George J. Carney, to visit at Concord, and if so, when, what fares you paid, and how much money you paid? Won't you state briefly?

A. Yes, I can give it in my own way.

Q. Give it in your own way, only as briefly as you can?

A. My brother asked me sometime in November. He sent to my house and sent me some money.

*Judge Bellows.* November—what year?

*Mr. George.* November, 1865.

A. He wished I should visit Concord. I did not see him, but I at once—I can't say whether he suggested that I should visit the Lunatic Asylum, but at any rate, as I was interested in it and had never visited it, I did. He told me to pay in the cars, and that was all that I knew about it. I took four with me—part of the way five—and started on Thursday, the 28th day of November, 1865, as appears from my note here which I wrote to him.

Q. (*By Mr. Mugridge.*) Are you testifying from a minute, a memorandum?

A. I am not; I am simply refreshing my memory.

*The Chairman.* I think the witness says he wrote it at that time?

Q. (*By Mr. George.*) Was it made at the time?

A. Yes, sir.

Q. You knew it to be true at the time?

A. I did.

Q. (*By Mr. Mugridge.*) That is the original one? It is not a copy?

A. That is the original. At the Boston depot I purchased five tickets to Nashua and handed those tickets to the conductor. At Nashua the conductor came through, and I purchased five tickets to Concord. I came to Concord with five persons, and visited the Lunatic Asylum.

Q. (*By Mr. George.*) Paid in the cars?

A. Paid in the cars—out of the money. [Objected to.]

Q. (*By Judge Bellows.*) Did you give the day of the month?

A. Twenty-eighth day of November—Thursday, the 28th day of November, 1865. I purchased tickets from Nashua to Concord for five persons. It was the first train in the morning from Boston. We visited the Asylum at half-past three in the afternoon. I got on board the train for Boston. The same conductor was on the train as was there in the morning. I purchased five tickets to Boston in the cars, and when I got to Manchester, as I had been asked to go over that road and reserve the tickets, I did so. We all got out at Manchester, and purchased from Manchester to Boston again.

Q. On the Lawrence road?

A. To Lawrence. The conductor told me at that time that he hadn't been on for a month, and he hadn't through tickets to Boston,

Q. (*By Mr. Mugridge.*) What did you say about buying at Concord?

A. When I started from Concord, at half-past three in the afternoon, I purchased five tickets in the cars, of the same gentleman—who was acting as conductor, I suppose, that I met in the morning coming up. And when I reached Manchester, he left the train, and I purchased of another conductor, in the car, tickets from Manchester to Lawrence, I could not buy through because he said he had no other tickets to Boston, as he said it was his first trip for a month.

Q. (*By Mr. Tappan.*) The tickets that you purchased here were from Concord to Manchester?

A. No, sir; from Concord to Boston.

Q. (*By Mr. George.*) Are these the tickets. [Handing tickets to witness]

A. I think those are the tickets.

Q. Won't you look at them on the back?

A. Yes, sir, those are the tickets. [The tickets were put in evidence by Mr. George.]

CROSS EXAMINATION.

Q. (*By Mr. Mugridge.*) What was the date, Mr. Carney, when you first bought tickets from Nashua to Concord of Mr. Clough?

A. The 28th of November—nearly.

Q. How many tickets did you purchase in the cars, of Mr. Clough, from Nashua to Concord?

A. Five, sir,

Q. And paid him in the cars for them?

A. Yes, sir.

Q. Now, when you went back again in the afternoon, you bought five tickets to Boston where?

A. I bought five tickets in the cars of this same conductor.

Q. Did they run from Concord to Boston—the five tickets that you purchased?

A. From Concord to Boston—yes, sir.

Q. And you paid him for them in the cars?

A. Yes, sir.

Q. These went by the way of Nashua?

A, Yes, sir.

Q. And that was the same day, the 23d [28th], and you only came up, went there and went back the next train?

A. That is it, sir.

Q. Now, Dr. Carney, did you visit Concord again after this?
A. I did, sir.
Q. With whom did you ride?
A. With friends of mine. Oh, you mean the conductor?
Q. Yes, sir.
A. I shall have to refer to my notes. On Monday, the 18th of December, 1865, I purchased two tickets for Newmarket Junction.
Q. Did you ride again with Mr. Clough and buy any tickets of him after the 23d of November?
A. Yes, sir.
Q. When was that?
A. That is—I say, yes,—I haven't seen these papers until this morning.
Q. Well, you can refresh your recollection and see?
A. That I cannot say. That is not fresh in my mind; and I notice here that I simply say that I asked for them—yes, I can say that I bought two tickets of Mr. Clough at half-past three on the afternoon following December the 18th—on the 19th.
Q. On the 19th of December, 1865?
A. On the 19th of December, 1865, at half-past three in the afternoon, I bought two tickets of Mr. Clough.
Q. Where did they go to?
A. From Concord to Nashua.
Q. Have you got those tickets?
A. No, sir.
Q. Where are they?
A. I don't know.
Q. What did you do with them?
A. I don't know that. My object was to buy a ticket to New York of Mr. Clough, as I understood that he sold to New York, and he was very much disturbed, and it was with much difficulty that he answered; and I think he knew that I was ——
Q. What did you think he thought you was?
A. I think he knew that I was a "spotter" as they are called.
Q. Did you say that you thought that he thought that you looked like a "spotter"?
A. I think very likely I may have said so.
Q. Will you please describe fully the embarrassment, the nervous excitement which Mr. Clough seemed to be laboring with at the time you asked for a ticket? Describe his trepidation?
A. I cannot describe it fully; but he was a man who had been set down as a thief for years.
Q. Go ahead, sir, from where you left off.
A. The first time that I saw Mr. Clough, that is, the first time I bought these five tickets, he acted as he had a great while, in an ungentlemanly manner, to say the least, towards the passengers, and in a haughty and overbearing way; and I had my doubts at the time that I paid the money where it was going.
Q. You did?
A. Yes, sir.
Q. (*By Judge Bellows.*) That was when you bought the two tickets?
*Mr. Mugridge.* The first tickets?

A. The second time when I saw him on the train, he came along and I handed him some money and asked him for some tickets to New York first. He turned pale, and trembled, and was very much excited. I cannot give you any further particulars of his appearance at that time; but he afterwards stammered out that he didn't have any to New York. I told him then that I would like to go to Groton Junction, which didn't highten his color at all; he looked, if anything, paler.

Q. Well, proceed.

A. I then asked him to sell me as far as he could in that direction —to sell me to Nashua.

Q. You say he was paler when you said to Groton Junction?

A. I could not swear. He was pale enough.

Q. Did he tremble a good deal?

A. He did, a good deal, for an honest man.

Q. You didn't look at him as an honest man?

A. I didn't.

Q. You regarded him as a thief?

A. Yes, sir.

Q. And you do now?

A. Yes, sir. I know it

Q. Now, sir, I want to ask you what brings you upon the stand here to denounce him as a thief? What do you see in his conduct which leads you to swear that he is a thief?

A. I don't say that it is anything that I have seen in his conduct.

Q. If you never saw anything in his conduct which leads you to believe he is a thief, what do you come here and swear that he is a thief for, if you never saw anything in his conduct which indicates it?

A. Because, I have information which leads me to regard him so.

Q. And you will swear that he is a thief upon representations made to you by other people?

A. No, sir, I think not. I have had some dealings—I can tell an honest man when I see him.

Q. Let me ask you, sir, what dishonest action you ever saw in Geo. Clough in your life?

A. Well, sir, I will tell you. His actions when I asked him for that New York ticket, were the actions of a thief.

Q. They were?

A. Yes, sir.

Q. Now, you mean to swear that he is a thief by reason of actions that you then saw on his part?

A. I mean to swear, sir, that I am satisfied.

Q. And you state here, that he is a thief, and you regard him as a thief, and are willing to swear and have sworn he is a thief, because of his actions when you asked him for that New York ticket?

A. Actions and subsequent representations from other parties.

Q. From what parties?

A. My brother.

Q. Who else beside your brother?

A. I could not give the name, sir, definitely.

Q. Did ever you hear another man in your life call George Clough a thief?

A. I never heard any man say that definitely.

Q. Name a man?

A. I cannot name one. I never heard a man mention his name but what said it.

Q. Name a man that you ever heard call George Clough a thief?

A. Every man.

Q. Name one man, sir?

A. I should not wish to name, for fear I might do injustice.

Q. Name a man that ever called him a thief, if you can?

A. Am I obliged to name?

Q. You are, sir.

A. One man that has called him a thief is George T. Comins, who has a contract here in the state prison.

Q. Give me another man?

A. I cannot do it, because my memory does not serve me.

Q. Name a man?

A. I cannot recollect the names.

Q. What is your business?

A. I am now surgeon general of the Travellers' Life and Accidental Insurance Company.

Q. How old are you?

A. I am over thirty-one.

Q. What business have you been engaged in during your life time?

A. I have been engaged in study, at school, and in college and medical school, and the practice of medicine, and as surgeon general of this company.

Q. Where have you practised medicine?

A. In Boston.

Q. How long a time?

A. I practiced first at Bridgewater, in the year 1861. Since then, and until a year and a half ago I have practiced in Boston.,

Q. How long since you occupied the position of surgeon-general of the Accident Insurance Company?

A. For a year and a half, perhaps.

Q. How long did you practice in Boston?

A. From 1862 until that time.

Q. Who got you this place as surgeon-general here?

A Myself.

Q. Where is the company located?

A. In Hartford, Connecticut. I will give you my card if you would like.

Q. No, sir, I don't want it, and I shouldn't call upon you. Now I want to ask you another question. Did you ever see an act of George Clough's that indicated dishonesty?

A. I have.

Q. What was it?

A. The second time I was on that train.

Q. And asked for a ticket to New York?

A. Yes, sir.

Q. Will you describe it?

A. I will. He was as mean a looking man as I ever saw under the circumstances.

Q. Will you describe the particular look of meanness?

A. I could not, sir. All that I could say is, he was perfectly pale, and trembling like a leaf—very much embarrassed.

Q. And you think his embarrassment was greater when you asked for a ticket to the Junction than when for a ticket to New York?

A. Yes, sir; it kept increasing.

Q. Now, I want to ask you how much Mr. Clough charged you for the fare of the five persons from Nashua to Concord the 18th day of December?

A. For the five? There were only two. Oh, yes; from Nashua to Concord, $8.

Q. Eight dollars?

A. Yes, sir.

Q. Is that the full fare?

A. I haven't the least idea.

Q. The fare is $1.50, and ten cents more. Let me ask you if there is anything dishonest about that?

A. I have nothing to do with whether he returned them or not.

Q. How much did you pay Mr. Clough for the three fares back again on the half-past three train?

A. Fourteen dollars and twenty-five cents.

Q. Is that the precise amount that he should have taken in the cars from here to Boston?

A. I haven't the slightest idea.

Q. (*By Judge Bellows.*) Fourteen dollars and how much?

A. Twenty-five cents.

Q. (*By Mr. Mugridge.*) How much did you pay him?

A. Fourteen dollars and twenty-five cents.

Q. Have you a minute of that?

A. Yes, sir.

Q. Where is your minute of that?

A. Right there [indicating.]

Q. Now, sir, did you pay him for the two fares that you paid the next day, or the 18th day?

A. Concord to Nashua, $3.20.

Q. How many fares?

A. Two.

Q. That would be exactly the fare, wouldn't it, in the cars?

A. I haven't the slightest idea.

Q. Now, I want to ask you how much money your brother, Major Carney, put into your hands to do this business with?

A. One hundred dollars.

Q. Have you made any account to him of the balance of the hundred dollars?

A. I have.

Q. How much did you spend?

A. Fifty dollars and fifty-five cents.

Q. Did you charge anything for your services up here?

A. I cannot tell you.

Q. Did you expect to get any pay for your services?

A. I haven't the least idea.

Q. Did your brother promise you anything?

A. No, sir.

Q. You have got in here a dinner $3.75, sundries, carriages, etc., $3.75, What was that "sundries" for?

A. I haven't the least idea.

Q. Did you drink anything?

A. I am not a drinking man, sir?

Q. Did your companions?

A. None of them drink.

Q. What were those sundries?

A. I cannot tell. That bill is just as I made it out honestly and fairly, with no intention to defraud the company. I don't see what that has to do with the case at all. If my brother gives me a hundred dollars I can spend it as I see fit.

Q. You have paid back the balance between the hundred and the fifty?

A. I have; yes, sir.

Q. Now, will you be kind enough to look at Mr. Clough's way-bill of the 23d day of the 3 o'clock train, and see if there are fares between Concord and Boston returned, and if so, how many?

A. Well, I don't understand the arrangement of these things.

Q. I will put it in this way, if you please. There are eight fares on the way-bill of the 23d of November, on the 3.30 train returned between Concord and Boston. Now you say that on the 19th day of November you paid Mr. Clough how many fares from Nashua to Concord?

A. Two.

Q. Mr. Clough upon that day has returned four fares upon his way-bill. And you paid him the full fare?

A. I suppose so.

Q. Now, we put in on the up train, on the 23d of November, five fares between Nashua and Concord.

*The Chairman.* I want to make this inquiry; if there has been any evidence put in—I don't remember how it is—tending to show whether it was the duty of the conductor to return on his way-bill the proceeds of the joint tickets which he sold?

*Mr. George.* Oh, yes.

*Mr. Rolfe.* Always.

*Mr. Stanley.* And they did so.

*Mr. George.* That was their duty.

*Mr. Mugridge.* These were joint tickets through to Boston, and Mr. Clough has returned them here.

*Mr. Stanley.* He accounted for all the money that he got to the Concord Railroad, and the Concord Railroad to the others.

*Mr. George.* The way-bills must show that he did it.

*Mr. Mugridge.* He has returned the fare in the cars from Boston to Concord. The fare from Concord was $2.85; and Carney swears that he paid him $14.25. That would be just the fare. Your honor will bear in mind that when he went down he bought five tickets to Boston; and he paid $14.25 for them. That is just exactly the fare that Mr. Clough should have taken in the cars, according to the way-bill.

*Mr. George.* The way-bill shows the fares of either the up or down train.

*Mr. Mugridge.* It was either the up or down train, and he shows five fares returned.

Q. (*By Mr. Mugridge.*) Did you buy any tickets of Mr. Clough the last time you came up to Concord? When was the last time you came to Concord?

A. The last time was on the 19th day of December, 1865.

Q. Who came up with you then?

A. I went by the way of Newmarket Junction, and then from Portsmouth to Concord with Mr. Kendrick; and when I got to Concord, I went back at half-past three with Mr. Clough. That was the time I had this conversation about the New York tickets.

Q. Then the last time you came up, you didn't have anything to do with Mr. Clough?

A. I didn't come on Mr. Clough's train.

Q. But when you went back—that was when you had the conversation you had with Mr. Clough.?

A. Yes, sir.

Q. (*By Mr. George.*) Did you know Mr. Draper, a man who was in your brother's employ?

A. I did not.

Q. Knew nothing about his being on the same train?

A. [No answer.]

Mr. Stanley offered in evidence a list of stock sales of the brokers' board, an organization doing business in the city of Boston, in the state of Massachusetts, said list including sales by auction, and being certified by Joseph G. Martin.

*Mr. George.* Mr. Stanley has added up the first 400 shares purchased by Mr. Clough, as appears by the stock record. The first 400 shares Mr. Clough swears he estimates at $40 a share. Now the first 400 shares as shown by this certificate were sold at the brokers' board for $47.50 per share. The price of the stock at the brokers' board since the 400,—that is, the average cost at the brokers' board was $64 within a very small fraction; or more accurately, it is $63 7-10. There were sixteen sales. The whole cost of the sixteen shares was $1019, which would be over $63 a good deal.

*Mr. Mugridge.* I don't know as I object to putting in this testimony in this way. If it is put in to contradict Mr. Clough, it is not competent. I do not understand the object of the testimony.

*The Chairman.* I understand that Mr. Clough has not testified here to the precise cost of his stock. I understand that, testifying now from recollection, he estimates that it cost him about so much. Now, for the purpose of showing that his recollection may not be exact in that particular, I understand that they have offered here a record of sales at that time; and then that they have made a calculation themselves, which of course is not evidence, except so far as it is verified. They offer the calculation to show that the average price of the stock, instead of being $40, would be $47, for the purpose of correcting what they think may be an inaccuracy in Mr. Clough's recollection.

*Mr. Mugridge.* It does not appear that Mr. Clough bought his stock in Boston at the brokers' board. The point we make is that, what the stock sold for at the brokers' board is not competent to contradict him as to the price paid for the stock that he bought up here in Concord.

*The Chairman.* As this has been laid before us, it is not necessary for the referees to rule now.

## TESTIMONY OF JOHN D. COOPER.

Q. (*By Mr. Rolfe.*) State whether you formerly owned estate on Warren street, that was afterwards purchased by Mr. Clough?

A. I owned a place formerly, on Warren street.

Q. Where Mr. Clough now lives?

A. Where Mr. Clough now lives.

Q. Who made the sale? That is, who was the negotiating party when it was sold to Mr. Clough?

A. I made the bargain with Mr. Clough, all the bargain that I ever knew anything about.

Q. State whether you know the exact sum that was paid for the house and furniture?

A. All there was there, both real and personal, was about $8000.

Q. Do you recollect how that was paid?

A. Yes, sir; there was a note given for $6000, and $2000 paid down.

Q. State whether you built that house, substantially, and whether you purchased the furniture, the same furniture that was sold to Mr. Clough?

A. I bought the premises when the house was partially built, of Mr. Clough.

Q. (*By Mr. Mugridge.*) Of whom?

A. Of Mr. Clough.

Q. (*By Mr. Rolfe.*) Joseph L.?

A. Joseph L., I should think. He was the young man that built it, or partly built it:—or rather I took it and quite a portion of the work that was done, and built it somewhat different from what he had calculated, and finished it up.

Q. What furniture was there in the house that you sold Mr. Clough with the house?

A. I don't know whether I can tell every article, but I can the most of them, I think. Mr. Clough had all the carpets that were in the house. He had all the carpets that were in five rooms. He had the mirrors; four small chairs, I think; two large ones; and two sofas that were in the parlor; two large mirrors.

Q. Four small chairs?

A. I think there were four of the small size, and two of the large ones, and two sofas; and the furniture that was in two chambers, with the exception of the bed and bedding. He had the bedsteads in one chamber. I want to correct that—there were six rooms that those nice curtains were in, instead of five.

Q. (*By Mr. George.*) You said that, with one exception, all the bedsteads in one chamber?

A. He had the bedstead and all in one chamber, but I took the beds out, leaving the bedstead for Mr. Clough.

Q. (*By Judge Bellows.*) He had no beds nor bedding?

A. He had no beds nor bedding; he had the bedstead and the rest of the furniture in the chamber where the bedstead was, consisting of, I think, about four small chairs and one large chair.

Q. (*By Mr. Rolfe.*) Was there a carpet on the entry stairs?

A. There was.

Q. Do you recollect how many curtains there were in the parlor?

A. I think four.

Q. Do you recollect what the expense of that was?

A. The curtains in the parlor, I think, cost $100 to the window.

Q. Do you recollect the cost of the carpet?

A. I don't—I think all the carpets I have I paid for in one bill. The man came up and took the dimensions of the rooms, and they were all made and brought to the house, and the man put them down.

Q. You recollect the entire cost?

A. I think I paid $400 for the carpets. I am quite sure I am right on that point. They didn't take any rugs that were on the floor—the mats; I am speaking of the carpets

Q. Do you recollect about the expense of the carpets in the other rooms?

A. It is impossible for me to tell now.

Q. Describe them. Take the two in the sitting-room?

A. Those in the sitting-room were paid for somewhat similar to those, only I should judge a good deal nicer than those.

Q. What they call trepanned damask carpets?

A. Those in the chamber were considerably more expensive than those in the sitting-room. In one chamber they were such curtains—somewhat similar to those; and then those that came part way down.

Q. That is the parlor chamber?

A. Yes, sir; in the parlor chamber.

Q. Now in the sitting-room, how many were there there?

A. Two. There were two in each of the three chambers. There were two in every room except the parlor, I think.

Q. Are you ready to state the expense of those in the parlor chamber?

A. I am not.

Q. Can you state what the two mirrors cost?

A. [Witness shakes his head negatively.]

Q. Were they high cost?

A. Yes, sir; they were very high cost. They went from the floor up as high as they could, and the scroll came over them. And the other was not as high, going from the mantle-piece, and reaching up to the ceiling, and was wider than the longer one was. I have a bill of them, but I don't feel at liberty to put the cost.

Q. At what you think they cost?

A. What they cost. I have got in mind that I gave $1100 for the two; but for fear that I am not correct, I don't want to say that positively.

Q. (*By the Chairman.*) The two mirrors, you say?

A. Yes, sir.

Q. (*By Mr. Rolfe.*) Do you recollect what you gave for the mantle-piece in the parlor?

A. I paid $500. You understand, the marble fire-place.

Q. Yes, sir.

A. Yes, sir; I paid $500 for that. I did not consider that personal with the furniture.

Q. Are you able to state the sum that you paid—the sum that the house that you sold to Mr. Clough, cost you?

A. Well, in telling the price of the house, I have to take my stable in with it.

Q. Very well, I mean everything?

A. And my fence?

Q. Yes.

A. I am quite sure that it footed to $22,000, or a fraction over.

Q. (*By Mr. Mugridge.*) That includes everything?

A. Everything there was up there.

Q. (*By Judge Bellows.*) That included the furniture?

A. That included the furniture that I have been telling you about, that went into the house. I had other furniture that was not carried there, that I made no reckoning on.

Q. (*By Mr. Stanley.*) You state the cost $22,000?
A. Yes, sir. I am quite sure that I am right on that. I have the figures somewhere now.
Q. (*By Mr. Rolfe.*) Are you able to say the consideration was put at $12,000?
A. All I can say of that —— [Objected to.]
*Mr. Rolfe.* He knew more about it than anybody else.
*The Chairman.* He can state about that, if he knows.
Q. Can you state at whose suggestion that was put in?
A. Mr. Corning was the man who mentioned that. He gave his reason why he wanted it done in that way. [Objected to.]

CROSS EXAMINATION.

Q. (*By Mr. George.*) Were Mr. Clough and Mr. Corning on your notes.
A. They were.
Q. To what extent?
A. I cannot possibly tell.
Q. State as near as you can?
A. At that time I cannot tell.
Q. State as nearly as you can?
A. Well, I should think $15,000.
Q. Was your property all put into Mr. Clough's or Mr. Corning's hands?
A. Into Mr. Corning's.
Q. It was all put into Mr. Corning's hands?
A. I sold to him
Q. Was it put into his hands because they were on your hands?
A. I cannot answer that.
Q. You cannot answer that?
A. I know I sold to them with the consideration that Mr. Corning was to pay up, as I supposed, everything I owed, and then the remainder was to go to two of my sons.
Q. What property did you put into Mr Corning's hands?
A. Well, what we call the foundry property, and I think this house that we talked of, and some other—I can't tell exactly now what.
Q. How many debts did you owe?
A. I could not tell you.
Q. How many did you pay of the insecured debts?
A. I don't know; I couldn't tell you.
Q. Did you pay anything?
A. I don't know; I didn't have a statement of them.
Q. Well, were there a large number of debts remaining outside?
A. I cannot tell.
Q. Did you know whether there were or not?
A. No, I don't know.
Q. Can't you tell how much you owed?
A. I could not tell you.
Q. Can't you give an estimate?
A. I cannot, because I did not have anything to do with the——
Q. Could not you tell somewhere in the neighborhood of how much you owed?
A. No, sir; I can not.
Q. Was Mr. Gilmore one of the parties into whose hands your property was put? That is, did it go into the hands of Mr. Gilmore?

*Mr. Mugridge.* If this is insisted upon, I think the deeds are the proper evidence.

*The Chairman.* It is pretty difficult for me to see the relevancy of the testimony.

*Mr. George.* I simply want to show what I suppose the fact is, that he did not know anything about it.

*The Chairman.* If it is for the purpose of testing the witness, I suppose it is admissible.

Q. Can you tell what debts were paid? Were Clough and Corning and Gilmore all together on your paper?

A. I think so.

Q. Didn't Corning take this property for the benefit of the three?

A. I don't know whose benefit they took it for. He bought it, and agreed to satisfy those notes that were due, and the remainder was to pay two-thirds to one son, and one-third to the other; and my son had a box and kept them all; I never knew anything about it.

Q. What became of the debts not included in those debts? What became of those debts?

*Mr. Rolfe.* The witness hasn't testified that there were any other debts that were not included.

Q. What was about the amount of your indebtedness?

A. I can't tell.

Q. Can't you give us the best estimate you can?

A. I don't want to, because the fact was, I might not be anywhere in the neighborhood. My son kept the books and managed the business, and settled up with them.

Q. Can you tell anywhere in the neighborhood of how much you owed—whether it was $50,000, or more or less?

*Mr. Tappan.* I submit, Mr. Chairman, that after he had stated that the books were all in the hands of his son, and he had nothing to do with the business, it is not competent for him to state.

*The Chairman.* If the witness says he don't know, it is not.

A. At the time I suppose I could tell a good deal more about it than I can now.

Q. Was Corning, or Gilmore, or Clough on your notes to the extent of $25,000?

A. I think not.

Q. Do you know anything about it, so that you can swear?

A. I think I can safely swear that.

Q. How much were they on for?

A. It is possible it might be eighteen, but I should think not.

Q. Where were those notes? Where was that indebtedness?

A. At the banks.

Q. At what banks?

A. The major part of it, I think, was here in Concord, and I think at the State Capital Bank. That I wouldn't be positive of, but think so.

Q. What particular property did you put in his hands? Did you have a particle of property that you did not put in his hands? if so, what was it?

A. I had two houses. One of them on Warren street, and one on the corner of Green and Warren.

Q. Where did these go to? Were these attached?

A. Yes, sir.

Q. These were attached before you had the conveyance made?

A. Yes, sir.

Q. Which was done first,—the attachment or the mortgage.

A. This foundry and some of that that I have told you of, was not attached at all.

Q. After the attachment the conveyance followed immediately?

A. I cannot tell you now.

## TESTIMONY OF GEORGE T. COMINS.

Q. (*By Mr. Mugridge.*) Let me ask you whether you are a contractor at the state prison?

A. I am.

Q. Do you know Sidney H. Carney?

A. Yes, sir.

Q. How long have you known him?

A. We went to school together from the time we were six years old.

Q. Let me ask you if you are acquainted with George Clough?

A. I am not acquainted with him.

Q. Let me ask you if you ever spoke of Mr. Clough as a thief.

[Objected to.]

*Mr. Mugridge.* This man Carney came in here and made the voluntary statement that Mr. Comins told him that Mr. Clough was a thief. I propose to show what statements he did make.

*The Chairman.* The testimony appeared to me, and, I suppose, did to the other referees, as being utterly irrelevant. Mr. Dr. Carney, I remember, blurted out something about his considering Mr. Clough as a thief; but it seemed to me that that was entirely immaterial and irrelevant, and having nothing to do at all with the matter. When you pursued the matter, I thought it seemed, and according to the rule of law it would seem that the testimony was inadmissible. It cannot be of the least possible competency in this case.

*Mr. Mugridge.* We supposed that exception would be made, but we offer to put in the testimony.

*The Chairman.* I remember thinking at the time that if the witness was rash enough to lay himself open to a slander prosecution, I did not see what it had to do with this case.

## TESTIMONY OF GRANVILLE REMICK.

Q. (*By Mr. Rolfe.*) Mr. Remick, where do you live now?

A. I live in Concord.

Q. How long have you lived in Concord?

A. About six years.

Q. Where did you reside previous to coming to Concord, and for how long?

A. I resided in Pittsfield and in the vicinity about 25 years.

Q. State when you were first acquainted with George Clough?

A. It was in 1837.

Q. What was he doing then?

A. He was driving stage from Pittsfield to Lowell.

Q. What was your business?

A. I was clerk in a grocery store.

Q. What store?

A. John L. Thorndike's.

Q. State whether you knew about Mr. Clough's business during the time you was there in Mr. Thorndike's store, and what it was, and if he did anything in the chicken and produce line, state it?

Q. (*By the Chairman.*) When was it that he speaks of?

A. From 1837 to 1839. He used to drive a stage from Pittsfield to Lowell, and he was in the habit of buying produce, and poultry, and most invariably carried a load, when he went out from Pittsfield, of such commodities.

Q. (*By Judge Bellows.*) Did you say butter?

A. Butter, eggs and poultry.

Q. (*By Mr. Rolfe.*) Mr. Remick, state what was your business in Pittsfield after Mr. Clough left? Up to the time you came to Concord were you a trader?

A. I have been a trader.

Q. You have been a trader here since?

A. Yes, sir.

Q. State whether you had occasion to pass over the Concord Railroad frequently or otherwise while you was at Pittsfield, and since you have been here?

A. Yes, sir; had occasion to, and did so.

Q. State whether you have dealt any in these coupon tickets, these tickets that were bought above here, and where passengers had ridden to this place, and got off. [Objected to.]

*Mr. George.* The fact that he dealt in them didn't give authority to anybody else to deal in them, and didn't give authority to him to deal in them; and if the conductors took them, that don't excuse them.

*Mr. Rolfe.* That is only preliminary. We propose to show that Remick purchased a considerable number of these tickets, and that he rode on them without objection.

*The Chairman.* With what view is that testimony offered? What is the purpose of it?

*Judge Bellows.* What is the object of this testimony, Mr. Rolfe?

*Mr. Rolfe.* One view is to show what kind of tickets were used over the road.

*Mr. Tappan.* The object is to show that the tickets sold here and used over the road were not spent tickets.

*Mr. Rolfe.* There is another view, which is to show the amount of travel over the road, and how they were returned. If the referees will allow me to suggest, these tickets from above here, from Montreal and Detroit—the amount of travel over the road is shown by the sales and not by the collections—so that if a ticket is sold from Montreal to Boston, the Concord Railroad counts that in as a passage, whether anybody passes over the road or not, and is entitled to its pay.

*The Chairman.* Then what you hold, if that is so, is that it is no injury to the road to have a man take a ticket over the road and sell it?

*Mr. Rolfe.* I hold it is no injury for anybody to live up to their agreement.

*The Chairman.* If you take the position here that during all this time that Mr. Clough was on the road, there may have been others selling coupons—if you take the position that there were no such tickets that were spent tickets, then this may be evidence to refute.

*Mr. George.* We take the position that there were tickets, and what were spent tickets subsequently. Subsequent to 1864 there were no such things as spent tickets.

*The Chairman.* If the testimony of Mr. Remick should relate to the time since October, 1864, since that regulation was made, it might be competent to show that these were tickets of that description out then.

Q. (*By Mr. Rolfe.*) I have inquired if you had purchased these coupon tickets; and I will ask you further, if you have ridden upon them, and at what time?

*The Chairman.* Perhaps you had better settle the time first, as that might be important.

Q. Whether after June 1st, 1864?

A. I should think I had since then; but I don't know as I could state positively when. I have bought several of them of Canadians who have stopped here, and taken them to pass over the road to Boston.

[Objected to.]

*The Chairman.* The evidence is perhaps pretty slight, tending to show that it took place since 1864.

A. Yes, sir, I know it was since 1864 that I had some of them; because I was in business for myself.

Q. (*By the Chairman.*) Since what time in 1864?

A. Since spring; since March, 1864, I bought them.

Q. (*By Mr. Rolfe.*) Up to what time, as near as you recollect? To how late a date?

A. I should think two years; two or three.

Q. Won't you describe those tickets?

A. Seems as though they were tickets that you tear off a portion. There were three parts when I bought them; from two to three.

Q. Was there anything on them that went to show that they were good for any particular time, or not good for any particular time?

A. My impression is that they were good for thirty days; though they passed them after keeping them longer.

Q. Had you ever had them refused?

A. I don't remember as I had.

Q. (*By Judge Bellows.*) You said they were marked good for thirty days?

A. Most of them said "good for thirty days," but I remember passing on them after keeping them more than that time, without being objected to.

Q. Over what road?

A. The Lowell—the Concord and Lowell.

Q. State whether you most invariably rode that way?

A. No.

Q. How was that generally?

A. When I had a ticket of that kind I usually passed that way.

Q. When it went the other way, you went that way? [No answer.]

*Mr. Tappan.* Did I understand the referees to rule out evidence that tickets of this character were used prior to 1864?

*The Chairman.* We only suggest that if does not contradict it is of no consequence. Col. George does not contend that they were or were not. The referees did suppose that it was material if tickets of this kind were used and received in the cars after June, 1864.

CROSS EXAMINATION.

Q. (*By Mr. George.*) Mr. Remick, I want to know if Mr. Clough owes you anything, or if you owe him anything?

A. He don't owe me anything, and I don't know as I owe him anything.

Q. Has there been an assignment?

A. Yes, sir.

Q. How much did you owe him?

A. He lost the same as the other creditors.

Q. How much did he lose by you?

A. About $100 in the settlement.

Q. Do you know how much you owed him?

A. I owed him about $200.

Q. And you paid fifty cents on the dollar?

A. Yes, sir.

Q. Now, sir, will you state whether you ever had any tickets of Mr. Clough, or had any tickets given you by Mr. Clough?

A. Never.

Q. Never bought any, or had any given to you?

A. No, sir.

Q. Never passed you over the cars?

A. No, sir; nor any other man. Never passed over the road without paying, with the exception of one instance.

Q. Who was that?

A. Mr. Plummer Whipple.

Q. Never bought a ticket or coupons of Mr. Clough?

A. Never.

Q. Directly or indirectly?

A. Never.

Q. You say you bought some tickets of some Frenchmen. From what place were they?

A. From Montreal, I think.

Q. Are you able to tell when it was? Are you able to fix any date?

A. It was during my time of business for myself; from 1864 to 1866; somewhere in that vicinity.

Q. When did you go into business for yourself?

A. In 1864.

Q. What date in 1864?

A. March.

Q. Are you certain about the year? What means have you of fixing the year?

A. I was in company with Webster a specified time.

Q. When did you go into company with Webster?

A. 1861.

Q. What month? Are you sure about the year? What means have you of fixing the date?

A. Because that was the date.

Q. How do you fix it? Have you any means of fixing it—that year?

It is all fixed. That is the year that I went in with them.

Q. Haven't you any means of fixing it?

A. Yes, sir; I have the books.

Q. You have looked at your books to find that, have you?

A. Yes, sir.

Q. When did you look at them?

A. I looked at them a great many times.

Q. How long did you remain?

A. Two years.

Q. That would be 1862. What did you do in 1863?

A. I was in with Mr. Cochran.
Q. How long was you in with him?
A. Almost two years.
Q. What time did you cease to be in business with him?
A. March, 1864, I guess it was.
Q. You guess it was. I want to know if you can tell?
A. It was March, 1864, that our copartnership ceased.
Q. Now, how did you fix the purchase of these tickets as the time in reference to your going into business for yourself?
A. Well, I have it on the cash-book.

## TESTIMONY OF MR. JOHN F. KIMBALL.

Q. (*By Mr. George.*) Will you be kind enough to state where you live, and what your employment is?
A. Live in Lowell; cashier of the Appleton National Bank.
Q. Will you state whether Mr. Raymond Kimball was your uncle?
A. Daniel Raymond Kimball.
Q. And you may state, if you please, what his business was, and what relation you occupied to him of a business character?
A. He kept a livery stable connected with the Merrimack House; moving there from Belvidere, about 1836. My father died in 1838. I was then about fourteen, and he occupied the position of guardian, and I went to him for any advice or assistance.
Q. Please state about your uncle being a man of limited accounts, and always keeping accounts, etc.?
A. He was a man who relied upon his memory mostly. He kept a ledger upon which he kept his charges. He was a man who disliked to use a pen or make figures.
Q. Who had charge of his business, and of his accounts—and of his business negotiations so far as writing, etc., was concerned, from 1838 on?
A. From 1838, for perhaps two or three years after that, there was a stage agent by the name of Lewis who used to write his letters, but his accounts were mainly kept by me from the time I was fourteen years old.
Q. And were you acquainted with his business matters?
A. Yes, sir; he used to make a confidant of me; more perhaps as I grew older.
Q. And you continued to occupy that relation until when?
A. Until the time he gave up business in 1857. He gave up business in 1857, and became insane; but prior to that I had the whole charge of that business.
Q. As far as money matters were concerned?
A. Without consulting with me about investments.
Q. Now let me ask you whether at any time from 1842 to 1846, or any time subsequent to 1842, there was any note existing, so far as you know, from your uncle to Mr. Clough, to the extent of $6000, or any other sum?
A. No, sir; I am not aware of any amount. No, sir; he was not a man who borrowed money.
Q. If there had been an indebtedness of $6000, must you have known it?
A. In all probability I should have known it. There is a possibility that I might not have known it; but I should, in all probability.

CROSS EXAMINATION.

Q. (*By Mr. Mugridge.*) Mr. Kimball, did you know Mr. Clough?
A. I knew him by sight.
Q. And what relations formerly existed between him and your uncle?
A. Mr. Clough drove stage there.
Q. Mr. Clough drove stage to Lowell where you uncle lived?
A. Yes, sir; he drove from the Merrimack House, that being his headquarters.
Q. Do you know that your uncle always took a particular interest in Mr. Clough, and was his particular friend and adviser?
A. No more than to many others.
Q. Did you know the fact that your uncle took a friendly interest in his affairs and assisted him from time to time with his advice and counsel?
A. Not especially; not more than the other drivers. Their relations were friendly.
Q. Don't you know that your uncle sustained the position of friend and confidential adviser towards Mr. Clough?
A. Not especially.
Q. Generally?
A. Yes, generally.
Q. Did you know the fact, that your uncle was accustomed to advise Mr. Clough?
A My uncle was a gentleman who had a pretty good idea of his own opinion, and was rather fond of ——
Q. Let me ask you if your uncle was not a man property?
A. Yes, sir.
Q. He was regarded as a perfectly safe man to entrust funds with?
A. He was.
Q. What was the amount of his property when he became insane?
A. The estate was settled at $37,000; but the property cost him about $50,000.
Q. His business responsibility was good, was it not?
A. His responsibility was never questioned.
Q. Did you know of a man in Lowell that was trusted by Mr. Clough any more than your uncle?
A. He wasn't a man likely to put out money.
Q. Wasn't he of that disposition and cast of mind that led him to do these friendly acts for young men?
A. Yes, sir; I think I can call others quite as likely as he.
Q. That was the tendency of the man's character, to befriend young men in this way?
A. Yes, sir.
Q. Wasn't he a prompt man in all his business?
A. Yes, sir.
Q. And liked to see other people prompt?
A. Yes, sir.
Q. Now let me ask you if you mean to say positively, that he could not have held that amount of $6000 for Mr. Clough?
A. I mean to say that he would not have been likely to without my knowledge,
Q. You don't mean to say that it could not have been entrusted to him by Mr. Clough?

A. I don't mean to say that it could not, but it was not likely that it would.

Q. How old were you in 1842?

A. I was eighteen.

Q. Had you any other business?

A. Yes, sir; I was clerk in the post-office.

Q. And that was your regular business?

A. Yes, sir; I worked for him evenings and Sundays.

Q. Did you do all his writing?

A. All that I know of.

Q. Were there any notes outstanding against him, so far as you knew?

A. I didn't know of any. I am not aware of any, except one.

Q. Whose was that?

A. That was one that I had.

Q. State whether young men were accustomed to entrust their funds to him for safe keeping?

A. I don't know of anybody but myself. It was my practice. He has said to me repeatedly, "I'll take this for you, John, but I wouldn't for anybody else."

Q. I suppose you wouldn't want to be understood as swearing positively to the fact that he hadn't that note?

A. No, sir.

Q. You say your uncle was of that character that you have described?

A. He was willing to lend a helping hand to any person in need.

Q. Do you know whether Mr. Clough was frequently at his stable?

A. I don't remember his being there so late as 1842. It is my impression that Mr. Clough was with him some earlier than that time.

Q. How long?

A. It is a long time; and I wouldn't be very particular about it.

Q. Was he not at the stable of your uncle very frequently from 1843?

A. At the time he drove stage, he was with him perhaps every other night.

Q. Where did he stop?

A. I think he stayed at the Merrimack House.

Q. That was your uncle's place?

A. That was where he lived.

Q. What length of time do you think Mr. Clough was accustomed to stop at the Merrimack House where your uncle was?

A. While Mr. Clough was driving stage.

Q. How long a time, as you recollect it?

A. I think it was two or three years. I shouldn't be willing to swear positively to that.

## TESTIMONY OF JAMES P. EATON.

Q. (*By Mr. Mugridge.*) Where do you reside?

A. Manchester.

Q. Were you one of the conductors that formerly run on the Manchester and Lawrence road, from Manchester to Lawrence?

A. I was.

Q. You are one of the gentlemen against whom a suit is brought?

A. I am.

Q. State whether you had any conversations with Joseph A. Gilmore with regard to the bringing of the suit against George Clough by the Concord Railroad? If so, you may state when and where the conversation was, and what it was? [Objection to what Mr. Gilmore stated.]

*Mr. Mugridge.* They undertake to show that there was a collusive combination between Gilmore and Clough with regard to that suit. It is competent in that view, if no other.

*Mr. George.* I suppose it is a question whether any statements made by Gilmore after these suits are competent evidence.

*The Chairman.* Probably it would not do much harm for them to state what they propose to prove, and then we can see it a little better.

*Mr. George.* If anything that Mr. Gilmore stated after this suit was commenced is not competent, I suppose there is no occasion for stating it.

*Mr. Mugridge.* We propose to show that Mr. Gilmore sent for Mr. Eaton to come to his office; that he said to him,—"your suit is commenced"—I dont undertake to use the exact words—"this suit against you, Mr. Eaton, don't mean anything, and won't come to anything, but with regard to Geo. Clough's case, I will learn him to refuse to sign my notes, and go back upon me in the way he has." That is what we propose to prove by the witness. It is certainly competent on the ground taken by the help of collusion between the parties.

*The Chairman.* The evidence being objected to, perhaps, Mr. Mugridge, you had better state fully your views.

*Mr. Mugridge.* In the first place, we offer it upon this ground: It has been suggested here that there was a collusive arrangement or combination between Mr. Clough and Mr. Gilmore with regard to this suit, and that there were also collusive understandings and arrangements, between Mr. Clough and Mr. Gilmore, antecedent to the time the suits were brought, while they were upon the road. We offer this testimony in connection with other testimony upon that ground, to show that there has been no collusion between them at any time or any where, with reference to this suit or subsequent to the time the suit was commenced. We then offer it as a statement of the superintendent of the road, as bearing upon the nature of the transaction here, and shall contend from it that this was a malicious suit, born in malice, and that it was maliciously undertaken by the officers of the road, so far as they had anything to do with it. And then again we offer it as a statement of the agent and superintendent of the road; and contend that we have a right to draw from that statement, made by him, whatever is legitimate as against the road.

*The Chairman.* We understand the proposition is to prove a statement of Mr. Gilmore while he was still the superintendent of the road, and pretty near the time of the transaction. There is evidence in the case tending to show that this suit was instituted to some extent by the transactions of Mr. Gilmore, acting as the agent of the road. And there is, perhaps some reason why the testimony should be admitted as statements of him in regard to that very particular matter in which he had acted as the agent. And under all the circumstances we think it is best to admit the testimony.

The ruling was excepted to by the plaintiff.

*The Chairman.* Of course you will understand that this is a pretty doubtful question, but you can put the testimony in. You can assume the risk of it.

*Mr. Mugridge.* Yes, sir; we will do that.

Q. (*By Mr. Mugridge.*) Now, you may state what conversation you had with Gilmore, and how you happened to go to see him, and state all that happened?

A. Why I happened to come up here, Mr. George A. Barnes wrote me a note in Manchester to meet him at his store; it might have been three or four days after the suit was commenced.

Q. After the suit of the Concord Railroad against Mr. Clough was commenced?

A. Yes, sir; also mine.

Q. You were sued at the same time Mr. Clough was?

A. Yes, sir.

Q. Now you may state where you saw Mr. Gilmore?

A. At his room.

Q. Where was that?

A. At his residence, where he lived.

Q. State all the conversation as it occurred between you and Mr. Gilmore at that time, in regard to the suit against Mr. Clough?

A. Mr. Gilmore says to me,—"Eaton, your suit don't amount to anything; don't mean anything."

Q. Well, you may proceed?

A. He says,—"Mr. Clough, I will learn him to go back on me, when I lay flat on my back, and want him to sign a note; I guess he wishes by this time that he had assisted me."

Q. Did he say anything else at that time in regard to this matter?

A. Various things were talked of, but nothing to any particular point. "Your suit," says he, "amounts to nothing, I will take care of that."

Q. And then in the same connection with this remark?

A. Made this remark. I think about the first thing is this that he said of Mr. Clough.

Q. Gilmore was superintendent of the Concord Railroad?

A. Yes, sir. He left word for me to come and see him; he telegraphed me to come from Manchester to Concord.

Q. Can you state how many days this was after the suit was commenced?

A. No, I cannot.

Q. State your impression?

A. I should think the suits were Saturday, and I think this must have been the middle of the next week. That would be my impression; I could not swear positively.

CROSS EXAMINATION.

Q. (*By Mr. George.*) You were one of the conductors upon the Lawrence road?

A. Yes, sir.

Q. You ran from Manchester to Lawrence three times a day?

A. Yes, sir.

Q. Will you state whether you had any conversation with Mr. Gilmore with regard to the amount of your returns on the road? Did you have any conversation with regard to the amount of your returns?

[Objected to.]

*Mr. George.* I suppose I have a right to ask him whether there was a conversation with regard to the amount of his returns.

*The Chairman.* I suppose they have a right to examine as to all the details of this particular conversation.

*Mr. Mugridge.* But this is another conversation.

*Mr. George.* I suppose we have a right to inquire as to all the circumstances of this conversation or any other conversation, and inquire what the conversation was, and all the relations.

*Mr. Mugridge.* We think it is only competent for them to inquire as to the particular conversation concerning which we have inquired.

*The Chairman.* It seeems to me that you will have to confine yourself to that particular conversation, Col. George.

Q. (*By Mr. George.*) Was anything said at this particular conversation about the amount of your returns?

A. At what time do you mean?

Q. You stated a particular conversation that you had, Was there anything said about the returns made by you?

A. I don't think there was; I couldn't say positively.

Q. Did Mr. Gilmore say anything to you about your returning ten cents a train on the average? Was the fact so? [Objected to.]

Q. Did you return on these trains to Lawrence the sum of thirteen dollars a month taken on the cars?

*The Chairman.* I don't see the relevancy.

*Mr. George.* The relevancy is this: that if that be so, it is pretty important as to whether Gilmore told him that it was bosh. Here was a suit commenced against him. Now when that suit was commenced, Mr. Gilmore sent for him and told him the suit was all bosh. I propose now to show certain facts, with a view of showing whether it was probable or not.

*Mr. Mugridge.* The point is this: Col. George proposes to show that Mr. Gilmore didn't make a certain statement to him by showing what Mr. Eaton's returns were.

*The Chairman.* I think we are all agreed that the testimony is not admissible.

Q. (*By Mr. George.*) Was there anything said in this conversation between you and Mr. Gilmore, that when a new conductor, (Mr. Kendrick) was put on the train from here to Portsmouth, he returned $2903 while you returned $548? Were there any suggestions of that kind? And didn't he show you these returns? Did you talk about it?

A. I can't say that he did.

Q. Can you say that he didn't tell you that in five months you returned $548 to a cent? Didn't he tell you that?

A. No, sir.

Q Did he tell you at that time how much you did return?

A. I told you he didn't, I believe.

Q. Did he say anything about Kendrick's return?

A. He didn't talk with me about matters in general, because I was there but a little while, and he talked with me about Mr. Clough's case. I haven't any particular recollection about it.

Q. What was the first word he said.

A. I think he said, "How are you, Jim?"

Q. Now, go on?

A. Well, I don't know what was said. There was a general preamble that was generally talked over.

Q. What was that?
A. I couldn't tell.
Q. Can't you tell anything that was said?
A. I have told you what was said about Mr. Clough's case.
Q. How long were you there? What did he send for you for?
A. Because he wanted to see me, I suppose; he frequently wanted to see me.
Q. What did he want to see you for? If you know, state. State every word that you can remember in that whole conversation?
A. I have no particular thing to tell.
Q. If you recollect anything else at all besides what you have stated, state it, from the time you went into Gilmore's house till you came out again—that is, in addition to what you have stated, won't you state it? You remember anything else?
A. I don't recollect.
Q. Do you recollect anything at all beyond what you have stated?
A. General conversation occurred, same as it would if I went into any man's house. It was about these suits, and you know very well, and the court knows very well, that it would be impossible to recollect the particular words.
Q. I don't ask the particular words. Give me the substance. Did you go into any explanation as to how your returns happened to be so small? Any conversation on the fact of your returns being so small?
A. No, sir.
Q. Any conversation as to your using tickets that had been used?
A. No, sir.
Q. Any conversation in regard to your manner of making returns?
A. No, sir.
Q. Any conversation in regard to the manner of Clough's making returns?
A. Not that I know of.
Q. Any conversation in regard to Mr. Clough's using tickets that had been used?
A. No, sir.
Q. Any conversation in regard to Mr. Clough's using coupon tickets?
A. No, sir.
Q. Any conversation in regard to giving return tickets?
A. No, sir.
Q. Any conversation had in regard to the Jim Whitcher tickets?
A. No, sir.
Q. What was the conversation about that you had? I don't understand. You say you was sent for by Mr. Gilmore to go down there. What was the conversation about? Won't you tell us?
A. I would ask if that is a proper question?
Q. Yes, sir. Answer it if you can.
A. I can't answer it. I asked him how he was. "Damn it, I'm sick; here I am swelled up; can't piss; big belly; I want to know if you wouldn't." That is all; general preamble. "I am swelled up pretty big; damn it, hand me some gin there." The waiter came in. It was a very short interview. The details of the road I didn't go into at all; we didn't go into it. I know that I was sued. Three years ago it was commenced, and I haven't heard much about it since.

Q. Did Mr. Clough come to you and the other conductors and get you to give a bond for loss, or sign a bond?
A. I think Mr. Clough didn't.
Q. Who did?
A. I can't tell now who presented it.
Q. How much did you pay?
A. I paid all I agreed to.
Q. As near as you are able to state?
A. I don't know; I couldn't tell to save my life.
Q. Won't you state as near as you can how much you paid?
A. I can't state anything about it.
Q. Can't you state as nearly as you can?
A. No, sir.
Q. Can't you state as nearly as you can?
A I don't understand the meaning of that word; it is beyond my dictionary.
Q. Do you mean to swear that you have no idea how much you paid?
A. I mean to swear that I couldn't tell.
Q. Are you willing to swear here that you haven't any idea?
*Mr. Mugridge.* Is it at all material how much he did pay or didn't pay? [Question waived.]

## TESTIMONY OF DANIEL S. WEBSTER.

Q. (*By Mr. Mugridge.*) Mr. Webster, where do you reside, and what is your business?
A. Nashua; baggage master.
Q. Baggage master for what road?
A. Concord Railroad.
Q. How long have you acted in that capacity?
A. Five years.
Q. And you are stationed, of course, at the depot?
A. Yes, sir.
Q. The Concord Railroad came in on one side, and the Worcester road on the other side of the depot?
A. Yes, sir.
Q. Will you state how many cars were run over the Worcester road in the hight of travel in the year 1864 and 1865? I suppose you observed the trains as they came in, didn't you?
A. Never more than two passenger cars.
Q. That is what I mean—passenger cars?
A. That is the regular trains.
Q. Now, Mr. Webster, what is your recollection as to the passengers being able to get their baggage without tickets, at the Nashua station since you have been there.
A. While I was there, notice came down from Mr. Winslow that we were to see the tickets.
Q. That is, any passenger who started from Nashua must show his ticket before he could have his baggage checked?
A. Yes, sir.
Q. Let me ask you whether that regulation has been lived up to?
A. I had to live up to it myself, and I do,—not so much as I did two years ago, I don't think, but I used to make the folks show their tickets before they took their checks, and so far as I know, all that came in over the Worcester road.

Q. Have you observed the conduct of Mr. Clough with regard to directing passengers who came in or on the Worcester road as to getting their tickets at the station? You may state what you have seen of his conduct in that direction, and state the extent of it, fully?

A. Mr. Clough used to have them get their tickets before they had their baggage checked.

Q. Used to have who get them?

A. The passengers.

Q. What passengers?

A. The passengers who came in over the Worcester road, he would have them get their tickets before they had their baggage checked.

Q. How frequently have you seen Mr. Clough do this?

A. Well, Mr. Clough used to do that most of the time. There was sometimes that the Worcester trains would come in the same time that the trains came in on the opposite side. He was most always up there. He would be on the other side. There were a great many that bought tickets only to Nashua, and would have their baggage checked to Nashua, and before they went on, he would send the passengers around to get their tickets.

Q. Where would he go?

A. He would go right in through the depot.

Q. He took the passengers as they came in from the Worcester cars?

A. Yes, sir.

Q. Won't you give the referees the best idea you can how long Mr. Clough practised in this way?

A. Mr. Clough practised that as long as he ran, while I was there.

Q. From 1863 to 1866?

A. Yes, sir.

Q. From what time to what time did he do it?

A. I couldn't say exactly; but Mr. Clough was there more than half of the time. There would be some times that he would not be there. He would be on the Lowell train, and the Worcester train would not be in first.

Q. But when he was there?

A. But when he was there, I should say that he was out there,—well, according to the best of my judgment, I should think he was there—well, seven-eighths of the time.

Q. Seven-eighths of the time?

A. Seven-eighths of the time, I should think.

Q. Giving the passengers these directions?

A. Giving the passengers directions.

Q. You may state how many cars the Concord Railroad ran, in the hight of travel, down from Concord in the years 1864 and 1865?

A. I should think that the number of cars that ran into Nashua was three; it might have been four.

Q. Do you mean three passenger cars?

A. Three passenger cars.

Q. And sometimes perhaps four?

A. Yes, sometimes; say, for instance, Saturday.

Q. But the ordinary general run was three cars?

A. Yes, sir.

Q. Did ever you see a train of sixteen or eighteen cars run into Nashua?

A. Not on the regular passenger trains. I have seen extras.

Q. Did ever you know a regular passenger train of sixteen or eighteen cars run into the Nashua depot?

A. No.

Q. What is the largest number of passenger cars that you have seen run into Nashua on the regular trains, that you remember of?

A. Well, the largest train I ever saw——

Q. I don't mean the extra cases?

A. Well, the largest trains I have seen run in there on the regular train was a year ago this last summer.

Q. What was that occasion?

A. It was when the mountain travel was going.

Q. How many cars were there on that train?

A. There were thirteen, I believe—but I don't know as there were but twelve passenger cars and one baggage car.

Q. That is the largest train you ever saw this last year?

A. A year ago this last summer.

Q. And you say that the ordinary train was three or four cars?

A. Yes, sir.

Q. Well, take the 3.30 train in the afternoon, how many cars ordinarily ran into Nashua on that train in 1864 or 1865?

A. Four passenger and two baggage cars, I think.

Q. You are describing now the ordinary train?

A. Yes, sir; there may be trains that there might be more.

Q. Did ever you see sixteen or eighteen cars run in there in 1864 and 1865 on the 3.30 train?

A. No, sir.

Q. Did you know anything about the substitutes who travelled during the war?

A. No, sir.

Q Let me ask you how it was during the war about soldiers being carried down over the road. Did they go in the regular or extra trains?

A. Well, most of them—if there was any amount of them, they went into an extra. But I don't recollect instances of that but once or twice.

Q. Then they put them on what?

A. Then they put them on the Worcester road.

Q. How many would there be?

A. Well, there might be a hundred.

Q. Was the soldiers' travel mixed up with the general travel in the cars during the war?

A. Well, some of the time there was a few; where they were travelling from Concord to Nashua, there would be a few down and back.

CROSS EXAMINATION.

Q. (*By Mr. George.*) There was an immense deal of travel up and down during the war. wasn't there, of substitutes and substitute brokers?

A. Well, I shouldn't think there was so much, but still there might have been more than I knew of.

Q. Wasn't Concord the general rendezvous?

A. Oh, yes.

Q. Wasn't there an immense deal of travel of substitutes and substitute brokers, selectmen and town agents?

A. Oh, yes.

Q. Wasn't there forty thousand of these that came to Concord to be examined?

A. I don't think there was. My judgment might not be very good?

Q. How many were there in this state? Any idea how many went from this state?

A. I believe they footed up fifteen or sixteen regiments.

Q. But they were recruited several times. Wasn't the regiments filled up?

A. Yes, sir.

## TESTIMONY OF GRANVILLE REMICK.

Q. (*By Mr. Mugridge.*) Whether you have got any record of cash paid out for tickets?

A. [Producing book.] April 24, 1865, $5.50 paid out; May 19, 1865, $5.

Q. Give us another one?

A. I don't know as I can give any more.

### CROSS EXAMINATION.

Q. (*By Mr. George.*) That book that you have there is your cash book?

A. Yes, sir.

Q. If you paid any expenses for any purpose, you put it on that, did you?

A. Yes, sir.

Q. Now, suppose you bought a ticket at the office and went to Boston, wouldn't that appear?

A. I would call it cash and expense, and it appears so all through the book. Sometimes I would buy tickets and count them at night as cash.

Q. Won't you turn to one of those cases?

A. [Reading.] Fare and expense to Boston, June 29, $4.85.

Q. Well, won't you read those two.

A. Fare and back from Boston.

Q. Won't you read exactly what the book charges.

A. Fare to and back and expenses.

Q. Well, read the whole.

A. Well, that is the whole of it. Fare to and back; then $2; then $2.35; and then $4.35.

Q. Now, was that fare to Boston and back.

A. I presume it was.

Q. What is that?

A. June 29.

Q. Of what year?

A. 1864.

Q. The fare from Concord to Boston in 1864 is $2.90. Now, sir, was that fare to Boston and back?

A. It says so on here. That is all I can tell you about it. It is fare to and back.

Q. You swear it is to Boston?

A. No, I don't swear it.

Q. Where does that mean? There is fare to and back. Where does that mean? Won't you tell me where the point is that you go from, $2, and come back $2.35? Then that don't include any fare and expense?

A. [Reading.] Fare to and back, $2, $2.35; carried out, $4.35; expense, .50.

Q. You have no recollection of where that was, apart from the book?
A. It must have been Boston; I don't know where else it should be.
Q. Now will you take the next one?
A. Fare and expense, $7.25.
Q. You don't know where that was?
A. That was Boston; because I paid bills that day—Boston bills. Fare and expense, $8.30, October 19, 1864.
Q. Doesn't say where that is to?
Q. (*By Mr. Mr. Mugridge.*) How do the bills show?
A. They don't show. There is a ticket that I paid $1.50 for, March 24, 1865.
Q. (*By Mr. George.*) Now you can describe that?
A. It was one of those coupon tickets.
Q. Have you any recollection about it, whether it was dated or whether it wasn't dated?
A. No, I couldn't swear to any date. Generally they were dated; nothing particular on that. Twenty-fourth of April, 1865, paid out for tickets, $5.55.
Q. Does it say what tickets?
A. No, sir.
Q. Have you any recollection about those tickets, and of whom you bought it, or whether at the ticket office, or anything about it?
A. It was one of those coupon tickets that I bought of those Frenchmen that came along; I bought a great many of them.
Q. Have you any more of them?
A. Yes, sir.
Q How many?
A. Perhaps I might have bought from twenty to twenty-five altogether?
Q. How many?
A. Perhaps from fifteen to twenty-five altogether. Fare $6.50, June 29, 1865.
Q. Fare, $6,50. Where was that?
A. To Boston.
Q. Does that mean fare, or fare and expense?
A. Fare and expense, $6.50.
Q. What conductors did you ride with when they took them tickets?
A. I can't tell; the regular conductors.
Q. Did you ride with Mr. Clough?
A. I think likely I did sometimes.
Q. There were only two of them, Mr. Clough and Mr. Noyes; you had to ride with one or the other?
A. Yes. I recollect riding with Mr. Clough.
Q. They both took these tickets without objection, from you?
A. I don't think any objection was made.
Q. Did you ever have any talk about those tickets?
A. Never was any objection raised but once, and that was when I went down on the other road, down through Lawrence. There they wouldn't take them; said they belonged over on the other road—the Lowell.
Q. Let me ask if you ever had any talk with Clough about these tickets?
A. Never.
Q. Any conversation with any one?

A. Never. I don't remember that I had any conversation about it.

Q. (*By the Chairman.*) Mr. Remick, do you remember now, any of these tickets—these coupons that you speak of having bought in this way—do you recollect anything printed upon them showing the number of days they were good?

A. I could swear that there were some of them that said, "good for thirty days."

Q. Do you remember whether any of these said nothing about it?

A. I think they all had a stamp on the back.

Q. Were there any that had nothing but the date upon them?

A. I think there were; because I called in question, in buying them, that very point.

Q. (*By Mr. Rolfe.*) [Handing book.] Won't you give that entry?

A. "September 7th, tickets and fare, $7.75."

Q. (*By the Chairman.*) Is that 1865?

A. 1865; yes, sir.

Q. (*By Mr. Clark.*) Is that tickets and fare?

A. Tickets, fare, and expense.

Q. Proceed here right along. That is all the fares and tickets you bought. What is that? [Indicating.]

A. Fare $5.50 and expense $2.50.

Q. What date?

A. October 12, 1865.

Q. Fare $5,50; where was that to?

A. Boston. Sometimes, you see, I would buy these tickets and pay whatever there was, and count that as cash, and then when I prid them out count them as tickets. Expense $4.50, 14th November. Expense and fare $10.25, Feb. 26. I have a book that is the same in 1864.

Q. (*By Mr. Tappan.*) Is that 1864?

A. Yes, sir; commences 1864, in June.

Q. Now will you be kind enough to state what you paid for tickets in 1864?

A. I have got them right along what there is, in this book; but I say I have a book back of that; six months previous.

### RE-DIRECT EXAMINATION.

Q. (*By Mr. Mugridge.*) Mr. Remick, I understand you to say that you purchased these tickets from Frenchmen that came down from Montreal?

A. Yes, sir.

Q. Did you always pay the same price for them?

A. No, sir.

Q. Do you remember how cheap you have got the fare?

A. From $1.25 to $1.75; I don't know but there are instances where I paid $2 for some.

Q. You think the smallest was $1.25?

A. Yes, sir.

Q. Do you recollect what year you paid $1.25 for them?

A. No, I don't know as I could state.

Q. I understand you to say that these were always good in the hands of every conductor, except in one instance when you attempted to go down over the Lawrence road. What was the objection made then?

A. It said Lowell.

Q. And you were on the wrong line?
A. Yes, sir.
Q. Was that the only objection you found?
A. Yes, sir.
Q. Let me ask you how it was in the hands of conductors on the Nashua and Lowell, and Boston and Lowell road, and whether they were recognized by those conductors?
A. In every instance, I think.
Q. Was there ever any objection made? [Objected to.]
Q. (*By Mr. Mugridge.*) How many of these coupons do you think you purchased from the year 1864 up to sometime in 1866? How many do you think you bought of these coupon tickets?
A. Perhaps a dozen or fifteen.
Q. And you bought them all of these Frenchmen?
A. Almost universally.

*Mr. Mugridge.* I will put in now the sum received by Mr. Clough from his stock from May, 1855, to May, 1866, inclusive. It amounts in the aggregate to $12,603.50. That is for eleven years. Now we propose, when Mr. Stanley comes in in the morning, to put in the amounts previous to 1855.

Mr. Rolfe offered in evidence the statement of Mr. W. P. Hills, secretary of the Newburyport Savings Bank, and the copy of the deposit book of the Institution for Savings in Newburyport and vicinity. Geo. Clough, Newburyport, No. 3,719, 1835, Jan. 21st, rec'd $109.00

| | | |
|---|---|---|
| | April 8th, " | 91.00 |
| | Interest acct., | 26.71 |
| | Footed up, | $226.71 |
| (On the other side.) | | |
| | 1837, Sept. 26, paid, | $100.00 |
| | Dec. 30, " | 126.71 |
| | | $226.71 |

## TESTIMONY OF JOHN L. COLBY.

Q. (*By Mr. Rolfe.*) Mr. Colby, where do you live now?
A. Franklin; that is my native place.
Q. How long have you known George Clough?
A. Well, I don't know just the number of years, but it was the commencing of the railroad running to Franklin first from Concord here. I could not state exactly, by any means.
Q. How did you make his acquaintance?
A. Mostly by trading.
Q. What trading did you have with him?
A. Well, I sold him various articles, meat and provisions.
Q. Specify, if you please?
A. Well, I could not. The first that we dealt was in veal—calves, and the skins.
Q. To what extent, sir, and for what length of time did you deal with him?
A. A few months during the year, as generally calves run—the trade of them.
Q. To what extent? How many a day or week?

A. Well, that I couldn't state exactly—the number of weeks; sometimes once or twice or three times a week, as they grew plenty or I could get them.

Q. Well, sir, what was the course of trade? Do you recollect what profit you got out of them—what profit you got out of the calf and the skin?

A. Well, we generally—I believe that we bought them, and he gave us a certain per centage. There was some over a dollar on the skin and veal, the heads, plucks, and rennets.

Q. Will you state, if you know, how that was about the profits that you realized on the calves?

A. Well, it was understood between us; nothing very particular.

[Objected to.]

*The Chairman.* That isn't anything more than that Mr. Clough told him so and so, I suppose?

Q. I will ask you if you know how that was, aside from what Mr. Clough represented?

A. I knew nothing only what the contract was. He was to give us a certain proportion.

*Mr. Mugridge.* What we propose to show—and we would like the ruling of the referees upon it—is what the contract was between Clough and this gentleman, under which this veal was sent to market. We propose to show that there was a distinct agreement between Clough and the witness as to what the interest of each should be under the contract.

*The Chairman.* That perhaps is evidence.

Q. (*By Mr. Rolfe.*) You may proceed and state what the contract was?

A. Sometimes the contract was high or low, and we sort of divided; as we traded from time to time, we sort of divided.

Q. (*By the Chairman.*) Was there any agreement between you and Mr. Clough, what you should do?

A. Yes, sir; there was an agreement this way, that he would give us as much as he got; and sometimes it would be low, and then again it would be a higher price; and take it upon the whole, it was satisfactory, generally. We could do better than we could to sell them out there.

Q. (*By Mr. Rolfe.*) If you recollect, what was the profit on a calf?

[Objected to as being hearsay testimony.]

*The Chairman.* I think you may show the fact what he realized.

Q. You may go on and show what you realized as profit on a calf on the average?

A. I think, as nigh as I can recollect, he divided with us some over a dollar average; sometimes more and sometimes less; it would average some over a dollar.

Q. (*By the Chairman.*) You mean that you got $1, and that you divided?

A. I mean that I got $1 myself; except the skin; that was divided afterwards.

Q. (*By Mr. Rolfe.*) How was that?

A. The skins raised after a while; and I let him have them; they raised about half that year.

Q. Now what were the profits realized out of the hides. If I understand you right, you retained these hides?

A. I did at the first of our dealing. But I think that I was having a dollar; but it came to something more than that; it about doubled, it strikes me.

Q. And you and Mr. Clough, if I understand, made a double profit on it?

A. Yes, sir; the skins seemed to be a separate contract from the veal.

Q. Now, sir, how was it? Did you trade in any other articles—for instance, poultry?

A. Yes, sir; pelts and poultry.

Q. Do you recollect any particular instance where you bought rather large lots?

A. I do.

Q. How was it with Greenleaf and Greenough?

A. We had some of them. We bought part of it from Greenough, and the rest he wouldn't let us have.

Q. You got all of it from Greenleaf?

A. Yes, sir. I should think we made a trade with Greenough. I know we got one or two horse-loads, as much as we could draw in a double sleigh like that.

Q. Anything else?

A. I don't know as I could state.

Q. Did you buy any produce of the parties?

A. We used to buy of others. Why I know this more particular was because it was a larger amount. We bought great amounts and could not sell them out there; didn't expect to.

Q. How is it about potatoes?

A. I don't recollect as I sold him any potatoes.

Q. Do you know of his trading with other parties?

A. I know, by his delivering stuff to the depot, that he traded there with other parties.

Q. How much did you do in pelts?

A. Well, I could not recall this, as to the amount, as well as I could the others.

Q. When you dealt in pelts and other sleigh-loads of property, how did you send it?

A. Sent it by the freight train, I think. All this heavy cold weather we sent by freight, and these other trades; and in hot weather it went by express—put on to the express trains.

Q. (*By Mr. George.*) Put on the passenger train you mean?

A. Yes, sir.

Q. (*By the Chairman.*) Do you mean that they went in the express car?

A. They were put into the passenger train.

Q. Whereabouts were they put? In the baggage car, or the express car, or in the passenger car?

A. I don't know certain as to that.

CROSS EXAMINATION.

Q. (*By Mr. George.*) When did the cars first run to Franklin? Do you remember?

A. I remember the time.

Q. Wasn't it late in the fall of 1847?

A. Yes, I think it was; sometime in January, I should think.

Q. What, sir?
A. Sometime in cold weather.
Q. Do you remember to where the road was next opened?
A. I couldn't state the year—not now, I couldn't; but perhaps I could remember. It was in the winter, I know. I couldn't state the year?
Q. And he ran on the train as long as the road stopped at Franklin?
A. Yes, sir.
Q. Now, where was it next opened? You were at Franklin?
A. Yes, sir.
Q. Do you recollect Mr. Daniel Webster's coming?
A. I do recollect something of it.
Q. Wouldn't that refresh your recollection so as to tell how long the cars stopped there, and when they ran up to Grafton?
A. Yes, I think so.
Q. Didn't they run up the following July?
A. I think not; I think the year after.
Q. Do you mean to say that they stayed at Franklin a year.
A. Yes, I think so. I should think it was all of a year. It was through one season any way, I should think.
Q. You say it was not the same season. Are you quite confident?
A. Well, it is only my impression that it is so; no records with me to show any such thing; only my memory.
Q. Mr. Clough didn't run to Franklin after the road ran beyond there at all, did he?
A. I don't recollect as he did.
Q. When did the Northern road begin to run the road themselves?
A. I couldn't tell the time.

[TWENTY-SECOND DAY. Wednesday, December 30th, 1868.]

## TESTIMONY OF GEORGE CLOUGH CONTINUED.

Q. (*By Mr. Rolfe.*) You testified in cross-examination that you had some sap-buckets of James Whitcher. Do you recollect the number?
A. I had 300.
Q. State whether you paid for them, or how that was?
A. I paid him cash for them; I paid him $33 for three hundred sap buckets, eleven cents apiece.
Q. With the exception of the last bill that was furnished, up to the time you left the road, have, you or not, paid for what merchandise you have had of Mr. Whitcher?
A. Everything.
Q. In what way?
A. Paid him the money.
Q. Haven't you never let him pass over the road in compensation for buckets and pails? [Objected to as a leading question.]
Q. I believe I did not inquire of you, when you were on the stand before, in reference to whether you ever paid back any money to Mr. Gilmore for fares taken in the cars?
A. Yes, sir; I have.
Q. To what extent and how it happened?
A. I have paid money to Mr. Gilmore and to George Sanborn by the order of Mr. Gilmore.

Q. What was the occasion of it?

A. There were some of his friends that went down that hadn't any passes, and he didn't mean to have them pay, and when they came back he went and got the money.

Q. Would it be before or after you made the return?

A. It would be afterwards.

Q In what way would you reimburse yourself?

A. I used to take it out on the next way-bill. The last two or three months, Mr. Gilmore used to come to me himself, and he would say to me—sometimes he would say that he had one or two of his friends that went down with Mr. Noyes, and back to Manchester or Nashua, or wherever they went, and he would say, "you may pay me, and take the money out of your next way-bill." I should think those times would average once a week certain. I should think it would average once a week for the last four weeks I was on the road. And before that we used to pay it back to Mr. Sanborn. When I went up to settle my way-bill, Mr. Sanborn would say to me that Mr. Gilmore directed me to pay back so much money for one fare, or two, just as it happened.

Q. Well, why did Mr. Gilmore claim that you should pay the money to him?

A. Why, he came the last two or three months himself, and I supposed that he paid it back to the persons. I knew nothing about that.

Q. I wish to make some further inquiry, Mr. Clough, about Raymond Kimball?

A. In 1835 —— ?

Q. You commenced driving at that time to Lowell?

A. Yes, sir; I commenced driving for Mr. Osgood at that time.

Q. Now, sir, when you commenced, state your course of business with Mr. Kimball?

A. I got acquainted with Mr. Kimball about that time,—in the year 1835. I used to put money into his hands that I made along from time to time; all amounts from $50 and more, just as I happened to have it on hand, until up to 1842.

Q. Mr. Clough, do you recollect where you got $1000 that you paid to Mr. ——?

A. I had that of Mr. Kimball.

Q. Was it money that you borrowed of him?

A. Money that he had in his hands. I had his note.

Q. State whether you had money in his hands beyond that?

A. I had money out in smaller amounts; and I think I had more money in his hands at that time.

Q. From the time you first knew Mr. Kimball until the time you went on to the Concord Railroad, what was his character as a man of wealth and responsibility? [Objected to.]

*Mr. George.* It is entirely immaterial whether he was a man of wealth and responsibility or not.

*Mr. Mugridge.* Our position is this: that Mr. Clough put his funds into the hands of Mr. Kimball to keep for him.

*The Chairman.* Haven't you got this particular fact proven well enough by the witnesses called on the other side? Is there any need of raising a question of that kind, when you have got testimony to the same effect?

*Mr. Mugridge.* Haven't we a right to show the witness's idea and knowledge of his responsibility and pecuniary standing?

*The Chairman.* We will consider that question, if you think, after the testimony on the other side, that it is worth the while.

The question is, perhaps, somewhat of a difficult question to decide; but the ruling will have to be for the defence. If you choose to take the chance, we can let you put in the evidence. [Ruling excepted to by the plaintiff.]

Q. You may state up to the time you knew Mr. Kimball, until you got your pay of him, whether he was reputed to be a wealthy and responsible man?

A. He told me himself, in 1840, that he was worth $30,000.

*The Chairman.* That, I take it, is not what we ruled in.

Q. From the time you knew him, until you got your money of him, was he reputed to be a wealthy and responsible man?

A. He was reputed a wealthy man. He was called rich, and his note was good for almost any amount.

Q. Won't you describe your relations with Mr. Kimball?

[Objected to as repetition.]

*The Chairman.* That has been pretty fully put in.

Q. Mr. Clough, when you testified before in relation to the amount of notes that you had, you testified, I think, before from memorandum. Did you have any notes that you had not on that memorandum? If so, what were they?

A. Notes given to me?

Q. No; notes that you had given, How was it with that land that you bought over on South street, and built the brown building on?

A. I have notes amounting to $450 that wasn't in—that was given this fall. One is $250, and the other $200.

Q. (*By Judge Bellows.*) That you gave for what?

A. For two houses that I bought of Mr. Thorndike.

Q. Have you got a memorandum of the Masonic Temple contract?

A. I have got the specifications and the articles of agreement.

Q. Won't you produce it, if you have it?

[Produces book.]

Q. That is it, is it?

A. It is.

*Mr. Rolfe.* If the referees please, I will read one article of the agreement: "And in consideration * * *

Mr. George read from the files of the N. H. Patriot to show that the railroad was opened to Franklin, Monday, Dec. 28, 1846; and to Grafton, Aug. 30th, 1847.

Mr. Rolfe continued the examination of Mr. Clough

Q. Mr. Clough, won't you state about the Prince street property?

A. The Prince street property—I helped my nephew buy it; bought it for him, I think, for $1600. The property was mortgaged for $1000 to the savings bank at the time we bought it. We hired $600. It was mortgaged at the bank for $1000, and we hired $600 to pay the balance of that; and I signed the note with him. He lived there about a year. But I am to pay the note to the bank. I have sold it, and given a bond for a deed, and something paid on it now; $400 or $500, I think. The note to the bank—my name is on it. That is the way that stands.

Q. (*By Judge Bellows.*) You are to pay the note?

A. The note is to be paid.

Q. And you are to pay the $600.

A. Yes, sir.

Q. (*By the Chairman.*) The note at the bank is not paid?

A. No, sir; I believe there is $800 at the bank; I believe my nephew paid $200 on it.

Q. (*By Judge Bellows.*) You sold the place for the same price?

A. Yes, sir.

Q. Did you say you had conveyed it and gave a bond for a deed?

A. I gave a bond for a deed.

Q. What did you say about the payment of the price by these persons to whom you gave the bond for a deed?

A. I sold to them for the same price that I gave.

Q. Has it been paid?

A. It has been paid, part of it.

Q. (*By Mr. Mugridge.*) That is, the person to whom you gave the bond?

A. Yes.

Q. (*By Mr. Rolfe.*) Mr. Clough, were you present when Dr. Carney testified yesterday afternoon?

A. I wasn't all the time. I went out to get some papers at my house when he was first put on to the stand.

Q. Have you ever seen Dr. Carney before?

A. I never saw him but once before. I met him on the street once. That is the only time that I ever saw him to know him.

Q. State whether you have any recollection of seeing him upon the cars?

A. Never; never saw him before, to know him, until that day you introduced me to Jim in Boston.

Q. Have you heard his deposition read,—the one that he gave in Boston?

A Yes, sir, I have.

Q. Did you hear what he stated yesterday in relation to your conduct when he inquired for tickets to Worcester and to Groton Junction?

A. No; I wasn't here then.

Q. Will you state whether you have any recollection of being agitated, trembling, or any realization of turning pale ——

A. No, sir; never.

Q. At any time?

A. Never.

Q. Has there ever been any occasion when you were on the cars, that you now recollect, that you were in a state of agitation when you have been inquired of for tickets?

A. Never any such thing ever happened.

Q. Will you state whether you knew, or had any intimation, that anybody at that time—about the 23d of November, 1865—that anybody was looking after you or trying to detect you upon the cars?

A. No, sir; I never knew anything of it until two or three days before I left the road, I heard somebody say that there had been detectives on the road. I think that somebody on the Northern road spoke about it. I don't recollect now who it was who said there had been detectives on the Concord road. That is the first time that I heard or mistrusted anything of the kind.

Q. Will you now state whether you ever, to your knowledge, intentionally withheld any money that you have collected on the cars from the railroad? [Objected to as being a repetition.]

CROSS EXAMINATION.

Q. (*By Mr. George.*) Mr. Clough, let me ask you this question: Did Mr. Gilmore within a month or six weeks—within a brief period of the time you left the road—send to you, underscored with red ink, a copy of the rules which are attached to your deposition?

A. I don't now recollect whether he did or not.

Q. Will you say that he did not send to you a copy of the rules underscored? That is, underscored with red ink the rule requiring the return of the ten cents, requiring you not to use the money?

A. I have no recollection of it. I would not say that he didn't.

Q. Did you hear Mr. Sanborn's testimony upon that point?

A. I don't know whether I did or not.

Q. You won't say you didn't receive from him a copy of the rules underscored?

A. No, sir; I would not.

Q. Were you in the habit of receiving rules underscored in that way?

A. He used to send rules around occasionally; I don't now recollect how often.

Q. What's your idea of his object in sending the rules around to you? If he told you to disregard them, and pay no attention to them practically, what was your idea of his sending them?

A. My idea was that when he sent the rules, if there was anything that I could not live up to, I would tell him so.

Q. Your idea was that he sent them around for you to see whether there was any rules that you could not live up to?

A That is the way I done it.

Q. You have received rules from him at various times, now was it your idea, when you received these rules, that you were to say what ones you could live up to?

A. I don't know what my idea was at the time.

Q. Didn't you just say what your idea was?

A. I don't know what my idea was at that time.

Q. There was a Mr. Draper, wasn't there, on the cars? One of Major Carney's men was Draper, wasn't he?

A. I heard there was a man by the name of Draper.

Q. He was a coach driver, wasn't he, down at the Elm House?

A. I think there was such a man there.

Q. He left immediately before the last trial?

A. He left.

Q. Do you know anything about his leaving?

A. No, sir.

Q. Did you ever have any communication with him?

A. When he worked to the Sherman House stable, I heard he was one of the "spotters"—one of the detectives; and I may have had some talk with him at that time, but I cannot tell now what that was. I think that I asked him if he was one of them, and he said he was. That was pretty much all that was said. I told him that I wasn't afraid of any of the detectives, of him, or any of the rest of them.

Q. Did you tell him that you could show that you bought tickets, and that your way-bills ought not to account for the tickets received in the cars?

A. No, sir; I never told him any such thing.

Q. What did you tell him?

A. I told him I wasn't afraid of him or any spotters that followed me.

Q. Didn't you have any conversation on the subject of his evidence?

A. That was a subject.

Q. Did you tell him how much money he paid you?

A. No, sir. I have told you the weight of the conversation with him.

Q. Did you ask how he paid you money?

A. No, sir.

Q. And immediately before the last trial he left?

A. I don't know when he left.

Q. When did you last see him?

A. I couldn't tell.

Q. State as nearly as you can.

A. I couldn't tell whether it was before the trial or after the trial commenced.

Q. Where did you see him then?

A. I saw him in a hack?

Q. Did you have any communication with him, directly or indirectly, or with any body else?

A. I don't think I have.

Q. Will you say you did not have any?

A. I don't think I did.

Q. Don't you know that you did?

A. I don't think I did.

Q. Will you say absolutely whether you did or not?

A. I don't think I did; I don't remember.

Q. Now, you made a very strong effort to obtain Major Carney's evidence?

A. We took his deposition, but couldn't get anything out of him.

Q. You were very anxious indeed to ascertain what Major Carney's evidence would be?

A. That was our object.

Q. You took a great deal of pains, and you went to Boston?

A. Took his deposition in Boston.

Q. Did you keep him here a fortnight?

A. Kept him, I don't know how long.

Q. Did you keep him here a fortnight?

A. Kept him, I think, a week or ten days.

Q. Your object was to ascertain what his testimony would be? Won't you tell me why the same objection did not exist with regard to Draper's testimony? [Objected to.]

*The Chairman.* I don't see why that is not well enough in cross examination.

A. I supposed that Major Carney had all the books.

Q. Won't you tell me why the same reason did not exist to get at the facts of Mr. Draper's testimony as Major Carney's testimony?

A. Well, I supposed that Maj. Carney was the headquarters; he was the one that conducted the job, and he was the man to take.

Q. Your object was to get at the facts against you, and what they were going to testify against you?

A. Yes, sir; I supposed they were in his possession.

Q. Will you explain, sir, if you proposed at that time to account for money not returned upon the way-bills, by showing that you took the money that you received in the cars and purchased fresh tickets, and punched these, how any evidence of that character could by any possibility tend to show that you didn't receive money that you didn't account for?

A. Well, I suppose that there were days that I took the ten cents extra, and didn't buy any tickets at the office. How did I know but that these detectives were on that day?

Q. What do you mean—that you didn't return so much money?

A. I mean to say that I took the ten cents extra, anddid return all the money on the way-bill.

Q. There were days?

A. Yes, sir; there were days when I took that.

Q. And therefore your view was that there might be evidence that you didn't return that money?

A. I suppose that if I had half a dozen men following me, it might be detected of course.

Q. If Mr. Draper was in the cars, and you understood that he was a "spotter"—you understood that Mr. Draper paid you money, didn't you?

A. He said he did. I understood you to say that he paid money in the cars, and I understood you to say that he bought some tickets too.

Q. Did you say that you didn't have any conversation on that subject, with him?

A. No, sir; I don't think I did.

Q. Very well, pass on to the next. Have you ever stated to anybody that as long as Joe Gilmore was living, you might have trouble, but that now that he was dead, you didn't care?

A. No, sir; never said any such thing.

Q. Never said it to your counsel?

A. No, sir.

*Mr. George* stated that, if the counsel for defendants would withdraw objection, plaintiff's counsel would show that the statement involved in the last question was made to Mr. Rolfe.

Mr. Mugridge and Mr. Rolfe each withdrew any objection.

Q. Have you a note given for the Prince street property? You spoke of the Prince street property that you did not allude to before. You say that you have given a bond for a deed. To whom was that bond given, and to whom was security given on the other side?

A. The bond was given for a deed to a woman by the name of Bush, I think it was.

Q. Did she give you a note, or give anything to represent it?

A. No, sir; I don't think there is a note; but she paid the money in to Mr. Rolfe, and Mr. Rolfe paid the money to him. She took some money out of the Nashua Savings Bank, and made the first payment.

Q. I am merely asking whether there is a note as consideration of the bond, or whether it rests upon agreement on your part? It rests on the bond, and you have no obligation from her?

A. No, sir; I think not.

Q. How was the $600 paid by your nephew, and this $1000 obtained at the bank; that is, $1600 if I understand you. Let me see if I state it correctly. Your nephew bought it for $1600, and paid $600, and borrowed $1000 at the bank? That is the first transaction?

A. Yes, sir; that is the first transaction.

Q. Then you bought the property of your nephew, paying him $1600?

A. Yes, sir.

Q. And as part and parcel of $1600, you assumed the note at the bank?

A. Yes, sir.

Q. Now, how did you pay the $600 to your nephew?

A. We borrowed the money at the bank.

Q. Who paid the $600?

A. I paid it, and received the pay of that woman.

Q. Do you know the exact amount you received of that woman?

A. My impression is that it is not far from $500.

Q. Then you are $100 out?

A. That is, if that is so. I think it is somewhere in the neighborhood of $500

Q. Then there has been $200 paid on the bank note?

A. Yes, sir.

Q. Who paid that?

A. My nephew paid that.

Q. Out of whose funds?

A. Oh, when he was tending here in the store.

Q. Who found the money?

A. He paid it out of the store there, I believe.

Q. Didn't you own the store?

A. I stood in for it.

Q. And have lost $1300?

A Yes, I have got a note to get that out of, if I get it. I don't expect to get it all. I have one note of $300, and have collected about $500 on the books.

Q. Do you know how your trial balance came out? You own the store?

A. Not all of it. I owned half the property at first, and then Piper and Clough divided, and I took the store. They divided the stock, I know.

Q. Lost $1300 or $1400 by the operation?

A. It depends.

Q. You have already testified?

A. I testified that I have paid these notes, but I have got a $300 note, and have collected some $400 or $500 on the books, and I don't know how it will come out.

Q. You haven't reckoned that on your assets?

A. No, sir; those notes that I assumed to pay, I have got to pay.

Q. You haven't reckoned among your assets these accounts that you spoke of?

A. No, sir, I don't own them.

Q. (*By the Chairman.*) Mr. Clough, I didn't understand you when you testified before to state when your practice commenced of taking money that you got in the cars and buying tickets with it, and returning them to the office instead of returning it on the way-bill?

A. Well, sir, I think it was about a year after Mr. Gilmore went on to the road. I don't know whether he went on to the road in 1857 or 1858; but it is about a year after he went on to the road. The first additional price they charged in the cars was five cents instead of ten, when Mr. Gilmore came on to the road; and about a year after, he put it up to ten; and it was more difficult to get the ten cents than it was the five. Soon after that I commenced; and we had a good deal of trouble about it, and I spoke to Mr. Gilmore about it.

Q. (*By Judge Bellows.*) That, you say, was what year that he came on to the road?

A. I think it was 1857 or 1858.

*Mr. George.* You recollect, Judge Bellows, I put in the record. Mr. Gilmore was not appointed at the beginning of the year. Mr. Spalding resigned. Mr. Upham had been the superintendent. The Concord, Manchester, and Lawrence roads had been united, and Mr. Gilmore was appointed between the years; that is, between the years 1856 and 1857.

Mr. Mugridge introduced in evidence tax bills showing amount of taxes paid by the defendent, Mr. Clough, upon the following property, for the years specified:

Two houses on Thompson street, 1851 to 1865, inclusive, $324.65; valuation in 1851, $1500; in 1852, $1500; in 1854, $2000; in 1855, $2000; in 1857, $2000; in 1858, $2000; in 1859–1865, $2000; amount of tax on Jefferson street property from 1856 to 1865, inclusive, $87.41; valuation each year from 1856 to 1865, $600.

*Mr. Mugridge.* Now we have a schedule of the other taxes paid in the town and city of Concord, from 1842 to 1865, inclusive, outside of the Jefferson street and Thompson street property. These are by themselves. We have also a general schedule of all the other taxes paid by Mr. Clough in the town and city of Concord; it amounts to $3219.01; there is another item: the old store on Pleasant street; the taxes from 1859 to 1865, inclusive, amounts to $44.94; the valuation of the old store for the first year, 1859, was $300; and for all subsequent years to 1865, inclusive, $400. That covers the entire amount of taxes that Mr. Clough has paid since 1842.

## TESTIMONY OF MR. C. A. STEWART.

Q. (*By Mr. Mugridge.*) Mr. Stewart, you may state whether you are the city clerk of the city of Concord?

A. Yes, sir.

Q. Whether you prepared this tabular statement of the taxes paid by Mr. Clough from the books?

A. Yes, sir.

Q. They are correct?

A. They are correct so far as I know. There may be a trifling variation from the fact that the per cent., as worked out, may vary a small fraction of a mill on a cent.

### CROSS EXAMINATION.

Q. (*By Mr. George.*) Mr. Stewart, in 1842—I wish to call your attention from 1842 to 1849—Mr. Clough was taxed on one poll, horses, carriages, land and buildings, $9.98.

*Mr. Mugridge.* It don't appear that he was taxed on horses, carriages, or polls by that.

Mr. George read the title of the schedule of taxes as offered.

*Mr. Mugridge.* I desire to take exception to evidence put in in that way. And I want to call attention to the paper offered.

*The Chairman.* We do not see how there can be any objection to reading this paper, or how it can convey this idea.

Q. Will you be kind enough to state again how the taxes were on the Masonic Temple?

A. One-half, I think, was charged to Mr. Clough, and the other half to Mr. Corning.

Q. Do you know how that was? For instance, we will take the tax of 1859; won't you be kind enough to look at the tax of 1859 and see what was the tax of Mr. Clough? See what property was taxed to Mr. Clough.

A. Some of that is on property. That is on the fly leaf.

*Mr. George.* I will put this in: the items of Mr. Clough's tax.

Q. That is the basis?

A. That is the basis on which the tax is made for that year.

Q. That is the basis from which you made this detail?

A. These items, taken down by the assessor.

Mr. George read the items from the record for the years 1842 to 1862 inclusive.

|  |  |  |
|---|---|---|
| 1863. | Value of buildings, | $21,600 |
| | 2 horses, | 100 |
| | 1 cow, | 18 |
| | | $21,718 |
| 1864. | Value of buildings, | $24,700 |
| | 2 horses, | 100 |
| | 1 cow. | 20 |
| | | $24,820 |
| 1865. | Value of buildings, | $21,200 |
| | 4 horses, | 400 |
| | 1 cow, | 34 |
| | 1 carriage, | 300 |
| | | $21.934 |
| 1866. | Value of buildings, | $24,900 |
| | 2 horses, | 130 |
| | 1 cow, | 40 |
| | 1 carriage, | 200 |
| | | $25,270 |
| 1867. | Value of buildings, | $25,300 |
| | 1 horse. | 100 |
| | 1 cow, | 40 |
| | carriages, | 200 |
| | | $25,640 |
| 1868. | Value of buildings, | $30,250 |
| | 1 horse. | 100 |
| | 1 cow, | 40 |
| | carriages, | 150 |
| | | $30,540 |

Q. (*By Mr. Mugridge.*) So far as the books show, what was the practice, or what was the fact, in the assessment of taxes from 1842 on, as to the itemizing of property, so far as the real estate and the personal property were concerned?

*Mr. George.* The books show for themselves.

Q. From any examination that you have made of the books, do you find any items of real estate, or any items of personal property distinct and by themselves, or are they all classed together?

*Mr. George.* I object. The books speak for themselves.

*The Chairman.* I believe the contents of these books have already been read to us.

*Mr. Mugridge.* A portion of them have. I did not know but that it was proper for me to call the attention of the clerk to a fact discovered in the examination of the books; but if that fact is already sufficiently shown, I will not consume further time.

## TESTIMONY OF ROBERT P. BLAKE.

Q. (*By Mr. Rolfe.*) Where do you live, and what is your business?

A. I have always lived in the county, and have worked for the Concord Railroad Company.

Q. How long have you been employed by the Concord road?

A. I have worked for the Concord Railroad more or less for the last ten years.

Q. How and where were you engaged during the war?

A. A part of the time I was at work in the depot.

Q. During what years?

A. During a portion of 1864 and 1865.

Q. And all the rest of the time in the car house?

A. No, sir; I went away from the depot in November, 1865.

Q. Previous to August 1864, how were you engaged?

A. I was at work shifting in the yard.

Q. Shifting about the passenger cars?

A. Yes, and the freight cars.

Q. During the time that you were engaged in the depot, state if you are able, the number of cars that were run on the several trains?

A. There were five on the first train—the half past five train in the morning.

Q. Do you include the baggage car in the five?

A. No, sir.

Q. On the quarter past ten train down, how was it?

A. There were from six to eight cars.

Q. On the half past three train in the afternoon?

A. From eight to eleven.

Q. Do you speak of this as being ordinarily the case, or in the hight of travel?

A. In the hight of travel there would be that amount.

Q. State how the trains were made up during that time. That is to say, over what roads did these cars go? Take the half past five train in the morning?

A. There would be three cars that went from here to Lowell, and two that went from here by Lawrence.

Q. How was it with the quarter past ten train?

A. There would be three passenger cars and a smoking car by Lowell, and from two to three cars by Lawrence. Two cars was the regular train.

Q. The half past three P. M. train?

A. There were five cars to go by Lowell, and three by Lawrence, and one by Portsmouth.

Q. Did you ever work in the car house?

A. I am working there now.

Q. When did you first commence to work in the car house?

A. About ten years ago.

Q. Up to what time did you continue to work in the car house—commencing ten years ago.

A. I worked in the car house one season before the old depot was built.

Q. Did you continue to work in the car house up to the time you went into the depot?

A. I worked in the car house some, and worked on the train some.

Q. Previous to 1864, are you to able to state how the trains ran?

A. I am not.

Q. What time in 1864 did you commence to work in the passenger depot?

A. About the middle of August.

CROSS EXAMINATION.

Q. (*By Mr. George.*) During the war was there an immense amount of travel over the road?

A. There was.

Q. During the war, how was it about the depot, when the trains went out or came in, was the depot crowded?

A. Sometimes it was.

Q. Was it usually?

A. As a general thing it was pretty crowded.

Q. It was so crowded that it was sometimes difficult to get from one part of the depot to another, was it not?

A. Yes.

Q. I suppose the substitute brokers, substitutes, soldiers, town agents, and people of that kind, were traveling at that time, that never traveled before, nor have since?

A. They were.

Q. Large numbers?

A. Yes.

Q. Was your attention ever called, during either the year 1864 or 1865, to the matter, so that you kept an account of the number of cars that went out on the ten o'clock train in the morning?

A. My attention was called to it by my being there all the time.

Q. Did you ever keep an account?

A. I never set an account down, but I have counted the cars several times as they went out of the depot.

Q. Did ever you see so much travel as there was in the hight of travel during the war, over the Concord Railroad?

A. Yes, I think I did.

Q. On this road?

A. Yes, I think there was as much this last year as there was then.

Q. According to your recollection, the cars were not much crowded during the war, were they?

A. Yes, they were crowded.

Q. Were the cars crowded right along during recruiting times, and during the war?

A. Some trains were crowded, and some were not.

Q. How as a rule?

A. As a general rule they were pretty well filled.

Q. During the war were there a large number of women riding back and forward on the cars?

A. Yes, there were quite a lot of them.

Q. Soldiers' wives, and soldiers' sweethearts?

A. Yes; and some that were not wives.

Q. There was a large number of them?

A. Yes, sir.

Q. Were they constantly riding?

A. They were more or less on the train.

Q. There never has been any occasion since, or before, as far as you have known, when there was so much of this class of travel,—substitute brokers, substitutes, soldiers, town agents, or drafted men, and this class of women?

A. No, sir; I do not think there has.

Q. (*By Mr. Rolfe.*) How was it about this crowd at the depot; what was it composed of?

A. Some were people going to see their friends away, and some were going.

Q. How was it with the people that were living about the town here, about rushing to the depot at the time the cars came in?

A. I do not know that there were any more folks who went to the depot then, than at any other time.

Q. How was it about this class of women that were not wives, as you say; were they pretty thick about the depot?

A. Yes; they were.

Q. What has been the practice in Concord, about people rushing to the depot when the trains came in?

A. There are a great many people that now go to the depot to see the trains come in.

Q. And it has always been so?

A. Yes; I think it has.

## TESTIMONY OF W. H. ALLISON.

Q. (*By Mr. Rolfe.*) Do you reside in this city?

A. I do.

Q. For how long have you lived here?

A. I came here in 1833, and went away a few years, and came back in 1841, and have been here ever since.

Q. How long have you known George Clough?

A. From the time he first commenced running on the cars. I think that was the first that I knew of him.

Q. Have you had any trade with him? Did you ever sell him any horses?

A. I sold him two, I believe.

Q. Do you recollect the price?

A. The first one I sold him was a large black horse, which I sold him, I think, for $210.

Q. Did you give it to him for a different price from that?

A. I do not recollect how that was; I think that I wanted more than that for the horse. I asked that for it, but whether I billed it at the price I asked, I cannot recollect.

Q. What did you sell the other horse for?

A. I think it was $187.

Q. Will you state what was actually paid you for what is called the "Allison Building"?

A. My impression is that it was $3700. I called it $4000; but whether I deeded it for $4000 or not, I do not recollect. I think the price paid was $3700.

Q. (*By Mr. George.*) Was it not $3800 exactly, that you paid?

A. I think it was $3700.

Q. Can you state that it was not $3800.

A. I could not without referring to my books, but my impression is that it was $3700.

## TESTIMONY OF E. M. SARGENT.

Q. (*By Mr. Rolfe.*) Where do you reside?

A. In Lowell.

Q. For how long have you lived there?

A. I went there in 1836.

Q. Do you know Mr. Clough, the defendant in this suit?

A. I know Mr. Clough.

Q. How long have you known him?

Q. I think the first I knew of him, personally, was in 1838 or 1839. He drove stage to Deering and Chester.

Q. Have you ever been in his employ?

A. Yes.

Q. State when and how?

A. I drove stage over the Mammoth road, from Lowell to Manchester, for him, for nearly a year, in the latter part of 1840 and the first part of 1841.

Q. When did you commence?

A. My recollection is that it was in August or September, 1840.

Q. Give to the referees the best idea you are able, of the amount of travel at that time on that line. State first what kind of a team he drove.

A. He drove a four-horse team into Lowell. He had two six-horse and two four-horse teams. I had a light coach,—a six-passenger coach, and three horses.

Q. Did you drive every day with him?

A. Yes, but on different hours. He left early in the morning, and I left later.

Q. On an average, state the amount of travel carried by you and Mr. Clough during that year.

A. I do not know the average. I have seen Mr. Clough on his coach with as many as twenty-four passengers: but that was not common. Fifteen or eighteen passengers would be an average load.

Q. How many did you carry on an average?

A. I could not well carry more than eleven. It was but a six-passenger coach, and eight passengers would be a load. Sometimes people would ride on the baggage. There was a great deal of travel.

Q. How was it generally. Was your coach generally filled?

A. Yes, sir; always. It was always filled to the capacity of the coach—six passengers, and sometimes there were eleven.

Q. Did you know Daniel Raymond Kimball, who was commonly called "Rem." Kimball?

A. I did.

Q. Did he keep a livery stable?
A. Yes.
Q. Were you engaged there at the stable at any time?
A. Previous to my driving stage I was in the stable. I was a hostler there, and took care of Mr. Clough's horses previous to my driving.
Q. During all the time that Mr. Clough was driving, and you were taking care of the horses, state whether you ever saw Mr. Clough let Mr. Kimball have any money?
A. Yes, sir; I have, a great many times.
Q. Did Mr. Kimball give him anything in lieu of the money?
A. He gave him what I supposed to be a note, but as I was only the hostler, I could not give any statement of the amount, or what it was that Mr. Kimball gave him for it. I have seen Mr. Clough a great many times count money out to him, and leave it with him; and Mr. Kimball would pass him a paper back.
Q. What was the fare from Lowell to Manchester?
A. One dollar and twenty-five cents, I think, but I could not tell positively. It was long ago, and I have not had my attention called to it.
Q. How frequently did you see Mr. Clough with this load of twenty-four passengers?
A. Not frequently, with as heavy a load.
Q. How was it generally?
A. I may say, perhaps, that generally, he had fifteen, sixteen, seventeen or eighteen passengers.

CROSS EXAMINATION.

Q. (*By Mr. George.*) I understood you to say that you attended to Mr. Clough's horses, and took care of them?
A. I did.
Q. Where did he keep them?
A. He kept them in the Merrimack House stable.
Q. Who had charge of that stable? Who conducted it? who owned it, and had charge of it?
A. I had charge of it.
Q. You were an employee only?
A. W. W Larrabee was the proprietor of the hotel, and the Middlesex Manufacturing Company owned the stable.
Q. Did Mr. Kimball run the Merrimack stable at that time?
A. He run one of them. There were two stables. They joined, but were two separate stables. The Middlesex Company owned the stables that I worked in, and Mr. Larrabee rented it.
Q. Mr. Kimball boarded at the Merrimack House?
A. Yes.
Q. And kept the livery stable adjoining the house?
A. Yes.
Q. And subsequently bought the livery stable?
A. Yes.
Q. He kept a very large livery stable, didn't he?
A. Yes.
Q. About how many horses?
A. He did not keep so large a number of horses to let. He boarded more horses than he let.
Q. He kept a large livery and boarding stable?

A. Yes.
Q. He kept the largest number of livery horses?
A. No, not the largest number.
Q. How many did he keep then?
A. At that time I think he kept fifteen or sixteen horses.
Q. Did he keep nice horses and carriages?
A. They were considered so.
Q. He was a bachelor then, wasn't he?
A. Yes, sir.
Q. And was never married?
A. Not to my knowledge,
Q. Did Mr. Kimball subsequently become insane, and was he afflicted with softening of the brain?
A. It was so reported.
Q. And he subsequently died at an insane asylum?
A. Yes.
Q. Do you remember in what year he first showed those symptoms of insanity?
A. No, sir. He went to Europe, but I could not tell in what year.
Q. Can't you tell *about* the year?
A. I should think it was in 1856.
Q. Do you remember when he first commenced to show symptoms of insanity?
A. No, sir; it was not before he went to Europe that there was any general talk about it. It may have been discovered before that by some of his friends.
Q. Who received his property? Who was the legatee, or heir, if you know? [Question waived.]
Q. In 1840, 1841 and 1842, how many stages were there running from Nashua and Lowell, north?
A. I do not know how many from Nashua. There were several.
Q. How many stages were running to Lowell?
A. Mr. Dudley run one stage half daily, and Mr. Clough ran another, half daily, over the Mammoth road.
Q. Did J. B. French run a stage there?
A. Not that ever I knew of.
Q. Did Mr. Walter run a stage?
A. Yes; but I think it was previous to 1840.
Q. How many more stages were there running?
A. The Darien stages, I think, were taken off previous to 1840, and I think that on the Mammoth road was the only daily stage direct from Lowell to Concord.
Q. In 1847 the cars ran to Nashua?
A. Yes.
Q. And most of the stages running north, subsequent to that time, ran into Nashua?
A. From Lowell?
Q. No; the stages ran to meet the cars, didn't they, all of them?
A. No; not all of them. A Mr. Tuttle and a Mr. Beales ran a stage from Lowell to Nashua for a long time after the cars commenced to run.
Q. The great bulk of stages ran to meet the cars?
A. Yes.

Q. How many lines of stages ran from Concord to Nashua, to meet the cars? There was a very large number, was there not? There was a line of stages from Nashua up to New Hampton, was there not?

A. Yes; and there was the Forest line that ran out through Hancock, and that way.

Q. Was there a line of stages that ran to Hillsboro from Nashua?

A. I don't recollect.

Q. There were a number of stages that ran from Concord to Nashua?

A. There were several; I knew several of the drivers; I don't know how many.

Q. (*By Mr. Rolfe.*) Do you recollect about the "chicken line"?

A. Yes, sir; it was the line that ran from Lowell to Dover.

Q. Who drove on that line?

A. There were two drivers; Mr. Clough was one of them.

Q. Were you ever at Lowell at the time that Mr. Clough drove in over that line?

A. Yes.

Q. Did you have an opportunity, when he came in, to see what he brought?

A. I did, occasionally.

Q. What was it?

A. There was poultry, and ——

Q. (*By the Chairman.*) Didn't he carry any passengers?

A. Yes.

Q. (*By Mr. Rolfe.*) To what extent did he carry poultry?

A. I cannot say. There were a great many Lowell people who depended upon him as much as they would upon their market in season.

Q. (*By Mr. George.*) Do you know Mr. Osgood, who owned that line?

A. Yes.

Q. He failed, and bursted, didn't he?

*Mr. Mugridge.* What has that to do with the case? He did not fail upon account of his staging. If it is competent, we will show the occasion of his failing; but we will not have anything put in by inuendo.

*The Chairman.* It is not very obvious to us how it can be very material whether he failed or not.

## TESTIMONY OF JAMES S. CHENEY.

Q. (*By Mr. George.*) Will you state whether, thirty years ago, you drove a stage, or were the owner of a stage, or were in any way interested in a stage that ran into Lowell? If so, what stage was it?

A. In 1838 and 1839 I used occasionally to drive an extra into Lowell from Groton.

Q. How was it in 1840, 1841, and 1842?

A. I went from there to Worcester, and drove out of Worcester until 1842.

Q. Where did you drive to?

A. To Brattleboro, Vermont.

Q. Were you acquainted with George Clough?

A. I was some acquainted with him. I used to meet him at Lowell.

Q. Did you room with him in town?

A. Not then.

Q. Did you subsequently?
A. I did a short time, when I commenced on the road.
Q. The Concord road was opened in 1842. When that road was opened, what was your business and employment?
A. It was opened to Manchester in July, 1842, and I then commenced running on the road as express messenger.
Q. Your brother, and Mr. White, and Mr. Walker had the express over the road; that is, they were the contracting parties?
A. Yes, sir.
Q. And you were in their employ?
A. Yes.
Q. For how long a period of time did you continue to run as express messenger after the opening of the road to Manchester?
A. About fifteen years, I think; I could not tell exactly.
Q. State how it was about your running on the trains conducted by Mr. Clough?
A. For the first three months that they run to Manchester, and before the road opened to Concord, Mr. Clough used to run down and stay over night, and I did the same, and we boarded at the Merrimack House, and roomed together for a short time, perhaps two or three months. After the road was opened through, his route was changed so that he stayed in Concord over night, and mine was changed so as to bring me in Manchester over night.
Q. But you run back and forward over his train?
A. Yes.
Q. You had sole charge of the express business on you trains?
A. Yes.
Q. State whether you were aware of Mr. Clough's carrying butter, cheese, eggs, and articles of that character, on the passenger trains?
A. I never knew that he did.
Q. At that time where was the express matter carried? Was there a seperate express car, or was it all carried in the baggage car?
A. We had a separate apartment generally, but sometimes they would fill that so full of matter going to Concord, that I was obliged to put my express with the baggage.
Q. It was all in one car?
A. Yes.
Q. If Mr. Clough had made a business of carrying articles of that character to any considerable extent on the passenger trains, how would it have been about your knowing it?

*Mr. Mugridge.* I do not know the object of this testimony now attempted to be put in by Col. Cheney. It cannot be put in for the purpose of contradicting Mr. Clough. Mr. Clough has not sworn that he used to carry produce by the passenger trains, but on the contrary the evidence is that shipments were all made by the freight trains, and that only occasionally was there some poultry or something of that kind sent by a passenger train. The bulk of the business, as has been shown by several witnesses, was carried on the freight trains.

*Mr. George.* It is useless to bandy words. We do not understand the testimony as Mr. Mugridge does.

*The Chairman.* My recollection of the evidence is that Mr. Clough testified that sometimes he carried it one way, and sometimes another; but that the heavy stuff went on the freight trains, and that stuff not so heavy went on the passenger trains. It seems to me that this testimony does tend to rebut Mr. Clough's testimony in some degree.

Q. (*By Mr. George.*) Let me ask you if there was any considerable amount of business of that kind done, you would have been aware of it?

*Mr. Mugridge.* You cannot ask the question in that shape. The witness can show what means and opportunities he had for seeing what was done in that line.

Q. (*By Mr. George.*) Were you pretty constant, or otherwise, in your traveling from here to Boston? [Objected to as leading.]

Q. How was it, Mr. Cheney? Will you state with regard to your constancy, or otherwise, upon the road when you went as a messenger?

A. Do you mean to ask me whether I was every day on my business?

Q. Yes.

A. I do not suppose I lost two weeks in a year, in any year that I ran.

Q. At the time you roomed with Mr Clough, or at any other time, did you have any conversation with him in regard to his means.

A. My recollection is that he gave me to understand ——

*Mr. Mugridge.* I object to what he "gave you to understand." If the witness will state the conversation, I will not object to it.

*The Chairman.* The rule is that the witness shall state what Mr. Clough said, if he can remember it; but if he cannot remember the words, he can state the substance of what he said.

*The Witness.* It would be impossible for me to state the conversations that I had with him at various times; but having been in the same business myself, and being friendly with Mr. Clough, I felt a little interest in knowing how he had succeeded in the two or three years that I had not seen him. I heard that he had bought on the Mammoth road of Mr. Dudley, and I had some conversations with him about it. I recollect distinctly of having a conversation to draw him out, and know what he had done; and he then gave me to understand ——

*Mr. Mugridge.* I object to his stating what he was given to understand.

*The Chairman.* I am not able to appreciate the difference between the substance of what he said, and what he gave him to understand. The witness can state the substance of what he said to him.

*The Witness.* My recollection is that he gave me to understand ——

*The Chairman.* The counsel don't want you to use the words that "he gave me to understand," but they want you to state the substance of the information that he gave.

*The Witness.* The substance was that he had made some $4000 or $5000 out of the stage operations.

*The Chairman.* When was it that you got that information from him?

A. That was during the time that we roomed together, in the month of July, or August, or September, 1842, when the cars first ran from Manchester to Boston.

Q. (*By Mr. Mugridge.*) Do you mean to state that Mr. Clough did not take down any produce, poultry, meat, berries, or the like of that, upon the passenger trains, the first fifteen years that he ran upon the road?

A. No, sir; I would not state that.

Q. You do not know the extent of his operations upon the freight trains in that line, do you?

A. No. sir.

Q. (*By Mr. George.*) You merely mean to state that you do not know anything about it?

A. I mean to say that he could not have done an express business and I not know anything about it.

## TESTIMONY OF GEORGE CLOUGH CONTINUED.

Q. (*By Mr. George.*) I see that you were taxed in 1859 for cows and oxen, and things of that kind, over eighteen months old. I want to know where they were. Were they down on your farm?

A. I think so. A portion of them were there, and a portion of them were where I lived.

Q. You had a stock of cattle on the farm that you run?

A. I had some stock there. I do not recollect how many.

Q. That is the farm that you swapped for the railroad stock with Captain Walker?

A. Yes.

Q. The next year, 1860, there was $3766 of stock appraised; that was down on the same farm, was it not?

A. I presume a portion of it was. I had some cows at my home, at Railroad square, where I then lived.

Q. You ran that farm down there in 1859, 1860 and 1861?

A. I put it away, I think, in the spring of 1861?

Q. You were taxed for it in the spring of 1861, were you not?

A. I do not remember about that.

Q. You do not remember how that was?

A. No, sir.

Q. You had no stock of cattle anywhere else that you were taxed in Concord for, had you?

A. None, unless I had it at my house where I lived, at Railroad square.

Q. You did have a stock of cattle down there?

A. I think I had some cattle there.

Q. You say that this book (the specification of the building contract with Mr. Mason) is not signed at all?

A. It is a copy of the original; it is not signed; I had it copied.

Q. Who built that building?

A. W. Mason.

Q. Were there any extra bills on it.

A. I do not know that there was.

Q. Who finished off the Masonic Hall? Did the Masons do that, or was it included in the contract?

A. I think the Masons did that themselves.

Q. How much rent do the Masons pay you?

A. I think they pay $400 per year.

Q. Didn't the Masons finish off the hall and take the money out of the rent?

A. I do not recollect how that was.

Q. Can't you tell whether or not the Masons finished off that entire suit of rooms in their own way?

A. I do not remember now. Mr. Corning did the whole business, and I cannot tell.

Q. How many stores are there in the Masonic Temple?

A. There are three stores: one on Main street, one corner store, and another one besides.
Q. Who occupy the stores?
A. Mr. Shattuck occupies one, Hoyt another, and Mr. Bond the other.
Q. What rent did you receive from them in 1864?
A. I do not remember.
Q. What rent did you receive in 1865?
A. I cannot tell. I can get it from the books.
Q. What yearly rent does Mr. Shattuck pay you for the use of the store occupied by him?
A. Three hundred dollars.
Q. What was the rent five years ago?
A. It was $300 dollars then.
Q. Who occupies the next store?
A. Mr. Hoyt. He pays $600, and Mr. Bond pays $400. Mrs. Jones, milliner, pays $300. Mr. Morgan pays $500.
Q. How is it with the other two stores on the other block? How many of those are occupied?
A. Mr, Rrown occupies one now.
Q. For an undertaker's shop?
A. Yes.
Q. What does he pay?
A. Three hundred. Mr, Cochran pays $300; and the other store is rented for $300, but the man has not come to take it yet; he pays rent however from the first of December.
Q. How long has that been unoccupied?
A. For a short time,—since the last tenant moved out. I think the rent was paid there until fall.
Q. How long has it been unoccupied?
A. Only for a short time.
Q. For four or five months?
A. No, sir; not more than three months, I think.

*The Chairman.* I wish to call the attention of the counsel to the suggestion made the other day, that we shall expect a note of the exceptions on both sides; and shall understand, as I said before, that the exceptions to which our attention is not now called, are not insisted upon.

## TESTIMONY OF JAMES H. EASTMAN.

Q. (*By Mr. Mugridge.*) How long have you lived in Concord?
A. About twenty years.
Q. How long have you known George Clough?
A. I have known him, I think, for as many as eighteen years.
Q. Where was Mr. Clough living when you first knew him?
A. He lived down in South street, in Concord.
Q. Let me ask you whether at any time you worked for Mr. Clough, and if so, when and for how long a time.
A. I worked for him fourteen years ago this winter. That would be in 1854.
Q. What time in 1854 did you go there?
A. In the fall,
Q. For how long a time were you in his employ?
A. I was there with him, off and on, for three years.

Q. How was it about Mr. Clough's keeping any cows while you worked for him? How many did he keep?

A. The winter that I went there he kept from three to four cows,—sometimes more and sometimes less.

Q. During the time that you were there off and on, for three years as you have stated, how many have you known him to have at one time?

A. I have known him to have six.

Q. Who milked those cows while you were there?

A. I milked them.

Q. What did Mr. Clough do with the milk of those cows?

A. He sold it to milk customers which he had.

Q. Who carried the milk out, and distributed it among his customers?

A. I carried it when I was there.

Q. Are you able to state the number of customers that Clough had during the three years that you were there?

A. I do not know as I could.

Q. If you cannot state the number, name as many of them as you can think off?

A. There was Mrs. Neil.

Q. Is she the mother of David Neil?

A. Yes, sir; and a man that lived in the house with her by the name of Kindalls; and Tim Flanners.

Q. Did he keep a boarding house?

A. Yes.

Q. How many quarts of milk did he take?

A. I think he used to take from four to six. Deacon Flanners, a brother of Tim, also took milk; and a man by the name of Babb.

Q. Are you able to give any sort of idea as to the number of customers that he had during the time you were there?

A. As near as I can tell, there were somewheres between ten and twelve.

Q. Where were these cows kept? Where did you milk them?

A. They were kept there at his house, on Railroad square.

Q. Mr. Clough was living in Railroad square at the time you were working for him?

A. Yes, sir.

Q. I have inquired about the cows while you were there. Let me ask you if you were familiar with Mr. Clough's premises before you went there,—from 1850 to 1854? Were you about the premises?

A. Occasionally I was.

Q. For how long a time prior to the time that you went there, had you known Mr. Clough's premises, by being there occasionally?

A. Ever since he built his house there; and I was there before it was built.

Q. Will you please state how it was with regard to his keeping cows before you went to live with him?

A. I should say that he did, as nigh as I could recollect.

Q. How many?

A. I could not tell you that.

Q. State whether you were familiar with his premises after you left his employ,—after the expiation of the three years?

A. Yes.

Q. For how long a time did you know of his premises and what he kept about them, after you left?

A. Until he moved up to where he now lives.

Q. Tell how it was about his keeping cows after you left him, and up to the time that he went on to Warren street.

A. As nigh as I could recollect I should say that he kept from perhaps one to three along. I could not tell.

Q. You say that you worked for Mr. Clough three years in the way that you have stated, off and on, and that you milked the cows and carried the milk out?

A. Yes.

Q. Let me ask you how Mr. Clough paid you for the services you rendered?

A. He had a driving horse and a working horse, and I worked the horses around the streets here and done job work, and got my pay in that way for the labor that I done.

Q. You received your compensation in what you got in doing job work with his horses, for taking care of his cows in this way.

A. Yes, sir.

Q. (*By Mr. George.*) Do you know anything about Mr. Clough's running this farm down on eleven lots?

A. He had a farm there.

Q. Had he a farm there at the time you are speaking of,—when you lived with him?

A. Yes.

Q. Who did he have to carry on that farm for him?

A. I carried it on sometimes, and Mr. Elliot's man, and a man by the name of O'Connor did sometimes.

Q. Did anybody live in the house, as you remember?

Q. There was nobody in it at the first of my going there.

Q. Mr. Clough built a new set of buildings on that farm, didn't he?

A. Yes, but it was after I got away from there.

Q. He built a new barn and a new house, or made a barn into a house, didn't he?

A. There was a barn there, and he made that into a house, and then built a barn?

Q. A nice set of buildings?

A. It was a nice barn.

Q. A very nice barn?

A. A very good barn.

Q. Have you any idea what that set of buildings cost?

A. I have not.

Q. You are not familiar with such things?

A. No.

Q. Do you remember in what year he built it?

A. No.

Q. Do you know whether he had a large stock of cattle there for two or three years,—twenty or thirty head of cattle, more or less.

A. I think he had some there, but I could not say how many.

Q. He run the farm, did he not, and all the operations connected with it? He did not let it out?

A. He had a man that lived there.

Q. A man that carried it on for him?

A. I do not know how that was,—whether he let it to him or not.

Q. You do not know anything about the profit or loss on the operations?

A. No, sir; I do not.

## TESTIMONY OF ROSWELL SILVER.

Q. (*By Mr. Rolfe.*) Where do you live?
A. In Thorndike street.
Q. Where did you live in 1849?
A. I think I lived on West street.
Q. In 1859 where did you live?
A. I lived on State street for a while.
Q. Where did you move to then?
A. I then went on to what is called the Galt lot, near Railroad square.
Q. State whether you had business with Mr. Clough?
A. I did have some.
Q. Commencing when?
A. It commenced about in 1849, I think. I lived then on West street.
Q. You had a horse of Mr. Clough, called the Silver horse, or "Old Bill"?
A. Yes.
Q. When did you have him?
A. I had him at the time I lived on West street.
Q. Where was he kept?
A. He was kept at Mr. Clough's stable.
Q. Did you run him from Mr. Clough's stable?
A. I did.
Q. For how long a time?
A. About a year and a half.
Q. What doing?
A. Jobbing.
Q. Do you know about the stock that Mr. Clough kept at that time?
A. He kept some cows and some horses.
Q. How many?
A. Sometimes three and sometimes four cows.
Q. For how long a time do you know about Mr. Clough's keeping cows?
A. I think he has kept cows all the time from that time up to this, more or less.
Q. Previous to his going up to where he now lives, I mean?
A. He kept cows there at Railroad square.
Q. To what extent during the time he was there at Railroad square?
A. I think he had from three to four cows.
Q. Where did he pasture them?
A. Down in the oaks, near the South end.
Q. On the land that belongs to the Concord Railroad?
A. Yes.
Q. Did you ever take care of his cattle and cows?
A. Only occasionally, when there was nobody else around.
Q. What kind of cows did he keep,—as to quality?
A. He had some pretty good cows.
Q. (*By Mr. George.*) How many cows does Mr. Clough keep now?
A. I guess he don't keep any.
Q. Do you know whether he does or not?
A. I could not say positively.
Q. Do you know how many he kept last year?

A. He kept one, I believe.
Q. Do you know?
A. I know that he had one a part of the year.
Q. How many did he keep the year before last?
A. I could not say that he kept any?
Q. Can you say whether he did or not?
A. I could not.
Q. How many did he keep the year before that?
A. I am not able to say.
Q. How many did he keep in the year 1860?
A. He had as many as three, I think.
Q. Do you know?
A. That is my best recollection about it.
Q. Do you recollect anything about it?
A. I recollect that he had some cows.
Q. Do you recollect that he had some cows in 1860?
A. I guess so.
Q. Do you know in what year he moved into his house on Warren street?
A. I could not say.
Q. Was 1860 before or after?
A. I could not say when he did move up.
Q. Have you any idea whether or not he was there in 1860?
A. I think he did not live up there in 1860.
Q. Do you remember whether he went there in 1860, 1861 or 1862?
A. I could not say.

## TESTIMONY OF PETER DUDLEY CONTINUED.

Q. *(By Mr. Rolfe.)* During the time that Mr. Clough was driving stage from Lowell to Concord over the Mammoth road, did you drive on the opposite days?
A. I did.
Q. For how long a time after he went on to the Concord road?
A. I drove until the cars commenced running.
Q. Whose days were the best? Do you know?
A. I considered Mr. Clough's days the best.
Q. You met him every day, did you not?
A. Yes.
Q. State whether you had an opportunity of judging of the amount of travel, and of the number of passengers he carried?
A. I do not know as I had any means of knowing that in any other way than we passed, as anybody would, when we met on the road. He was pretty generally filled up. As a general thing he was heavily loaded.
Q. Do you recollect what the fare was from Lowell to Manchester?
A. One dollar and fifty cents.
Q. And what was it from Manchester here?
A. Sometimes we charged $1.25 and sometimes $1.50.
Q. What was the fare through?
A. Two dollars.
Q. During the time that you ran, did you run an extra on the days that you drove?
A. I do not remember about that, I presume that I did run an extra, but do not recollect very clearly about that.

Q. Did Mr. Clough run an extra during the last year?

A. I should say that he did.

Q. Do you recollect, during the first year that Mr. Clough ran, how much you made net?

A. No, sir. I do not know what he made, nor can I say what I made. I know that the days that I run, after I sold to Mr. Clough, were not so good as they were before. That, all the stage folks will tell you.

TESTIMONY OF G. G. SANBORN CONTINUED.

Q. (*By Mr. Mugridge.*) Who ran upon the corresponding trains, after Mr. Clough left the road, to those that he ran on while upon the road?

A. I think Mr. Gale took it first for a short time, and it was finally put into the hands of Mr. LeBosquet.

Q. Has Mr. LeBosquet since died?

A. Yes.

Q. Who has run his train since that time?

A. Mr. Alexander.

Q Have you here the returns made by the conductors that have run Mr. Clough's trains to the general ticket office since he left the road?

A. I have. [Witness produced returns.]

## TESTIMONY OF HENRY P. ROLFE.

Q. (*By Mr. Mugridge.*) How long have you known Mr. George Clough?

A. Since the fall of 1848.

Q. Where was he living at the time you formed his acquaintance?

A. I do not know where he did live. The first time that I knew where his residence was, he was living on Railroad square.

Q. Will you be kind enough to state to the referees whether you had anything to do with regard to negotiating the price, or whether you had anything to do with the transaction connected with the purchase of the Cooper house? If so, state to the referees what you knew or did, in connection with that transaction.

A. I did a part of the business, and was knowing to all of it. I acted as a sort of "go between," between Mr. Clough, and Mr. Corning, and Mr. Cooper, and Mrs. Clough.

Q. What was the price paid by Mr. Clough for all that he had in that purchase? If you do not know from your personal knowledge, you need not state it. What was the price, including the real estate and the personal property?

A. I know about it, because I paid the $2000 that was paid down, and I think that I wrote the note for the balance, in the State Capitol Bank. I was present when the matter was arranged between Mr. Corning and Mr. Cooper, and Mr. Chandler.

Q. State what the price was, including the real and the personal property?

A. It was $8000.

Q. Be kind enough to state whether the consideration in the deed was fixed at another price than the $8000 that was paid; and if so, at whose suggestion was it done?

Q. My recollection in that particular is a little different from Mr. Cooper's. Mr. Corning desired the price to be put at $12,000; and

Mr. Cooper also desired it. Mr. Cooper claimed that the property cost him a great deal more than that, and that it was worth $12,000 and should be sold for that, but that as Mr. Clough must necessarily lose something from having signed his notes, he wanted this property to go into Mr. Clough's hands at $8000 or so, in order that Mr. Clough might not be the loser by the transaction. That is my recollection about it.

Q. And upon consideration the deed was fixed in that way?

A. Yes.

Q. You say that you paid the $2000 that passed through your hands. What was the amount of the note that you wrote?

A. Six thousand dollars.

Q. And this $6000 note, and the $2000 cash payment, made the amount that was paid for the property?

A. Yes.

Q. Did you have anything to do with the purchase of the Jamaica Plains property? If so, state in your own way, your connection with the transaction, and what you know about it.

A. I did nearly the entire business in relation to obtaining the title to that property. I went to Boston with Mr. Joseph Palmer, and was present when the first mortgage was executed, which was made to secure the debt. I afterwards went three times to Worcester to negotiate with Amos Binney; and I went to Boston two or three times to negotiate with Amos Binney Merrill, who was, I think, a relation. The Amos Binney who lived at Worcester, was a Methodist Presiding Elder. At the time the mortgage was executed to secure Mr. Clough, the property was under an attachment, and Mr. Palmer took me into an office where there was a lawyer by the name of Parker, I think, to have him explain to me that the attachment was not good for anything; which Mr. Parker tried to do; but I failed to see it; and at Mr. Clough's request I then immediately commenced to negotiate with Mr. Binney, who was the attaching creditor.

Q. State the result of that negotiation?

A. I issued an attachment, and procured the property to be set off. The mortgage did not include all the lots, and I think I procured the whole of the property to be set off on his execution. I made an agreement with Mr. Merrill and with Mr. Binney (whom I afterwards met in Boston,) that they should assign the execution to Mr. Clough at a certain time, and that they should retain the property in Mr. Binney's hands until the equity of redemption should expire, so that we should have a complete title in that direction; and in order to have no legal controversy about the matter Mr. Binney discounted, on his execution, for the cash, a note; so that what was paid, in addition to the Palmer note, amounted to between $1700 and $1800.

Q. You mean to say that the Jamaica Plains property cost Mr. Clough between $1700 and $1800?

A. Yes; then after that, Mr. Clough and myself went to New York to obtain a release from Mr. Palmer and his wife; which we obtained; and Mr. Clough there gave Mr. Palmer $100; so that the whole property cost Mr. Clough at that time, a very few dollars over $1800.

Q. The title that he got to the property cost him that?

A. Yes.

Q. And his title was a perfect title?

A. Yes, sir.

Q. Did you attend, as counsel for Mr. Clough in the taking of the deposition of Martin D. Starkey?

A. I think I did. I attended the taking of a part of it.

Q. State whether, since that deposition was given, you have had an interview with Mr. Starkey, and if so, what he said to you at the time, and where you saw him? And give a detailed account of the interview in your own way.

A. Some time after the referees had been in session, and the tickets had been produced, on which were found several dates after Mr. Clough left the road (I mean the Whitcher tickets), I went to Westmoreland, N. H., to see if I could find Mr. Starkey. I had understood that that was his place of nativity. I ascertained there that there was a man by the name of White living at Hinsdale, Mass., who was related to Mr. Starkey by marriage; and I went from there to Hinsdale and found Mr. White, and he informed me that Mr. Starkey was in the ticket office, as night ticket agent, at Port Jarvis, on the New York and Erie Railroak. I took the cars from there and went to Port Jarvis, where I found Mr. Starkey. I reached there Saturday evening, and on Sunday morning, having ascertained where Mr. Starkey lived, I went to his house. His wife came to the door, and as soon as he heard my voice (he was in an adjoining room,) he stepped to the door and called me by name; and wanted to know if I had "come to take him on to New Hampshire." After shaking hands and passing the compliments of the day, I said "No." He then wanted to know what I did want of him. I told him "nothing that would do him any injury." I then stated to him the dates of the tickets that were said to have been found upon the person of James Whitcher; and I said to him that from what James Whitcher had told me, I inferred that he (Starkey) could tell me all about the Whitcher tickets. At the time he first stepped into the room, he was looking very flushed, and immediately he turned pale, and said to me, "I shall have to sit down; you frightened me very much; and when I get over my fright, I will tell you all about it." He then said, "Mr. Rolfe, I have suffered a great deal. I have had my leg cut off by the cars. Put your hand on my shoulder; my shoulder-bone has been broken all to pieces; three of my ribs were broken, and I was rolled over and over on the track; and I have suffered so much that trouble greatly affects me. Mr. Clough was always kind to me; and I thought, while I was lying sick and suffering, that I had done him a great deal of wrong by a deposition that I had given in Concord." I asked him if he would be willing to give his deposition again. He said that he did not want to. I inquired "why not?" He said, "If I should state differently from what I stated in my deposition, they would shut me up, would they not?" I asked him what he meant by that? And he said, "They would put me in the state prison." I told him that they might, but I didn't think that they would. He said that he was not yet quite dressed; and I asked him if he would come down to the hotel and see me, and he said that he would, and that he would be there in a few moments. I went back to the hotel, and he soon came down; and I then told him what Whitcher had told me.

Q. State what the communication was that you made to him. What did you tell him?

*Mr. George.* I object.

*Mr. Mugridge.* It is a part of the conversation. We propose to put in the entire conversation between Mr. Starkey and Mr. Rolfe.

*Mr. George.* What he proposes to state is incompetent.

*Mr. Stanley.* The conversation that has been already given is entirely immaterial.

*The Chairman.* It does not seem as if there could be any controversy about the proper way to examine this witness in this manner. I understand that the testimony is put in for the purpose of contradicting Mr. Starkey. Of course, nothing can be admissible except what tends to contradict Mr. Starkey; and this conversation can only be admissible as being in point of fact a part of what Mr. Starkey said.

*Mr. Mugridge.* We do not offer it to prove the fact of what Mr. Whicher said to Mr. Rolfe.

*The Chairman.* It does not prove Mr. Whitcher's statement; and it can only be put in by Mr. Starkey's speaking in reference to it, and to what Starkey said contradictory to his deposition.

Q. (*By Mr. Mugridge.*) Proceed and state what you said to Starkey that Whicher said to you?

A. I told Starkey that Whitcher said to me that he had most of the tickets that were found on his person, at Manchester, of him—that is, of Starkey. I said "of you." I asked him how that was. He said; "I stated in my deposition, that I let Whitcher have tickets once, but," said he, "I do not recollect the number that I said I let him have, but I did let him have tickets two or three times; I cannot tell the number." Said I, "Whitcher says that you claim that you let him have $150 worth of tickets." He said, "I do not remember of claiming any such thing." I asked him what tickets he let him have; and he said: "I let him have tickets over the Concord road, and coupons from over the Boston, Concord & Montreal road, and over the Northern road, and over the Concord & Claremont, and the Manchester & North Weare, and the roads below." I asked him if he let him have any Boston tickets or any tickets that were good to Boston. He said "No." I told him that there were several found, and he said that Whitcher must have got them of him when he was not noticing it, or he must have got them of somebody else, for he didn't let him have any; "for" said he, "the Boston tickets were counted out to me, and I have no recollection of my count not coming out right." I asked him if he ever let Whitcher have any tickets before he went on as conductor; he said that he had,—a good many. I asked him then ——

*The Chairman.* What part of Mr. Starkey's deposition does the evidence that is now going on, contradict?

*Mr. Tappan.* The part in relation to his having the tickets of Mr. Clough. His statement to Mr. Rolfe contradicts the statement in his deposition relative to the number of tickets that Starkey says that he had of Mr. Clough.

*Mr. George.* I do not think that it does.

*The Chairman.* I believe that in his deposition he says that Mr. Clough let him have twelve or fifteen tickets.

*Mr. Mugridge.* I will read from Starkey's deposition. Interrogatory 7th and its answer:

"Int. 7. State whether or not, during the time you were acting as conductor, you received or disposed of any tickets. If so, state from whom you received them, and to whom you disposed of them, giving all the circumstances in detail. Ans. I did dispose of some tickets. I did not receive any. I gave some to Jim Whitcher; I think there was a dozen or fifteen, or in that vicinity. They were coupon tickets between Manchester and Concord, and Concord and Manchester perhaps. It was somewhere about the seventh of February. He asked me for them.

*The Chairman.* It seems to me that the established practice in such cases as this would require that there must be some particular thing in the deposition indicated, that this conversation runs contrary to. I do not think it is necessary that the whole conversation should be stated; but only that part which is contradictory of some particular thing in the deposition,

*Mr. Mugridge.* The particular point we want to contradict is the fact that he states in his deposition that he only let Whicher have tickets once, and then some ten or twelve tickets; and we also propose to contradict his statement that he got tickets of Mr. Clough.

*The Chairman.* Any testimony as to that is admissible.

Q. (*By Mr. Mugridge.*) State what he said in relation to the tickets?

A. He said that he had let Whitcher have tickets a good many times.

*Mr. George.* That has not been ruled in.

*Mr. Mugridge.* He denies in his deposition that he let him have them more than once.

*Mr. George.* The deposition speaks for itself.

*Mr. Mugridge.* In another answer from that which has been read, he says that he let Whitcher have tickets but once.

Q. Go ahead, and state in regard to the tickets.

A. I asked him if he had tickets of Mr. Clough.

*Mr. George.* He swears in his deposition that he did not receive them of him.

*Mr. Mugridge.* He swears in his deposition that he did have them.

*Mr. George.* The only time that he swears in regard to giving tickets to Jim Whitcher, is in reply to the 7th interrogatory, which has been read. He then goes on and swears expressly that he took his tickets and was punching them at the rear end of the car, and Whicher came along, and he let him have some tickets. Then he swears that he said that he had received tickets from Clough and from all the other conductors; and Mr. Clough swears to the same thing. That statement was in reference to these tickets that he gave to Jim Whicher, because he goes in and tells what he did with them. With the leave of the referees we would like to read the whole deposition.

*Mr. Mugridge.* We object to its being read.

*The Chairman.* In that deposition he states that he had a certain uumber of tickets, or tickets, from Mr. Clough. Can there be any impropriety in contradicting that?

*Mr. George.* Not if they contradict that.

*Mr. Tappan.* The whole purpose of the deposition was to show that these Whitcher tickets came from Mr. Clough. That has been the object all the way through; that was the object of Starkey's deposition; and it is that statement in his deposition that we now wish to contradict.

*Mr. George.* I suppose that you must first point out the statement of Mr. Starkey that you propose to contradict.

*The Chairman.* He says in his deposition, "I received tickets from George Clough." I suppose Mr. Rolfe's testimony is directed against that statement. If he has anything that bears against that he can state it.

*Witness.* He said that for about a year and a half before he left the road, Mr. Clough did not let him have any tickets. Then I said: "You let White, and Gilman, and others have tickets; where did you get them? Did you get them of the other conductors?" He said no; he did not; that he asked Mr. Clough for tickets, and when he had given him one or two, he asked him for others, and Mr. Clough refused to give them to him, and told him that if he had anybody that he wanted to pass, to let them get into the cars. He said that he had been in the habit, when the cars left Nashua, of helping the conductors take tickets in the cars. And when he took up any considerable number he said that he saved out one or two; and in that way, he said, he had any quantity of tickets; and he let Jim Whitcher have them, and let others have them. I asked him how he managed to get tickets over the entire length of the road, from Nashua up to Concord.

*Mr. George.* Is this competent evidence?

*Mr. Mugridge.* It relates, as we say, to these same tickets that he let Whicher have.

*The Chairman.* His letting Whitcher have them is of no consequence. The question is, how it is connected with Mr. Clough.

*Mr. Mugridge.* He says that Mr. Clough did not let him have any tickets for a year and a half before he left the road; while he states, in reply to the 7th interrogatory, that he let him have twelve or fifteen tickets,

*The Chairman.* Was that read in the case?

*Mr. Tappan.* I think so. The latter part was ruled out, but I think the first part was ruled in.

*The Chairman.* My recollection about the matter is that the defence very strenuously, and the referees very properly insisted that no part of that deposition should go in which was not connected with Mr. Clough.

*Mr. Mugridge.* If there is any testimony in your mind that applies to the testimony of Starkey, as you recollect it, you may state it.

A. I have stated about all. There is nothing more than the manner in which he got his tickets over the whole length of the road.

Q. After having this interview with Mr. Starkey, you came home?

A. Yes.

Q. If you know anything in regard to George Clough keeping cows at any time since he has lived in Concord, state what you know about it.

A. He has kept cows more or less, but I know more particularly from 1854 to 1858 or 1859. In 1854, he had one cow—that I recollect—a large red cow. He bought another at Hopkinton—a very nice white-faced cow. He bought another of Wyatt—a white cow; and I had a very nice cow that had twin calves, and he bought that cow of me. He rode up with me to Bellows' when the cow calved, and on the way back he bought the cow and calves of me, and they were driven down here. At that time I recollect of his having these four cows. The white cow he sold that fall to a man by the name of Mooney, and the cow that he had of me he also afterwards sold to Mr. Mooney. Mr. Mooney asked me about her, and I saw the cow when he was leading her through the village to take her away. In 1859 I bought a cow of him, and he had several others left after I picked this out. That is all I recollect about his cows.

Q. Do you know of his selling milk at any time between 1854 and 1859?

A. Only from what I have heard.

Q. (*By the Chairman.*) How many cows had he when you bought one?

A. Three or four.

CROSS EXAMINATION.

Q. (*By Mr. George.*) How many visits have you made to Mr. Whitcher and Mr Starkey since these suits were commenced ?

A. I have made one to Mr. Whitcher and one to Mr. Starkey ?

Q. Did you get Starkey to sign any sort of writing ?

A. Yes, sir.

Q. Have you got that ?

A. Yes.

Q. How long did it take you to get him to sign that writing ?

A. Not long.

Q. How long ?

A. He signed it very soon.

Q. Will you let me see it ?

*Mr. Mugridge* If the gentleman desires to put it in evidence, he is entitled to it; if not, he is not.

*The Chairman.* If there is no objection made we cannot rule upon it.

*Mr. Mugridge.* We object to his seeing it unless he wishes to put it in.

*Mr. George.* Will you let me see it ?

*Witness.* I have no objection to your seeing it, but if Mr. Mugridge and Mr. Tappan think I had better not let you see it, I will not.

*Mr. Stanley.* I submit that we have a right to see it. Mr. Starkey has sworn to a written statement, and we have a right to see if his written statement compares with his oral statement.

*The Chairman.* You will let the referees see it, if they wish ?

*Witness.* Certainly.

Q. (*By Mr. George.*) You are liable to misunderstandings as well as other people, are you not ?

A. I presume I am.

Q. You signed a certificate that you had no interest in, or connection with Mr Clough's insurance ?

A. I did.

Q. Was that true ?

A. I do not know that I had.

Q. Were you not an agent, and did you not make a charge for services in that case. and receive pay ?

A. I do not that I did.

Q. Do you know whether you did or not ?

A. I cannot tell; I will look at my books and see.

Q. How many justices of the peace, to your certain knowledge, lived nearer Mr. Clough than yourself ?

*Mr. Mugridge.* Is this an important matter to inquire into ? Mr. Rolfe certified upon an application made for payment of loss in accordance with certain rules of the Insurance Company, that he was the justice living most contiguous to the fire. Mr. George now proposes to inquire of Mr Rolfe whether he was the justice living nearest to the fire. I would like to know the competency of the evidence.

*Mr. George.* The paper is a paper in this case ; it is signed by Mr. Rolfe, and sets forth certain facts. Upon the cross examination of Mr. Rolfe, I certainly have a right to ask him whether he knew, at the time he signed the certificate, that the certificate was erroneous ?

*Mr. Mugridge.* The paper was introduced by Mr. George, who now proposes to contradict his own evidence in the case, or in other words, to show that the testimony he has introduced is incorrect, and to assault it in some way by Mr. Rolfe. I object to the inquiry in that view.

*The Chairman.* What strikes us about it is this: That the paper was introduced here as a paper which had been made out as it was said partly through the agency of Mr. Clough for the purpose of showing that he made a claim to a certain amount of damages on account of his loss by the fire; and for the purposes of this hearing I do not know whether it is of any consequence whether it was proved or not. The cross examination is entirely irrelevant to anything that Mr. Rolfe has testified to; it does not tend to test his recollection at all, and as we think, is entirely irrelevant to anything in the case.

*Mr. George.* My point is this; That it is very fair to presume that Mr. Rolfe's recollection cannot be very accurate if he would certify that he was the nearest justice of the peace, when there were two others that lived within two rods.

*The Chairman.* The referees are of the opinion that it is not a legitimate way of testing the recollection of the witness to examine him about totally irrelevant matters. Perhaps Mr. Rolfe had a different understanding from that of the insurance company, as to the meaning of the word "contiguous." The term *contiguous*, I suppose, could hardly mean anything else than that the property occupied by him was adjoining the property insured.

Q. (*By Mr. George.*) Did ever you state to Mr. Stanley in substance, that as long as Mr. Gilmore lived, was uneasy, or trouble, and that when he died ——

*By Mr. Mugridge.* Is that material? Suppose I make a statement to Mr. Rolfe or to Col. Tappan; does that statement affect Mr. Clough, or bind him in any way?

*Mr. George.* Mr. Rolfe occupies the position of counsel here; he is now a witness; I understand that he has made such a statement; if he has not, I would like to have Mr. Rolfe deny it under oath.

*Mr. Mugridge.* You shall have that privilege then; we will withdraw our objection.

Q. (*By Mr. George.*) Did you ever have any conversation of that character with Mr. Stanley, or in his presence? Was your statement to Mr. Stanley this: that while Mr. Gilmore lived, Mr. Clough was in a good deal of trouble, for he did not know what he would swear to, but that after he was dead, he felt better?

A. No; I never said anything of that kind to Mr. Stanley in my life.

Q. Nor in his presence?

A. Nor in his presence. I may have said to Mr. Stanley that Mr. Gilmore was the most depraved man I ever knew, and that he would swear to anything; but I never said that Mr. Clough was in trouble about it, for I never heard him express any opinion of that kind, one way or another.

Q. How much has Mr. Clough paid you for the last twenty years?

*Mr. Mugridge.* Is that material?

*The Chairman.* Inquiry has been made as to Mr. Clough's expenditures in various ways, and I do not know why this is not as relevant as anything else.

*Witness.* For actual services, since I have first known Mr. Clough, he has paid me less than $400.

Q. How much for money paid out for expenses, &c.? How much has he paid you for any purpose or for any reason?

A. Up to the time when Rolfe & Marshall dissolved partnership, in 1859, there was an account upon the books against Mr. Clough, of one hundred and twenty odd dollars. Mr. Marshall desired me on settlement, to take his part of that in pay, which I did,—sixty odd dollars. That bill Mr. Clough has never paid, dollar nor cent; and one-third of it was for cash paid out. Some four or five years ago Mr. Clough asked me to make out my bill against him. I looked the account over as well as I could, and I had less than $400 on my books. Whenever I have made any disbursements for Mr. Clough, he generally has paid me the cash for them. That is the way the account stands now.

Q. (*By Judge Bellows.*) When were you called upon to make out the account?

A. Perhaps three or four years ago.

Q. (*By Mr. George.*) Do you mean to say that you have never received, directly or indirectly, any money from Mr, Clough, beyond $400?

A. I have never received, directly or indirectly, for services, more than $400. I have borrowed money of Mr. Clough in small sums,—not large, all of which I have paid him except that I borrowed of him some two months ago, $50, and since this hearing commenced I have borowed $14—which sums I have not yet paid him.

Q. You have not received from him directly or indirectly any property, in the shape of presents or otherwise, or anybody else, for you, beyond the $400?

A. I have never received anything in the shape of presents, property or money, in any manner, shape or form, directly or indirectly, from Mr. Clough, except as I have stated.

Q. Nor has any member of your family?

A. Nor has any member of my family received a dollar or a cent.

Q. Has ever any allowance been made to you in trade, or in anything of that kind?

A. Not a cent. I sold him a pair of buckskin horses that I went to Canada and bought, at a direct loss to myself of $107 in cash.

*Mr. Mugridge.* I believe, Mr. Chairman, that this our case.

## TESTIMONY OF JONATHAN L. PICKERING.

Q. (*By Mr. George.*) You are the deputy sheriff who served the writ against Mr. Clough?

A. Yes, sir.

Q. Will you state whether at the time you attached Mr. Clough's property, Mr. Clough had in his possession and you attached the property named in that receipt. [Presenting receipt.]

*Mr. Mugridge.* Well, now, so far as any attachment is concerned, of actual property, I do not object: but the idea of putting in that receipt in this way I do not think is competent.

*Mr. George.* I do not ask about any receipt; I simply ask whether that property named in that receipt was there and attached by him as the property of Mr. Clough.

*The Chairman.* It seems to me that, until you in some way or other make the receipts admissible by some preliminary examination, you cannot put it into the case. If Mr. Pickering knows what property he attached, perhaps he can state that.

Q. (*By Mr. George.*) Mr. Pickering, you may say what property there was that you did attach.

A. The furniture that was named in that receipt that was in the house. I went around through the rooms. There were some horses that I didn't see.

Q. That furniture you saw?

A. That furniture I saw.

Q. I want you to state how the receipt was made out?

A. I commenced to write the receipt, but Mr. Rolfe said he could write faster than I could, and he wrote. There is some property in that that I didn't see, that Mr. Clough said I might put in.

Q. Mr. Clough gave the names of articles that you didn't see?

A. He said there were some horses, and carriages that I could put in.

Q. Now will you state the articles?

*The Chairman.* Is there any objection to that being used.

Mr. Tappan waived objection.

Q. You may state?

A. [Reading from receipt.] 12 carpets, 2 stair carpets, 8 damask stuffed chairs, 2 large damask stuffed chairs, 2 sofas, 4 curtains, 2 mirrors, 1 centre-table, 4 pictures, 1 secretary, 1 piano stool, 1 sick room chair, 1 book case, 1 safe, 1 lounging chair, (I should think it was,) 1 black-walnut stand, 2 chairs, 5 pictures, 2 statuetts, library, 1 piano cloth, 1 parlor rug, 4 entry mats, 4 mats in library, 1 secretary, 8 dining chairs, 1 rocking chair, 3 chairs, 1 clock, 2 mirrors, 1 pitcher and salver, goblet and basin, 1 rug, 1 hat tree, 6 mahogany chairs, 1 rocking chair, 1 easy chair, 1 mahogany sofa, 1 centre-table, 1 stand, 1 ottoman, 3 pictures, 1 glass, 1 crumb cloth, 1 entry carpet, 6 rugs, 1 chamber set, 1 somnole, 6 beds and bedding, 1 chesnut chamber set, 2 damask curtains, 1 other chamber set, 1 crumb cloth, 1 what-not, 1 cloth, 1 rocking chair, 3 chamber sets, 3 horses, 1 three years old colt, 3 carriages, 2 sleighs, and 4 harnesses.

Q. Was Mr. Clough present at the time that was made?

A. Yes, sir, he was.

Q. What was the value fixed upon that property, at that time, of Mr. Clough's

A. The receipts show $4000, [Objected to.]

*The Chairman.* The receipt probably is not evidence of the value of the property, unless Mr. Clough is made a party to it in some way.

Q. How was that arranged with regard to the value of the property?

A. I think we jumped at the amount of the property.

Q. Did Clough agree to it? [Objected to.]

*The Chairman.* What occurs to me about it is this, that if Mr. Clough stated his belief that the property was worth that, a simple acknowledgement of its value, it might be evidence of its value. I don't suppose it is binding; I understand the officer to say that he jumped at it himself.

Q. Any conversation in regard to it?

A. I think that after the receipt was made out something was said about the amount. I said that I guess it was about $4000, and I think Mr. Clough said,—"It is immaterial what you call it."

*The Chairman.* I should think that could be evidence.

Q. You are accustomed to the buying and selling of that kind of property, are you, occasionally?

A. Yes, sir.

Q. How many occasions do you attend the sales of horses and carriages in a year?

A. Well, I could not tell; probably—well, I used to be in that business altogether.

Q. How is it now?

A. I don't know as I could tell now what the value was. If I had been there with the intention of buying—there were some articles that I don't know the value of. I didn't go there with the intention of buying, and I couldn't tell as to the valve of that property as I should have if I had been there with the intention of buying.

[Objected to the opinion of the witness as to value, witness not being an expert.]

Mr. George read from the deposition of George Clough, taken Nov. 24, 1868, and offered in evidence the 1st cross interrogatory and answer, as contradicting the testimony of Mr. Clough.

Mr. Mugridge then offered in evidence and read the whole deposition.

Mr. Stanley offered in evidence the certificate of J. G. Martin, of Boston, broker, showing the highest and lowest monthly prices, at the brokers board, of the Concord Railroad stock, from 1860 to 1868, inclusive, and the dividends on the said stock declared from 1860 to 1868, inclusive. [Marked D.]

Mr. Mugridge offered in evidence a memorandum of the dividends paid George Clough on the Concord Railroad stock from 1846 to Nov. 1868, inclusive, amounting in all to $22,769. [Marked B.]

[Adjourned.]

## DEPOSITION OF JACOB F. SMITH.

Int. 1. (*By counsel for plaintiff.*) State your age, residence and occupation. Ans. Age, 32 years; residence, Epsom, N. H.; occupation, shoemaker.

Int. 2. Have you ever been passed free, or been furnished with free tickets or passes over the Concord, Manchester and Lawrence Railroads, or connecting roads, to Boston, or any portion thereof? If yes, state the circumstances fully. Ans. I never have been passed over the road to my knowledge without its costing me the full amount of fare. I did once ride up from Newmarket Junction to Suneook without paying anything. I think Mr. Langley was conductor, and passed me. I got aboard without any ticket, and he said nothing to me.

Int. 3. Had you ever purchased any tickets except at the ticket office, or of the conductors in the cars, or been furnished with any, or been knowing to others being furnished with any over the Concord, Manchester & Lawrence Railroad, or any portlon thereof. Ans. I purchased two of a young fellow who said he bought them at the office, I can't tell who it was. They were tickets from Suncook to Boston, I paid him $4.50 for the two tickets. He said he bought them to go down to Lynn and did not use them. I never had any other tickets of any one except ticket masters. The conductors had taken up my tickets and given me checks frequently. The checks were taken up the same trip always, excepting once. I could not tell exactly the time nor why he did not take the check. I purchased my ticket at Suncook to Portsmouth when this check was given. I think Mr. Clough was conductor; could not swear positively. I sold Mr. Clough some hay which was drawn here the 20th and 21st days of January, 1865. He paid me the money for it at the time. That is all I know about it. I also bought three of a Frenchman from Concord to Boston, at a discount.—I don't know his name.

Int. 4. State whether or not you ever received any tickets or passes either directly or indirectly from said George Clough. Ans. I never did.

Int. 5. State whether or not you ever told any person that you had received tickets from him? Ans. I never did to my knowledge. I don't recollect that I ever did.

Int. 6. State whether or not you ever told Levi Case of Epsom, that you had received tickets of said Clough? Ans. I think I never did. I have no recollection of it.

Int. 7. Will you swear that you did not so tell him? Ans. No, I would not swear that I never did tell him so, and still I don't recollect of it.

Int. 8. Will you swear that you did not so tell him, the day before yesterday? Ans. Yes, I will so swear. I told him day before yesterday that I did not receive tickets from him.

Ins. 9. Will you swear that you never told either Levi Cass, or any one else, that you sold hay, or any other articles to said George Clough, and received pay or part pay, in tickets or passes, or words to that effect. Ans. I never did to my recollection. I have no recollection of any such conversation.

Int. 10. Did you not say, when making your answer to Int. 3, that you supposed you was summoned here to testify with regard to hay furnished said Clough? If so, why, and explain what connection there was betweeen said hay, and the question which asked you whether or not you had purchased tickets other than at the ticket office? Ans. I did so say. Mr. Cass drawed the hay at the time, and said he thought I received my pay for it in tickets. I don't know as there was any connection between said hay and the question, only I did not suppose I should have been summoned here if it had not been for Cass, and I supposed he told them I received my pay in tickets. I have no doubt he told them to summon me.

Int. 11. State whether or or not you know of any persons having had tickets or now having tickets over said roads, or any part thereof? If so, state who? Ans. I don't know any one who has tickets over the road. I don't know positively of any one having had tickets. I saw that a young fellow of the name of Johnson in my shop, had two. I don't know where he got them, I don't know of any others.

Int. 12. State whether or not, during the war, you were engaged at at all in the business of substitute broker, and if so, whether or not you traveled in the cars? If so, to what extent, and whether or not with substitutes. Ans. I was somewhat engaged in that business. The first year I was in the business, I was in Boston. I did travel considerable—quite often, I can't tell to what extent. Sometimes I had substitutes with me.

Int. 13. State how many times, as near as you can, you went over said roads with substitutes. How many substitutes you had with you at each of said times, and from what points to what points you took them? Ans. I could not give any estimate how many times. It is impossible to tell. I can't tell whether it was 20 or 50 times. I started at one time from New York with four substitutes and got here with one. At another time when I came from New York, three or four of us were connected together, and we had, perhaps, a dozen men with us. I recollect that case in particular, because they all got away from us. We did not get to Nashua with a man of them. I should think likely

I have carried a hundred men in all over the Concord road and branches, in the years 1864 and 1865. There might not have been as many and might have been more. I used to make them pay their own fare when I could, when I could not, I paid their fare myself. When I paid, I usually paid the fare to the conductors in the cars. What men I paid for I paid in the cars, as a general thing. I might sometimes have had tickets. I can't say as to that.

INT. 14. State whether or not you have a memorandum which will enable you to state more definitely and particularly in regard to these matters. If so, will you produce that memorandum? ANS. I think I have. I may have carried 200 men over the road. I had a book, and think I have it now, which will give an account of every dollar I received and paid out in 1864. I will produce it if I can find it. I think I have got it.

INT. 15. State whether or not, since the adjournment, you have made diligent search for said memorandum, and with what result? ANS. I did search for it, and the result was I could not find it in the house, and I don't know where it is. It may be there.

INT. 16. State when you commenced the substitute business and when you ceased the same? ANS. The first business I kept any record of was the 21st of December, 1863. That was about the first; and closed about the time Lee surrendered, about April 3d, 1865.

INT. 17. State, if you are able, where you were on Thursday, Dec. 24th, 1863, and with whom you were? ANS. I went on that day from Boston to Concord—so my memorandum says. There was a man with me by the name of Downs, A. B. Downs. He used to work at the American House.

INT. 18. State how you passed over the road that day. If with tickets, state where you obtained said tickets and of whom? ANS. I passed over the road with tickets that day. Mr. Downs gave them to me. He gave me three tickets, two from Boston to Concord, and one to go back.

INT. 19. State whether or not while you traveled as a substitute broker, either with or without substitutes, any conductor ever passed either you or any of your substitutes, or made any discount or deduction from the regular fares. If so, state all the circumstances? ANS. Never to my recollection but once. Once when I came from Philadelphia to New York the conductor wanted me to pay for four men I had with me. I told him I would pay for them if I got them through. I got two through and paid for. The others I didn't get through or pay for.

INT. 20. State whether or not, George Clough ever passed you, or any substitutes you had, free over the Concord, Manchester and Lawrence roads, or connecting roads to Boston, or any part thereof, or carried you or any of your substitutes at any less than the regular fares, or whether you ever knew him to pass anybody free or carry anybody at less than the regular fares? ANS. He never passed me free over said road or any of my substitutes at less than the regular fares. I have known Mr. Clough to omit to take up tickets some times. I never did but once. I don't know the persons name whose ticket he did not take.

INT. 21. Will you make further search for your memorandum for 1864, and if you can find it will you produce it? ANS. Yes, sir.

JACOB F. SMITH.

## DEPOSITION OF JOSEPH K. COLBY.

INT. 1. (*By counsel for plaintiff.*) Please state your age, residence, occupation, and what it has been for the past few years. ANS. I was 25 years of age the 18th of last October. I reside in Concord, but do business in Boston. I am in the wholesale dry goods business. I have been in the same business since 1859; and my residence during that time has been as stated above.

INT. 2. What is your father's name, residence and occupation? ANS. Timothy Colby. He is a house carpenter by trade. He resides in Concord.

INT. 3. How often have you passed over the railroad between Boston and Concord during the last five years, and which line of road? State as nearly as you can. ANS. Sometimes I have once a week for two or three weeks in succession. And I didn't come once for about six months, —between five and six months. I have been west and been gone from four to five weeks. I should not think the average would be more than once in three weeks. For the first two years I came to Concord three or four times a year. I used to come by way of Lawrence. The last year I have come by way of Lowell. Lawrence three or four times.

INT. 4. State whether or not you paid your fares over the Concord or Manchester and Lawrence Railroad. If so to whom you paid them, or if passed over said roads free, by whom you were passed? ANS. I have bought tickets over the Concord and Manchester and Lawrence Railroads, and have had tickets sent me to come to New Hampshire and vote, sometimes. I have bought tickets at the Boston and Maine ticket office. I have passed over the Concord and Manchester and Lawrence Railroads free, sometimes. George Clough has run that road. I have been up with him and with Mr. Eaton. I don't know whether it was up or down. They, Clough and Eaton, have passed me free.

INT. 5. How many times has George Clough passed you free over the Concord, Manchester and Lawrence roads, or any portion thereof, during the last five years? State as nearly as you can. ANS. I could not tell how many times, I have kept no account. I have been over the road perhaps once a week for a month or two, and then again might not have been for a month. I should not think it would average more than once in three weeks, each way. Some of the time I have had tickets, and sometimes not. I have had tickets sent me sometimes just before election.

INT. 6. Please answer directly, how many times in your best judgement George Clough has passed you free over the Concord, Manchester and Lawrence Railroads, or any portion thereof, during the last five years? ANS. That is a question which is hard to answer, as I have kept no account. Sometimes I have not seen George Clough and sometimes I have had tickets. It would be all guesswork if I should say. You might call it fifty times, as near as I could say. But I can't swear to it.

INT. 7. Upon your oath do you state that fifty times is the nearest estimate of the number of times George Clough has passed you free over the Concord, Manchester and Lawrence Railroads, or any portion thereof, during the last five years? ANS. I do so state.

INT. 8. Did you ever pay George Clough a fare over the Concord, Manchester and Lawrence Railroads, or any portion thereof? ANS. I don't remember that I ever did.

INT. 9. Did you ever ride upon a train over said railroads or any portion thereof conducted by George Clough, when you were not passed free? If so, state how recently and when, and how often. ANS. I have. I have last fall and this winter, when I had tickets. I could not say how often. I don't know.

INT. 10. Where did you get your tickets over the Concord, Manchester and Lawrence Railroads, alluded to in your last answer? ANS. Starkey gave me some tickets once, several of them, and I had tickets sent me to come to Concord to vote, and I have bought tickets. Starkey, the brakeman, gave me some tickets. I don't know whether it was those I used, or whether I bought them.

INT. 11. Will you state upon your oath that you bought a ticket over the Concord, Manchester and Lawrence Railroads, or any portion thereof, during the year 1865? If yes, state how often, of whom and when you bought such tickets, as near as you are able. ANS. I don't know, and could not swear that I bought a ticket over said roads in 1865.

INT. 12. How many times, in your best estimate, on your oath, did you pass over said roads in 1865? ANS. I don't know and could not guess how many times. You might say thirty times. It might have been more, and might have been less.

INT. 13. How many tickets, if any, has George Clough directly or indirectly given to you or provided you with during the last five years? ANS. I don't remember that he ever gave me a ticket except once from Lawrence to Boston, which was a Boston and Maine check. I don't remember that he ever gave me a ticket directly or indirectly over the Concord, Manchester and Lawrence Railroads.

INT. 14. Do you mean by your last answer to be understood upon your oath that George Clough has not directly or indirectly provided you with any tickets over any portion of the Concord, Manchester and Lawrence Railroads, within the last five years. ANS. Not to my knowledge. I don't know that he has.

INT. 15. Have you ever had any conversation with George Clough with regard to tickets furnished you by Starkey? ANS. Never.

INT. 16. Have tickets been furnished you by any one else than Starkey? If so, by whom? ANS. No tickets have been furnished me. I have had passes. Mr. Henry Eaton or James P. Eaton might have given me tickets sometime ago. Not recently.

INT. 17. When did you receive the tickets from Starkey? How many did he give you? What was his occupation at the time, and upon whose train as conductor was he running as brakeman? ANS. I don't remember how long ago it was. I don't remember the quantity. He was running as brakeman at the time, with George Clough's train.

INT. 18. State as nearly as you can when it was, how many tickets you had of him, and between what points the tickets extended? ANS. It might have been a year and a-half ago. I might possibly have had ten tickets. The tickets were between Concord and Manchester and Manchester and Nashua, I think. I don't think he has given me tickets more than three or four times, and not over one or two at a time, except the time when he gave me the ten spoken of.

INT. 19. State how it was about the tickets furnished you by Starkey, being tickets which had been once used, or whether they were new tickets and never before used? ANS. The tickets never had been punched. They might have been used once, but not punched.

INT. 20. Describe said tickets as nearly as you can? ANS. I don't know that they were anything more than common tickets, signed by G. G. Sanborn. Good for this day only. Think they were Concord Railroad tickets, and not coupon tickets from other roads. Think they were not dated so far as I noticed.

INT. 21. What did you do with those tickets? ANS. I used them on the road.

INT. 22. How happened Starkey to furnish you with these tickets, and what consideration, if any, did he receive therefor? ANS. I don't know, only he was a friend of mine, and gave me the tickets. I never gave him anything in return to my knowledge. I sold him some goods once and hain't got my pay. I don't know whether he expects me to allow it on that account or not; there was never anything said about it. It was a year ago last summer, I think.

INT. 23. State all the conversation, negotiation, and all the circumstances you are able to state pertaining to the furnishing of tickets by Starkey to you? ANS. At the time he gave me the eight or ten tickets I was going down on the train, and was in the baggage car. I did not agree to give him anything in return. I don't remember that we had any conversation at all, only he took the tickets out and gave them to me.

INT. 24. How many tickets in all, according to your best estimate, has Mr. Starkey given to you? ANS. It might have been sixteen.

INT. 25. Has ever George Clough passed you or obtained tickets for you over any of the roads connecting with the Concord road? If yes, to what extent, and when? ANS. Yes, he has given me his own passes from Boston to White River Junction and return; and also over the Boston, Concord and Montreal Railroad to Littleton and return. Between Boston and White River Junction, he might have given me two, within the last six months. The one over the Montreal was in the last six months. I got a pass through him once for a young man to go from Boston to Hanover to vote. These were blank passes signed by him, and I filled them out. He has given me no other passes over these roads that I know of. Those are the only ones that I remember during the last five years.

INT. 26. Have you been in the habit, during the last five years, of paying fare over the roads between Concord and Boston, connecting with the Concord, Manchester and Lawrence roads? If not, through or from whom have you obtained passes or tickets over said roads? ANS. I have sometimes paid my fare over these roads, and sometimes I have had tickets or a pass. I have been over the Maine road quite a number of times with Starkey, when he has asked the conductor to pass me. Mr. James P. and Henry Eaton have given me tickets once in a while over the Boston and Maine road.

INT. 27. Have you paid any fare, either over the Boston and Maine road, between Lawrence and Boston, or the Boston and Lowell road, during the year 1865? If so, when and how often? ANS. I think I have. I cannot state when.

INT. 28. Has George Clough, directly or indirectly, within the last five years, furnished you with tickets or passes, or procured conductors or others to pass you over either the Boston and Maine or Boston and Lowell roads, or any part thereof? If yes, to what extent? State fully. ANS. Not to my knowledge.

INT. 29. Will you state upon your oath that he has not done so repeatedly? ANS. I will, that he has not to my knowledge.

INT. 30. Of what number does your father's family consist? ANS. My father, mother, brother and sister.

INT. 31. State how it has been with regard to your father's being in the employment of Mr. Clough, giving the details, so far as your knowledge extends? ANS. I hardly think he ever done a job for Mr. Clough until he built this house for him. I mean the house in which he now lives.

INT. 32. State how it has been with regard to your father's family passing free over the Concord, Manchester and Lawrence Railroads? ANS. I don't know anything about their passing free.

INT. 33. Do you state upon your oath that you never knew any member of your father's family to pass free over said roads or any portion thereof? ANS. I went once with my mother on the Portsmouth train from Concord to Mr. Eaton's house, about three miles below Manchester. We went free, I think. It was sometime ago. I don't remember of any other instance; at the same time there might have been. I know that my father, when he was building Mr. Eaton's house, had a pass to go and come when he pleased.

INT. 34. Please state if you have known any member of your father's family to pay fare over the Concord, Manchester and Lawrence Railroads, or any portion thereof, during the last five years, and prior to January, 1866? If yes, state when and where. ANS. I know that they have paid fare within five years. I can't say whether they have paid within five years prior to January 1, 1866.

INT. 35. Can you state an instance where any member of your father's family has paid fare over said roads, or any portion thereof, prior to the first day of January, 1866? If yes, please state it. ANS. I don't know that I could give the date. I think my father went down with some workmen and paid fare for the whole of them, or they all paid their own fare, I don't know which. I think it was while he was at work for Mr. Eaton, and before he got his pass. I remember no other instance.

INT. 36. Please state how many tickets were sent to you at any one time to come to Concord to vote? Who sent them and what was done with them, stating the different times? ANS. I don't know as more than one ticket at a time up and back was sent me. James Fellows, who was then a leading republican, gave me one or two at Boston, when he was down. It is my impression that Mr. Burnham, who is expressman in Cheney's Express, sent one down for me to come up to vote; and I got one free from some railroad employee to go back with.

INT. 37. Have you any knowledge of the procurement or purchase of tickets from the conductors or employees of the railroad, or persons not connected with the railroad, except as before stated in this deposition? ANS. I have no knowledge except as before stated.

INT. 38. What was the amount of your dealings with Mr. Starkey as above referred to? What goods have you let him have, and have you furnished any other conductor or employee of the railroad within the last five years with any goods or merchandise? And if so, what? ANS. I think the amount of that bill was somewhere about ten dollars. I believe one item was some linen handkerchiefs. There was something else. I don't remember when it was; think it was a year ago last summer or fall. I have furnished Mr. Clough with some goods, cotton goods, to furnish his new house, received my pay in greenbacks, at the regular rates or prices, to the amount of about $100. There might have been some others, years ago. Mrs. Eaton has bought goods of me in Boston. I don't remember the amount. It was small, not over $10. Got my pay in money. Never took any tickets in pay for anything, of anybody.

CROSS-EXAMINATION.

INT. 39. State whether or not you had any tickets at any of the times when you rode over the road, when Mr. Clough was conductor, prior to the last fall and winter, as stated in your answer to interrogatory 9? ANS. I don't remember as to that.

INT. 40. Have you any means whatever for now making any estimate of the times Mr. Clough has passed you free over the Concord, Manchester and Lawrence Railroad, or any portion thereof, as required in interrogatory six? ANS. I have not.

INT. 41. Is the number fifty, stated in the answer to said interrogatory, anything more than mere guess work? ANS. I should think it was all of fifty. It is just guess work. It might be more and might be less.

JOSEPH K COLBY.

## DEPOSITION OF HERMAN STRAUSS.

INT. 1. (*By counsel for plaintiff.*) State your age, residence and occupation. ANS. Age, 21; residence, Concord; occupation, merchant. Have been in Concord six years—that is, made it my home here. For the first three years I peddled. I was with Mr. Charles Morse about two years, and been in business for myself, at Morse's old stand, since last September.

INT. 2. Have you ever passed free, or been furnished with free passes or tickets over the Concord, Manchester and Lawrence Railroads, or connecting roads, or any portion thereof, to Boston? If yes, state all the circumstances fully. ANS. I have frequently been passed free from here to Nashua by Mr. Clough, and once in a while back from Nashua to Concord. I should think I had been passed from Concord to Nashua about eight times, and about six times from Nashua to Concord, within the last year. Mr. Noyes has passed me from Nashua to Concord about twice within a year. I was acquainted with Mr. Clough, and asked him every time if I could go down with him. He said I could. I asked Mr. Noyes to pass me, and he did so.

INT. 3. Have you ever purchased any tickets save at the ticket offices, or been furnished with any tickets or passes over the Concord, Manchester and Lawrence Railroads, or any part thereof, or been knowing to others being so furnished? ANS. I have purchased tickets very often. Frenchmen came into the store with tickets from here to Boston, which they told me they had purchased in Montreal through to Boston, and they stopped here, and sold me their tickets the rest of the way. That is, some of them told me so. I bought two of John P. Haven from here to Boston last August. They were regular tickets from Concord to Boston. I used the tickets myself, as I had occasion. Mr. Clough gave me, perhaps, three tickets from Concord to Boston—regular tickets. They were coupons from tickets above. They were given to me within six months I asked him for some tickets to go to Boston with, and he gave me these. Nobody else ever furnished me any tickets in any way, nor have I known of tickets furnished by or for anyone else. Mr. Clough has bought goods at my store. I sold them just as I sell to anyone else. I charged him the same profit I did other people.

INT. 4. Did you charge said Clough any less for goods furnished him for his family in consideration of his furnishing you free passes or passing you free? ANS. No, sir.

INT. 5. State what goods, and at what prices, you sold to George Clough, and the cost of such goods. ANS. I sold him clothing and furnishing goods. Paper A, annexed contains bill of items sold him up to January 1st, 1866. There may be three or four items since. The profit, as near as I can state, is stated against each item under the head of "profit" on "Schedule A," hereto annexed. He paid me the bill in full in cash.

SCHEDULE A.

| | | | | | PROFIT |
|---|---|---|---|---|---|
| 1865 | March | 27, | 1 umbrella, | $2.50 | .50 |
| | Oct. | 12, | Overalls and frocks, | 3.50 | .50 |
| | | | Cap, | 3.00 | .75 |
| | | | 3 collars and tie, | 2.50 | .62 |
| | | 24, | 1 shirt and drawers | 4.00 | 1.00 |
| | | | 1 pair gloves, | 2.50 | .50 |
| | | | 1 " | 1.00 | .25 |
| | Nov. | 3, | Overalls and frock, | 3.50 | .50 |
| | | | Overcoat, | 35.00 | 7.00 |
| | | | 1 pair drawers, | 3.00 | .75 |
| | | 22, | 1 pair gloves, | 1.00 | .25 |
| | | 30, | 1 pair drawers, | 2.00 | .37 |
| | | | Shirt, | 4.00 | .75 |
| | | | Scarf, | 1.50 | .37 |
| | Dec. | 27, | 2 pairs socks, | 1.50 | .33 |

INT. 6. Did you ever, in any way, directly or indirectly, allow said Clough anything for passes or tickets that he ever furnished you? ANS. No, sir.

INT. 7. Do you swear that you never furnished him goods at lower rates for any such reason? ANS. Yes, sir.

CROSS EXAMINATION.

INT. 8. State, as near as you can, how many free passes over the road, or any part of it, you have had when Mr. Clough has been conductor, within the last six years. ANS. About eight times from here to Nashua, and about six times back.

INT. 9. Were the three tickets which Mr. Clough let you have, as before stated, all of them coupons on tickets from the north? ANS. I think they were.

HERMAN STRAUSS.

## STOCKS IN CONCORD R. R. BOUGHT AND SOLD BY GEO. CLOUGH.

Being an abstract from the Treasurer's books.

|  |  | |  | SHS |
|---|---|---|---|---|
| 1846. | January | 30, | Enoch Plummer, | 12 |
| | February | 2, | J. W. Elkins, | 12 |
| | " | 10, | " | 16 |
| 1854. | Decemb. | 23, | Defau & Perkins, | 15 |
| 1855. | Sept. | 4, | Baldwin, | 10 |
| 1857. | January, | | Defau & Co., | 7 |
| | " | 19, | R. A. Richards, | 15 |
| | " | 19, | Cabot, | 3 |
| | July | 27, | Bailey, | 50 |
| | Sept. | 16, | C. Arnold, | 50 |
| 1858. | Feb. | 8, | M. Bowles, | 5 |
| | " | 8, | T. P. Shaw, | 25 |
| | " | 9, | Foster, | 20 |
| | March | 23, | Williams, | 7 |
| | " | 27, | Dodge, | 5 |
| | " | 29, | Gilman, | 20 |
| | " | 30, | C. Hill, | 4 |
| | August | 4, | J. L. Perley, | 20 |
| | Nov. | 12, | Ballard, | 1 |
| 1859. | Feb'y | 12, | E. S. Fowle, cashier, | 150 |
| | March | 18, | Defau & Co., | 35 |
| | " | 18, | C. Arnold, | 5 |
| | May | 7, | T. H. Perkins, | 7 |
| 1860. | March | 10, | S. P. Clark, | 20 |
| | " | 19, | Defau & Co., | 40 |
| | April | 12, | G. W. Joy, treasurer, | 4 |
| | " | 12, | Wilson, | 13 |
| | " | 6, | Richardson, | 10 |
| | " | 11, | Merrill, | 3 |
| | " | 26, | Defau & Co., | 10 |
| | May | 19, | Foster, | 10 |
| | " | 19, | Defau & Co., | 20 |
| 1861. | March | 19, | N. P. Lovering, treasurer, | 50 |
| | May | 24, | Clough, | 100 |
| | July | 12, | Wingate, cashier, | 75 |
| | " | 30, | " " | 50 |
| 1862. | January | 14, | Amoskeag Savings Bank, | 50 |
| | July | 26, | New Hampshire Savings Bank, | 50 |
| | Sept. | 30, | Union Bank, | 50 |
| | Nov. | 21, | J. A. Gilmore, | 60 |
| | " | 21, | Defau & Co., | 7 |
| | " | 25, | " | 6 |
| | " | 28, | " | 9 |
| | Nov. | 29, | " | 18 |
| 1864. | April | 18, | Amoskeag Bank, | 100 |
| 1867. | March | 9, | J. G. Lincoln, | 14 |
| | " | 9, | H. P. Rolfe, | 100 |
| | " | 28, | B. F. Martin, | 18 |
| | " | 19, | J. Henard's estate, | 100 |
| | " | 26, | L. Shaw, trustee, | 20 |

| | | | | |
|---|---|---|---|---|
| 1867. | March | 26, | L. Shaw, attorney, | 10 |
| | April, | | Perley Clough, | 100 |
| | May | 1, | James R. Hill, | 150 |
| | " | 20, | Natt Head, | 50 |
| 1868. | " | 20, | N. G. Ordway, | 6 |
| | October | 14, | C. F. Low, | 6 |
| | " | 31, | George Clough, | 100 |
| | Nov. | 2, | Merrimack River Savings Bank | 50 |
| | | | Total shares, | 1976 |

TRANSFERRED FOR HIM.

| | | | | |
|---|---|---|---|---|
| 1846. | February | 2, | Mechanics' Bank, | 12 |
| | " | 5, | " " | 12 |
| | October | 16, | C. A. Evans & Elkins, | 16 |
| 1858. | February | 1, | E. S. Towle, collateral, | 150 |
| 1859. | May | 11, | J. A. Gilmore, | 7 |
| 1860. | March | 1, | J. C. A. Wingate, collateral | 100 |
| | May | 14, | J. Kenrick, | 30 |
| | Sept. | 15, | Wingate, collateral, | 75 |
| | Dec. | 26, | N. P. Lovering, treasurer, | 50 |
| 1861. | January | 6, | J. Colby and others, | 50 |
| 1862. | " | 12, | Defau & Co., | 50 |
| | July | 12, | " | 50 |
| | Sept. | 29, | " | 50 |
| | October | 9, | " | 100 |
| 1863. | " | 12, | Wilbur, treasurer, collateral, | 100 |
| 1866. | May | 15, | Union Bank, collateral, | 150 |
| 1867. | March | 18, | Amoskeag Savings Bank, collateral, | 100 |
| | May | 1, | Merrimack River Savings Bank, collateral | 64 |
| | " | 3, | " " " " " | 100 |
| | June | 12, | C. H. Carpenter, collateral, | 64 |
| 1868. | January | 21, | " " | 30 |
| | October | 30. | Merrimack River Savings Bank, | 50 |
| | " | 30, | George Clough, | 100 |
| | Nov. | 2, | C. H. Bartlett, collateral, | 50 |
| | | | Balance, | 380 |
| | | | | 1976 |

SUMMARY.

| | |
|---|---|
| George Clough has | 380 |
| Amoskeag Savings Bank, | 100 |
| Merrimack River Savings Bank, | 200 |
| C. H. Carpenter, | 94 |
| C. H. Bartlett, | 50 |
| Union Bank, | 150 |
| | 974 |

The defendant offered to show in the trial the amount of money returned in the cars by John Le Bosquet and J. H. Chandler, the conductors who succeeded him on the same trains formerly run by the defendant, from January, 1864 to January, 1866, for the purpose of comparing the amounts returned by said conductors with the amounts returned by him, for the two years immediately anterior to his leaving the road; and offered the following tables, made up from returns made by him and said conductors to the general ticket office of the railroad. The last two years that defendant run are compared with the first two that the new conductors run:—

MEMORANDUM OF CONDUCTORS' WAY-BILLS.

| | Clough. | Noyes. | Clough. | Noyes. | J. B. Le Bosquet. | Chandler. | Alexander. | Chandler. |
|---|---|---|---|---|---|---|---|---|
| | **1864.** | **1864.** | **1865.** | **1865.** | **1866.** | **1866.** | **1867.** | **1867** |
| January, | $255 70 | $320 05 | $337 95 | $343 40 | $459 60* | $569 33* | $251 23 | $239 65 |
| February, | 303 65 | 262 75 | 254 00 | 275 60 | 326 22 | 389 61 | 226 83 | 234 72 |
| March, | 359 40 | 377 40 | 254 10 | 292 05 | 462 49† | 353 61† | 270 21 | 288 67 |
| April, | 395 20 | 402 45 | 277 35 | 230 30 | 346 00 | 299 29 | 291 79 | 271 97 |
| May, | 393 15 | 354 15 | 225 90 | 235 45 | 293 25 | 263 57 | 329 54 | 298 82 |
| June, | 367 80 | 410 50 | 253 60 | 258 60 | 347 07 | 287 92 | 349 73 | 284 40 |
| July, | 276 75 | 252 40 | 252 30 | 262 90 | 292 79 | 262 80 | 316 85 | 327 80 |
| August, | 334 30 | 357 40 | 277 90 | 290 75 | 355 27 | 316 39 | 314 48 | 285 37 |
| September, | 355 60 | 51 65 | 267 10 | 251 70 | 380 99 | 351 65 | 316 15 | 325 58 |
| October, | 251 95 | 262 05 | 274 45 | 297 65 | 288 70 | 356 67 | 270 75 | 317 88 |
| November, | 244 60 | 233 25 | 414 25 | 355 60 | 305 00 | 271 48 | 275 20 | 320 02 |
| December, | 343 60 | 292 20 | 422 25 | 368 20 | 290 50 | 319 27 | 330 55 | 250 87 |
| | $3,881 70 | $3,576 55 | $3,511 15 | $3,462 20 | $4,147 88 | $4,041 59 | $3,543 31 | $3445 75 |

* A portion of January was run by Messrs. Clough and Noyes.
† March 20, 1866, the ticket office was closed.

| | |
|---|---|
| Amount returned as taken in the cars by George Clough, during the year 1864, . . . . . . . . | $3,881 70 |
| Amount returned as taken in the cars by John Le Bosquet, during the year 1866, . . . . . . . | 4,147 88 |
| Balance in favor of Le Bosquet, . . . . . | $266 18 |
| Average per train, allowing seventy-eight trains per month, | 29 |
| Amount returned as taken in the cars by George Clough, during the year 1865, . . . . . . . . | $3,511 15 |
| Amount returned as taken in the cars by J. H. Chandler, during the year 1867, . . . . . . . | 3,543 31 |
| Balance in favor of Chandler, . . . . . | $32 16 |
| Average per train, as above, . . . . . . . | 03 |

## REFEREES' REPORT.

On the 5th day of January, 1869, the referees reported as follows:—

"That the plaintiff recover of the defendant the sum of five thousand six hundred and thirty-five dollars."

The referees find the following facts, and award as the court shall determine the law, viz:—

"The defendant, in certain instances, received of passengers paying in the cars less than the full fares, and in order to conceal this irregularity from the plaintiff, did not return the fares so received on the way-bills filed in the ticket office from day to day, and pay over the same to the ticket master, as by the rules of the plaintiff, of which he had knowledge, he was required to do.

"The defendant's testimony tended to show that, with the money so taken, he bought tickets at the ticket office, and after punching them so as to show that they had been used, returned them to the ticket master, with the tickets taken up in the regular course of business, and in this way the money all came to the plaintiff's possession. This was done with the knowledge and consent of Mr. Gilmore, the superintendent of the plaintiff's road, but with the understanding and agreement between him and the defendant that the whole transaction should be concealed from the plaintiff and its directors, and it was so concealed.

"The plaintiff, ascertaining that money received by the defendant of passengers paying their fare in the cars had not been returned on the way-bills and paid over to the ticket master, and having never had any knowledge or given any assent to any other mode of paying over, commenced this action in good faith, and thereby changed its position in a material matter.

"If the defendant is estopped from showing, by way of defence, that the money came into the plaintiff's possession as above mentioned, or if such facts being shown would not constitute a defence, then, and not otherwise, the referees further award that the plaintiff recover of the defendant the additional sum of five thousand five hundred and nine dollars.

"The plaintiff had contracts with other railroad corporations, by virtue of which such other corporations issued what they call joint tickets, entitling passengers to be carried on the plaintiff's road and the roads of other corporations.

"To these tickets coupons were attached for each road, and the conductor on each road took off the coupon belonging to his road when the holder passed over such road. It often happened that holders of such tickets from northern and western roads to Boston, over the plaintiff's road, left the trains before passing over the plaintiff's road, and sold the balance of their tickets to other persons, who used them.

"It was a part of the defendant's duty, as a servant and agent of the plaintiff, to sell joint tickets from Concord, and other stations on the plaintiff's road, to Boston and other places, and he was furnished with tickets for that purpose.

"The defendant, for his own profit, purchased such joint tickets from northern and western roads, from which the coupons prior to the plaintiff's had been taken, and sold them to passengers in the cars who would otherwise have bought the joint tickets of the plaintiff, entrusted to the defendant for sale.

"This was injurious to the plaintiff's business entrusted to the defendant, and deprived the plaintiff of the whole or some part of the profits which should have been received from the sale of its own tickets, the plaintiff deriving a larger profit from its own tickets than from those of other roads.

"This was done with the knowledge and consent of Mr. Gilmore, the superintendent of the plaintiff's road, but the plaintiff's directors and the plaintiff had no actual knowledge of it.

"If the plaintiff is entitled by law to recover in this action the profits made by the defendant in thus buying and selling the joint tickets, the referees further award that the plaintiff recover of the defendant the additional sum of two thousand dollars.

"The referees further award that the plaintiff recover of the defendant his costs of reference, taxed at two hundred and eighty-one dollars and fifty-six cents, and referees' fees and expenses paid by the plaintiff, amounting to one thousand nine hundred and ninety dollars, and costs of court to be taxed by the court.

"The referees append to their report, and make a part of it, a statement of the exceptions taken by the defendant.

EDMUND L. CUSHING,<br>WILLIAM HAILE,<br>HENRY A. BELLOWS, } Referees."

# INDEX.

The following testimony of Martin V. B. Edgerly, of Manchester, N. H., was accidentally omitted in printing the case, for the reason that it did not appear in the notes of the reporter, which were furnished the printer as copy.

It was introduced before the referees on the forenoon of December 17, 1868, when the hearing was resumed after a long adjournment, before the reporter had arrived in town, and consequently it was not reported by him with the other evidence.

We give it from the notes of Judge Bellows, one of the referees, as taken by him at the time.

## TESTIMONY OF M. V. B. EDGERLY.

Am in life insurance business for ten years; removed to Manchester in 1863; before in Pittsfield, and had an office in Manchester; had frequent occasion to pass on Concord Railroad; knew Mr. Clough; had occasion to buy tickets and coupons of him; can't say how many, but both tickets of the Concord road and coupons. I rode mostly between Manchester and Concord; occasionally outside. I may have paid $3 or $5 at one time, and had tickets at half price, as I supposed; can't say as exactly half fare. I bought one or two years, and perhaps longer. Can't say as the tickets had been once used; don't remember whether they had been once stamped. In 1861–2–3 I went home every Friday and returned on Monday. I purchased of him several years; can't say to what time. At first, when he passed me at half fare, he gave me a return ticket, I paying in the cars. At first I paid full fare. Did not know Clough at first.

### CROSS EXAMINATION.

I was introduced to Mr. Clough by True Garland, in the cars. I paid full fare for three or six months, and ten cents extra when I had not got a ticket. I went with him after a while at half fare, Clough saying that he had made arrangements to let me go. This was some months after I was introduced to him. I rode at half fare for two or three or four years.

CONCORD RAILROAD CORPORATION v. GEORGE CLOUGH.

# CLOSING ARGUMENT

OF

# HON. MASON W. TAPPAN,

COUNSEL FOR DEFENDANT,

DELIVERED

## BEFORE THE BOARD OF REFEREES,

AT CONCORD, N. H., JANUARY 1, 1869.

CONCORD:
THE PEOPLE STEAM PRESS, STATE BLOCK.
1869.

# ARGUMENT.

---

Mr. Chairman and Gentlemen:

I feel very much the weight of responsibility which devolves upon me, on this occasion, and I would gladly have had it fallen upon other shoulders. I know how important the case is to my client; I know how much is depending upon the result, both to him and all connected with it; I am conscious, also, that even with my best endeavors, there will probably be many imperfections and omissions; and I fully appreciate the fact that I am before a tribunal competent to weigh all the evidence that has been put into the case, and to detect any mis-statement or fallacy of argument. I shall endeavor to argue the testimony fairly, and to present our view of the case, on the facts and evidence before you, with such ability as I possess. I am aware, gentlemen, that many points have been fully argued and discussed from day to day, as we have gone along, and I shall therefore endeavor to be as brief as possible in this closing appeal which I make in behalf of my client.

Permit me, at the outset, to congratulate you sincerely, gentlemen, that this protracted trial has at last reached a stage when we can at least see the beginning of the end.

But there is no one connected with these proceedings that I can so much congratulate as my client, Mr. Clough. For three long years, gentlemen, he has stood here, a target for all the abuse and detraction and suspicion that these proceedings have brought upon him; but at last the hour has come when this agony of suspense will be ended, and I hope it is not too much to add, Mr. Chairman, that the hour has come for his complete and entire vindication. God alone knows, or can know the suffering, the mortification, the anxiety that have been inflicted upon him and his family for this long space of time, growing out of proceedings which, I think the evidence shows, were commenced and have been prosecuted for other motives than those which appear merely on the face. I am sorry to be obliged to say this, but I believe the evidence bears me out in the assertion.

And it is proper, gentlemen, I think, for you to look a little at the surroundings of this case. We certainly desire to have the motives of this proceeding inquired into, and I think we have a right to ask of you to look at this case in the light of some of its surroundings.

Now if these proceedings were commenced with honesty of purpose; if they were commenced in good faith for the purpose of having an investigation into the conduct of Mr. Clough, and the other conductors; if they were commenced deliberately with this purpose, why, gentlemen, that is one thing; if, on the other hand, it is shown that these suits and this investigation were commenced hastily, inconsiderately, without any proper deliberation, but from motives such as have appeared in this case, I think these facts and considerations ought to have weight before this tribunal or any other.

Well, gentlemen, what was the initiation of these measures?

The evidence discloses that the first that was heard of any movement against the conductors was the declaration of Mr. Gilmore, before the board of directors, that the conductors on the Concord Railroad had been stealing fifty thousand dollars a year in the cars. Whereupon Gen. Butler was applied to, and Maj. Carney was employed and put upon the road, and other detectives were put upon the track of the conductors, Mr. Clough included.

But, gentlemen, it appears from the evidence in the case that Mr. Gilmore afterwards endeavored, as the gentleman on the other side says, to do what he could to squelch these proceedings and to stifle the investigation. He did this because he found that his own proceedings in connection with the road would be dragged to light, and this charge against the conductors, which he had made to cover his own misdoings, would be the very means of his exposure. It also appears that Mr. Gilmore had other motives than those of a sincere desire to bring Mr. George Clough and the other conductors to justice, and other motives that induced him to make this declaration.

It appears that for some reason or other Mr. Clough had refused to sign his notes; and Mr. Gilmore said,—"I will learn George Clough to go back on me, and refuse to sign my notes when I am lying here sick." And it seems to me, gentlemen, that the whole proceedings have been conducted, from beginning to end, and especially at the outset, with something of the spirit that is manifested in this declaration of Mr. Gilmore. I think the whole prosecution—I may be allowed to say—has evinced, in its carrying on, a bitterness of spirit, an uncharitableness—and, I had almost said, a malignity—such as I never saw evinced in any other case. I do not wish to impute wrong motives to anybody. I do not wish to charge my brother, or to charge Mr. Gilmore,—who has been summoned before another and a higher tribunal—with improper motives. I know the old maxim is—"Tread lightly on the ashes of the dead." I do not propose to disturb, ruthlessly, the ashes of Gov. Gilmore; I do not propose to say anything against him or about him except what the evidence in this case warrants; I do not propose here, gentlemen, to enter into a harsh denunciation of the course pursued by anybody; but I do not intend to spare a word that I think to be necessary to vindicate the cause of Mr. Clough before you, whether that word shall reflect upon the living or the dead.

Now, then, gentlemen, I have said that it seems to me that these proceedings were instituted from some other motives than those merely of a sincere belief that these men had been stealing and plundering in the cars to the amount of fifty thousand dollars a year, or to any other amount. And I believe it. I believe the facts and the evidence in this case show it. The evidence discloses that this whole concern of the Concord Railroad was grossly mismanaged, and was reeking with corruption in every department. And there were omnious mutterings everywhere,—by the public, before the legislature, and through the press. The stockholders were taking the alarm, and everywhere the atmosphere was freighted with rumors of mismanagement and corruption in the affairs of this road. And it appears, gentlemen, that there was an annual meeting to come off at no distant day. Somebody had got to be elected directors and clerk and president again. And you will see that the initiation of these proceedings, and the dragging them before the annual meeting, disclosed the fact that it was the intention of somebody, if possible, to raise a smoke and a smudge, and, by prosecuting these conductors upon some imaginary charges, to screen themselves from the guilt which justly attached to themselves. This is the way it looks to me, and this, gentlemen, is what I believe. As I said, I am not going to impute improper motives to anybody any further than it seems to me the evidence warrants. My brother on the other side has seemed very sensitive in regard to this case, from the very day that the trial opened down to the last moment of the closing of the testimony. And I must say this: that the gentleman has seemed bent upon putting himself upon trial in the matter. I have not put him upon trial. The counsel associated with me have not put him upon trial. But, for some reason or other, it has seemed to me that he has felt the necessity of relieving himself from improper connection with this case, and the proceedings under it. In his opening argument, before any charges or insinuations had been made against him by anybody, he made the declaration to you—I quote his language—that "Gov. Gilmore had got him into this scrape, and after making these declarations before the board of directors, and after directing the prosecution, had turned around," as the gentleman says, "and tried to ruin *him*."

Well, gentlemen, we have nothing to do with that; and the fact as to whether Col. George is injured or not injured, ruined or not ruined, is a matter of no particular concern, perhaps, in this case. But it has something to do with the motives that have impelled this prosecution. It does disclose something of the *animus* that has pervaded it all the way through.

I fully appreciate, and commiserate, what it seems to me I may characterize as at least the rather unfortunate position of my brother on the other side. Gov. Gilmore, by whom, through evil and good report, the gentleman stood as counsel and shield and adviser through his entire connection with the Concord Railroad, has passed to his long home. The old board of directors, with whom he used to meet, at Parker's, and at the American, and at the Revere House, where they used to have such jolly good times, and where the bills for liquors and cigars alone were, one year, fifteen hundred dollars, have all been swept away by the fiat of the stockholders, and the places that once knew them shall know them no more forever; and when the new board came into power, they found, gentlemen, this elephant on their hands. It was no elephant of theirs; and I don't think, from the manner that they have given this case a pretty wide berth on this trial, that they care to have much to do with it. So that the animal has been left pretty much in the care of my brother George, to feed and manage and get along with the best way he could. There has been rarely any one here on this trial representing the Concord Railroad. The gentleman has once in a while shown us the rubicund face of Mr. Weld, of Boston. He has occasionally been trotted in here, and would stay perhaps ten or fifteen minutes, and then, presto! he is gone. So that the whole thing seems to me to show that this is, after all, about as much a trial between John H. George and Mr. Clough as between the Concord Railroad and George Clough.

Now, I say, gentlemen, I am not going to impute motives to anybody; but it seems to me, and I think I am warranted in saying — I know that the gentleman on the other side is enthusiastic, I know he is zealous—but it seems to me that there has been a bitterness and a vindictiveness imparted to the proceedings in this case, and in the manner in which George Clough has been pursued, such as nothing that has appeared in the testimony would at all warrant. And this much I feel it my duty to say.

But gentlemen, I want to allude to another matter here, because I propose to say at the outset, before going to the testimony, all that I propose to say at all in reference to these extraneous matters.

Now, gentlemen, it has been said and charged and reiterated, over and over again, that there has been a combination, on the part of the conductors and somebody else, for a new administration of the Concord Railroad, and to stifle investigation and to stop this suit. I assert, gentlemen, that there is no evidence of anything of the kind.

I do not deny, gentlemen, the feelings that these conductors would be likely to have, after these proceedings were commenced from the motives in which they believed they had their origin. I do not deny that they had feelings in regard to it. I do not deny that they did everything in their power, and that Mr. Clough did everything in his power, that he could legitimately and honorably do, to have the old board of directors turned out, and a new and better and honester and more capable board put in. And if the clerk went by the board, with the board of directors and superintendent, nobody is to blame but himself. He helped to raise the wind, and they all reaped the whirlwind. Or, in other words, to use a quotation which I once heard the learned chairman of this board quote in an argument at the Sullivan county bar:

"He digged a pit;
He digged it deep;
He digged it for his brother;
For this his sin
He did fall in
The pit he dug for t'other."

That is all there is to this charge, so far as this is concerned. The proceedings were commenced, as I think, with the view and the hope and the expectation that by this extraordinary zeal and great spasm of virtue, and by parading through the press and before the public and the stockholders their great and strenuous efforts to ferret out these monstrous frauds of the conductors, they themselves would be retained in power another year; and the thing would go on in its old tracks and its old grooves in the same way it had gone on before. And, gentlemen, if the conductors have done nothing else,—if it be through their instrumentality that reforms in the Concord Railroad have been brought about—if it shall turn out that they were instrumental in relieving the corporation from its former management, and putting it into the hands of new men, of whom nobody complains, I think, instead of being pursued here for fifty or a hundred thousand dollars, when no cent can be proved against them, they are entitled to receive from the corporation fifty thousand dollars for their service; and I think the public will so regard it. Gentlemen, do you believe this charge of fraudulent combination to stifle this investigation—do you believe that the men who at present constitute the board of directors of the Concord Railroad ever entered into any matter of that kind? Do you believe that such a man as Judge Minot is capable of it? Do you believe that Nathan Parker, or J. Stephens Abbott, or the others, are capable of it? Do you believe, gentlemen, that the members of the new board went into a combination, or that there was ever any expectation that anything that was done at the annual meeting would have any reference whatever to the prosecution of these suits? As I have said, the conductors, in conjunction with the stockholders, or a large majority of the stockholders, were anxious for a change. The public clamored for it. And they got it. And this very prosecution that was got up for the purpose of screening themselves was the very act that finally carried them all away with the besom of destruction.

Now, gentlemen, something has been said in reference to an attempt to stifle an investigation before the legislature. The evidence in regard to that matter has come out before you. Mr. Clough had no counsel before the legislature at all for himself. The investigation, as I understand it, had nothing whatever to do with these suits. The only question there—and I was counsel for the road, with others—the only question there was whether after measures had been taken by the new board of directors to investigate the affairs between the corporation and Mr. Gilmore, and all the business of the corporation, whether the legislature, at the instigation of anybody, should step in and take it out of their hands. And it was on that ground, and on that ground alone, that the investigation was postponed, in order that the new directors might go on with it themselves. And they have gone on with it, gentlemen; they have gone on with it thoroughly; and have probed the affairs of the Concord Railroad to the bottom, and spread the thing before you and before the public, and shown, in their last report, the corruption and rottenness which existed there; and they have demonstrated that while the superintendent was charging the conductors with stealing in the cars, of which there has been no proof, he was himself a defaulter to the amount of more than fifty thousand dollars!

Well, gentlemen, so much for this matter of the charge which has been so frequently made here, that somebody had been trying to shirk an investigation. Why, gentlemen, after the proceeding that had been had at that annual meeting, the new board of directors could not stifle this proceeding if they would, or if we had wanted them to. On the other hand, there never has been one moment from the hour when this prosecution first started—from the hour when this suit was first commenced against George Clough—there never has been one moment when he would have had it stifled, or would have had it suppressed. He could not afford to have it, gentlemen. I appeal to one of the honorable members of this board, who knows George Clough, if, up to the time these proceedings were commenced, he did not sustain, in this community, and everywhere that he was known, the character of an honest and upright and good citizen. I believe that his reputation stood as fair among those who knew him best, and who had occasion to know all about his business transactions, as any man in the city of Concord. And I say, when these charges were trumpeted forth—when these charges were bruited through the press, and this man and that man was button-holed at the corners of the streets and shown these little papers of computations and calculations and comparisons, to prove that Mr. Clough had been a plunderer for these long years that he had been on the road—I say that George Clough could not afford to have this investigation suppressed. And never for a moment has the idea been entertained that they could be suppressed in any way, except before a tribunal competent to probe this thing from beginning to end. And, gentlemen, we understand that we are before that tribunal. We understand, too—and we knew full well when we agreed to submit it to this honorable board—that an adverse decision would fall with greater weight than if from a hap-hazard trial before a jury.

And, on the other hand, we know that if we are vindicated, and if the charges are proved to have been false, (as I believe they have, every one of them,) a verdict from such a tribunal—a vindication from such a source—would come with all the more strength in our favor. Aye, gentlemen, there is another thing. Talk about George Clough wanting to suppress this investigation! On the other hand, in all this proceeding, he has courted the utmost publicity; he has courted the most searching investigation into all his affairs; his whole life has been laid open before you, from the time that he started from his father's roof, a poor boy, with all his worldly effects done up in a pocket handkerchief, down to this very moment. Every business transaction, everything that he has done, every dollar that he has made, and where he has made it, the way that he has done his business, have all been laid open before you. Every cent's worth of property—aye, gentlemen, the service of silver plate that he won by getting up subscriptions to a newspaper, has been paraded before you; and even the silver spoons that came down to his wife from her mother—everything has been inquired into. And we have been willing that it should be so. And I submit that the testimony and deposition of Mr. Clough show that he has been most anxious to tell the truth, the whole truth and nothing but the truth, and to keep nothing back in reference to his whole affairs. And I say that this ought to weigh something in his favor. I assert that on the stand here, and when for those long weeks he was under examination, when his deposition was taken, that every answer that he has given shows that Mr. Clough—let it cut where it might, no matter whether it is made against him or for him—Mr. Clough has told the truth in regard to everything. Yes, gentlemen; more than this. The counsel on the other side has sought to vindicate himself from any improper motives in prosecuting this suit, or any improper connection with it, or any improper feelings in regard to it. I have this to complain of in regard to that. I say that, at least after these proceedings were commenced, after suit was brought, it was a little in bad taste—to use no harsher expression—to follow this man in the way he has been followed, through the press, in the stockholders' meetings, and on the corners of the streets, and everywhere. That is what I complain of, gentlemen. If counsel were obliged to be present at the stockholders' meeting, for instance; or if counsel were obliged to be present at the state house—it was because, gentlemen, they did not want Mr. Clough's case prejudiced in advance of a trial. They did not want it tried by outside pressure.

And this leads me to say what I was going to say in regard to another matter, that Mr. Clough has carefully provided for. These suits, these charges were rung throughout the length and breadth of the land. He could not walk these streets without feeling that the citizens with whom he came in contact every day regarded him as a plunderer; his family could not feel that the house they lived in, and the clothes that his wife and his children wore on their backs would be regarded as their own; but that somebody would say: "All this has been stolen from the Concord Railroad."

And so, gentlemen, we have been obliged to meet all these prejudices, and to repel them as far as we might. And Mr. Clough has been here at large expense on this trial, to have the whole proceedings taken down. Every word that has been said, every particle of proof that has been put into this case, whether it bears for him or against him, has been faithfully reported, and is to be printed and circulated, no matter what your decision may be. If he shall be convicted — if, in your judgment, the charges are sustained, it shall at least appear, gentlemen, upon what evidence that decision rests. And if, on the other hand, it shall appear that Mr. Clough has been fully vindicated in your minds, then we want to show to the world — we want to show to those who have heard these charges, which have been rung from one end of the country to the other — we want to show to everybody, by a faithful report of this case, just the miserable farce that has been played, from beginning to end, in regard to the whole matter.

Well, gentlemen, what was the first step in the proceedings in this case?

Mr. Gilmore makes his charge before the board of directors. Whereupon — I say, with an indecent haste — Gen. Butler was employed as counsel. And Gen. Butler recommended the redoubtable George J. Carney, "Assistant Quartermaster, and afterwards Quartermaster in the army of the James," to come here, with his associates, and to go on to this road with his detectives.

Now, gentlemen, I want to say this: If there had been, as it seems to me, this purpose to have a quiet investigation,; if there was doubt in the minds of the directors of the Concord Railroad in regard to the honesty of George Clough, in how much more quiet and proper and effectual a manner, as it seems to me, could all these proceedings have been carried forward. How much easier it would have been — and it seems to me, gentlemen, that this course must have already been taken, and I think you must infer and set that fact down in favor of the defendant, that it has been pursued by the gentleman on the other side,—to put somebody (some one or more citizens of New Hampshire who would not be suspected), at any one or more of the stations on this road, running through some weeks or months,—into these cars, and quietly pay money, and see if they could "spot" or detect these conductors. I think that would have been the way that would suggest itself to anybody that did not want to make a great sensation and a great splosh. I don't know but it has been done. If it has been done, then there is no evidence of it before you. It would be the most natural thing to be hit upon by men who wanted quietly to ascertain whether these men were honest or not. But instead of that, we have Major Carney, and we have "Surgeon-General" Carney, and we have Mr. Draper, and Mr. Batchelder, and other detectives, put upon the track of these men for two or three months. We have them here under false and lying pretenses of carrying on a large "claim agency," traveling daily over the road, and paying some fifteen hundred dollars in money in the cars, but no farthing of which has been found sticking to the fingers or traced to the pockets of George Clough.

Well, gentlemen, I am not going to comment on their testimony now. I shall feel constrained to say a word or two about it before I get through, but I will only allude to it in passing. But, gentlemen, it was some little time (and I have no wonder that it was,) before these suits were brought, after Major Carney's reports were made. I do not wonder that no suits were instituted on the strength of that report. And I don't think that you will wonder at all after having these reports before you.

But it was the "WHITCHER TICKETS," gentlemen, which finally produced these prosecutions, or caused them to be brought. To use again the expression of the counsel on the other side, in the course of this trial, they "accidentally dropped a hook into this hole, where they didn't expect to get anything at all, or at most only a nibble, and they brought out a four-pounder!" And they made a great sensation over it! Two of the best citizens of Manchester were arrested as accomplices in this crime, but were soon discharged. Jim Whitcher was arrested. Extra trains were run on Sunday. The whole world was in a ferment. Sensation despatches were written (mostly by the counsel on the other side,) and sent off to the Boston Journal; and everything was done to make this appear to be one of the most atrocious and (to quote the language of the gentleman himself,) the "most astounding" peculation that ever was heard of in the annals of crime! And the sensation was kept up; and all, gentleman, on account of the finding of these three hundred and forty-two tickets on the person of Jim Whitcher. Why, as was well said by my brother Rolfe in the opening, if they had not been intent upon making this great noise, and keeping up this sensation for some ulterior purpose, ten minutes examination of these tickets would have shown them that they could not have been obtained from George Clough, not one of them. I undertake to say, the evidence here before you will show that not one of these tickets probably came from George Clough at all. On a part of them the dates show that they could not have come from him; and if they had made even the most cursory examination of the tickets, when they were discovered, they would have found that in all cases they must have come from other sources than George Clough.

Why, gentlemen, as I said before (and I now come to this matter of the Whitcher tickets)—I say, upon careful analyzation of this testimony there is not one of those Whitcher tickets that could have been obtained from George Clough, or that Whitcher ever got from him in any way. Isn't it a little remarkable, Mr. Chairman and gentlemen, that the witness most competent to give proof, if it exists,—is it not a little remarkable, gentlemen, that the man of all others who knows most about it, and is supposed to know most about it, Mr. Whitcher himself, is not put upon this stand at all? He is the man that occasioned all this uproar. The reports of the detectives amounted to nothing, for nothing was discovered. These prosecutions had their inception right here with these Whitcher tickets. Whitcher knew where he got them. They could have produced him; and yet they didn't put him before you at all. Do they say, on the other hand, that we might have brought him? Why should we? We have had Lanes and Curtises enough on this stand; and it is no business of ours to go fishing around in their dirty pools, when there is no evidence against us. It is incumbent on them to produce Mr. Whitcher; and they have not produced him. And it is because they knew that Mr. Whitcher, whatever he may be, would not have the hardihood to stand here before you and say he got these tickets of George Clough. Why, gentlemen, if there were no other sources from which these tickets could come, then there would be some suspicion that the conductors, or that Mr. Clough might have let Whitcher have them. Here these tickets were flying about everywhere. Everybody had them. Tickets were purchased in Concord. Tickets were purchased in Manchester by White and Weston and others—these coupon tickets

Gilmore was having tickets by the hundred. Other men were going to the ticket office, and getting tickets whenever they wanted them. Starkey was on the cars taking up tickets. And there was a hundred ways that tickets could get into the hands of James Whitcher, without presuming that every one of these were furnished him by George Clough. And here I desire to call your attention to the testimony—because I want to have it known just how this matter stood in regard to this question of tickets—I want to refer to the testimony of Mr. Sanborn, the ticket master. And he swears that Whitcher used to go to the ticket office and get tickets; that Natt Head used to get tickets there; that Joseph Goss used to get tickets there; and that Tom Wattles used to get tickets there—all of them living in Hooksett, where Whitcher lived; and he says that Captain Harrington and others used to get tickets there. And he swears to you also that Whitcher was one of Gilmore's fuglers, one of his men that he relied upon to do anything that he wanted done in Hooksett and vicinity, and who had just as many tickets as he wanted. This is what he says:—"*He, (Gilmore,) would come into the office and get tickets, and give me orders not to make any record of them; tickets were taken out of my package and no record made of them; one year three or four hundred. Whitcher said he had so many men, whom he was keeping for town meeting, and wanted some tickets for them. I should think the days when men didn't come there to get tickets, for four weeks prior to an election, were the exceptional days.*" And Mr. Sanborn says, when these men came for tickets, he would deal them out in numbers from one to fifty. And he says further:—"*I won't say that I have not dealt out to Gilmore a hundred at a time; there has been no year but that Gilmore would come and get tickets to give or send away. I recollect one time, when Gilmore was a candidate, carrying in twenty-four or twenty-five tickets to him, from Manchester to Nashua; I was directed to make no account; he repeatedly ordered me to keep tickets out; it was very common to get tickets both before and after he was governor; it would average three or four times a week; they went to all stations; I took some of these tickets back again; I took them up and made no account of them.*" This is the testimony of Mr. Sanborn, the ticket master on the Concord Railroad, their own witness. And the report of the directors, which is in this case, (the last report,) shows that Mr. Gilmore was indebted to the road one thousand dollars *for tickets!* Now wasn't this matter known to Mr. Gilmore and to Col. George? They took the initiatory steps to have these suits brought, and brought too, not on the report made by "Major Carney of the Army of the James," or of "Surgeon General Carney of the Providence Accident Company," or of any other detectives—for they *detected* nothing—*but because of the accidental finding of these tickets* on the *person of James Whitcher!* Didn't they understand all this? Didn't they know that these hundreds and thousands of tickets were flying about everywhere like autumnal leaves? And yet, for some motive or another, either to plume themselves upon some investigation that they were to make, or some great saving which they were going to make to the road, for the purpose of subserving their own interests and ends, and getting a new election, they charge this whole matter upon George Clough without ever looking to see whether these tickets were dated before or after he went off from the road! They made hot haste to jump to the conclusion of his guilt without making the slightest examination! And my brother, in the exuberance of his fancy and the intensity of his zeal, exclaims that he has got a "four-pounder" where he didn't expect a nibble!

I think then that I have a right to characterize the proceedings in this case not only as hasty, but as vindictive and as malicious. And I feel and have felt all through this thing that there was some motive pervading this matter that I was entirely unable to account for. And I think the referees will be of opinion that there was some motive somewhere with somebody, to push this thing on, other than that of a sincere belief that this amount of money had been stolen in the cars, as they have charged.

And now, gentlemen, I want just to analyze these Whitcher tickets. And I will only detain you a very few moments. There are three hundred and forty-three of them in all. Of these three hundred and forty-three tickets, Mr. Sanborn, the ticket master, has sworn that *two hundred of them could not by any possibility—and nobody pretends here to-day that they could—have come from George Clough!* There was then only a hundred and forty-three of them that could by any possibility come from his hands at all. That is the testimony of Sanborn; that is the testimony of the ticket officers, which shows that barely a hundred and forty-three of the three hundred and forty-three could by any possibility have been obtained by Whitcher from George Clough. And such is the evidence.

Well then in regard to the one hundred and forty-three that possibly might have come from him. Mr. Clough left the road in January, 1866. Thirty-three of these tickets bear dates (which every man can read) after George Clough left the road. That disposes of thirty-three of them.

There were three tickets to Boston that could not have been obtained from Clough, for the reason that those tickets were joint tickets, and were charged to Mr. Clough, and he would have them to pay for; and it is not very likely that Mr. Jim Whitcher would get these from him or any other conductor. That disposes of three more. Sixteen of these one hundred and forty-three bear date of "February;" some in 1866; and some "February 5th" only, without the year being named. But it is perfectly apparent that the February 5th could not be February, 1865; for it appears from the almanac (of which you take judicial notice,) that February 5th, 1865, would be on *Sunday*. So that of these tickets of which I spoke as bearing date February 5th, the year could not have been 1865. It must, therefore, gentlemen, have been February, 1866, the February after Mr. Clough left the road. I say February, 1866, because although, Mr. Chairman, you will bear in mind that these tickets have the year 1865 printed on the end in that way, it does not by any means follow that these tickets were only used in the year 1865. On the contrary, there is one of these tickets (here it is) the date of which is very plain, and which bears the date of 1865 printed on the end, like the others, but on the back you will see that it is issued February 10th, 1866. The practice was, as the ticket master swears, to have a large quantity of tickets printed, use those tickets until they were gone, and then get a new batch; and when they were issued, the year was put on in which they were printed. So here were sixteen more of these tickets which were for February 5th, which have the year marked 1865 on the back side, which must have been dated February 5th, 1866, after Mr. Clough left the road; because, as I have said, if it was February, 1865, it would come on Sunday, and the cars did not run on that day. There are nine more of these tickets which are Montreal coupons; two dated February 9th, 1866; three dated February 5th, which was undoubtedly February, 1866. I believe the year does not appear on these tickets; but they are new tickets, as Mr. Sanborn thinks from the face. The tickets were very clean and were undoubtedly tickets that Starkey let him have. Five more of them are dated February merely, without any figure; new and clean, Mr. Sanborn says. And there is one dated February 27th, which it is evident must have been issued February, 1866. There are three old ones found at his house, neither of which, an examination of the tickets will show, Clough could have had, because they came from points above Concord. That disposes of all the tickets except seventy-two, which bear no date at all. Showing, therefore, that every one of these three hundred and forty-three tickets, with the exception of seventy-two tickets, could not possibly have been received from Mr. Clough. And of these seventy-two tickets, all I have to say in regard to them is, what Mr. Sanborn said himself, when he examined them, that they "are all new and clean, and there is nothing to show handling and nothing that would indicate that they were probably issued before February, 1866." They are all new and clean. I say, then, in regard to these seventy-two—looking at the testimony that we have from Starkey himself—that these tickets show on the face of them, that every single one of them were tickets that Whitcher received from some other source besides from George Clough. The tickets themselves prove that every one but seventy-two must have been received or got somewhere else than from Mr. Clough; and of these seventy-two, as I have said, the indications are, and the testimony of Mr. Sanborn is that they must have been issued after Mr. Clough left the road, after January, 1866; because they bear upon their face the fact that they are new and clean tickets; and there is nothing to indicate that they were not so issued, but on the other hand every presumption that they were issued after Clough left the road!

But, gentlemen, we have testimony that disposes of this whole matter. And that is the testimony, as you will recollect, put in by the evidence of Mr. Rolfe, of what Starkey himself stated in regard to this matter, and the circumstances under which he stated it. Starkey says that *he* let Whitcher have these tickets, and let him have them at two or three different times; that he let him have them when he was on the road as conductor; and also when, as brakeman, he picked them up in the cars before he went on the road as conductor. And Weston and White both swear that not a single one of the tickets which were found in their possession, did Mr. Clough let them have. They both swear that they never purchased one of Clough, nor that Starkey ever got one of Clough to pass them. And White says when he got tickets of Starkey he was "boss and all hands on the road." So that neither the Whitcher nor the White nor the Weston tickets ever came from George Clough. Besides all this testimony, gentlemen, which I have referred to on the subject of the Whitcher tickets, we have the testimony of Mr. Clough himself; and he swears to you that for a year and a half prior to the time he left the road, he never let Mr. Whitcher have any tickets at all; and he swears to you further that although he has let him have a few tickets, by direction of Gilmore and to give to poor people, not more than ten or fifteen during all the time that he has been on the road; he swears that he never let him have any tickets for more than a year and a half before he went off the road.

So, gentlemen, that disposes, I think, of the Whitcher tickets. And I may say, in dismissing this part of the case, that there is not the first particle of testimony from any quarter whatever, which shows or has any tendency to show that any single one of these Whitcher tickets ever came from the hands of George Clough.

And, gentlemen, it was upon this flimsy foundation, in the finding of these Whitcher tickets which came from Starkey, that Whitcher had picked up from the conductors everywhere, or stolen, or got, nobody knows when or how,—it was upon this flimsy foundation that this prosecution was commenced, and this great sensation inaugurated!

Well, gentlemen, I come now to the great charge which is made in this case. The gentleman said in his opening that all this talk about the report of the detectives and other small matters was of no account whatever. They only related to the subject of a few thousand dollars; that was of no account; they were small matters in comparison with the great subject that he was going to present to you. And he said that you would be perfectly "astounded" at the evidence. The "most astounding" and "astonishing" evidence was to be produced to show the wholesale plundering of these men in the cars. And now, gentlemen, I suppose that is the great thing that is in this case, if there is any stealing anywhere. I suppose the public think, and I suppose the stockholders thought, that if there had been any considerable amount stolen by these men, it had been done, not by selling western coupons, or peddling peanuts, but by receiving money in the cars and putting it in their own pockets. And I think, gentlemen, when this evidence comes to be laid before the public, as it is now before you, that the most "astonishing" and "astounding" part of the whole case, considering how this charge has been put forth, and with what confidence it has been proclaimed and asserted here that this wholesale plundering was going to be proved—I think the most astonishing part of the whole case will be, when you come to consider it, that it rests entirely in the fertile imagination of the counsel on the other side; and that there is not a particle of proof to sustain it. I think, gentlemen, when the testimony of the redoubtable Major Carney and the other detectives was introduced, when every breath in the room was suspended and every ear open to hear what was coming, that you must have been slightly "astonished" as step by step that report was put in; *and in no single instance anywhere was Mr. Clough found to have pocketed a single dollar or a single cent that had been paid to him by these detectives in the cars!* I say this is the most "astounding" part of this case, considering the expectations that had been raised by reason of these reports. And when this evidence falls, when the foundation slips out from under the feet of my brother, and Major Carney falls, and Surgeon General Carney falls, and Draper falls, how much is there left? They say that Mr. Draper was sent away by somebody, and therefore he is not here to testify. The fact was that he got so much in debt that he could not stay. And I think it is not to be imputed that he has been driven away by us. I wish he might have been here, and that we might have had the benefit of his report, as well as Carney's and Batchelder's. When this evidence of wholesale stealing in the cars, which has been charged over and over and over again everywhere, slips away, and there is not the shadow of a shade of a foundation to sustain it, then the gentleman is left to fall back upon the question of "estoppel," and to charge George Clough for the fares of the "peanut boy."

Now, gentlemen, how is it in regard to this "astounding evidence"? In the first place, (before I come to comment on the testimony of these detectives) let me consider the question of the number of cars and the size of the trains. The "astounding" proposition of the gentleman is, that although, according to the records of the road, there were only passengers enough riding in the trains to fill two or three cars, and that Clough only took, as his waybills shew, some two or three dollars a trip (I forget the exact amount), yet the fact was, all through the war it took sixteen or seventeen cars to carry the passengers, and that instead of two or three dollars a trip from passengers riding in three or four cars, he should have returned a sum corresponding to the amount that would be paid by passengers riding in seventeen cars! In other words, they were going to show you that hundreds and thousands of dollars had been taken in this way by the conductors. They were going to show it by the size of the trains; by the amount of passengers that rode in the cars; and that, instead of passengers enough, as I have said, to fill up two or three cars, they were going to show you that seventeen cars ran over this road, filled with passengers; and that substitute brokers and substitute runners, and soldiers and town agents, and soldiers' families rode in these cars daily, filling them up, and paying immense sums of money in the cars; *and that George Clough stole the whole of it, ex-*

*cept his beggarly returns that he made to the road!* Well, gentlemen, this proposition, except to a man, it would seem to me, of very exuberant fancy, would be sufficiently absurd on its face. The bare statement of the proposition is its best refutation. The idea that when Mr. Clough or any other conductor only returned the amounts that they show they did return in the cars,—the idea that they only returned this small amount in comparison with what the gentleman says was paid to them—these thousands of dollars paid by soldiers and brokers and others—the pretense that all this additional money should be pocketed by George Clough, right in the face and eyes of the public, and under the very noses of the directors and the superintendent and other officers of the road,—I say the very statement of the proposition stamps it with absurdity and impossibility on its very face. Nobody believes it. And I should not suppose that even with the fertile imagination of the counsel, when he comes to sit down and look at it calmly, could believe it himself!

But how are the facts, gentlemen, in regard to this testimony? What is the evidence which they have brought forward to sustain this charge of wholesale plundering in the cars?

How is it, gentlemen, in regard to the size of the trains and the number of passengers who rode therein? What is the testimony?

It would seem that the records of the road ought to weigh *something* on these points, and that the officers of the road would also be likely to know a little something in regard to both of these particulars,—the number of cars and the number of passengers. But my brother proposes that it should rest on the loose guesses of Mr. White, the occasional counts of Mr. Perkins, and upon the pure, the intelligent, the *sober* and disinterested testimony of Henry P. Lane and Sam Curtis—about whom I shall have something to say before I get through.

You have the testimony of Mr. Perkins and Mr. White, introduced by them; and what do they say? Why, Mr. Perkins states that sometimes there were sixteen or seventeen cars on a train; and that he counted them. But he states, gentlemen, on his cross-examination, that he never observed particularly in regard to the size of the trains *only when he did count*, and he counted because, he says, they were *extraordinary* occasions and *extraordinarily* large trains. That is what he says. And Mr. White says in regard to it: "From the month of July to September, I have *sometimes* counted sixteen cars; at this time there was a lively recruiting business going on; it was at the season when the travel was the highest; soldiers were recruited and marched off immediately to the field; it was about the time of the battle of Gettysburg, in July, 1864; at this time the soldiers were sent right away; in July the squads were sent away most rapidly; if a squad of a hundred and fifty, they would have to put on three cars; if two hundred, four cars, etc." That is what he says in regard to the number of cars; showing, gentlemen, that when these extraordinarily large trains of cars were run, they were exceptional cases. I do not deny that there were times during the war when there have been sixteen or seventeen cars upon the road. I deny, gentlemen, that that was usually the case. I deny that there is any evidence which shows that so many cars as that ran usually. But, on the contrary, they are mere exceptions, counted, as

Mr. Perkins says, because they were extra large trains. Now, gentlemen, it seems that Mr. White and Mr. Gilmore had some dispute in reference to the size of these trains. Mr. White and Mr. Gilmore were talking about it, and Gilmore denied there were so many. And it was said on the other side that Mr. Biddle kept a precise account. And, gentlemen, they were going to show by Mr. Biddle that there were these large trains. Mr. Biddle was sent for in great haste; he was trotted up in a great hurry, and he was put upon the stand, and was to be examined; but, gentlemen, for some reason or other, he gave way to another witness. And it has happened—probably upon further consultation with Mr. Biddle—that Mr. Biddle, who kept this exact account, and whose account, for some reason or other, my brother says, Mr. Gilmore has destroyed—(the Lord only knows for what)—Mr. Biddle, who made the account, and who made the record, *has not been produced upon the stand at all, to swear to that matter!* Now I do not know about the loss of his record. That is the assertion of the gentleman. We have not any proof of it. We have not a word from Mr. Biddle in regard to it. But I know this: that my brother is not apt to forget any testimony that bears favorably upon his side of the case, and if they could have shown that there were sixteen or seventeen cars here, by Baruch Biddle, who knew as much about it as anybody, except Mr. Blake, whom we put upon the stand, I know that Mr. Biddle would have been produced here. But they have not produced him. The fact is not as they have stated. There never were such trains run on the road at that time, unless they were exceptions to the general rule, or were filled with soldiers who had government transportation. The size of the trains is testified to by Mr. Upham, the ticket master at Nashua, and by Mr. Robert Blake, who has charge of the car house at Concord, and who knows more about it than anybody else, except Mr. Biddle, and he says there never was any such trains. Another thing they were going to show was that the soldiers paid in the cars. They were going to show this transportation, and that these soldiers paid in large numbers in the cars; that the brokers and substitutes paid by the thousands of dollars in the cars. Well, it turns out, like all the other large assertions, that instead of large numbers paying in the cars, *the soldiers were furnished with transportation by the Government,* and that they were furnished with extra tickets, and day by day large squads of soldiers were being sent off to the field to recruit the regiments that had been depleted. They were sent off in squads of twenty, or fifty, or a hundred, or a hundred and fifty; and they rode in the passenger cars, and they increased the size of these trains from the ordinary run of travel four or five cars, on the Concord Railroad, to from six to eleven. *But they were not paying their money to these conductors in the cars.* They paid no dollar to them; but the government furnished them with separate tickets. The government paid their transportation; so there was no chance for Mr. Clough to steal any money from the soldiers.

It should also be borne in mind that at this time the cars were filled with "dead heads"— everybody, almost, was riding on "free passes"—so much so, that when the list of bank directors, lawyers, politicians, insurance agents, substitute brokers, merchants, clergymen, &c., all over the state, who rode free, was being put in, the Hon. Chairman remarked, in view of the vast throng that was presented, that it would probably save time to put in a list of those who *paid* their fares! leaving the mass of the great public to ride free! And a statement has been referred to in the course of the trial, which the new directors authorized, that they found in existence, when they took charge of the road, more than ten thousand of these free passes! And yet, when the size of the trains were swelled in this way, you are asked to pronounce that George Clough stole thousands of dollars in the cars, because the trains were larger than the returns from paying passengers indicated!

Well, gentlemen, a great parade was made over the matter to which I have referred, about the thousands of dollars that were being paid by the substitute brokers in the cars. This has been asserted over and over again; but who did they put on to prove this, and what is the testimony upon which this charge against George Clough rests? Why, they put upon the stand Henry T. Lane and Sam Curtis! And, as has been said, if they had raked the purlieus of perdition over with a fine tooth comb, they could not have combed up two such infernal wretches as came here upon the stand and testified in regard to that matter. Both, when they were before you, were so drunk that they could hardly articulate correctly. Both of them admitted themselves to be the basest of human characters, and showed that their whole course of life was such that no decent man would put a particle of confidence in the truth of the testimony they might give. And this was the "astounding testimony" that was going to come before the referees in regard to the matter of wholesale plundering by George Clough in the cars!

These two runners, gamblers, drunkards, debauchees—who admit that they get their living by gambling and liquor-selling—these are the immaculate witnesses on whom the gentleman relies to furnish this "ASTOUNDING EVIDENCE!"

But, gentlemen, let us analyze this testimony. Lane starts off as a man of very large consequence. He is put upon the stand as a "substitute broker," a man who brought large numbers of recruits from New York, and recruited our regiments here! Mr. Chairman, you recollect what Henry P. Lane said. He had always paid in the cars; never bought any tickets, but always paid. He "brought seven or eight thousand men here, and put them into our regiments." Well, if my brother on the other side was content to rest this charge upon such statements, he has the right to do so.

But how did it turn out? Mr. Chairman, when you come to test and analyze the testimony of Curtis and Lane,—men who, if they could keep sober enough,—when any recruits were on board the cars, were employed merely to turn them this way at Worcester instead of having them go to Boston—it is found that neither Lane or Curtis ever brought a recruit from New York themselves without the intervention of a broker. They were sent merely to turn recruits this way; and Lane himself swears he had nothing to do with the payment of their fares, but that the company that he was to work for took the delivery of the men here in Concord from the brokers who came through with them. This he was obliged to say upon cross-examination. And I want to call your attention to it. You will remember it; but I want to have it appear that I state, as I go along, what is exactly true in regard to this evidence. I have told you how he started out—that he brought on seven or eight thousand men, and that he paid in the cars mostly; carrying the idea that for these seven or eight thousand soldiers he had paid George Clough and other conductors, in the cars, the entire amount of their fares! And he swears to another thing which you know to be false: He always came, he says, on the early train, and George Clough was always on this early train. Clough, as you know, could not be there only every other day at any rate. He carried the idea, in his direct examination, that he had paid this large amount, not one dollar of which was returned by Clough in his returns to the road; and therefore the whole of that paid by Lane—and Curtis swears pretty much the same thing—was pocketed by Mr. Clough! And yet, upon his cross-examination, see how he collapses. He says: "sometimes it was the practice to ticket their men through to Concord, because it was half a dollar less." Mr. Upham, who ran upon the road about the same time, you recollect, swears distinctly that every time he was on the road those men were ticketed through. And he gave you the reason: that it was for their interest, because their fare would come fifty cents less than it would if paid in the cars. Mr. Upham swears to this distinctly; and he is their own witness, and a man whose word will not be gainsaid anywhere. He swears *that every time he ran over the road as conductor, these substitute brokers had tickets, instead of paying in the cars.* And I beg that the referees will make a note of that fact. Lane says farther: "I used to make bargains for these men at the Worcester depot; I simply took their names, and the brokers took them through to Concord. Ordinarily the men remained in the hands of the broker until they arrived at Concord. The broker had the entire responsibility until they were delivered in Concord." Again: "Generally the broker paid the fares of the men." Although he would have you infer that he paid the fares of these men in the cars, yet he says afterwards: "*The broker paid the fares of the men; I had nothing to do with the paying of the fares; the broker had charge of them, and paid their fares until they got to Concord; I cannot state how many fares, nor for how many men, were paid to George Clough; I cannot tell how many men; I cannot give any sort of an idea to the referees, whether it was one or a thousand.*"

And then look at the animus of their testimony, too. Both of these men had had trouble with George Clough, who had been compelled, on account of the disturbance they made in the cars, to confine one or both of them in the baggage car, because they were unfit to ride among decent people.

Curtis starts out in the same manner, with a large number of men, and swears substantially the same as Lane in regard to the brokers paying their fares. You will recollect that one of these men was put in the baggage car; and the other wanted to go free, and Clough would n't pass him free. And this accounts for the swift and wholesale manner they came in here to testify against him. And these are the men—drunken, miserable, gambling wretches—that are fished up and brought in here to support this charge of fraud and wholesale plundering on the part of George Clough in the cars. And this, let me repeat again, is the "ASTOUNDING EVIDENCE" that the gentleman has got and laid before you, in support of this monstrous charge!

Well, now, what does Mr. Perkins say? I will read the substance of his testimony: "I do n't remember about Mr. Clough's trains in particular." That is his testimony. He swears that he cannot distinguish particularly in regard to Clough's trains. "I never counted the up trains," he says. "The reason why I counted that going down was because it was an extraordinary long train. I never noticed only when I counted." And again, on page 20 of my notes: "I counted because they were extra long trains; a great many soldiers in the train." (Corresponding exactly with Mr. White's testimony.) "The whole of our company had a pass. I don't recollect any particular instance of brokers paying to Mr. Clough." So you see there is not a particle of reliable testimony; the whole charge rests upon inferences and assertions alone. He don't recollect of any particular instance of brokers paying to Clough. "*Ordinarily*," he says, "*the broker came through on the trains; we had nothing to do about paying their fares.*" Occasionally, it was undoubtedly true that these runners might have picked up a man, and then they paid their fares, if they did not purchase a ticket. But ordinarily, Perkins says, the broker came through with the men, and he had nothing to do with paying the fares. He speaks about two car-loads of substitutes that paid in the cars. And this was a matter that was referred to by the gentleman on the other side—that they "paid by the car-load." But Perkins, although he says he knows that two car-loads of substitutes paid in the cars, he also says: "*I know it was not paid to either Clough or Noyes; it was paid to a conductor on the early express train.*" And again: "*It was no part of our business, or of our men, to see to the payment of fares;* they were delivered to us here; when we got a squad of fifty or a hundred men they were sent off to a regiment; sometimes we would put in a hundred men in a day; have seen them march off under guard." Showing that while these large trains were being run, gentlemen, it was at a time when recruiting was going on briskly here. Regiments had been depleted in the field, and it was necessary to send the men off immediately; and they went by the passenger trains. But they did not pay their fares in the cars, and neither Mr. Clough or the other conductors could have stolen a cent of money from them, for the government paid their transportation. But there is another omission here, that seems to me to be a little bit remarkable, in regard to this, as well as in regard to Biddle's testimony. This case has been pending some three years or more. There has been ample time to get evidence from every quarter of the globe almost, if it was necessary. It was well known exactly how these facts were, to the gentleman who was managing this case on the other side. And if they could have shown that these brokers—(these men, Lane and Curtis, did not rise to the dignity of substitute "brokers" even; for the "brokers" were of a much higher grade than Lane and Curtis, however low a broker's position might be; they were mere subagents and runners, as I have said;)—and now, if it could have been shown that these "brokers," as Col. George says they did, paid invariably in the cars, let me ask you why no single broker has been put upon this stand? Not one has been brought forward. Aye, gentlemen, they *have* taken the testimony of a few brokers. That has come out in evidence before you. They have taken the depositions of one or two or three of these brokers. That you understand, and that is in evidence before you. But have the depositions of these brokers, gentlemen, been put into this case? Not one of them. Perhaps I am wrong when I say not one; because I believe Jacob Smith, in his deposition, says that he sometimes brought on a few substitutes. It was attempted to be shown, in order to implicate Clough, that he had at one time passed Smith free, and received pay for doing so in hay. But, unfortunately for my brother, Smith, when he came to testify, swore that for the hay which he let Clough have Clough paid him in money. That ended that matter. But Smith, although stating that when he had substitutes he usually paid in the cars, yet swore that he did not recollect of ever paying any money to Clough—and Smith is the only person in the shape of a substitute "broker" who has testified in this case, out of the hundreds of "brokers" that were running men in here! So that I am right when I say that there is not a particle of proof here from these brokers from New York, Concord, Manchester or Boston or anywhere, to show that they paid a single cent in the cars to either of these conductors.

Now, gentlemen, I ask again, is not this omission a little significant? If the fact were so, when we have the names of twenty or more different men in New York who were constantly coming on here to see the town agents, and who were constantly bringing on men to fill up our regiments, why, let me ask again, has not some one of them been produced? They were "brokers;" they were men known in New York, and probably, most of them, known to people here. It does not appear that the least inquiry has been made to find where these substitute brokers were, by whom they could support this charge that has been made of plundering thousands of dollars from day to day and from week to week by George Clough. And it rests only upon the unsubstantial and false and malignant testimony of such men as Henry P. Lane and Sam Curtis.

When it is shown by Perkins, and even by Lane and Curtis, that the "brokers" had charge of the men they brought on—that they had the oversight and paying of their fares, or purchasing their tickets—that they delivered their men in Concord, and that the parties with whom they dealt here had nothing to do about paying their fares—when Upham swears that these men always had tickets when he ran the trains,—is it not singular, gentlemen, that not one of them has been put upon the stand to show what amount of money they paid to George Clough? So much, gentlemen, for this "astounding testimony!"

And, gentlemen, this brings me to consider the testimony of these detectives. And, as I said before, I think everybody, including this honorable tribunal, feels that here, after all, is the great hinge of this case. If the fact be that these conductors have been plundering this road in this way, and they have had three or four detectives upon their track for two or three months, more or less, of course they must have ferretted this thing out, and it must appear from this evidence, if at all, that these men have been false and fraudulent and dishonest plunderers and thieves, and entitled to the execration of every honest person in the community. Now then, I believe this was what we all expected. When Major Carney, "of the Army of the James," was put upon the stand with such a parade, and his testimony referred to as of such paramount importance, it is no wonder, although we were firm in the faith of his innocence, that we should expect, if there was any evidence against him anywhere, that the disclosures in this report of the chief spy would furnish that evidence. But what is the fact? Let me state it, gentlemen, and let me proclaim it so that everybody may hear and know that although Maj. Carney was on this train as detective,—this Carney, who swore that his pay depended upon the amount that he found against the conductors—who was thus interested in convicting Mr. Clough,—and his testimony and his manner and his words showed that he got, from some source or other, imbued with the same spirit that has been pursuing George Clough all through these proceedings,—although he went on there with this prejudice—although it was for his interest to get all against Mr. Clough that he could, and although he swore that his pay depended, as I have said, on the amount that he found against him, *not one single dollar or cent does he report as having been found against George Clough! Every dollar, every cent that Major Carney paid in the cars, day after day, to George Clough, is returned on George Clough's way-bill at night—every dollar, every cent!* But on the cross-examination of Major Carney, "of the Army of the James," it is shown that he was put on to the Concord Railroad by General Butler, the senior counsel in this great humbug; and he was obliged to state, when he took the returns of George Clough in his hands and looked them over and compared them with his report, that George Clough had returned every dollar and every cent. "On such a date I paid him thirteen dollars in the cars for so many fares between Concord and Boston, perhaps, or between Concord and Manchester, or Concord and Nashua;" and that was reported. We took George Clough's return for that date, put it into his hands to read, and you find every one of these fares is returned just exactly as he had reported them. I have got them analyzed here, gentlemen, but I don't propose to take up your time in reading it. I went over them carefully last night, after one o'clock, to see that I was not mistaken in regard to this matter; and I say *that in every single instance Major Carney's report corresponds with Clough's returns!* Why, gentlemen, I said this was the hinge—the pivot—upon which everybody will consider this case to turn; and we, as counsel for George Clough and his family in this matter, naturally felt anxious concerning this redoubtable gentleman and his famous "report." And we tried to get hold of it and find what this man, who, it appears, took a thousand dollars to make a report in another case, would swear to; and so we made anxious search for "General Butler's box" containing it. We traced it to Boston and to Lowell, and then, as we believed, into the custody of the gentleman who is here as counsel on the other side, at Concord; and then we followed it to Boston again, and then we heard of it at Washington; and we followed it everywhere, all the time pumping the terrible "Major Carney" to see what he knew about it. But, like an *ignis fatuus* that it was, it everywhere eluded our grasp! And when at last the box came before you, and the key was turned, I admit that my heart palpitated a little; I didn't know what Major Carney might say in that report. But when, with trembling hand and anxious eye, we came to open the box and see all that it contained, I must say that the mountain which, you recollect, went through that terrible agony of labor and brought forth only the smallest kind of a mouse, didn't compare with the abortion that jumped out of that box! And there is nothing that I remember in history that compares with the entire fizzle of this testimony and the report of this puissant "Major of the Army of the James" and the other detectives, but the redoubtable exploit, down South, of General Butler's "powder boat!" Now, gentlemen, I don't stand here to disparage the military ability of General Butler. But whatever may be his fame as a soldier or a statesman, it must be conceded on all hands that his "powder boat" was a failure. His name, and his great reputation as a lawyer, have been paraded and blazoned before you, as if, by their aid, this "Major Carney" and "Surgeon General Carney" and Colonel George were going to turn the universe all upside down and explode at least half of Concord, and particularly George Clough and the other conductors! And so, I say there has been no fizzle in history that at all compares with it, but just that "powder boat" fizzle to which I have alluded. And it seems to me, gentlemen, that this same "Major Carney, of the Army of the James," must have been the very fellow who touched that boat off! There, you recollect, was a terrible noise, a great deal of smoke and smudge, and a few clams and considerable sand went into the air. But after the smoke and smudge had cleared away, there stood Fort Fisher, with not a single brick or stone displaced. And it is just exactly so here. There has been a great noise, a terrible stench, an awful smudge and a dreadful sensation, but nothing in the world to it—a complete fizzle! Why, what a tremendous man this "Major Carney, of the Army of the James," was. You recollect what airs he put on when he testified; and you recollect, also, the appearance of his brother, "Surgeon General Carney, of the Accident Company," who appeared later upon the scene of action and testified as a detective! Yes, you recollect how the Major was struck with the guilty appearance of Mr. Clough. I beg you to remember this, because it is of the same piece exactly with "Surgeon General Carney" who was on the stand day before yesterday. You recollect the Major went down one day, and paid George Clough, he says, thirteen dollars in the cars. Well, he thought, he said, that he would "try a little game of his own." And this was the "little game":—this is what he says—I read from his journal:—"This afternoon I took the 3.30 train for Nashua. I had only three men with me, and I intended at first to give to Clough just $7.50, the fare of us to Nashua, 3, and Boston, 1; thus:

| | |
|---|---|
| 3 *a* $1.50, | $4.50 |
| 1 *a* 3.00, | 3.00 |
| | $7.50 |

But as there were a number of soldiers on board, *I determined to try a little game of my own*, and accordingly, when Clough came round I told him I wanted to pay for six to Nashua and one to Boston; thinking that if required to identify my men I could do so for three of them, but hoped to allay suspicion by stating that very possibly they went down to Lawrence by mistake." In this way, by this nice "little game," he hoped to detect Mr. Clough. So he gave him a ten dollar bill, a two dollar bill, and a one dollar bill, making thirteen dollars in all. He says, after they left Manchester Clough came along and asked him how many he paid for. He told him *six*; and that Clough then told him he would be back and see him soon. In a short time, he says that Mr. Clough came back, and approaching "stealthily from behind," he slipped something into his hand. He looked, finally, when he dared, to see what it was, and there were *fifty-five cents in money!* He said, after Mr. Clough passed out he (the Major,) was in great trepidation, to use his own language; that he was in considerable excitement; and he said he took out a book to keep from showing his "nervousness;" and when Mr. Clough came up behind him

in that "stealthy" manner and thrust these fifty-five cents into his hand, then his nervousness was almost beyond control, and he didn't know hardly what to make of it,—(this was on the direct examination)—and he concluded that as this was not any "multiple of six or seven," these fifty-five cents "must have been intended as a present!" And Mr. Clough looked guilty in all his lineaments. Well, it turned out, when my friend, Mr. Mugridge, inquired about this matter and punctured this bubble, that these fifty-five cents *were exactly the change to which Carney was entitled!* He paid Clough six fares to Nashua, *a* $1.60 each, (the regular fare and 10 cents extra added,) $9.60
and four to Boston, *a* $2.85, (extra 10 cents added,) 2.85

$12.45

$13.00
12.45

.55

Twelve forty-five, from thirteen dollars which he gave Clough, leaves *fifty-five cents*, the exact change to which he was entitled and which Clough paid him back! And it was this perfectly proper transaction that Carney undertakes, in his direct examination, to magnify into evidence of Clough's guilt! This paying him back the fifty-five cents which belonged to him as change between the price of the fares and the thirteen dollars, was distorted by this supple tool of his employers into a "stealthy" approach from behind to give him a *bribe* of fifty-five cents! But Carney himself, when followed by Mr. Mugridge's "sharp stick," on cross-examination, was compelled to say *that he "knew he was testifying to what was not true in regard to Clough's giving him the present,"* and that *"Clough returned him the exact change for the thirteen dollars, and that he did not see the slightest impropriety in Mr. Clough in reference to this transaction.*"

You see here, gentlemen, the character of the man who took the thousand dollars in the Leighton case, for making his report in a particular way; and I don't think you will wonder that, according to his own admission, (although he was "with General Butler, in the Army of the James,") that the treasury department promptly dismissed him from its service on account of that Leighton affair! He is, however, just the pimp for those who set this ball in motion against the Concord Railroad conductors!

Gentlemen, the entire evidence upon which this case rests —I don't care where you place your finger—is not one whit more substantial than the testimony that I have just detailed; and I thank God, after these charges have been blazoned forth to the world, that the evidence on which they rest is to go too, so that everybody will see whether George Clough is the thief and plunderer that they charge him to be, or whether the whole case does not rest upon testimony just as frivolous and unreliable as this I have detailed here, of "Major Carney, of the Army of the James."

Well, gentlemen, they had another detective, and that was Bachelder; and he rode frequently in the cars. He tells you the number of times that he rode, and the number of tickets that he bought, and the amount of money that he paid in the cars. And so far as his appearance was concerned, he appeared tolerably well on the stand compared with Carney, although he was in this unfortunate office of spy and detective, and hoped to get a place as conductor if he succeeded in getting the old ones off. He appeared well compared with Major Carney, and he puts in his report here.

He tells you that November 23d he purchased three tickets of Clough to Boston for $8.55, and one fare to Nashua. Taking up George Clough's report for that date, every one of his fares is returned. And so it is through the whole of Mr. Bachelder's operations upon the cars. November 23d, 25th, 28th, December 13th, 15th, 16th, 25th—all these dates he rode in the cars and paid money, more or less, to Clough; and every single time, gentlemen, these fares are returned by George Clough in just the way (with the exception I will make in a moment,) that they were paid!

Now then, gentlemen, when it appears from the testimony of these detectives who were on these trains following in the track of this man, that not a single cent of stealing or misappropriation has been discovered, I ask you if there is not an end of this case,—if substantially that is not all there is to it,—if substantially that is not all there was charged, and that the charges have all been dissipated into thin air?

But we have another detective here, and that is "Surgeon General Carney," to whom I have alluded. And, by the way, gentlemen, George Clough swears before you that he never had the least intimation that any of these men were on his track. He knew nothing about it until two days, I think he says, before the vote of the directors. And, gentlemen, how could he know? Here were not only one or two, but four different detectives, all of them on George Clough's track, because Mr. Gilmore had said he was the chief offender in this matter. And not one of them succeeded in "spotting" him at all, to use the language of Surgeon General Carney. And I think, gentlemen, that it is remarkable that neither by mistake—because I suppose that is not pretended or contended here at all but that Mr. Clough might make a mistake in his returns.——

Mr. Rolfe.—He did make a mistake against himself in making change with Bachelder.

Mr. Tappan.—Yes, sir; and I think it is remarkable that it did not happen that by mistake even, there was one solitary instance where Mr. Clough has taken a dollar or a cent wrongfully! And I say, when this fact became known to the men who had this investigation in charge—I say, when this fact was apparent on Carney's report to those who set this ball in motion, it is evident it was pursued afterwards from some motives other than the honest motives of having this thing investigated, or from the belief that George Clough had been plundering to the extent to which he had been charged, or to any extent. When these reports came in they stood as they do to-day. There was nothing upon them to show or to warrant the fact that he had been stealing in this wholesale manner, or in any manner, in the cars. The whole thing was dissipated and shown to have been false and idle and without foundation, and there the matter should have ended.

And now I want to refer a moment to "Surgeon General Carney." And, gentlemen, I need not say much in regard to him. He came here very much as his brother did, and seemed to be very much inflated with his own consequence; and he undertook, wholly without warrant, after he had sworn that there was nothing in the transactions from which he could infer any such thing, he volunteered the assertion, or the declaration, that he regarded Clough for years as a thief. He knew he was a thief. After his report had shown that Mr. Clough had returned every dollar—when he had himself sworn that there was nothing in any transaction that he had detected to inculpate Mr. Clough, or which showed that he had been dishonest at all—still he went out of his way,—(and it shows the *animus* with which he testified,)—he told you that he had regarded Clough for years as a thief, although he never knew him! Well, gentlemen, if the character of that man and of his brother, so far as they have any character, was not depicted in their own lineaments, and if the true character that that man bears is not carried in his face, and it is not that of a sneak, I never saw that word written upon the face of any man on earth! But "Surgeon General Carney" comes here like his brother, and he was put upon the track as a detective, and he paid money in the cars to George Clough; and he swears to you how and when he paid it. It seems that on November 23d, 1865, he paid five fares from Nashua to Concord: $8. Clough has returned five fares, Nashua to Concord, at $1.60 each, making exactly $8. And in the afternoon of that day he bought five tickets to Boston: $14.25; and Mr. Clough has returned five tickets to Boston, $2.85 each, which is the regular fare and the ten cents extra added, making $14.25. At this point the gentlemen on the other side supposed, I presume, that they had got enough of Surgeon General Carney, and they stopped. But it seemed, on cross-examination, that he admitted that he was here once after this, and he testified that he came up again, and on the 19th of December he bought two tickets from Concord to Nashua, and taking George Clough's report in his hands, he found that four tickets were returned between Concord and Nashua on that day; so that Surgeon General Carney, no more than his brother, Major Carney, detects Mr. Clough in anything that is dishonest. On the contrary, everything taken on the cars was returned by Mr. Clough in the same way and manner that the detectives paid him.

And, gentlemen, for all this work what has the Concord Railroad got to pay? For this immense labor and these immense services which have produced such "astounding" results, the Concord Railroad, Major Carney says, has got to pay him $25,000! That is what the Concord Railroad are in to pay for the services of Major Carney, of the Army of the James, and Surgeon General Carney, of the Provident Accident Company, I suppose, $25,000 more! and I should like to see the counsel on the other side institute a suit to recover that fee for Major Carney, for he seems to be set on this particular sum, and he swears *that if he could have* $24,999.99 *he would not take it!* And he says, further, that he thinks his services are worth that sum. He thinks he has been of that benefit to the road. And, gentlemen, I hope in your report, when you come to consider Major Carney, you will recommend that the Concord Railroad should pay to him at least $25,000 for the services which he has rendered. He ought to get a great deal more, but he would compromise, gentlemen, for $25,000! If he gets, as he suggests, only a "percentage on the increased revenue of the road," by means of his operations, I do n't think it will come to quite that sum! But, gentlemen, I suppose the counsel on the other side will argue, although not a dollar has been detected or disclosed by these reports to have been wrongfully taken, that they do disclose one astounding fact that nobody ever heard of or thought of before;—and the gentleman is obliged to fall back from his $50,000 a year plundering in the cars—to fall back from his terrible bombshell, that Major

Carney and the other detectives were going to throw into our camp, and undertake to charge him because he reported—perhaps he will say shows—that in a few instances he failed to take the extra ten cents! That is all there is to it, gentlemen; all there is in those reports that by any possibility can be tortured into anything favorable to the other side of this case. And yet, although we have shown to you how this matter of ten cents extra has been carried forward all through, ever since the rules were promulgated, although that is a fact that must be in the knowledge of everybody that has anything to do with the Concord Railroad, let me just say that out of forty-one fares as paid in the cars, as shown by this report, there are but five instances where it appears that only the regular fare was returned. And from the fact that because in five instances only, out of forty-one, Mr. Clough, although receiving only the regular fare, without the ten cents, returned them with the ten cents added, therefore he did not return them at all! And although Mr. Clough shows you exactly how it was done, and although out of these instances, as I have said, there are only five in forty-one, the position is taken that because Mr. Clough, receiving from Dr. Carney, for instance, or from Mr. Bachelder, $1.50 for a fare in the cars from Concord to Nashua, returned on his way-bill $1.60, which would be $1.50 with the ten cents added, therefore he failed to return their fares at all, or that he is *estopped* to say that he did.

And that brings us to the famous doctrine of ESTOPPEL; and I may as well consider that, and say what I have to say now on that subject as at any other time. Now, what is the testimony? I think that you will have no doubt, gentlemen, that Mr. Clough has told, in regard to this matter, just exactly the truth, with no attempt to extenuate anything that he has done, because he has done nothing except what he was instructed to do; and so he has told you that in some instances, after this ten cent rule was promulgated, he could not always get the ten cents extra in the cars. The tickets were obliged to be taken up between each station, the cars were filled with "dead-heads," with free passengers, everybody was going free and having Gilmore's passes, the cars were filled up; and many of the passengers objected to paying the ten cents extra. And so it turned out that not only on the Concord Railroad, but on all roads, (such must be within the knowledge of the referees themselves,) it became utterly impossible to enforce this rule at all times. Well, you will see, gentlemen, how often Mr. Clough did enforce the rule. You will observe that this was only an exception. When money was paid to him in the cars, in the large majority of cases, as he reports and these detectives show, the ten cents extra was collected and received; and in the exceptional cases Mr. Clough was not able to take the ten cents extra, simply because, being obliged by the rules to take down the names of those who had free passes, if he stopped to quarrel with everybody that objected to paying the ten cents, he could not get through the train, and take up the tickets, before he arrived at the next station; and therefore he was actually obliged, when they gave him change, to take what change was given, and not to stop and have a quarrel. That is all there is about it. And, under the circumstances, what does he do? Why, he did just exactly what every other honest man, who is desirous of doing his duty and doing nothing else, would do; he went to the fountain head to know what course he should take. Now, what was the fountain head? Where could he go? When he found it was impossible to enforce this rule and collect this ten cents extra, to whom did George Clough turn for information as to what his duty was under the circumstances? Why, he went to the agent and superintendent of the road. He went to the man who was the executive officer of the road. He went to the man who had the entire power over these trains. He went to the only person who had any authority to speak for that corporation. This, gentlemen, was the man who goes to the platform and says, "Conductor, pass that man free," and "Let that man go for half fare," and carried this wood at such a price, and that wood at another price, and whose word in all these things—at least so far as those under him were concerned—was the supreme law; and all this was within the scope of his authority to do. Suppose Mr. Clough had seen Judge Upham upon the platform—suppose he had seen Mr. Spalding—suppose he had seen any other of the directors. They had nothing more to do with this than you or I have, and their authority in such a matter would be of no sort of effect whatever, because Mr. Gilmore, the "general agent and superintendent"—the man who runs the road—the man who is the eyes and ears and mouth of the corporation, was the mouthpiece through which the corporation speaks and takes notice, and there can be no other avenue of communication with or from the corporation. Now I am not aware that there is any judicial decision anywhere—I have not been able to find any in the books—in regard to the powers of the superintendent of a railroad. But it is apparent from the very nature of the case, what the powers and duties of a superintendent are and must be. George Clough cannot stop to call the board of directors together. They have nothing to do about it. What they resolve or undertake to do is one thing; but when they put forth an executive officer, or their mouthpiece who manages their concerns, they are bound by what he does, and by what he says.

But I take another position. I say the act to which I refer, and that is the assent of Mr. Gilmore, the direction he gives Mr. Clough in regard to returning the ten cents and in regard to his course of conduct when he was unable to take it, was an act fully within the scope of the general agent's authority. As I said, George Clough could not go to anybody else and ask what he should do. He could not go to the other directors to ask them what he should do, but he went to Mr. Gilmore and asked instructions from him. Mr. Gilmore tells him to get the extra ten cents when he could, and when he could not, then buy a ticket and punch it and return it to the general ticket agent, and that would make it all right. And that is the course that Mr. Clough says he pursued, and that is the course that he was obliged to pursue; and I say that Mr. Gilmore ordering him to transact the business in that way, that act was an act of the corporation and within the scope of the agent's authority, and so within their knowledge. But I take the further position here, that even if the act in question was beyond the scope of Mr. Gilmore's authority, (which I do not admit, only for the sake of the argument,) the corporation have held him out, or enabled him to hold himself out, as having more enlarged powers than the strict letter of their rules would indicate, and are therefore bound by his acts. The way in which Mr. Gilmore conducted the business of the road—the departure from the fixed tariffs in the matter of certain kinds of freights — the constant issuing of free passes — his directions to the conductors to pass this man and that without any ticket or pass — the taking of tickets from the ticket-master and conductors without any account being made of them, and other acts outside of the "rules and regulations," were matters of public notoriety, or if not of public notoriety, they were well known and understood by everybody connected with the operation of the road — were known to the directors, for they, themselves, were constantly violating their own rules, and they cannot now be permitted to say that Mr. Gilmore acted in excess of his authority, especially as against this defendant, with whom he came in daily contact, and who was cognizant of the powers which the superintendent exercised. Nor will it do to charge corrupt *collusion* between the agent and the defendant, for this will apply as well to Mr. Sanborn, the ticket-master, to Mr. Spalding, one of the directors who had the benefit of the irregular tickets, and to every other officer and employee of the road, wherever the rules were varied from, as they constantly were.

I say then, if these directors, having constant knowledge that Mr. Gilmore was setting aside the rules every day, if they held him out as having more enlarged powers than he actually possessed, or if they allowed him to hold himself out as having more enlarged powers, then they are bound by his acts; and that is the doctrine of *Hatch* v. *Taylor*, in 10 New Hampshire Reports, as I understand it. So then, I say, when Mr. Gilmore directed Mr. Clough to purchase tickets and punch them and return them to the general ticket agent, that is all the authority that Mr. Clough needed, and he did exactly what was in the line of his duty, when failing to collect ten cents extra, as he must fail in a few instances; he did as he was directed, and purchased tickets and punched them and returned them to the ticket office in that way.

Now, then, the evidence in reference to wholesale plundering in the cars fails, and there is no evidence, as I undertake to say, upon which you can lay your finger, that George Clough has taken one cent's worth in the cars, and the gentleman is obliged to fall back on something else. He sets up now the technical doctrine of "estoppel." And so this case dwindles down and fizzles out! Starting with the charge of immense wholesale plundering to the tune of fifty thousand dollars a year, at last, gentleman, the counsel is obliged to stand before you, contending that you shall charge Mr. Clough for money taken in the cars, although he has returned every cent so taken, *because he has not been able in every instance to collect the ten cents extra!* Well, gentlemen, he gets a little lower even than this, before he gets through.—He not only wants to charge him on this fraud and "estoppel," but also for the fares of the "pea-nut boy" who rode back and forth on the train! And so ends this great sensational bubble of wholesale plunder on the part of George Clough. Gentlemen, I do not propose to detain you very long on this matter of "estoppel," which has been very fully argued by myself, and particularly by my friend Mugridge, when the point came up in course of the trial. The proposition here is, even admitting the facts to be as we have found them, even admitting that everything we say is true, *even admitting that Mr. Clough has paid into the corporation by purchasing a ticket and returning it to the general ticket agent, every dollar he thus received in the cars, but failing to take the ten cents extra—that even though he has done all this, he is "estopped" now to say so, and that he is to be charged in this suit, although he may have paid over every dollar!* And I say here, as has been said before, and as has been said by one of the learned referees in a case that is reported, that the doctrine of estoppel might well have been termed "odious," if it were allowed to work such a wrong as this!—Again, I contend that the rule of the road has been substantially complied with, and that everything has been done that possibly could have been done under it, and, if the conductor has not in all instances returned the ten cents extra, he has done the next best thing that he could.

If you should find, gentlemen, from the evidence, (and I think that not only the evidence, but your own experience and observation will convince you of this,) that it was impossible for Mr. Clough to collect this ten cents extra, and perform the other duties, of taking up tickets between the stations, &c., and that the road would have lost more by his attempting to do it than it would by sometimes omitting it, then I hold that there has been a substantial compliance with the rule. I say further that the doctrine of "estoppel" cannot apply here, because the corporation have in no way changed their position. There is not a particle of evidence in the case from Carney's report, or from the report of either of the other detectives, or from any other source, that Clough had ever received a cent of money in the cars that he had not returned, prior to the bringing of this suit. The evidence that he did receive money and did not return it in the regular way because of his inability to take the extra ten cents, comes wholly from Clough, when he gave his deposition, since this suit was commenced. I repeat that there has not been pointed out, and there cannot be, a scrap of evidence of this fact *before* the suit was instituted. How, then, has the corporation changed its position, and in what way can the doctrine of estoppel *en pais* apply to this case? It was from Mr. Clough's mouth alone, after the suit was brought, that the corporation has any knowledge that he received a dollar in the cars that he did not return on his way-bills; why shall he not have the benefit of the explanation that he gives, and the reasons that compelled him to return it in another way? The evidence from Mr. Upham, the president, and from Mr. Spalding and Mr. Kidder, two of the directors, shows that they did not know, nor did the other directors, so far as they are aware, know before the suit was brought, that Clough received money and did not take the ten cents extra, or that he returned it in the way he did—thus completely negativing any knowledge on the part of the corporation that the defendant had received money that he did not return on his way-bills, except the knowledge they had through their general agent and superintendent.

And, as has been before urged, the knowledge, to Mr. Gilmore, of the way this money was paid over, was knowledge to the corporation, and was therefore done by and with the consent of the corporation, and they are *estopped* to say it was not so. Arrangements made by him, directions given by him were of supreme authority to Mr. Clough, and such must be the rule. Railroads cannot be operated upon any other basis. The agent or superintendent for the time being has the power which governs and directs the whole machinery of the running of the road. The directors can turn him off if they please, but those who are under him, and the public can know no one else but the general agent and superintendent that they hold forth to the public and to their own employees as the man who is to give them direction, and the man from whom they are to receive instructions. And I think, gentlemen, that it comes with a bad grace, I think that it comes with an exceedingly bad grace, after this great flourish of trumpets in regard to this wholesale plundering, and the burdening of the public ear through the public press and everywhere with the charges against these conductors of the amounts of money they have stolen, taken in the cars fraudulently and appropriated fraudulently to their use; I say that it comes with an exceedingly ill grace from the gentleman on the other side when he now undertakes to charge Mr. Clough on the ground of this technical "estoppel;" that although the facts that we allege are true, and although every dollar has been paid, yet he cannot be allowed to say so! And if there is a decision against us on that ground, I am really glad that everybody will know the position that has been taken and the reasons for such decision. But, gentlemen, there is no fear of it. I don't believe that this doctrine of estoppel can be sustained before this tribunal, or that there is any law which warrants setting up such a ground as this when it works such gross and manifest injustice! But I want to allude to one or two other things, gentlemen, in regard to this charge which I am now answering of wholesale "plundering in the cars," for that is the great charge we are called upon to meet. Now, then, I said the other day, that it seemed to me if I was sitting in the place of this honorable board, or one of them, when these charges were put forth in this manner with an assertion that Mr. Clough returned only one-quarter part of what he ought to have returned, that the evidence that would carry most weight to my mind would be the returns of these conductors who succeeded George Clough in running these same trains. Now that is the way it strikes me, gentlemen, and I think that if the conductors who succeeded Mr. Clough, either Mr. Chandler or Mr. Le Bosquet or anybody else, had been produced and he had stated that he ran the same train that Mr. Clough did, and it appeared from his report in the two years that he had run that he had returned a very much larger amount than Mr. Clough had returned—I think, myself, gentlemen, that that testimony would have carried considerable moral weight with it; and I think, also, although the testimony was not legally competent, and could not be put in except by agreement, I think if there had been an honest purpose to do only one's duty, as the gentleman says he wants to do in this prosecution—no more and no less, and no disposition to charge Mr. Clough except upon all the evidence and the facts in the case—I think that he would have allowed that testimony to have gone in when we offered and desired to put it in. And if, gentlemen, on putting it in, after this great braying of trumpets in regard to this plundering and stealing, it had only turned out—and I am now only *supposing* it to be the fact, (the gentleman has been full of his "suppositions" and "illustrations" all through the case;) and I have no right to say what that report did show, but I will suppose what it showed—that these new conductors took only thirty-two cents a trip more than Clough returned, would it not have had considerable weight in your minds as tending to show that Mr. Clough had made honest returns? Aye, gentlemen, more than that: would it not have been entitled to great weight, especially, as Mr. George Blake swears, the travel has been quite as much since Mr. Clough left the road as before? The trains have been quite as large in 1866 and 1867 as they were in 1864 and 1865. And when, too, it appears in the testimony that all these hundreds and thousands of free passes, upon which everybody rode over the Concord Railroad, had been cut off, under the new direction and management—if, under such a state of circumstances as that, the returns of Mr. Le Bosquet and Mr. Chandler, the conductors that succeeded Clough, only showed *thirty-two cents more a trip*, would not that be pretty strong evidence to show that no dollar had been stolen by George Clough? And if the facts were not so, the gentleman would have allowed us to put in these returns. I appeal to you if it would not be so. And I have argued this omission; and I have the right to insist upon it, and to have it make all the weight in favor of Mr. Clough that it can possibly be made to have; because it would be the most satisfactory and convincing testimony, aside from the positive testimony of these detectives who were in the cars, that any tribunal or the public could possibly have as touching the innocence or guilt of George Clough. But we have been offered to put in this testimony if we will put in the evidence of the returns of the conductors on the Portsmouth road, or the returns of another conductor on the Concord road. Now, as I told you the other day, I do not know what the returns of Mr. George Noyes will show; I have not looked into that; but I infer from the fact that the counsel on the other side is willing to put in the returns of Mr. Noyes, that his returns are, for some reason or other, favorable to them, Mr. Noyes, perhaps, having returned a greater amount than Mr. Clough. And I ask you if there is any fairness when the gentleman offers to put in evidence the returns on the Portsmouth road. If he has looked them over and seen that there was a larger amount returned, from the fact that there are more way stations where tickets cannot be furnished, and finds that he is willing to put them in in comparison, I say that those returns have nothing to do with this case; and I say you had no more to do with the Portsmouth road than you had with the Grand Trunk road or the Erie road. But it would be more satisfactory, I think, if you were to compare the returns of Mr. Clough and the returns of Mr. Le Bosquet, who succeeded him on the same train. But to show that, we have not been allowed. We have shirked no responsibility; we have dodged behind no covert. We have tried to meet this thing knee to knee, foot to foot, and breast to breast; and if we have not, gentlemen, we have not done what we intended to do. If we have failed in a single instance, we have not done what we intended. And if we have not followed every step, and answered every charge with evidence that is irrefragible, then we greatly mistake the evidence that is before you.

But, gentlemen, I want to call your attention to another thing. Over and over again has the counsel reiterated the assertion thas there was a very meagre amount (I do not remember the exact figures,) returned by George Clough in the cars; that it was "astonishing" and "astounding," with all this travel that he should only return the pittance that he has returned.

I want to have you remark another thing. There is no doubt of the truthfulness of Mr. E. W. Upham; and I am glad that this testimony has got into this case. I am glad that, although the gentleman is unwilling to institute a comparison where it would seem that a comparison ought to be instituted, I am glad that he put in the testimony of Mr. E. W. Upham. Because he is an honest man. There is no charge against him that he has stolen a cent of money in the cars. And yet he ran on the train a few times, that same year; and his returns are before you. And now I want to show how they compare with Mr. Clough's.

On the 24th day of October, 1865, gentlemen, Mr. Upham was on the train as conductor, and his entire way-bill on that day is $17.40, according to my minutes.

Mr. George.—How many trains, Mr. Tappan?

Mr. Tappan.—I do not know. I have it lumped together. Mr. Clough ran on the 24th of October, and his entire returns for two trains are $17.90, a little larger than Mr. Upham's. Mr. Upham was on the road on the 25th as conductor, and he has returned as his whole way-bill $8.70 on that day. Mr. Clough was on the road on the 25th—the same day—and he has returned $17.40 on that day, on the same number of trips. On the 26th day of October Mr. Upham was again on the train, and his whole way-bill is $18.05. Mr. Clough was on the road on that day, and his whole way-bill is $22.15. Mr. Upham was on the train on the 27th, and on that day his return is $7.65; and Mr. Clough's is $4.45. On three of the days that Mr. Upham was on the road Mr. Clough's returns exceeded those of Mr. Upham, and on only one does that of Mr. Upham exceed that of Mr. Clough. And all the way through, you may look through the returns of Mr. Clough that year, and compare them with the returns of Mr. Upham, and you will find that just about the same average amount is returned all the way along by both of them. And I suppose that if Mr.

Le Bosquet's and Mr. Chandler's returns had been put in, even after the free passes were all cut off, you would have seen just exactly the same state of things.

And, gentlemen, I believe this is all that I care to say upon this point of stealing in the cars, except to state to you the fact, which of course you know, that Mr. Clough himself comes before you, and with an oath upon him, swears that to his knowledge, either directly or indirectly, intentionally, he has never wronged the Concord Railroad out of a dollar, or ever taken a cent. You can judge whether in this respect, or in any other respects, where Mr. Clough has testified, he appears before you in the character of an honest man, and whether as a witness he means to testify truthfully and honestly to everything. And I submit to you whether, even in some instances where the truth has seemed to make against him, as perhaps, apparently, it has, he has not come up and told everything to you, with an honesty and straightforwardness of purpose, such as few men exhibit on the witness stand.

And now, gentlemen, all I have to say is, that Mr. Clough is known to this community; he is known, too, to one member of this board. And if he is a man, if his character entitles him to be regarded as a man that can stand up before this tribunal and wilfully burden his soul with perjury, I never have heard any such intimations in regard to George Clough. Nor do I believe that a man who has come forward in the way he has—(dragged before a magistrate as he was, day after day, week after week, and month after month,)—and made a clean breast of everything—stating facts that were within his knowledge alone, many of which he might well have excused himself from stating, and which, as in this matter of the extra ten cents, was susceptible of being tortured against him, and which he might have denied, or withheld, if he intended to be false,—I say I do not believe that such a man has deliberately perjured himself before you. And I submit to you whether he has not manifested a disposition to meet all these charges, and state the truth in regard to them, according to the best of his knowledge and ability. And I say, then, if George Clough stands uncontradicted—if this thing rested on his testimony alone—and there is no evidence to convict him upon his own testimony, he should go free. But, as I have said before, there is no single scrap of testimony upon which a man's finger can be placed—nothing but these violent inferences and presumptions and suppositions, that have been made over and over again in the course of this trial, upon which there can be any pretense to charge Mr. Clough.

But there was an attempt, gentlemen, to show, and one of the charges of the famous "synopsis" of the gentleman on the other side is, that Mr. Clough received and was cognizant of bogus tickets sold and used by other conductors. And that was put in, gentlemen, with a great flourish of trumpets. Now there is not a particle of testimony upon which to substantiate that charge; and all the testimony that has any tendency to show it, was the testimony of Carney, who said that he got four tickets of Kendrick, and he undertook to describe the tickets which Mr. Clough took up. Now, gentlemen, I undertake to say that when you come to examine these tickets you will find that every one of them (I believe I have not examined them particularly) were tickets that were not good over the Concord Road. I believe that Mr. Clough and Mr. Sanborn state in their testimony that it was the practice of the conductors, if they had no ticket which was good, to put in a ticket of any kind, and then, when they arrived at the place of their destination, to replace it with a good one; or sometimes they would send forward a scrap of paper and write on it—"Good for a passage to Nashua," or whatever the station might be. These might have all been got in this way. There is no evidence to show that it was not so. And this evidence was ruled out, unless in some way Mr. Clough was shown to have partaken of the fraud, if fraud there was. But there is no fraud. There is no evidence that Kendrick perpetrated a fraud, nor that George Clough was cognizant of a fraud.—Therefore there is no evidence, not a particle, in regard to these "bogus tickets."

Well, gentlemen, I will just say a word or two in passing in regard to this matter of free passes. It is not worth while for me to take up any time in regard to them, because by the decision of the referees all the evidence in regard to that subject has been ruled out. Whatever has been done in regard to free passes the court here hold that Mr. Clough cannot be charged with it, inasmuch as he has received nothing, or unless in cases where he has passed persons free and has received some benefit from it. Now, gentlemen, everybody, as I have said, was passing free over this road; almost every lawyer and physician and prominent man in Concord and Manchester, and everywhere; bank officers, politicians, and myriads of people were riding free over the Concord Railroad, and it would be strange indeed if, under these circumstances and under this loose rule, Mr. Clough, having not only the authority of the superintendent, but having, at various times, been ordered by every director in the board save one, to pass people free, had not, now and then, let one slip by him. The counsel on the other side is now a great stickler for *rules.* But it was all very well for the *directors* to tell Clough to pass their friends free—all very well for my brother George, as clerk of the road, with his family and friends to pass free in violation of his famous "regulations" and "rules," but George Clough is now to be "*estopped*" from making a defence to this suit, because *he* has "violated the *rules!*" Consistency, thou art a jewel! That the directors did violate the rules by ordering Clough to pass people free has been proved. Mr. White says that he heard Gilmore tell Mr. Clough to pass persons free, and Mr. Sanborn swears to the same thing; and, therefore, I don't think it is becoming in these gentlemen to make such a great noise about Mr. Clough occasionally passing persons free, when every officer was passing in the same way, including the clerk, and their families! I don't think the public will give them much credit for sincerity when these facts are considered.

Well, now, gentlemen, there has been an attempt to show that in some instances Mr. Clough has received some benefit from persons having been passed over the road by him. As I understand the testimony, there is no evidence in the case to show that Mr. Clough has ever received a dollar of benefit by passing persons free; that is to say, that anything has gone into his pocket. One of the grounds stated in this "synopsis" is that Mr. Clough sold tickets and passages for boots to J. Grier, and that he improperly passed his tenants and workmen, and others, for his private benefit. Now, gentlemen, in regard to Grier: it doesn't appear, except from the testimony of Mr. Grier himself, and I submit when you view that in the light of the bills that he has produced in connection with Mr. Clough's own testimony—that there is not a particle of evidence by which it can be shown that Mr. Clough ever received a cent's worth of benefit from Mr. Grier. Mr. Grier was a poor man. Mr. Clough has not denied that he occasionally passed him over the road,—and Grier says that he *supposed* he made consideration enough when he worked for him to pay him for what passes he had. On the contrary, Mr. Clough swears that he never had any understanding that he was to receive any consideration, directly or indirectly, when he passed him, but he passed him merely on the ground of charity.—And I submit, when you consider the bills paid, and the prices shown in those bills for the work that Grier did for Clough, whether he ever made any consideration to Mr. Clough for passing over the road. But this deposition was made in time of great excitement; people were taken up here and their depositions taken; and all they got from Mr. Grier was that he supposed he made some allowance and did his work enough cheaper to pay for passing over the road a few times. But the receipted bills, which were paid in cash, show that full prices were paid. And then they brought on Dr. Blaisdell, and Dr. Blaisdell swears in substance—I need not repeat what he swears,—but he says that when he suggested the making of repairs that Mr. Clough never gave him any intimation that he would pass him over the road; and he got down there once to go to Boston and asked him to pass him free, and Mr. Clough was obliged to send him back; and this was after he had got part way over the road. And then they put on Jacob F. Smith; and the object I suppose was to show that Jacob F. Smith had been passed and that Clough had received some pay for tickets in hay; but it turned out, when the gentleman came to testify, that he swore that Clough paid him the money for the hay; so that ended that. And they put in Herman Strauss' deposition, which however was ruled out, I believe, as not tending to show that Mr. Clough received any benefit. However that may be, he says he never in the world furnished him goods at any lower rates by reason of having been passed free over the road. The evidence from Strauss is that he never gave Clough a cent's worth of benefit in any way for any time that Clough may have passed him over the road. Mr. Clough, I believe, swears himself that he passed one or two of his workmen, but it do n't appear that he received any particular benefit one way or the other, or that he would not have passed them if they had worked for Mr. Gilmore, or anybody else. And I submit, then, that this attempt to charge him with having received any benefit from free passes, is an utter failure.

Now, gentlemen, there is another charge that has been made here with a great deal of parade by the other side; and you recollect on a certain day, I believe it was December 19th, when I requested you to make a note of it, the gentleman stated what great things he was going to prove. I told you then that we had had assertions all the way along that at this time the counsel on the other side said he was going to prove "most overwhelmingly,"—these were his words—that Mr. Clough had taken fares repeatedly in the cars, not once or twice, and then had slipped return tickets into their hands for the purpose of inducing them to pay in the cars; that it had been not once, but many times. Now gentlemen, what I have to say in the first place in regard to that is this: unless there is some evidence to show that money has been received by Mr. Clough in that way and appropriated to his own use, he cannot be charged in this form of action. Suppose he has — suppose the very worst that is possible — suppose he has given tickets and taken half-fare, if it is not shown that Mr. Clough has taken that money and appropriated it to his own use, he cannot be charged in this form of action for taking half-fare and returning half-tickets. And there is no evidence that he has ever appropriated a dollar, or a cent. But how are the facts in regard to it? — The evidence is that this whole practice of Mr. Clough — perhaps it may turn out so in reference to the other conductors — of giving return tickets and taking half-fare was a matter that was directed by Gilmore himself. You have the testimony from Sanborn as well as from Clough; you have the direct testimony from G. G. Sanborn — and the gentleman on the other side will not attempt to detract from the credit to be attached to him. It does n't depend upon the testimony of Mr. Clough alone, although there has been nothing put in to detract at all from Mr. Clough's testimony, but it appears from the testimony of Mr. Spalding and Mr. Sanborn, as well as of Mr. Clough, that Mr. Gilmore was in the frequent habit of altering these rules and directing how everything in regard to the Concord Railroad should be done. Why, Mr. Sanborn has just as much disregarded his duty as Mr. Clough. He has just as much liability as Mr. Clough. He has let Mr. Gilmore have tickets out of the ticket-box and failed to return them. He has recognized the "governing power of the road" as in Mr. Gilmore, and he knows and has sworn that he was in the practice of setting aside the rules of the road. Now, gentlemen, how was it? Why, here was, as the evidence shows, this multitude of free passes—everybody had a free pass. Mr. Gilmore became delicate about giving his passes. Here were some few of his fuglers, lawyers and friends; here were my friends Stanley and Clarke, associate counsel in this case with Col. George, and my other friend, Topliff, of Manchester, who was a warm friend of Mr. Gilmore, and they did n't get Mr. Gilmore's pass, and they felt that they had a right to ride free quite as well as others who had Mr. Gilmore's pass; and they complained to Mr. Clough, and Mr. Clough would say to them, "I cannot pass you without authority from the higher powers;" and so Mr. True Garland went to Mr. Gilmore, and through the agency of Mr. Garland Mr. Clough was finally directed to pass these gentlemen by taking fare one way and giving them a return ticket. Now, gentlemen, that is all there is to that; but yet the counsel says that he is going to prove, "most overwhelmingly," that it was not once nor twice, nor thrice that this was practiced, but that Mr. Clough gave a return ticket into the hands of these persons "*for the purpose of inducing them to pay in the cars.*" So the gentleman gets up a considerable bluster over the matter.—There was Edgerly, Stanley, Clarke, French, Topliff, and others that were riding by the instructions of Mr. Gilmore.—And then Mr. Clough was asked to state whether he had ever passed Mr. L. B. Clough; and Mr. Clough testified that he had. It appears, gentlemen, that he had no especial authority, but he said, "If one passed him, I have generally." "Well, what did you do it for?" "Why, I knew he was a particular friend of Gilmore, and I knew that Stanley and Clarke were friends of his, and I did n't know that there was anything wrong in that." I merely bring it up to show the force of his testimony. "Did you pass Obadiah Clough in that way?" *Ans.*—"Well, I would not state positively." The gentleman is going to argue that he did once pass him. What is the explanation? "Why, there were a good many persons that he was passing for Gilmore, and he would tell me to save out so many tickets sometimes, and I would do it; and he desired that a great many people might go for half-fare, receiving tickets back. And it is quite likely that men that I don't now recollect did pass in this way." There was LaFayette Robinson, and Clough was asked the same questions in regard to him; and he says—"he thinks he has sold him tickets." Mr. Clough was also asked if he ever took fare of O. B. Hardy and gave him a ticket to return on. Mr. Clough says, "he always had tickets—I never sold him any." He was inquired of in regard to Mr. True Eaton—he don't recollect in regard to him, or in regard to Mr. David Carr and John S. Carr—he may have passed some of these persons whom Mr. Gilmore was in the habit of dead-heading. Now, then, I believe here are all the persons that Mr. Clough was inquired of as passing in this way for half-fare and giving back return tickets. Now if Mr. L. B. Clough, or Mr. Eaton, or the Carrs, or anybody else were in the habit of riding in this way, unless by the express instructions of Gilmore, why have not these men been produced?

If these practices had been general; if Mr. Clough had been in the constant habit of receiving fare in the cars, and then going up and slipping tickets into their hands "for the purpose of inducing them to pay on the cars," why havn't some of these men been produced here to contradict Mr. Clough, when he says that he never passed them in that way and gave them tickets for that purpose? And from anything that has been shown, gentlemen, in regard to these half-fare tickets, has there been a look as though this was done by Clough to *"induce people to pay in the cars?"* It was a very silly and unnecessary device, one would suppose, when he had such ample opportunities for stealing, if he was so disposed, without resorting to any such mode as this. We never should have had any display of eloquence over such matters as this if they had made out anything on the main charges in their case.

And now, gentlemen, I submit to you that all this talk in regard to this matter of giving a return ticket, to induce people to pay in the cars, so that he might steal the money, rests on no foundation whatsoever; and the "overwhelming testimony" that was to be put in here has not been forthcoming, and there is no evidence upon which, when it is fairly understood, Mr. Clough can be charged as having made any departure from his duty in this regard.

Mr. George.—Mr. Tappan, I don't care to contradict when you make statements of this kind, but I simply desire to have it understood that I don't, by my silence, acquiesce in their correctness.

Mr. Tappan.—You will of course correct me if I am wrong, and the referees will correct me. What is the mistake?

Mr. George.—Your mistakes have been in reference to misstatements in regard to persons to whom he gave a return ticket. I simply wish to call attention to this. Either you are wrong or else Mr. Stanley's minutes are wrong.

Mr. Tappan.—Very well; it is barely possible that I may be mistaken, and it is barely possible that Mr. Stanley's minutes are mistaken. I have stated it as I have it in my minutes, and as I believe the testimony to be.

Now, gentlemen, a great deal has been said in regard to this matter of "using tickets over and over and over again;" and I should really be glad, gentlemen, if you have time, as a matter of curiosity, if you would just look over your minutes and see how many times the counsel on the other side has repeated that language, of "over and over again," in this connection. They were going to show how Clough had taken up these tickets and sold them "over and over and over again;" and we have had it *ad nauseam* during the twenty-seven days since this trial began. I admit that it was possible for Mr. Clough, or any other conductor, to have taken up these coupon tickets and used them "over and over and over again," if they pleased; but what I say here in the first place is that there is not a particle of evidence except what came from the testimony of Mr. Clough himself, as I understand it, that any such tickets have been used in any such way. And then it has been explained to you how and why they were so used. They could be used to a limited extent, I admit, but it is one thing to charge and it is one thing to show how these tickets *could* be taken up and used "over and over again," and another thing to back with proof that they *have* been so used.

Now, in the second place, gentlemen, you will recollect the testimony of Mr. Sanborn, who swears distinctly in reference to this coupon ticket matter, and the practice of the road under it. Mr. Sanborn swears expressly—and he shows you how and why—*that no considerable peculations or operations of this kind could be transacted without the whole thing being brought to the knowledge of the road.* All the tickets issued by the upper roads are charged to the Concord road; these coupons are taken up by the conductors and returned to the general ticket agent, and the Concord road settles with the upper roads at the end of every month. The coupons returned by the conductors should correspond with the number charged to the road, and if they did not so correspond—if any were missing, kept back to use "over and over again," that fact would be known at the end of the month.

And how can Mr. Clough know, supposing that he had any desire or wish to use these tickets over again—how could Mr. Clough know that the tickets he gives out, or, if the inference is that he gave out tickets in large masses, as the Whitcher tickets, which were "used over and over again"—how can Mr. Clough have any assurance that those tickets would not be kept out, or that they would come around by the end of the month? And Mr. Sanborn swears if there was any material discrepancy between the number of tickets reported by the upper roads and the account at the office, they would discover it at once and inquire into it. It is absurd to say that Mr. Clough, who is a somewhat shrewd man in his business matters, if he was doing anything fraudulent, would have been likely to have laid himself open to detection in this way. These tickets were merely local tickets and could not be used off the Concord Railroad. I think Mr. Clough states that he may have perhaps given one of these tickets in a few instances, when he happened to have no other local ticket. You will see, therefore, even if Mr. Clough were so disposed, he could not abstract these tickets and use them "over and over again" to any considerable extent, or put any considerable amount of money into his pockets in that way. It could not be done. But what necessity was there, gentlemen, for doing it? As I have said, they were merely local tickets—they could be used only on the Concord railroad; there was no necessity for giving such tickets. When a man was in the cars and had no ticket, why should Mr. Clough, or any other conductor, when he paid his fare, give him one of these tickets? Why couldn't he, instead of giving him a ticket, take the money and put it in his pocket? Every ticket that he kept out in that way would be the ready instrument for his detection. There was no necessity for his doing it if he wanted to steal. All he had to do, and which was more safe for him to do, was to take the money and put it in his pocket without giving any ticket at all.—And I don't see what benefit it could be to Mr. Clough or anybody else, the practice of stealing these tickets in that way.

Now a word as to the "joint tickets." My brother has the reputation of being an astute lawyer, and what he don't know, especially in regard to the law and practice of railroads is scarcely worth knowing, and I was therefore a little surprised at the nature of his inquiries, and the pertinacity with which he urged them in regard to the matter of these "joint tickets." You will recollect that he pressed Mr. Sanborn, the ticket-master, and the other gentleman, very hard indeed, and tried to get out of them that Mr. Clough could have retained these joint tickets, and he put this question "over and over again:"—"Supposing that you sold ten tickets from Concord to Boston, and suppose having sold ten tickets, he put the money in his pockets, and supposing he took up ten tickets and didn't return them, now I want to know, what means you could have of knowing that?" That is the exact question, as taken down by the short-hand reporter; and that is the question that he crowded on these men, and as often as it was asked, Mr. Sanborn repeated to him in the most natural manner that could be, "*Why, sir, this could not be done without collusion with the conductors below, and with the passengers.*" And I have heard something said before with regard to the testimony of these lower conductors. I should be glad, if there is any testimony from them, that they would put it in. But at any rate it is enough to say that it could not be done without collusion with the conductors below, and also with the passengers. And if it was so, they could have had these conductors on the stand. I understand that their testimony has been taken in depositions. Can any one tell me why they have not been used? It thus appears that the stealing of these "joint tickets" (i. e., tickets that were good over other roads,) could not have been done at all, because they were charged by the ticket-master directly to the conductors, who had to account for them dollar for dollar. And this disposes of another of the gentleman's long-winded "suppositions."

[At this point a recess was taken until 2 o'clock, P. M., after which Mr. Tappan continued.]

There was one fact, may it please the referees, when I was touching, before the adjournment, on this wholesale plundering in the cars, (which is of course the main charge on which this prosecution rests,) which I accidentally omitted, and it is this: the practice (which has been proved by the testimony of Mr. Upham and of Daniel S. Webster,) of Mr. Clough, in every instance where he could, to have the passengers purchase their tickets at the office. You recollect the testimony of Mr. Upham, that it was frequently the case, when the passengers came on to Nashua without a ticket, that Mr. Clough would send the passenger or a brakeman to purchase tickets for them at the ticket office, so that they might be furnished with them in the cars. The testimony of D. S. Webster, which was put in only yesterday, I think, on this point, was, that it was the invariable practice of Mr. Clough, when passengers came from the Nashua and Worcester road, to assist them and direct about purchasing their tickets. Now, this is, perhaps, a small matter, but after all it is a significant one, and I put the question to you if the conduct of Mr. Clough in this respect indicates that he wanted to induce people to ride without buying their tickets at the ticket office, so that he might steal their fares? On the other hand, does it not indicate that he was an honest man, and that in every instance where he could, he desired that the passengers should be furnished with tickets; and does it look likely that a man who was constantly—every day—taking this course to have the passengers supplied with tickets, would be taking half-fare and giving a return ticket to such gentlemen as had these tickets, *to induce them to pay in the cars?*—And that is all I propose to say in regard to that. You will bear in mind, however, that the testimony is that passengers could not get their baggage checked till they showed their tickets.

Now, there is another matter, gentlemen, that I will refer to very briefly, and I feel that I ought to hurry along as fast as I can, for I know you are anxious that these arguments should be closed to-day. And, I should almost be ashamed to allude to it, if it had not been brought forward here with so much parade as the third article in the famous "synopsis;" which is, that Mr. Clough didn't punch the tickets, simply making a pretence, and making them susceptible to a second use. It is charged that Mr. Clough didn't punch the tickets. Now what is the evidence upon which that third charge of the "synopsis" rests? What is it based upon?—Why, just the simple testimony of Mr. Wm. Roby, who rode in the cars with his wife and gave his tickets to Mr. Clough; and Mr. Clough, as he thinks, failed to punch his ticket. And what did he say? Here is the exact language he used, showing you how the thing presented itself to Mr. Roby: "Mr. Clough took the ticket and went to punch it, but the punch *didn't happen to hit it.*" This is what old Mr. Roby says; and that is all there is to this, gentlemen; that is all the evidence upon which this charge rests. "He didn't *happen* to strike it; it went by the end and didn't strike the ticket, and he handed it back to me and afterwards took the ticket off." Who knows but what Mr. Clough afterwards punched the ticket; and what benefit could it be to Mr. George Clough to take a ticket in that way? Who knows but that it *did* "happen" to hit, and that Roby is mistaken? But at most the evidence shows that this failure to punch Roby's ticket was purely accidental—the punch slipped by the end and didn't *happen* to hit it! And if he sold any such ticket to any man, where is the witness that has been before you to show the fact? Why, gentlemen, while he was finding a purchaser for a ticket taken in this way, he could have stolen, if he had been so disposed, a hundred dollars in the cars. The thing is simply ridiculous and absurd, and only shows the straits to which they are driven to find something to substantiate the charges which they have made against George Clough. And there I dismiss the matter. Now, gentlemen, we find ourselves against another "estoppel," and that is the buying and selling of these coupon tickets. Failing, as I think I have shown in the examination of the testimony that I have gone over, to show where Mr. Clough has fraudulently appropriated a single dollar, or a single cent, and failing to produce any evidence which tends to charge him in that direction, they are obliged to stand upon all sorts of shifts, and upon all sorts of positions and "estoppels" in order in some way to charge Mr. Clough, when everybody understands that if there is anything against him it is this wholesale stealing that he had been carrying on for years, on the train.

Well, now, how is it about this matter of buying and selling western coupons? The evidence upon which that rests is the testimony of Mr. Clough, himself, which does n't deny but that he purchased these tickets, and we do n't deny but that he made something out of them. It is a matter that he might have concealed, if he had been so disposed.— I put it to this board, that, if he was disposed to commit wholesale perjury, he might have well concealed this matter. Instead of that, he comes forward and tells us exactly how it was; and he tells you he bought and sold these tickets.— The evidence shows, and I believe (with great submission to what the opinion of the honorable board before me will be,) that the law will warrant him in doing everything that he did in regard to it. He says that he purchased and sold western coupon tickets, that he made somewhere in the vicinity of $2000 out of that transaction. Well, gentlemen, it was notorious, (at least the evidence before you shows,) that White and Weston, and traders in Concord and everywhere, were dealing in these western coupon tickets. Nobody thought there was any harm in it, nor any wrong.—

They came here in that shape through the instigation of the Concord Railroad themselves. They were issued for the purpose of inducing travel to come this way instead of going in other directions; and hence it was just as cheap to go all the way to Boston as to go to Concord or Manchester, and you could buy a ticket from Detroit and all points west cheaper to go to Boston than you could to go to these other places. And so it happened that hundreds of Frenchmen, coming from Canada, and others, coming to Concord and Hooksett, or wherever they were going, purchased their tickets through to Boston, which gave somebody a right to ride to Boston. That was the contract which the Concord Railroad made when they issued those tickets, and that, purchasing a ticket to Concord or to Manchester, after they had traveled as far as they chose upon it, if they saw fit to transfer it to somebody else, was no fraud on anybody. But what did Mr. Clough do in regard to this matter? Before he did anything he went to his superior officer and told him what the others were doing, that they were buying and selling these tickets, and he asked him if he could not have the liberty to do it. And Mr. Gilmore informed him that he didn't know why he could not make something out of it, as well as anybody. And here Mr. Clough was only following his natural instinct, such as he has shown everywhere, to improve every occasion where he could, to turn an honest dollar, and to accumulate the property that he has accumulated. Mr. Gilmore, as he said, told him he did not know why he might not make something out of it as well as anybody else. And with his approbation he bought some of these tickets to the extent of what he says he bought, though he don't undertake to tell or to know the precise amount.—

Now I don't propose to go over my views again as to how far the road would be holden by the actions of the general agent; but this I believe, that so far as the fact of notice to the corporation is concerned, that notice could not come officially through any other source than Joseph A. Gilmore. And my position is that if Mr. Gilmore, the general agent, had knowledge that Mr. Clough was dealing in these coupons, that is the only way that this notice could be brought home to this corporation. They had notice of what he was doing. They continued him in his place; and they are "estopped" now from showing that he had no right to deal in these tickets.

But I will state here my position in regard to an agent's going beyond the scope of his authority.

I do not know whether or not Gilmore, in giving instructions and saying to Mr. Clough that he might do this, was going beyond his authority. I have no doubt that it was notice to the corporation. But my position is that the road held him out, and enabled him to hold himself out in such a manner, as possessing a more enlarged authority than the letter of instructions would give him; and therefore they are bound by his acts, by the adjudicated cases in this state. And therefore I say, if this was an act beyond his powers and the scope of his authority that the directors of the road enabled him to hold himself out as having all the powers that he claimed, and that they cannot now turn round and repudiate what he has done.

But, gentlemen, there is another thing about this. As I understand it—I do not pretend to be a very good lawyer, not so good as either of the legal gentlemen before me—but it has seemed to me, in looking over all the evidence touching this point, that there is not any evidence here that could properly go to a jury, to charge Mr. Clough with a single dollar of this two thousand as against the Concord Railroad. I cannot see that there is any evidence that a jury has a right to hear under the instructions of the court as to how they shall assess damages or compute the damages growing out of this transaction. Mr. Clough made this money by selling these coupons, not out of the Concord Railroad altogether, but also out of the Lowell road, and the Nashua and Worcester road. Other roads were entitled to some portion of these tickets that he purchased. Mr. Sanborn swears that the amount which the Concord Railroad was entitled to from these western coupons averaged all the way from twenty-eight cents up to fifty-two cents, at different times. Now how has it been shown by the other side how many of these tickets were purchased when their proportion was twenty-eight cents, and how many when their proportion was fifty-two cents, or when they had not a right to a single cent, when the tickets were taken up at Nashua, going to New York, or from Nashua to Boston? I say that the burden of proof is upon them. It is incumbent upon them to put before this tribunal such data as will enable you to compute and to say how much Mr. Clough has received which belongs to the Concord road, out of that transaction. It is but a mere fractional part of the whole, at best; and there is no evidence upon which you can say that there is a dollar.

But, in accordance with the suggestion that has been made by the chairman, suppose the Concord Railroad have the right to adopt his acts, and claim the money that he has received. I think that must go upon the assumption (with all deference to the suggestion) that the corporation would have sold a regular ticket every time that one of these coupons was sold; or in other words, that that coupon would not have been ridden upon. As I understand it, in order to charge Mr. Clough on that ground, it must appear (and the suggestion goes upon that ground) that that ticket would not have been used at all. Well, now, my position is exactly the reverse of that; and I submit that the presumption is exactly the other way. I think the fair presumption is that somebody would ride on these tickets, and have a right to ride upon them, whether Mr. Clough sold a regular ticket or not. And if Mr. Clough had not bought them, somebody would have bought them, and would have had the right to have ridden upon them. And I submit that there is no wrong to anybody; nor should Mr. Clough be charged upon these technical grounds of not having brought home notice to the directors of the corporation, although he had the sanction of its general agent.

I confess that I am a little puzzled over these rules over which such a display and parade have been made. This rule adopted in 1860 for instance:—"All tickets over the Concord, Manchester or Lawrence Railroad shall be dated on the day of sale; said tickets shall only entitle the holder thereof to a passage on the day of their date; provided that joint tickets shall be good for such further time as may be necessary to enable the holders, by the regular trains of the road, to reach the stations to which such tickets are sold." I think that rule will fairly bear the construction that it is to apply only to tickets sold by the Concord road. But suppose it applies to tickets sold on the western roads as well. What was the practice under it, gentlemen? What was the practice under it, as shown by the tickets that have been produced before you, and as shown to you by the testimony of Remick and, I believe, of one other witness who has testified in the course of this trial? What was the practice under that rule? Why, it is evident at once that the western roads, if they had assented to that rule, and every road that was bound to comply with that rule, could not comply with it literally. It was impossible for a person to ride on a ticket of that kind, in compliance with that rule. There must be *some* time within which such tickets should be good, and some time beyond which they should not be good. The testimony of Mr. Sanborn is that these tickets were marked as being good for "thirty days," or "fifteen days," and at one time for "ten days." But there was a time limited, and a ticket had been produced of that kind, showing that it was good for "thirty days." And Mr. Remick swears that the tickets that he had, some of them were marked good for thirty days, and that they were tickets that had not spent their force, but were good for that length of time. I cannot see, therefore, how any rule of the road has been violated at all.

And now, in regard to the instructions of 1864, it seems to me to exactly confirm, by the very letter of the instructions, the construction that I am contending for. The fourth and last clause is:—"Tickets and coupons over other roads shall not be sold, * * * * when they believe, or the date indicates, that they were not received in the "regular course of business."

If they were received in the "regular course of business" of the road, nobody can complain. The complaint was in the case of Johnson *v.* The Concord Railroad, that the ticket was not received in the "regular course of business." It was so far beyond its date — being some four months after date — that nobody could contend that it was good. And the point was not made there, as I understand, in relation to tickets that were not spent.

Well then, the instructions of the road did not prohibit anybody from riding on such tickets; and if it did not prohibit anybody from riding on such tickets, it did not prohibit Mr. Clough, or anybody else, from taking a transfer of such ticket. The "regular course of business" would be within thirty days, if that was the time, or within fifteen days, if that was the limit of the ticket, or within ten days, if that was the limit of the ticket. And I say, not only as a matter of fact, but as a matter of law, that when Mr. Clough purchased these tickets with the full knowledge of the agent of the road, which was the full knowledge of the corporation, and when these tickets were not spent tickets, but tickets that somebody had a right to ride upon, and within the regular course of business,— I contend that Mr. Clough did just exactly what he had a right to do. And I think everybody would feel that it was an outrage, after all this parade of virtuous indignation, and all this great show of doing something to uncover the frauds and peculations of these conductors, that the road should now plant themselves on this position and only attempt to charge him in this roundabout way. And that, gentlemen, is all, I believe, that I care to say in regard to that matter. As has been suggested to me, it is not what the Concord Railroad might have got, but the question is what they have lost. And what evidence is there to show that they have lost a single dollar? They have, in doing this, only carried out the letter of their own contract; a contract beneficial to them; a contract with the western roads to bring travel this way. And it would be monstrous for them to undertake to turn around now and repudiate a contract with a man who has purchased these tickets with the full knowledge of the "governing power" —I undertake to say—of that road.

Now, gentlemen, I come to the last topic upon which I propose to speak in this case. And that is the question of Mr. Clough's property. I am reminded, however, that I have overlooked the "peanut boy!"

Well, gentlemen, as I referred to the peanut question casually, as I went over it, I do not suppose it is necessary to refer to it again. It only shows the frailty of the other side, in fastening to every little twig in order to save themselves. Their powder-boat having blown up, and failing to show any stealing in the cars by Mr. Clough, they are obliged to cling to every little thing to save themselves; and they have fastened at last on the *fares of this peanut boy!* And they say that Mr. Clough is liable for the fares of that boy, because he set him up and furnished him capital and paid him so much a day. Well, gentlemen, I am not going to say much about that. That only shows Mr. Clough's character in doing everything he could to make a cent. And I think it was certainly quite as much of an advantage to the boy; his gains were sure; he had somebody to stand behind him, and somebody to advise and direct him what he had better sell; and he had somebody to assist him and furnish him capital. And it was for the advantage of this road that these boys should go on the trains. Some carried water, and some carried lemonade, and some carried newspapers, and some carried peanuts. And who ever heard, gentlemen, that any of these roads exacted fare? And does it make any difference in this, whether he was hired by Mr. Clough or not, that because Mr. Clough made something out of the transaction, the road can turn around now and recover of him the fares of this boy? Well, I suppose that is going to be seriously argued. And the gentleman is welcome to all he can make out of it.

Now, gentlemen, I come to Mr. Clough's property. And this is another great charge in the "synopsis" of my brother; that is one thing that has been paraded on all occasions, in public, at the corners of the streets, and in the course of this trial, "over and over and over again;" that Mr. Clough commenced running on the cars when he had but very little property, and that now his property has swelled up to the enormous amount of one hundred and fifty thousand dollars or more! And, gentlemen, I am very glad to say and to know that George Clough has, by honest accumulations and industrious habits, got some property. And I am very glad that we have been able to show you that he came honestly by every dollar. Three years he has stood here with this brand upon him, followed and hunted by the hounds on his track, and feeling as if every man, woman and child that looked upon him regarded him with suspicion, and as though he had not a right to live in his own house; that even his silver ware, that he had got by getting subscriptions to a certain paper; and that the very spoons that came down to his wife from her mother, had been stolen from the Concord Railroad. I thank God that the knowledge is going to the public, sanctioned, as I believe it will be, by the verdict of this tribunal, that the property which Mr. Clough owns he has a right to own—that it is his—and that if he has got a house, he has a right to live in it; and, if that home has been made beautiful and attractive, and surrounded by the evidences of taste and refinement, that it shall redound to Mr. Clough's credit; and that he has a right to be proud of his home, because his own honest industry has earned it. And I can stand here and say this, feeling it to be true, in the face of every aspersion, and every charge that has been made.

Why, gentlemen, look at George Clough. See him start out from home, at fifteen years old, a poor boy, with all his worldly effects on his back and tied up in his pocket-handkerchief! That was George Clough when he started out at the age of fifteen to make his own way in the world! But, gentlemen, that poor boy had about him a capital that was worth more than twenty thousand dollars left him, in the absence of the habits that he carried from his home with him. The capital that he started with, when he left the paternal roof, with his earthly effects tied up in that little bundle, I repeat, was worth more to him than if twenty thousand dollars in money had been left him that day by his father—without the habits that he possessed and the resolution he had formed! Now, follow him. Follow him to all the places that he went to. Look at him at Newburyport; a chore boy, working incessantly, by day and by night, wearing the cast-off clothes of the boarders at the hotel, saving every dollar that he made, blacking boots, tending stable, doing all he could to turn an honest penny; never for once—no, not for once—going into a grog-shop; never for once visiting a saloon; never for once going to a dance; never for once allowing himself a single day or evening of recreation; but saving his earnings and laying up, in two years, three hundred solid dollars of his own! I tell you, gentlemen, that instances of this kind in a boy of that age and under these circumstances are rare. And I tell you further, gentlemen,—what you know as well as I know—that you never saw a boy—there is not an instance on record where a boy of that kind starts out in that way, with those habits, with that determination to save what he earns, and to be a rich man, but that he always succeeds; and he always will succeed, and always ought to succeed.

Well, gentlemen, he told you that he took all these hard earnings of two years—these three hundred dollars—and placed two hundred dollars of it in the Newburyport Bank, keeping a little for himself and giving the rest to his mother.

Well, gentlemen, here happened to be something that did not die. And we corroborate the fact of the deposit of his first earnings in the Newburyport Bank. It would have been counted on the other side only as one of those lies and fabrications, if we could not have brought the cashier of the bank where he deposited that money. It would have been one of Mr. Clough's fabrications. But it so happened, although some of those with whom George Clough has had dealings have passed to that "bourn from whence no traveler returns,"—but for which, notwithstanding the innuendo of the gentleman, I don't see as Clough is to blame,—that the bank was in existence, and we could bring the cashier of the bank and verify Mr. Clough's statement. He put two hundred dollars in the bank, and went home and visited his mother, and gave his mother fifty dollars, and started out to try his fortunes again. We find him next at Raymond; never losing a moment, saving all that he earned, receiving his wages, gradually increased when he came to be hostler; increased again when he came to be stage-driver—and so he went on, until finally he came to be owner of the line of stages on the Mammoth road. And we find him, when he came to be stage-driver, with these same habits, contriving every way to increase his earnings, and the result was, that while on that staging, out of the perquisites he got by the carrying of eggs, butter, poultry, veal, and everything that he carried, he made more than his wages. And it was so notorious that Clough carried all these things, and was trading in these articles, that his line became known all the way from Pittsfield to Lowell as the "Chicken Line!"

Well, gentlemen, I say that habits such as these will insure success and prosperity. I say that habits such as these will insure wealth. And it is shown here that from the time Mr. Clough started out from his home, in the way that I have described, until he was made conductor on the Concord Railroad, he never lost but twelve days of his time; and those were ten days when he was sick with the measles, and two days when he stopped to go and get married! These are all the days that George Clough lost. And I thank God, too, that he got a wife that was a help-meet, and who co-operated with him, and who was content for many long years "to labor and to wait" in an humble position and an unpretending residence—who was content to labor and struggle with him—who was not ashamed to do her own work with her own hands—who was content to make her own children's clothes out of the cast-off clothes of their father. And in this way their property grew, until to-day they have a competence, and it is property for which he owes the Concord Railroad no dollar; and the Concord Railroad has no right to try and wrench it from them!

Well, gentlemen, the counsel on the other side says, or gave you the idea, that Mr. Clough had two or three thousand dollars when he started as conductor. But we have shown you (and that you at least will not gainsay,) that he had property to the amount of eleven thousand dollars at that time. That was his cash capital. That is the capital that he had to start with before he ever went on to the Concord Railroad. And that is a pretty good capital for a poor boy starting out in life at the age of fifteen, his worldly effects all told not worth nine shillings! And it is a pretty good capital in the hands of such a man as George Clough to accumulate and to build on; and it has accumulated until at last it has reached the sum of seventy-two thousand dollars. Why, you may take that capital at the time he went on the road, and putting it upon compound interest, that capital to-day will amount to nearly fifty-two thousand dollars! You all know how money accumulates; and you know, as well as I, how property increases in the hands of such a man as George Clough—such a man as he has shown himself to be in all his business transactions.

It was only the other day that an old gentleman that I, and probably the members of this board (or some of them,) have seen—a man who had charge of the baggage at the Astor House steps, New York, a "friend of Daniel Webster," as he was proud to call himself—a man that I have been in the habit of always seeing at the stairs at the Astor House, whenever I have been in New York, who went there twenty-eight or thirty years ago, with no property at all, who had no salary, and no pay there, only the perquisites that he could get from carrying the baggage up and down stairs and taking care of it. And the other day he died, and his remains passed through Concord on the way to their last resting-place; and he left behind him a property of a hundred and forty thousand dollars! Is there anything extraordinary in this? Why, another man might have gone to the Astor House, and in three years been a confirmed drunkard.—George Clough would not. And Mr. Jones was not. Some men would have squandered their earnings as they went along, and always have been poor, but not such a man as George Clough has shown himself to be. Instead of spending any portion of his time in haunts of pleasure or of dissipation, he was always, early and late, attending to his business; instead of squandering, he *saved*, and put at interest all he earned, and he has a right to be rich!

Is there any doubt, Mr. Chairman, and gentlemen, that George Clough possessed eleven thousand dollars when he went on to the Concord Railroad? Do you doubt it? Is there any chance to doubt it? He swears to you distinctly how he made it. You have seen it, and you know it. We have corroborated his testimony as far as we could. And the only point where I suppose it is to be contradicted is the fact that Mr. Clough says that he had six thousand dollars in the hands of Mr. Raymond Kimball of Lowell. Gentlemen, have you any doubt of that fact? Why, they would have you believe, perhaps, that Mr. Clough fabricates this story, and states this fact now because Mr. Kimball is dead. The fact that Mr. Kimball is dead is not so unfortunate for them as it is for us. And they bring Mr. Kimball, his nephew, here to testify in regard to his uncle's business. And even Mr. Kimball testified on the stand to the character of his uncle, and told you that he was a man who, when he took a liking to another, he could not do too much for him. He took a liking to George Clough, and he acted as George Clough's friend, and as his banker. And George Clough, acting under his advice, whenever he got fifty dollars, put that into his hands, and Kimball invested it for him, or held it himself. And when he wanted it, after he came to Concord, and it became necessary for him to use it, he drew upon him for the money, until he got it all. We did not, of course, expect, after this lapse of time, to be able to corroborate Mr. Clough's testimony in regard to many matters. We can, of course, in the absence of such corroborative testimony, ask you that Mr. Clough's testimony shall be believed when there is no evidence to contradict it. But we have thought it our duty to substantiate the testimony of Mr. Clough as far as we have been able to do it. But we had no idea that we should be able to corroborate the fact of this money in Kimball's hands. That we are able to do so seems almost providential.

We called Mr. Sargent particularly in regard to his knowledge of Mr. Clough's staging, and were not aware, until he came here, that he knew anything about Clough letting Kimball have his money. But it seems that Mr. Sargent used to be there in that stable, and he states that there was a desk there which Mr. Kimball had, and that he had repeatedly seen George Clough give him money, and that Mr. Kimball would go to his desk after taking the money, and hand him a piece of paper, which he supposed to be a note. So that we corroborate the testimony of Mr. Clough as far as we can in this particular. And if there was no other evidence in this case, I believe that you can find that George Clough had this sum of eleven thousand dollars. And with his habits, and with this sum of eleven thousand dollars to start with, I might well ask you, and I might well ask everybody else to believe, that that was sufficient to account for the amount of property that he has to-day.

But, gentlemen, we follow him further, and see him on the cars. We see him taking advantage of everything that turned up in his favor to make a cent of money. We see him carrying berries, poultry, all kinds of farming produce; we see him dealing in potatoes; and he has stated to you the amount which he thinks he gained in that way. If any of you are familiar with the trade of these men who are speculating in and dealing in these articles every day, you would not be astonished that Mr. Clough made what he did in carrying on this business. He carried on this business over the freight trains mostly; and it is some four or five thousand dollars that he says he made in this way. The people of Lowell relied upon him for their marketing articles more than they relied on the market. Such is the evidence. We have shown you by witnesses from Pittsfield, from Franklin, from Concord, and elsewhere, the extent of his operations in this particular. He had persons buying for him in large quantities; and every day, each way over the road, for years together, he was transporting these articles and selling them at a profit, and in this way adding to his property.

Now, gentlemen, I propose to do little more than to give you some facts and data upon which we rely in regard to the amount of Mr. Clough's property. It has been incumbent upon us, by the course that the investigation has taken, to show how much he had, so far as we could, and how he got it. Of course, you understand that this is not a matter that could be reduced to mathematical demonstration. Nobody could do it. But if we show you how this probably could have been, and give you a probability as to how this property could be acquired, we have met and answered all that was required of us. But, fortunately, we have been able to go further than mere probabilities and conjecture; for we think we have shown you, step by step, how all his property has been acquired.

Now, gentlemen, our theory, if it be a theory—(but it is not a theory, it is a fact,)—our theory, our position is this: We start Mr. Clough off with eleven thousand dollars of property. We have shown you the gains that he made upon that property. And we have simply kept that property at simple interest. I mean the original capital of eleven thousand dollars, and the gains that he has made upon it, which amount to the sum of his property on the 14th day of February, 1866, when these suits were commenced. We have not deemed it competent to show what his property is valued at to-day. That is not a point in this case. The great question which you want to know is how much the property which George Clough has got to-day cost him. That is the question, as I understand it.

That property cost him, in February, 1866, as he has testified, and given you the details—you will find it not only in the testimony which you have, but more elaborately in the deposition which he has given, and more fully explained, I think,—that property cost him ninety-one thousand nine hundred and thirty-five dollars and sixty-seven cents ($91,935.67). Deducting his debts and liabilities at that time, nineteen thousand six hundred and seventeen dollars and twenty-one cents ($19,617.21); and you have seventy-two thousand three hundred and eighteen dollars and forty-six cents ($72,318.46) as the actual cost of his property in February, 1866.

I will leave all my computations and figures with the referees, if they will be of any service to them, and perhaps save the trouble of taking fuller notes. They will do, however, as they please about that.

Now, gentlemen, as I have said,—and I desire the referees to understand the position that we take; and if there is any fallacy, or anything that is wrong, or anything that will not bear the test of examination, I am sure the referees will understand it; I cannot see the fallacy of it;—now then, we have taken this eleven thousand dollars and put it at interest; and the interest on that, for that time, is fifteen thousand eight hundred and forty dollars ($15,840).

Now that this matter shall be exactly right, and that all proper deductions shall be made, Mr. Clough at this time owned two houses; and we have cast simple interest on the cost of these houses, and that amounts to two thousand one hundred and sixty dollars ($2160); and we have deducted that from the interest on eleven thousand dollars, and we make the balance of the interest thirteen thousand six hundred and eighty dollars ($13,680) on the eleven thousand dollars. Or in other words, we do not think it is extravagant to say that a man with Mr. Clough's business habits would keep his capital at simple interest. The fact is, that in many instances he has received a great deal more; but our theory now is at simple interest only. Then, gentlemen, we take his salary that he has received; of course out of that we take his cost of living. We take his salary, and that salary for twenty-four years and a half is fifteen thousand two hundred and seventy-five dollars ($15,275). Then there is the Brown building on Pleasant street. That property he has on hand now. And we have put in a table here showing the net gain or net profits, deducting all outs and expenses, that he has made on that Brown lot. I think there is one thing that you will think is somewhat surprising; though, perhaps, not surprising in a man of Mr. Clough's business habits. And that is, that out of so many transactions that he has been able to lay before you, he has, in such a large majority of them, been so fortunate. Everything seems to have turned to his advantage. Take his horses: the gains which he made, the prices which he gave, and the prices which he got, would seem to be extravagant almost. But we have substantiated these in many instances, and we have given names and dates in all, and if they are not true, witnesses could be brought to contradict them. So it is in other particulars; take this property; it has grown on his hands; take this property, for instance, on Pleasant street; that cost him, originally, five hundred and twenty dollars; he has received, for seven years rent, the amount of fourteen hundred dollars ($1400); the interest on that rent is four hundred and sixty dollars ($460). The gross receipts from that purchase amount to eighteen hundred and sixty-four dollars.

Now we take the water tax, forty-two dollars ($42), and the interest on that. We deduct the interest on the original cost, two dollars and forty cents ($2.40); we deduct insurance, seventeen dollars and forty cents ($17.40); we deduct interest on that; we deduct the taxes; we deduct the interest on the taxes; and the whole outs amount to three hundred and ninety-eight dollars and seventy-eight cents

($398.78). Deducting that from $1864, leaves $1,465.34 as the absolute net gain on that one purchase!

Now, then, there are the two houses on Thompson street. These houses cost originally twenty-two hundred and seventy-five dollars ($2275). The rents that he has received from 1849 to 1862 are forty-four hundred dollars ($4400). And he has detailed all this in his testimony, and shown from whom he received it. The interest on that sum is twenty-five hundred and fifty-one dollars and sixty-eight cents ($2551.68). The rents from 1862 to 1864 were six hundred and eighty dollars ($680); the interest on the same, one hundred and two dollars ($102). The rents from 1864 to 1868 were eight hundred dollars ($800); making in all the total receipts $8,533.68. And now we take from that the interest on the amount invested, $2068.56, water tax, interest on that, repairs, insurance, interest on insurance, and taxes, making in all $3366, which, deducted from the total receipts, leaves $5167 as the actual gain on the two houses on Thompson street.

In the same way, taking the original cost of the Jefferson street property at six thousand dollars, and taking his rents and receipts from the same, and deducting the taxes and all the outs, we find that he has made on the Jefferson street property $1925 actual gain, and legitimate gain, and not a dollar of it stolen.

Well then, gentlemen, here is the Sanger house. The gain on that, as his testimony shows, is $596, and on the Washington street property, $612.

As I said, all these matters are fully explained in detail in his testimony. I will not undertake to detain you with the details. The referees will bear in mind that these houses that I am naming now are houses that he bought and has sold. Then there are the other houses still in his possession. But on these other houses I have made no computation, but only on property that he has bought and sold, and where he has actually made these gains which we have shown in proof. Then there is the Jamaica Plain property, etc. I will not detain you with the details of that, but he has made on that purchase at Jamaica Plain $4798.81, the facts in regard to which you fully understand. And on the Masonic Temple property we have gone into the same estimate, and will furnish you with a detailed statement of rents received and taxes and outlays that have been paid out. And I do not count now the rise in the value of property; but merely what he has actually received from rents, over and above what he has paid out, and over and above the interest on the money. And he has received $4122.01.

On the Manchester Print stock he made by absolute sale $1500.

And on the transaction with Mr. Elkins, as he has shown you, he made $5000. On the Boston, Concord, and Montreal stock he made $2000. On the old stock, $200. On the Orange farm, $1000. On the steamboat stock, $2630, more than he gave. On his Manchester land he made $500. On the sale of western tickets he made $1500. The gain on the Page farm was $100. His profits from cows he estimates to be about $40 a cow. Here was a great splurge made, as if a man could not make such a profit on the keeping of cows. Why, gentlemen, if you know anything about the keeping of cows, you know that every farmer will say that the most profitable business of the farm is this matter of dealing in milk. But the gentleman thought he had got us pretty tight when he found the tax bill. But he kept the cows, and sold the milk; and you can easily see how much the milk will amount to in a year, and see how much Mr. Clough would be likely to make. And I submit that Mr. Clough had cows taxed. But however that may be, it was very evident that the counsel on the other side didn't believe they were there, and that consequently the facts stated by Clough could not be shown. But, gentlemen, we happened to be verified here. We could get the very man who took care of them. We put before you the very man who milked the cows, the very man that peddled the milk. And, gentlemen, I think we have made that matter pretty tolerably certain in your minds.

He got $210 from securing deserters. He says, by estimate, he made these years that the boy ran on the cars, $3276; he estimates these profits at only about nine shillings a day. That is for you to judge. It is a mere matter of estimate. But it all shows that Mr. Clough was not afraid nor ashamed to do anything that was honorable to increase his means. I would like to know where there was another conductor, where there was another man that had a chance to steal a hundred dollars a day without detection, as they assert, that would go into these little speculations for the sake of making a small amount of money, and thus by little and little adding to his gains?

And, gentlemen, I might ask you another thing here, for fear I may forget it. I ask you if a man of ordinary shrewdness, (and I think Mr. Clough has turned out to be a man of ordinary shrewdness,) if a man such as Mr. Clough has exhibited himself before you, and if he had been plundering by the hundreds and thousands—if he had got his property by stealing from the Concord Railroad, whether he would be likely to parade these ill-gotten gains before the public? This is not the way that tolerably shrewd men usually act, and I don't believe that George Clough would have been likely to have lived in a house (though he may have bought it for less than half its cost,) that would excite the envy and jealousy of those not so fortunate as himself, if the money that went into its purchase had all been stolen!

Now, gentlemen, I suppose a good deal is to be said about the good house that Mr. Clough lives in. I think you will acknowledge that even a man who has a good deal of money would be glad to have got a house that cost $22,000 for $8000, all furnished. But I ask you again, as reasonable men, if George Clough had been cognizant that that money taken to purchase this house had been stolen, would he have been likely to have thus paraded his ill-gotten gains? I tell you that men don't generally do that. But George Clough and his wife and family had lived in a poor house for many years. And they had toiled there, as I have said, together. And they had a right when the property accumulated, when the money had grown upon their hands from the honest gains that he had made in real estate, and in every other department that he turned his hands to—they had a right to live in a little better house. And if they could get it at a bargain, they had a right to live in the best house in Concord—if it be the best house; I do not know but that it may be; and nobody has a right to call him in question about it. And I ask you, again, if any man believes, who has common sense, that if George Clough had been conscious that he had done the Concord Railroad any wrong, and taken money which did not belong to him, he would have paraded himself before the public in this ostentatious way? Nobody can believe it.

On the chicken and produce trade he realized $5200.— That he could not go into fully; could not tell to the dollar, and did not pretend to. But we show the amount of business that he did, and we show his persistency. And although Mr. Spalding came here and had heard that Mr. Clough had carried a "*fish or two*"—not merely a few fish, but a "fish or two"—I think you will find that he had been dealing in other kinds of produce as well as a "fish or two," and that the amount which he must have made cannot be far from the sum which he says he has made.

And then he says he has made $850 in the tobacco speculation. Mr. Eaton, who was connected with him, is a man well known to the counsel on the other side, and this deposition was taken two years ago; and if Mr. Eaton could have contradicted Mr. Clough, they would have had him here. But he stated the truth in regard to this, as he has in regard to everything in relation to his property.

And then his horse trades; he gave you his horse trades in detail, he shows you the amounts that he has made; and they foot up to the sum of $1543, and many of these we have shown by other testimony than his own.

And the whole of his legitimate gains then, gentlemen, without any interest whatever, where we show you absolutely, and put our fingers upon the very spot where he has made this money, amount to the sum of fifty-six thousand nine hundred and ninety-three dollars ($56,993.)

And now, gentlemen, we have taken that sum—I mean the gains; we have not taken it in the aggregate; we have not averaged the interest, but we have taken every gain that he has made for a new principal, and cast interest upon that. And we think, as I have said, that it is not an extravagant nor an unreasonable theory to say that a man like George Clough, with his business habits, has merely kept his gains at simple interest, with the exception of the railroad stock, on which we have counted the actual dividends received up to 1866.

And, by the way, just look at that. Just look and see how it was in regard to that one item of railroad stock. Where is there a man that can be shown to have made so much on a legitimate rise of the stock. He bought it low. It did not average more than forty dollars per share. He has realized on that stock more than twenty-two thousand dollars of dividends. And yet, notwithstanding we show all these legitimate gains—notwithstanding he is able to tell you how he has made the amount which his property cost him when he was sued, you are asked to believe that he has stolen it all from the Concord Railroad!

Well, reckoning in then these gains on the Jefferson street property, on the Thompson street property, on the Pleasant street property, on the Concord Railroad stock, and others not here enumerated but shown in the proof, and casting only simple interest on each gain, and that interest amounts, with the principal and all the gains, to one hundred and twenty-two thousand nine hundred and twenty dollars and thirty cents (122,920.30); from which we have deducted his cost of living. We have put that in as he has sworn to you, from 1842 to 1848, at $300 a year, making $1800. And we have shown the habits of this man and his wife, and how they were determined to save all they could, and that they brought themselves within that sum for that length of time. And from 1848 to 1860, the time he moved from Railroad square, he lived within his salary; and since that time, up to the time he was sued, he has estimated his cost of living at $2500 a year. Adding to it the amount of losses, taxes paid, etc., it makes $41,203.44. Deducting that from the amount previously given, and we show where he could have made legitimately $81,716, which is $9398 more than his property foots up on the fourteenth day of February, 1866.

Now, gentlemen, as I have said, we could not bring this to a mathematical demonstration. It is very probable that he may have under-estimated the cost of his living. But we have approximated to the result. We have shown that his property cost him the sum of seventy-two thousand dollars, and a little more; and we have shown by calculations from the testimony, which is on record, where he made legitimately the sum of eighty-one thousand dollars, without being obliged to steal or plunder a single dollar from the Concord Railroad. Why, gentlemen, something was said here in regard to the amount of clothing and the under-valuation that Mr. Clough put upon his expenditures in regard to clothing; and I do not know for what other purpose the testimony of Mr. Keyes, who made the insurance report, was put in, unless it was to contradict the testimony of Mr. Clough in this particular. Well, the counsel on the other side may make all he has a mind to out of that. But, gentlemen, I have this to say, that Mr. Clough had but little to do with making up that insurance inventory. And that, too, was made in the time of the war, when everything worn was considerably higher in price than now. The insurance agent made it in his own way, and Clough assented to it.—

The gentleman may make all he can out of that, for, making liberal allowance for any under-estimate of his expense of living, &c., we still have a margin of more than nine thousand dollars above the actual cost of his property at the time this suit was brought.

Now, gentlemen, I have troubled you long—longer than I had intended. In the argument I have felt called upon to make, I have been considerably embarrassed from various considerations. In the first place, I cannot help feeling, as I have gone along, that much that I have been saying has been talked about in the course of this long investigation.—And I could not help remembering, also, the character of the tribunal before which I stand—that you, gentlemen, understand just as fully in regard to this evidence as I do; and that, therefore, there was little need to argue the matter before you. And another thing is the fact that you are anxious to get through with the arguments to-day. And I have, perhaps, discussed the case in a more rambling and desultory manner than I otherwise should if I had been freed from these embarrassments.

In conclusion, gentlemen, I think I am warranted in saying of this proceeding and the evidence under it—though it may be contrasting a comparatively small case with one that is world-renowned—as was said by a celebrated English barrister in his great argument in behalf of the queen:—"In this case such is the evidence before you—evidence inadequate to prove a debt—evidence impotent to deprive of a civil right; ridiculous to convict of the lowest offence; scandalous when brought forward in support of a charge the highest known to the laws; monstrous to ruin the honor and blast the name of an English queen." And I say in regard to the testimony brought against the defendant here, that it is inadequate, impotent, scandalous! It is indeed monstrous, and, I might almost add another term—infamous, when we take into consideration the magnitude of the charge with which the case set out, and the character of the witnesses by whom it has been attempted to support it! The attempt upon the vaunting, but flimsy testimony that has been put in—upon the assumptions that have been made with no testimony to back them—upon the suppositions that have been put forth, without the shadow of a fact on which to base them—upon the inferences that have been attempted to be drawn from matters in themselves the most innocent, to ruin this man and his family, and to blast his name and reputation forever, is, I think, something unheard of in the history of judicial trials in New Hampshire. I do not undertake to know who is to blame for all this, and, as I said at the outset, I am not going to impute motives. But this much I know, that Mr. Clough has been pursued in the most unrelenting manner—the fact of his guilt has been taken for granted from the start, and even when the evidence fails the charge is still continued! But I thank God that the time will come, no matter what the decision here may be, when all the facts growing out of, and surrounding this remarkable proceeding, shall be read and known of all men, when the whole truth shall be made to fully appear, and George Clough shall stand vindicated before the whole world!

I thank you most sincerely, gentlemen—and these are no merely formal words—for the kindness and courtesy which you have exhibited to myself and my associates, and to us all, throughout this protracted trial.

And now, gentlemen, I leave Mr. Clough, I leave his family, I leave his reputation and good name, which is better than riches; I leave the decision of this case confidently in your hands, for I know that your breasts are inclined to justice, and simple, naked, exact justice is all I ask, all I desire, and all I claim for my client.

---

[*From "The People" of January* 28.]

In February, 1866, suits were brought by the Concord Railroad against George Clough and other conductors—damages being laid, in Clough's case, at the sum of one hundred thousand dollars. Mr. Clough's case only has been tried, and the session of the referees lasted twenty-seven days. The proceedings against the conductors grew out of a declaration by Mr. Gilmore (who was himself, as appears by the annual report of the present board of directors, a defaulter in a very large amount, and who took tickets without accounting for them, by fifties and hundreds at a time, and who scattered his free passes by the ten thousand,) before the board of directors, that the conductors had been stealing fifty thousand dollars a year in the cars! Detectives, unbeknown to the conductors, were at once put upon the trains, and continued upon them some six weeks or two months, paying for fares in the cars to the conductors about fifteen hundred dollars in money. When the report of the detectives was put in the case, it would show, for instance, that a certain amount of money on such a day was paid to George Clough in the cars for so many fares between such and such stations on the road. Clough's way-bills, or returns to the road, which he was obliged to make every day, were then put into the hands of the detective who was testifying, and he was asked to state whether it appeared from said way-bills that the money he paid to Clough on that day was duly returned by him. And in every single instance it appeared that the money so paid him by the detectives was returned to the road by Clough. Not a wrongful appropriation of money paid him in the cars, to the extent of one cent, was discovered against Clough; and the only "irregularity" shown was the failure in five instances only to take the ten cents extra. It appeared on the trial that it was impossible for the conductors, in all cases, to get the ten cents extra, as required by the rules, for the conductor, if he was obliged to stop and have a quarrel with every passenger who objected to paying it, could not get through the cars and take up his tickets before the train reached the next station. Mr. Gilmore, therefore, instructed Clough to get it when he could, and when he could not, to buy tickets, punch them, and return them to the ticket master with his other collections. Mr. Clough himself testified that this was the way he did in cases where he did not get the extra ten cents, and in this way he received considerable sums for fares, which he paid over to the road by purchasing and punching tickets, as he had been directed. And there was no other evidence but that of Mr. Clough himself, as to his practice in this respect. But the counsel for the road contended that even though the fares might all have been paid over in this way, yet it being done in violation of the rules of the road, he was *estopped* to show this; and the referees held that, if in the opinion of the court he would be "estopped," or if the facts would not constitute a *defence*, that he would be liable in the sum of $5509.

It also appeared that in order to induce travel to come this way from the west, the Concord Railroad authorized through tickets over their road to be sold from points west to Boston at a less rate than to local points on their road. For instance, a passenger could purchase a ticket from Detroit to Boston cheaper than from Detroit to Concord. The result was that when the passenger arrived at Concord, or other points on the Concord Railroad, he had the balance of his ticket left, which was good for a passage to Boston. These tickets were sold; anybody and everybody at Concord, Manchester and elsewhere purchasing and selling them. Mr. Clough, by express permission from the superintendent, bought a good many of these tickets, and made, in the years that he was on the road, according to his own testimony, some two thousand dollars from this source. The counsel for the road claimed that what he had thus made belonged to the road, and the referees decided that if, under the circumstances, it did belong to the road, then it would have a right to recover of Clough this sum of two thousand dollars.

It appeared, also, that Mr. Clough had been in the habit, ever since he had been on the road, of carrying produce and other articles, in which he speculated and made considerable money. It was contended that all this inured to the benefit of the railroad; and whether the award of the five thousand six hundred and thirty-five dollars was on this ground, or upon what ground, does not appear by the report.

Other exceptions were taken besides the questions raised by the referees themselves, and we understand that these are all now before the court for consideration, and will undoubtedly go to the full bench. JUSTICE.

[*From the "Republican Statesman" of January* 8.]

In the case of the Concord Railroad against George Clough, the referees rendered judgment against defendant for $5635, and on questions contingent to decisions of the court the further sum of $5509, for not returning reduced rates of fare taken to the ticket master, as required, and for joint tickets over roads purchased of ticket agents by defendant, $2000. This result, while not as satisfactory to Mr. Clough and his friends as they hoped,—because they have believed, and now believe, that the railroad has no legal or moral claim whatever upon him,—still is of such a nature as not to reflect upon his character for integrity.

Mr. Clough testified before the referees, and frankly stated that he had been in the habit, openly, of carrying articles of freight over the road without paying for their carriage; from which transactions, in a series of years, he made considerable profit. Also, that he had been in the habit of purchasing through coupon tickets at a reduced rate, from passengers stopping off at way stations, and selling them again at full rates to other passengers; and that from this system he derived considerable sums of money, the amount of which he estimated as well as he could. That he had also allowed refreshments and other articles to be carried and sold on the cars, and had received a portion of the profits for many years. The carriage of articles and the sale of coupon tickets, Mr. Clough proved, was with the knowledge and consent of the officers of the road; but the referees were of the opinion that a conductor ought not to carry on such transactions on his private account—that they must be deemed to have been carried on for the benefit of the road, and that Mr. Clough was liable in this suit for the profits.

We do not understand that the referees have found that Mr. Clough retained any of the money paid to him for fares while conductor. This was the only grave or serious charge that was made—was the one upon which the action was originally founded, and which was pressed with the most zeal by the counsel for the railroad. We regard the report of the referees as a substantial acquittal of Mr Clough on all such charges. Whether, on the other grounds, he should be held responsible for profits made by him in a long series of years through transactions in which he engaged by the consent of the higher officers of the road, the public can judge.

---

[*From the "Boston Journal" of January* 12.]

CONCORD, N. H., Jan. 9, 1869.

*To the editor of the Journal:*

In relation to the late railroad "conductor case," which has excited so much general interest, it is due to Mr. Clough, the defendant, to say that with four detectives upon his trains for about two months—they paying in the cars some $1500—not a single instance was shown where Mr. Clough had failed to return every cent of the money so paid to him. It does not appear, therefore, by the report of the referees, upon what ground he is charged for the $5635—whether for carrying freight and articles on the passenger trains, which he had been in the habit of doing for some years, and which the plaintiffs contended should enure to their benefit, or otherwise. It also appeared on the trial that it was impossible in all cases for Mr. Clough, with other conductors, to collect the ten cents extra in the cars, which the rules required when the passengers paid fares on the train, and that Mr. Gilmore, the superintendent, directed Mr. Clough, when he could not do this, to buy tickets, punch them, and return them with his collections to the general ticket agent. On this ground he was sought to be charged for the fares so received when the ten cents extra was not taken, although the money collected was accounted for in that way. The referees award conditionally the sum of $5509, if the court shall hold that the defendant is estopped to show the above facts, or if, being shown, they would not constitute in law a good defence. It appears also by the report that it was the practice of western roads, by virtue of an arrangement with the Concord road, in order to ensure travel to come this way, to issue through tickets to Boston at a cheaper rate than to Concord and other points on the Concord road, so that passengers from the west desiring to stop at Concord, Manchester, and other points, often bought through tickets. These tickets, good for a passage to Boston, when the passenger arrived at his point of destination, were sold to whoever would buy them. Mr. Clough, by the express permission of the superintendent, purchased many of these tickets, and made in the operation, during the years he ran as conductor, some $2000, and the referee award to the Concord Railroad this sum of $2000, provided that the court shall hold that the road is entitled to the profits thus made by Mr. Clough on those tickets.

The case goes to the full bench on the above questions, and upon other exceptions taken by the defendant's counsel in the course of the trial. F. F.

www.ingramcontent.com/pod-product-compliance
Lightning Source LLC
LaVergne TN
LVHW021057110826
845150LV00001B/92

* 9 7 8 1 4 2 5 5 6 6 8 2 1 *